D0164223

Biology of Fishes

CARL E. BOND
Oregon State University
Corvallis, Oregon

SAUNDERS COLLEGE PUBLISHING
Philadelphia

ders College Publishing
Washington Square
delphia, PA 19105

Biology of Fishes ISBN 0-7216-1839-1

 Made in the United States of America. Library of Congress catalog card number 77-84665.

0123 147 9876543

PREFACE

This book is intended as an introduction to the study of fishes for the general reader as well as for the college student, and can have application as a text for undergraduate courses in ichthyology or for fish sections of vertebrate biology courses. It is a beginning book designed to provide an appreciation of the diversity and importance of fishes, or to provide a minimum background for the study of more advanced works on fishes. The sequence of subjects places a section on structure first, as the basis for the following section on relationships and systematics of fishes. These two sections form the background for the chapters on biology and related subjects in the third section. Sections I and III can be used, with reference to section II, as the reading material for a shorter course or as the fish segment of a general vertebrate course.

A functional approach has been followed throughout most of the chapters, and I have made an effort to keep all coverage simple. In order to avoid breaking the continuity of passages, I have inserted only a few literature citations. Most of the works used in compiling the text are given in reference lists at the ends of chapters.

The sequence in which bony fishes are listed in Chapter 7 was chosen to emphasize the concept of "lower", "middle", and "higher" fishes but the main elements of the more modern classification systems are discussed in appropriate places. Hopefully, the individual professor can arrange reading assignments and lecture coverage to reflect a chosen classification scheme.

I must acknowledge the assistance, influence, tolerance and forbearance of a number of persons who, in various ways, furthered the production of this book. My scientific interest in fishes is a result of the teachings of Professor R. E. Dimick, and I have become increasingly grateful to him over the years. Rich Lampert encouraged me to start this book, and kept me going through some critical points. Robert Lakemacher took over where Rich left off. Charles Warren and Richard Tubb helped me find time to work.

Many colleagues and graduate students at Oregon State University read portions of the manuscript critically, or otherwise made valuable suggestions. Included are Charles E. Warren, Lavern Weber, Michael Barton, Kevin Howe, Fred Bills, and George Putnam. Joseph Wales furnished photographs and specimens. Richard Casteel of the National Park Service made photographs available. Gerald Smith of the University of Michigan made important and much-appreciated suggestions. Robert Morris of the University of Oregon named a chapter.

Bonnie Hall deserves special thanks for her patience and cooperation in interpreting some often difficult subjects and converting them into clear drawings. Marge Jackson is thanked for her help with typing.

My wife, Lenora, was of inestimable importance to this book. She gave many hours of her time to typing and proofreading, and suffered a few years of changed routine with characteristic good humor. Without her sacrifice of part of the house to the papers, books and general clutter necessary to the writing of this book, the job couldn't have been completed.

CONTENTS

Section One

INTRODUCTION AND STRUCTURE

1 INTRODUCTION

This book is designed to provide some general knowledge of fish and fish-like vertebrates in a form which can be utilized in a college ichthyology course. It will treat those vertebrates occupying a position in phylogeny between the lancelets and the amphibians. Truly enough, this assemblage includes some diverse forms, and throws jawless animals in with gnathostomes, but some features are held in common. All have crania, live in water, possess gills that are used throughout the life span, and have fins that lack the five-digit characteristics of higher vertebrates. Scales or other dermal armor are typical features of most living groups, and those living forms with naked skin have fossil relatives that were equipped with scales or armor.

General groupings of the living animals which fall into this definition of fishes are lampreys, hagfishes, sharks and rays, chimaeras, lungfishes, and bony fishes (Fig. 1–1). The first two are agnathous and represent a level of vertebrate development which is quite widely separated from that of the living gnathostomes. However, their inclusion in a study of fishes may be justified by their aquatic habitat, their general similarity in structure to higher vertebrates, and the idea that by including them we might better appreciate the lineages of aquatic vertebrates.

There are about 35 species of lampreys; these are eel-like and secretive, and are seen mainly when they are on their spawning migrations. Adults range in length from 75 mm to more than a meter. Larger species are parasitic, feeding upon fishes and, at times, marine mammals. The smaller kinds are usually nonparasitic and feed only in the larval stage. Although all spawn in fresh water some of the parasitic forms grow to maturity in marine waters.

Hagfishes are blind, eel-shaped marine predators and scavengers, with about 20 species in temperate seas, ranging in length from 300 mm to a meter. Like lampreys, they have a cartilaginous skeleton and lack scales and paired fins.

Sharks and rays are more familiar than the lampreys and hagfishes. They are widely distributed in marine waters and have been of interest for many reasons. Many are edible or capable of furnishing useful items such as leather, ornamentation, or sharp points for spears and clubs. Others are considered to be harmful; for instance, their venomous spines can cause injury to swimmers or fishermen.

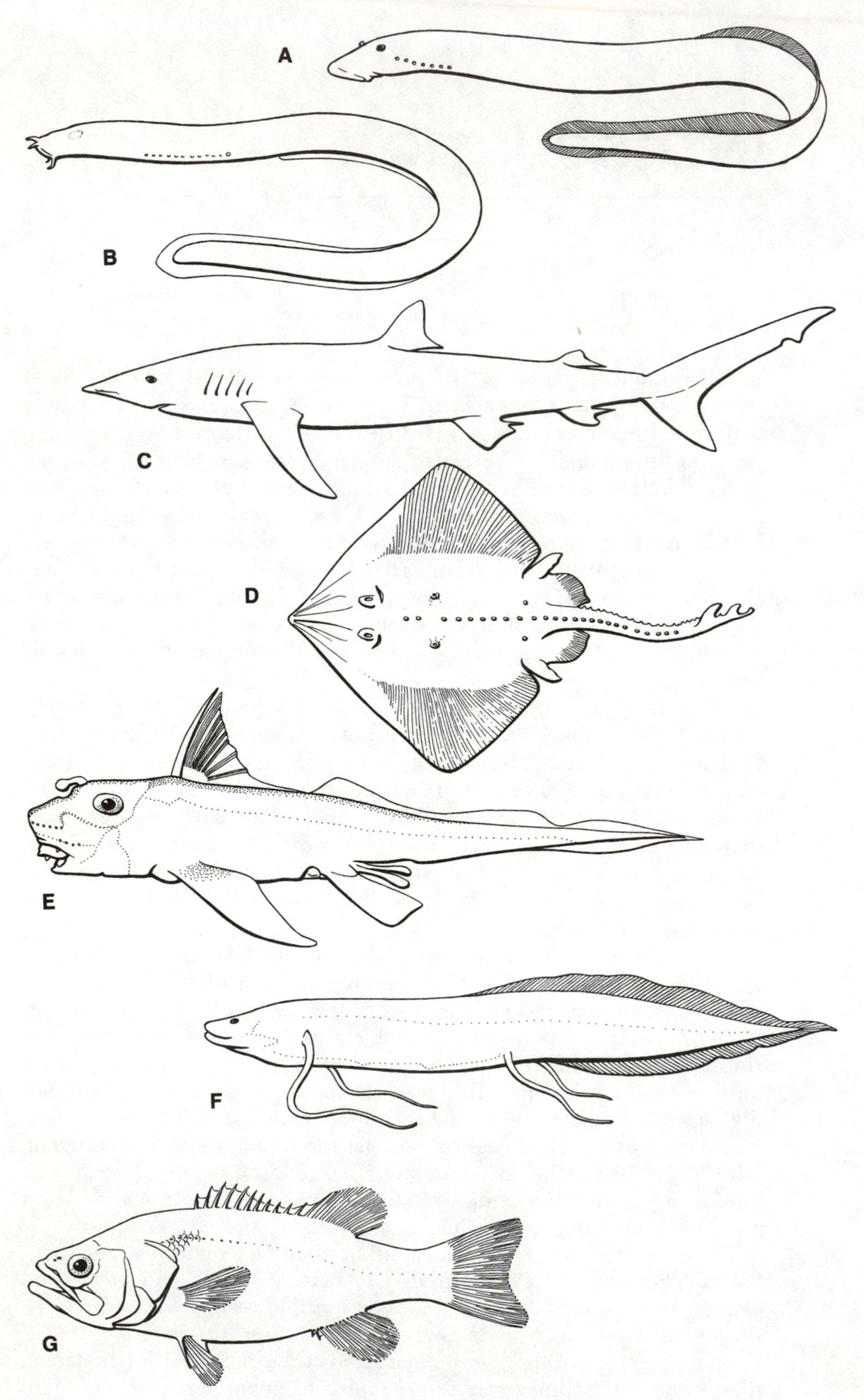

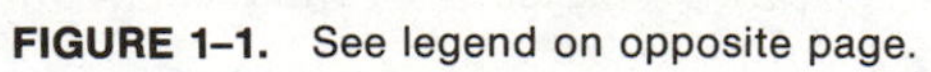

FIGURE 1–1. See legend on opposite page.

Of course some sharks habitually feed upon large prey, and may consider man fair game. Sharks are rather large animals — the largest fish known is the whale shark, a titan of 12 meters or more. An average shark is about 2 meters long, but the dwarf shark (*Squaliolus*) matures at about 200 mm. There are perhaps 220 species of sharks; all but a few are confined to salt water. Only a few such as the bull sharks, including the Nicaragua shark, enter fresh water.

Rays and skates are somewhat more numerous, with more than 300 species. These flattened relatives of sharks are nearly cosmopolitan in the oceans, and, like sharks, are represented by numerous species in tropical areas. One species of sting ray lives in the Amazon. Rays range in size from about 10 cm wide up to the giant manta, with a "wingspan" of over 7 meters.

Chimaeras share certain features with sharks and rays — they have cartilaginous skeletons and are confined to the oceans. There are probably no more than 30 species, and only a few are commonly caught. Their strange appearance elicits such vernacular names as rabbitfish, ratfish, and ghost sharks. Despite their bizarre appearance some chimaeras are used as food, and their oily livers produce a fine lubricant.

Fishes with bone in the skeleton (Osteichthyes) are numerous and varied. Various estimates of the number of living species have ranged from under 15,000 to nearly 40,000. Actually, tallies of known species tend toward the lower figure and recent estimates are usually between 19,000 and 21,000 bony fish species. New species are constantly being described, and the number of subspecies is truly a matter for speculation. Bony fishes constitute about 40% of the living vertebrate species.

Although lungfishes have ossified tissue in the skeleton, they are usually separated taxonomically from the other fishes because of anatomical peculiarities. Only six species of lungfish remain, one each in Australia and South America and four in Africa. African and South American species range from about 60 cm to a meter long, and the Australian lungfish may reach almost 2 meters. They are often referred to as choanate fishes because they possess external nares that open into the nasopharynx via the choanae or internal nares.

Latimeria, a coelacanth or lobefin, represents a group related to the ancestors of amphibians. These fishes were thought to have been extinct for millions of years until one was captured near the coast of South Africa in 1938. Subsequent specimens have allowed zoologists to ascertain that this remarkable animal retains many features that were accurately deduced from fossils of its extinct relatives, although it shows specialization for its particular mode of life.

Figure 1–1. Examples of groups of living fishes. *A*, Lamprey (class Cephalaspidomorphi, subclass Hyperoartii); *B*, hagfish (class Cephalaspidomorphi, subclass Hyperotreti); *C*, shark (class Elasmobranchii; *D*, ray (class Elasmobranchii); *E*, chimaera (class Holocephali); *F*, lungfish (class Osteichthyes); *G*, teleost (class Osteichthyes).

Other bony fishes proceeded along evolutionary lines that were distinct from those trends that produced terrestrial vertebrates, and there is general agreement that they represent three levels of organization in which a general progression from primitive to derived characteristics may be identified. These assemblages are Chondrostei, Holostei, and Teleostei — groups that are defined for convenience's sake in dealing with fish evolution, because no absolute characteristics can distinguish all the living and extinct forms of any one of these groups from their closest relatives in another.

Sturgeons and paddlefishes of the Northern Hemisphere and bichirs and reedfishes of Africa constitute the Chondrostei, and are more primitive in some important respects than the other groups. All are found in fresh water, with the sturgeons capable of entering the ocean. There are about 23 species of sturgeons, two species of paddlefishes, 10 of bichirs and one reedfish.

Only eight species of Holostei remain extant; all of these are found in North America — one bowfin and seven gars (garpike). At least one of the gars can enter salt water. Holosteans have numerous fossil relatives distributed in other areas.

The Teleostei comprise the remaining 19 or 20 thousand bony fish species, of which diverse types are found in all oceans and most land areas. They are found from the abyss to the spray zone, from thick swamps to the rushing torrents of the Andes and Himalayas, and from hot springs to freezing bog ponds and marine waters so cold that antifreeze is required in the blood. They swim using many methods, walk and wriggle both in and out of the water, leap, glide, and nearly fly. They range in size from miniscule, 1-cm gobies to giant tunas, marlins, swordfishes, and the arapaima of South America, all of which reach a length of about 4.5 meters. Their modes of life and some of their anatomical and behavioral adaptations for feeding and breeding are quite fantastic.

Teleost evolution has generally been from soft-rayed fishes of rather generalized anatomy to the more specialized spiny-rayed forms. Usual examples of lower or more primitive teleosts, sometimes known collectively as Malacopterygii, are herrings, tarpons, trouts, and the bony-tongues or osteoglossomorphs, although carps, catfishes, eels, and their relatives qualify for inclusion on this level. Some of their characteristics grade into the features of higher or more derived groups, and some scientists recognize an intermediate level of teleosts that consists of groups exemplified by lanternfishes, flyingfishes, cods, topminnows, and others. These show certain affinities to both the lower and higher levels. Higher fishes are the perches, basses, scorpionfishes, and many other spiny-rayed or otherwise specialized related groups.

Because arranging the teleosts into lower, middle, and higher levels of organization may be a gross oversimplification that tends to obscure phylogenetic relationships, the terms "lower" and "higher" fishes in this book will be used only for the most general purposes. Phylogeny of the teleosts is intricate and difficult to trace. Despite the excellent efforts of systematists who are making progress in unravelling the evolutionary puzzle, evolution is for the most part a continuum, and our invention of categories for the stages and levels recognized in known fossil and living fishes is an artificial one. The great diversity of all the fish groups, living and dead, the plasticity of some groups, and the conservatism of others shows the artificiality of any schemes for classifying fish and their structures. Therefore, although it can be seen that there are three or more lines of teleosts leading from the Holostei, the interpretation of these lineages differs widely among ichthyologists. One of the most useful modern concepts of these lineages is that not all "lower" teleosts belong to the same line, and another is that not all "middle" and "higher" teleosts belong on the same evolutionary track.

As might be expected, the bony fishes will receive the most attention in this work. Not only is this due to their numbers, but also to their importance from many standpoints. Living species will receive most coverage, but the important extinct groups will not be ignored. Of the vast numbers of fishes that have become extinct — which evolved, perhaps flourished, and then failed to make some adjustment along the way—some are known in fossil form. Many have been studied by those clever detectives, the palaeontologists, whose painstaking work has provided a better evaluation of the relationships of groups of fishes than could ever be gained by attention only to living species. The often fragmentary, accidentally preserved remains of a random fraction of the extinct forms give a tantalizing glimpse of way points and end points of fish evolution.

The first known fish-like vertebrates occur in the fossil record of the late Cambrian Period, over 500 million years ago (Table 1–1). These were jawless pteraspidomorphs known as "ostracoderms" because of their armor. By the end of the Silurian, over 400 million years ago, the bony fishes coexisted with several lines of both jawless and jawbearing (gnathostomatous) fishes that were destined to disappear. All major groups of jawed fishes presently living were in existence by the middle of the Devonian. Although probably of great antiquity, the soft-bodied lampreys and hagfishes have left a meager fossil record; only one lamprey is known from the Carboniferous.

Man is interested in fishes for diverse reasons. They afford sport for the angler, provide food for millions all over the world, and have other commercial uses as animal food and raw materials. Many species are kept as pets and have made advanced amateur ichthyologists out of an army of hobbyists. Some students concentrate on the structure of fishes, while others study their systematic arrangement and great evolutionary significance.

Probably the most interesting aspect of the study of fishes is what they do, and how they do it. Even if a person could ignore all else around, and observe and record only what an individual fish

Table 1–1. RANGE OF MAJOR GROUPS OF FISHES THROUGH TIME

Era	Period	Epoch	Dur*	BP*	Major Group
Cenozoic	Quaternary	Recent			
		Pleistocene	2	2	
	Tertiary	Pliocene	13	15	
		Miocene	15	30	
		Oligocene	10	40	
		Eocene	15	55	
		Palaeocene	10	65	
Mesozoic	Cretaceous		70	135	
	Jurassic		60	195	
	Triassic		130	225	
Palaeozoic	Permian		55	280	
	Carboniferous		70	350	
	Devonian		50	400	
	Silurian		50	450	
	Ordovician		50	500	
	Cambrian		100	600	
					Osteostraci, Anaspida, Petromyzonida, §Myxinoidea, Heterostraci, Thelodonti, Acanthodii, Actinopterygii, Crossopterygii, Dipnoi, Placodermi, Elasmobranchii, Holocephali

*BP = approximate millions of years before present at beginning of epoch or period; Dur = approximate millions of years duration.

§Fossil examples and age unknown.

was doing the study would be of interest and importance. But that type of observation is only the beginning. The behavior and activity of the individual fish is at once directed by its environment and influences it. And the place of a particular species in an environment depends upon its structural and physiological capabilities for taking advantage of food sources and finding suitable facilities for reproduction — however far from the feeding grounds — so that perpetuation of the species is ensured.

The fish, busy in its habitat and interacting with other members of its own species or with other species, is a complex of structures, organ systems, and behavioral patterns that have evolved over the millions of years that fishes or their ancestors have been swimming around in the abundant waters of this planet. Ichthyologists face many interlocking problems in studying what fishes do, because it is necessary to know with what equipment the fish works, how this equipment evolved, how the fish uses it, and, what may be of prime importance, the reason for differing behavior in closely related forms in spite of a similarity in equipment. In studying these matters, the ichthyologist generates more than knowledge for its own sake; because fishes are used for various purposes, precise information on many aspects of fish ecology, behavior, reproduction, population dynamics, and so on is required to manage wild stocks or to culture tame ones.

The required knowledge of fishes is not at all complete, even though laymen, naturalists, scientists, and others have been recording facts for centuries. Thousands of species are known only by their structures and systematic relationships, with little or no information available on living organisms. Many species that enter fisheries over the world have only the basic features of their life histories known. Others that have been studied in detail continue to present investigators with mysteries, as expanding knowledge allows the importance of peculiarities of physiology and behavior to be evaluated and appreciated. So there is much yet to be done, much to be discovered, analyzed, and reported. The labors remaining for ichthyologists are great and compelling.

Almost all ichthyologists are concerned with more than pure science and are involved to some extent in practical aspects of fishery science, even if this goes no further than studying the systematics of economically significant groups. Those involved in pure science often see the results of their investigations being put to practical use by others. Because a knowledge of anatomy is basic to these studies, this work begins with a section on the structure of fishes that describes various parts or regions of the body, and which mentions function briefly. Using this as a foundation, the second section describes the kinds of fishes, their evolution and relationships, their world and ecological distribution, and their importance to man. Then, armed with the vocabulary of structure and phylogeny, a section on some peculiarly fishy subjects can be approached, and the matters of what fishes do, and how and why they do it, can be explored. A final chapter will deal with fishes as a resource to be used and managed by man.

References (These general references pertain to Chapters 1 through 4.)

Alexander, R. M. 1975. The Chordates. London, New York, Cambridge University Press.

American Society of Zoologists. 1977. Symposium: Recent advances in the biology of sharks. Am. Zool., 17:287–515.

Bridge, T. W. and Boulenger, G. A. 1904. Fishes. *In*: Harmer, S. F., and Shipley, A. E. (eds.), The Cambridge Natural History, Vol. 3. London, MacMillan. Reprint ed. 1958, Codicote, England, Wheldon and Wesley Ltd., and Weinheim, Germany, H. R. Engelmann (J. Cramer).

Brodal, A., and Fange, R. (eds.). 1963. The Biology of *Myxine*. Oslo, Universitetsforlaget.

Carter, G. S. 1967. Structure and Habit in Vertebrate Evolution. Seattle, University of Washington Press.

Daniel, J. F. 1934. The Elasmobranch Fishes. Berkeley, University of California Press.
Dean, B. 1895. Fishes, Living and Fossil. New York, MacMillan.
Goodrich, E. S. 1909. Cyclostomes and Fishes. *In*: Lankester, R. (ed.), A Treatise on Zoology, Part 9. First Fascicle. London, Adam and Charles Black. Reprint ed. 1964, Amsterdam, A. Asher.
———. 1930. Studies on the Structure and Development of Vertebrates, London, Constable and Co. Reprint ed. 1958, 2 vols., New York, Dover.
Grassé, P. P. (ed.). 1958. Agnathes et poissons: anatomie, ethologie, systematique. *In*: Traité de Zoologie, Vol. 13, 3 parts. Paris, Masson et Cie.
Grizzle, J. M., and Rogers, W. A. 1976. Anatomy and Histology of the Channel Catfish. Auburn, Alabama, Auburn University Agricultural Experiment Station.
Hardisty, M. W., and Potter, I. C. (eds.). 1971. The Biology of Lampreys, Vol. 1. London, Academic Press.
———. 1972. The Biology of Lampreys, Vol. 2. London, Academic Press.
Hyman, L. H. 1942. Comparative Vertebrate Anatomy, 2nd ed. Chicago, University of Chicago Press.
Jollie, M. 1962. Chordate Morphology. New York, Reinhold Publishing.
Jordan, D. S. 1905. A Guide to the Study of Fishes, Vols. 1 and 2. New York, Henry Holt.
Lagler, K., Bardach, J. E., Miller, R. R., and Passino, D. R. M. 1977. Ichthyology, 2nd ed. New York, John Wiley and Sons.
Lineaweaver, T. H., III, and Backus, R. H. 1969. The Natural History of Sharks. Philadelphia, J. B. Lippincott.
Marshall, N. B. 1954. Aspects of Deep Sea Biology. London, Hutchinsons.
———. 1966. The Life of Fishes. New York, Universe Books.
———. 1971. Explorations in the Life of Fishes. Cambridge, Harvard University Press.
Norman, J. R., and Greenwood, P. H. 1975. A History of Fishes, 3rd ed. New York, John Wiley & Sons.
Parker, T. J., and Haswell, W. A. 1962. A Text-book of Zoology, Vol. 2, 7th ed. (Revised by A. J. Marshall.) London, MacMillan.
Romer, A. S. 1970. The Vertebrate Body, 4th ed. Philadelphia, W. B. Saunders.
Smith, L. S. 1973. Introductory Anatomy and Biology of Selected Fish and Shellfish. Seattle, College of Fisheries, University of Washington.
Torrey, T. W. 1962. Morphogenesis of the Vertebrates. New York, John Wiley & Sons.
Webster, D., and Webster, M. 1974. Comparative Vertebrate Morphology. New York, Academic Press.
Weichert, C. K. 1965. Anatomy of the Chordates, 3rd ed. New York, McGraw-Hill.

2
EXTERNAL FEATURES AND INTERNAL ANATOMY

General External Features

Although the overall structure of a fish is arranged to present a more or less streamlined unit (Fig. 2–1) the head, trunk and tail can be distinguished. There is usually no difficulty in recognizing the extent of the head in most bony fish, but the true distinction between the trunk and tail is internal (see p. 72). Because boundaries of the head may be obscure in sharks, rays, the lampreys and hagfishes, and bony fish that have hidden opercula, systematists and anatomists studying these groups use other points of reference.

The Head

Regions of the head (Fig. 2–1) include: the snout, between the eye and the anterior tip of the upper jaw; the operculum, or gill cover; the

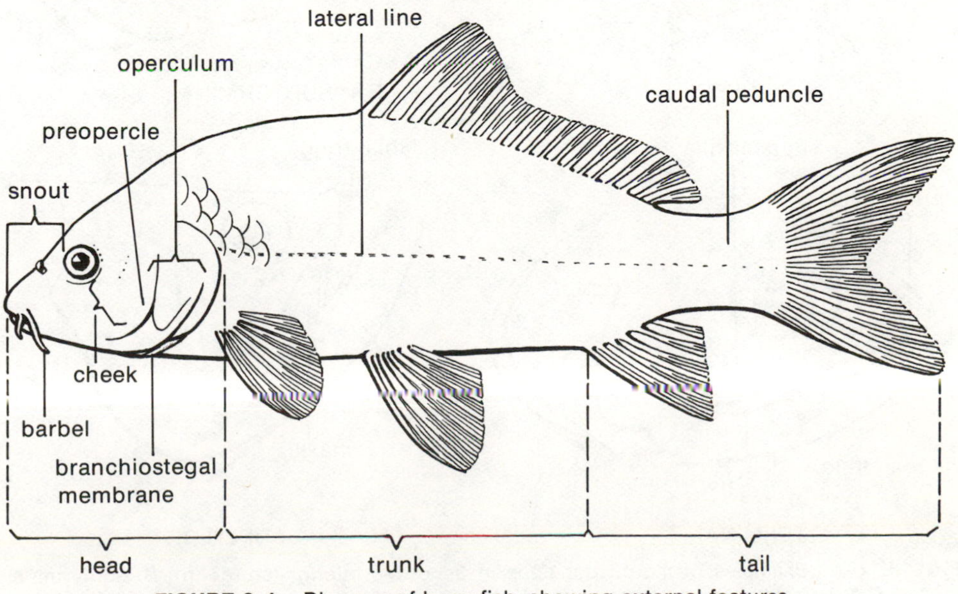

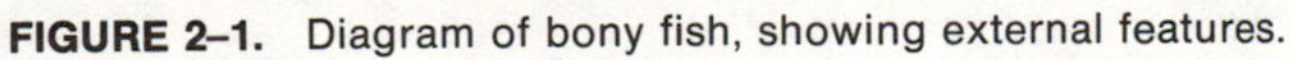
FIGURE 2–1. Diagram of bony fish, showing external features.

cheek, between the eye and the angle of the preopercle; the branchiostegal membrane, below the operculum; the chin, or mentum; and the interorbital. The lachrymal region is below the front edge of the eye. Special features of the head are numerous, and a knowledge of some of these can aid in recognizing species, in studying systematics, or in assessing the mode of life of the fish.

The position of the mouth may be inferior, as in the sturgeon, subterminal as in dace, terminal as in trout, and oblique, or even superior, as in sandfish (Fig. 2–2). The mouth parts usually visible are the lower jaw or mandibles, the premaxillae, the maxillae and, in some species, the supramaxillae, which seem to have little function. Often the mouth is protrusible, the ascending processes of the premaxillae sliding down and forward from their resting place in the nasal region. The lips or jaws of many fishes are bound to the snout or chin by a continuous bridge of skin, or frenum, so that the mouth is nonprotrusible.

Barbels may be present (Fig. 2–1). These are sensory structures which carry tactile and chemical receptors; they may be minute and simple, or conspicuous and sometimes complex, as in some catfishes. They take the designation of the structure that bears them, such as maxillary, mandibular, nasal, rostral, and mental. Similar to barbels but

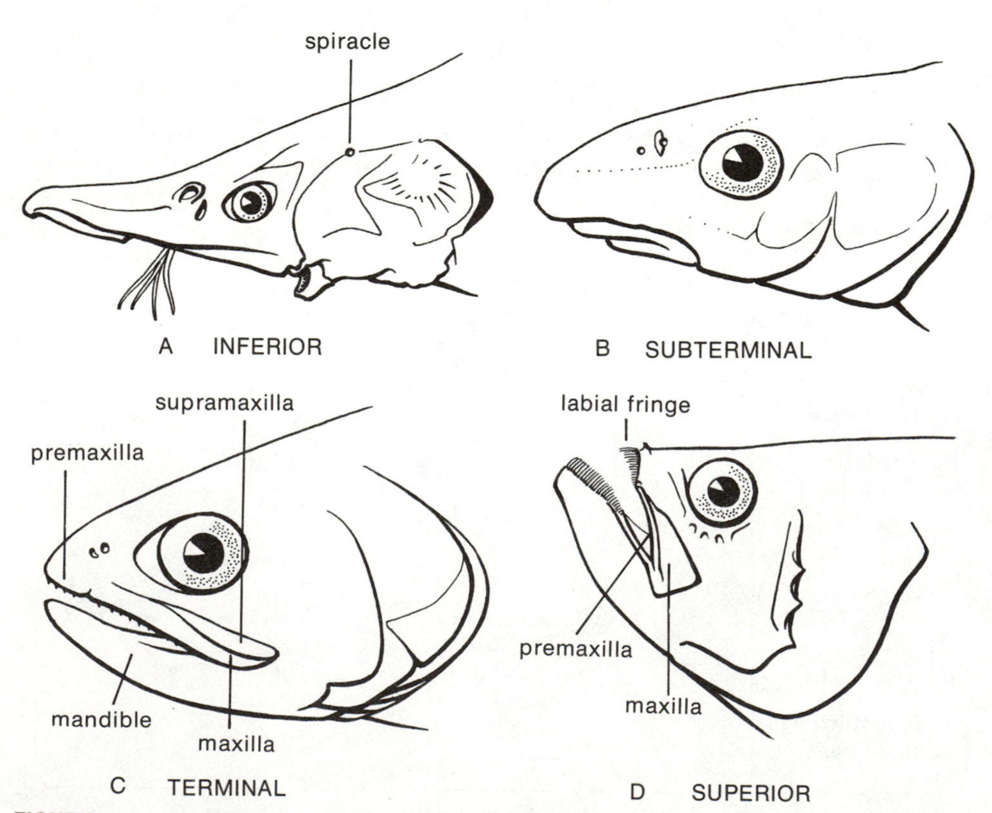

FIGURE 2–2. Examples of mouth positions in fishes. *A,* Inferior (sturgeon); *B,* subterminal (dace); *C,* terminal (trout); *D,* superior (sandfish). (*D,* based on Jordan and Evermann, 1900.)

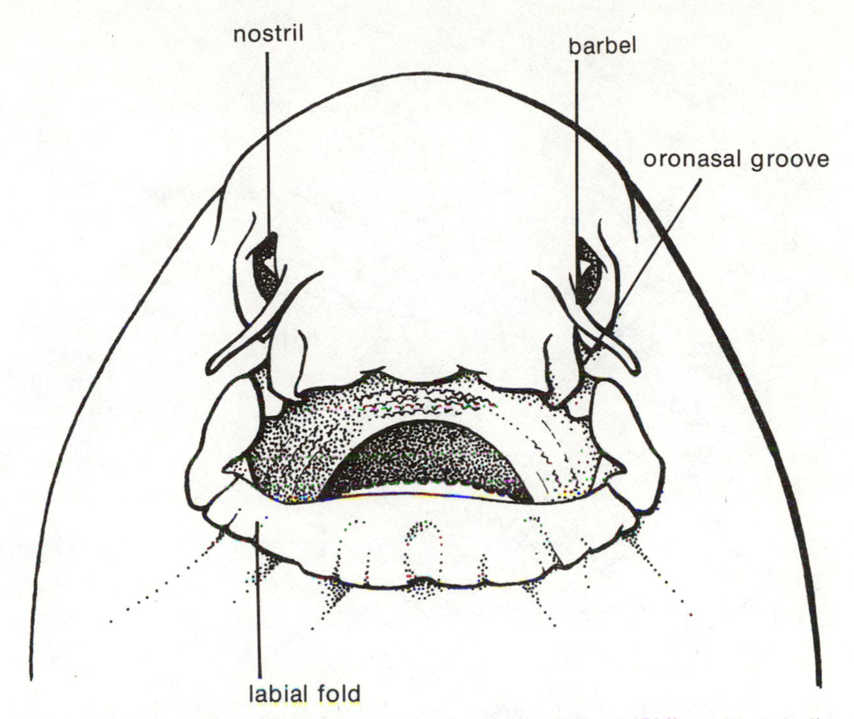

FIGURE 2–3. Ventral view of mouth and rostrum of shark (*Chiloscyllium indicum*) showing grooves and labial folds.

usually without special sensory functions are cirri, or various flaps of skin that may decorate the lips or other parts of the head. Many of these projections seem to blend the fish with its surroundings or otherwise make it less conspicuous, at least to the human eye. Man is only a Johnny-come-lately predator of fishes, but it seems reasonable to assume that the cirri and flaps serve to help conceal fish from their prey or from predators of long standing, as well as from man.

Sharks may have oronasal grooves and labial folds in the mouth region (Fig. 2–3). Lampreys have a series of fleshy fimbria surrounding the mouth, which is a jawless sucking disc (Fig. 2–4).

Some other prominent features of the head are spines on various bones (Fig. 2–5). These are commonly found on the preopercle or opercle, and make some common fish such as yellow perch (*Perca flavescens*) and sculpins (*Cottus* spp) hard to handle. Head spines usually take their names from the bones that bear them, but sometimes from their location.

Sensory canals on the head can be recognized by rows of pores in the skin (Fig. 2–6). Placement of the rows and the number and placement of pores are sometimes used in artificial keys to aid in fish identification. Sensory organs in the canals respond to water movement and other stimuli.

Nostrils of living fishes, except for lungfishes and some specialized

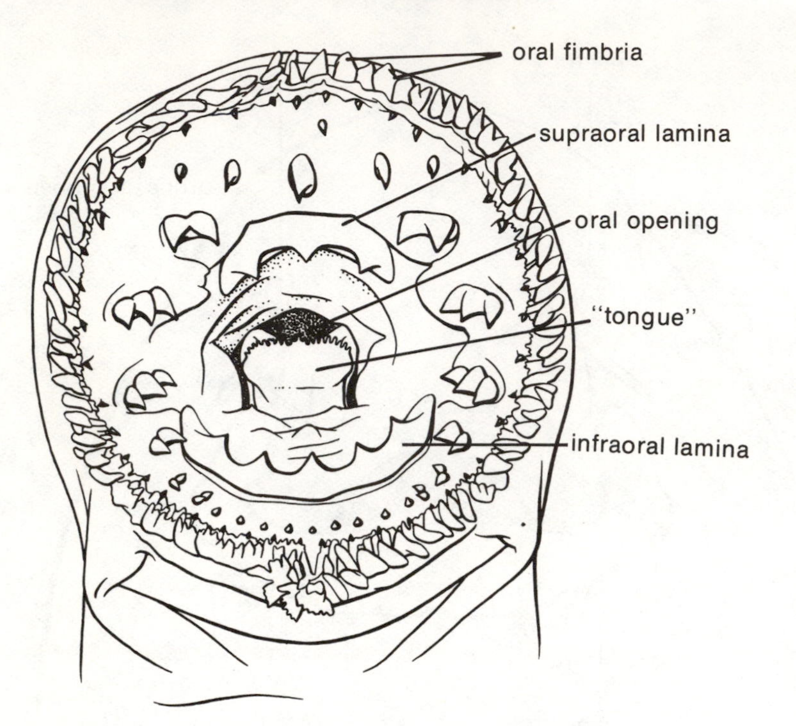

FIGURE 2–4. Oral disc of lamprey (*Lampetra minima*).

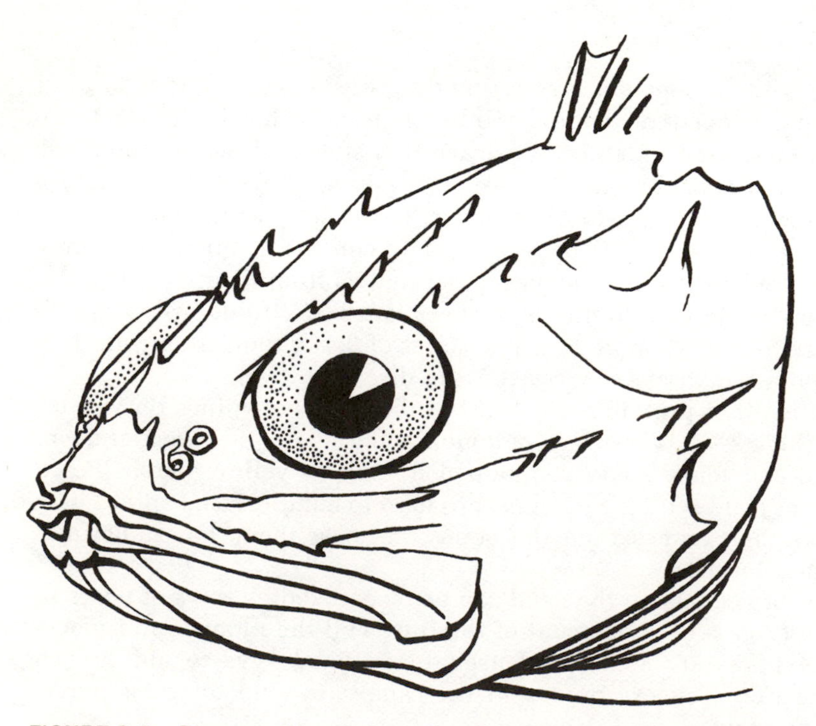

FIGURE 2–5. Diagram of head of rockfish (Scorpaenidae), showing spines.

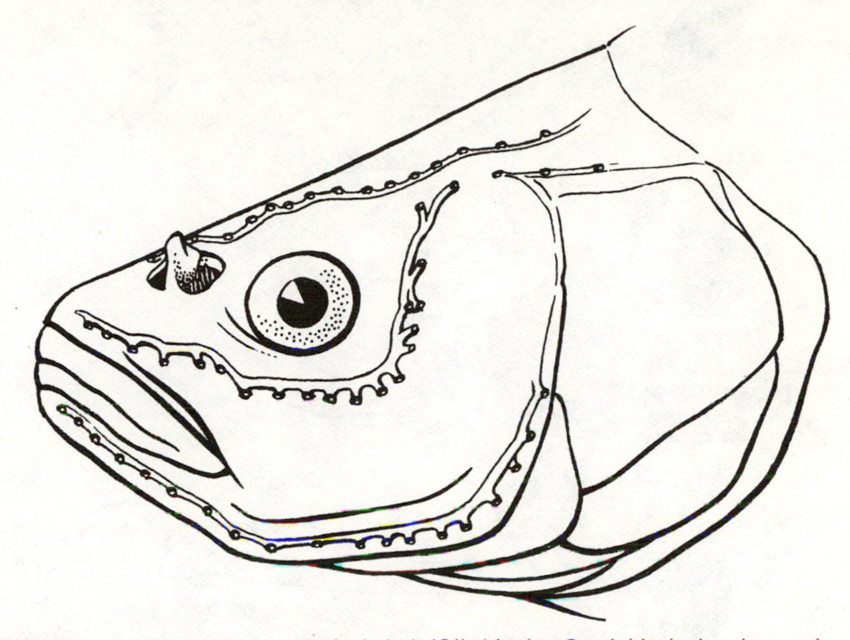

FIGURE 2–6. Diagram of head of tui chub (*Gila bicolor*, Cyprinidae), showing position of cephalic sensory canals and pores.

bony fishes, have no internal openings. There may be a blind sac on each side, with its single opening separated into incurrent and excurrent portions. More usually the sac has two openings barely separated from each other. In some fishes, such as eels, the olfactory organ is in an expansion of a tube-like cavity with the external openings widely separated (Fig. 2–7).

In rays, many sharks, and some primitive bony fishes (*Polypterus*,

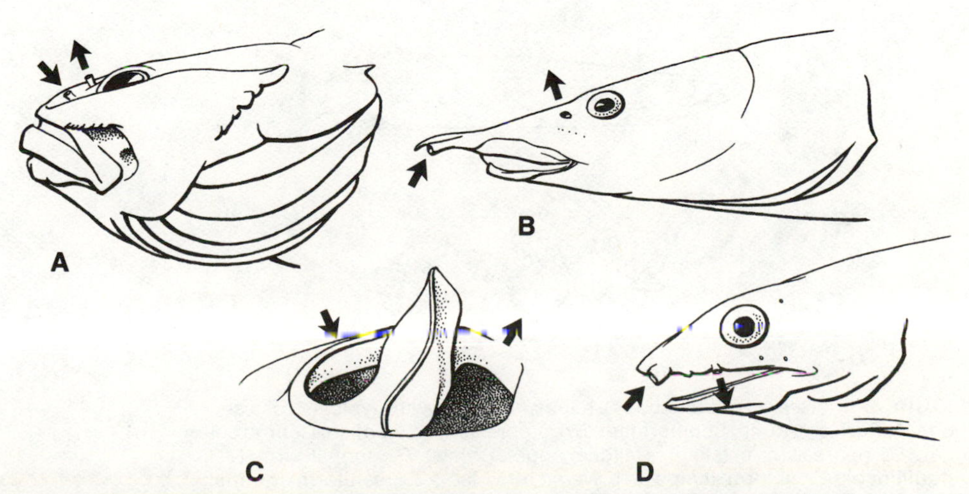

FIGURE 2–7. Nostril positions. *A*, Sculpin (Cottidae); *B*, spiny eel (Mastacembelidae); *C*, typical bony fish nostrils divided by flap of skin (Catostomidae); *D*, eel (*Myrophis*).

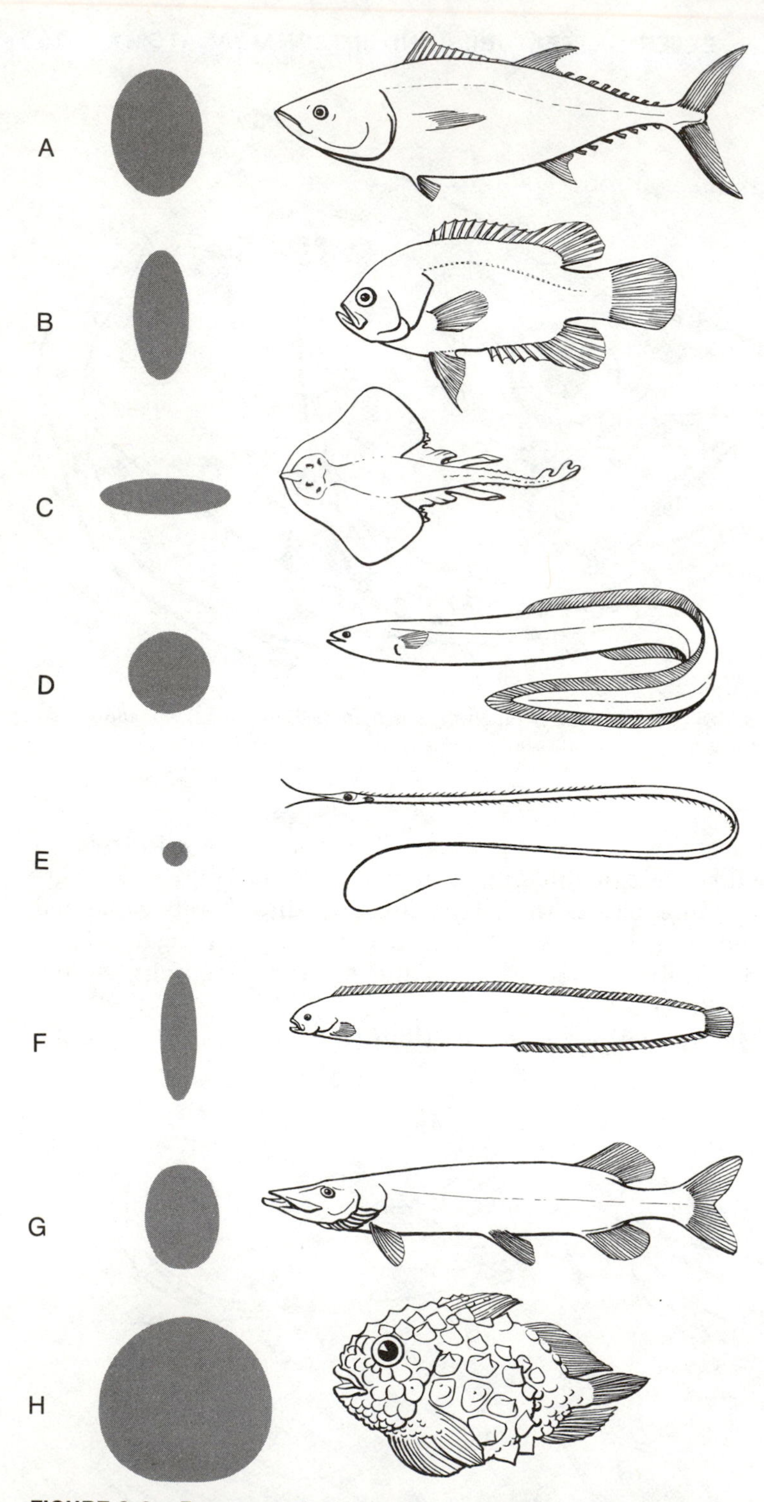

FIGURE 2–8. Representative body shapes in fishes, with typical cross sections. *A,* Fusiform (tuna, Scombridae); *B,* compressiform (sunfish, Centrarchidae); *C,* depressiform (skate, Rajidae), dorsal view; *D,* anguilliform (eel, Anguillidae); *E,* filiform (snipe eel, Nemichthyidae); *F,* taeniform (gunnel, Pholidae); *G,* sagittiform (pike, Esocidae); *H,* globiform (lumpsucker, Cyclopteridae). (*H* based on Jordan and Evermann, 1900.)

Acipenser, Polyodon), an opening called the spiracle is found behind the eye (Fig. 2–2). This aperture is the remnant of a gill slit that has been lost through evolutionary processes. In bottom-dwelling rays respiratory water is brought to the gills through the spiracles.

Body Form

The body form (Fig. 2–8) of a fish can be used in quick appraisal of the fish's way of life. A common body form of fast-swimming, open-water fishes is exemplified by the tunas and relatives (Scombridae). This ultra-streamlined configuration, with an elliptical cross section and narrow caudal peduncle just in front of the tail fin, is called fusiform. This term is often applied to the body shapes of fishes that are considerably more laterally compressed than the tunas, such as trout (*Salmo*) and black bass (*Micropterus*).

Many fishes that are not constantly moving, but which may be capable of quick bursts of speed, are markedly compressed laterally, and are called compressiform. Familiar fishes of this shape are sunfishes (Centrarchidae), snappers (Lutjanidae), porgies (Sparidae) and flounders (Pleuronectidae). Fish that are flattened dorsoventrally are termed depressiform. Depressiform fishes include skates (Rajidae) and their relatives, angel sharks (Squatinidae), toad fish (Batrachoididae), and anglers (Lophiidae). Obviously this shape suits the fish for life on the bottom, but the greatly flattened mantas (Mobulidae) and eagle rays (Myliobatidae) have adapted to a flight-like swimming above the bottom.

Eel-shaped fishes are called anguilliform, from the name of the genus that includes the American eel, *Anguilla*. Other descriptive terms used in connection with body form are: filiform, for thread-shaped fishes such as snipe eels (Nemichthyidae); taeniform, for the ribbon-like shape of gunnels (Pholidae), pricklebacks (Stichaeidae), hairtails and cutlassfishes (Trichiuridae), etc.; sagittiform, for the somewhat arrow-like shape of pikes (Esocidae), gars (Lepisosteidae), and others; and globiform, exemplified by the rotund lumpsuckers (Cyclopteridae).

Of course, not all fishes have body forms that can be described by these convenient terms. Boxfishes and cowfishes (Ostraciidae), sea-horses (Syngnathidae), and sea moths (*Pegasus*) (Fig. 2–9) are some examples of odd shapes. A familiar freshwater fish, the brown bullhead (*Ictalurus nebulosus*), is an example of a fish constructed from a combination of shapes, having a depressed head, a body of round cross section, and a laterally compressed caudal peduncle.

A body form often encountered in marine fishes, many from considerable depths, is that exemplified by the chimaeras (Fig. 1–1) and grenadiers, which have a large head and forebody with a tapering afterbody and tail. This "chimaeriform" body, or one resembling it, can be seen in some seapoachers (Agonidae), spiny eels (Halosauridae), and a few others. Some of these fishes can hold the body straight and swim by undulating the pectoral fins, while others swim by undulations of the body.

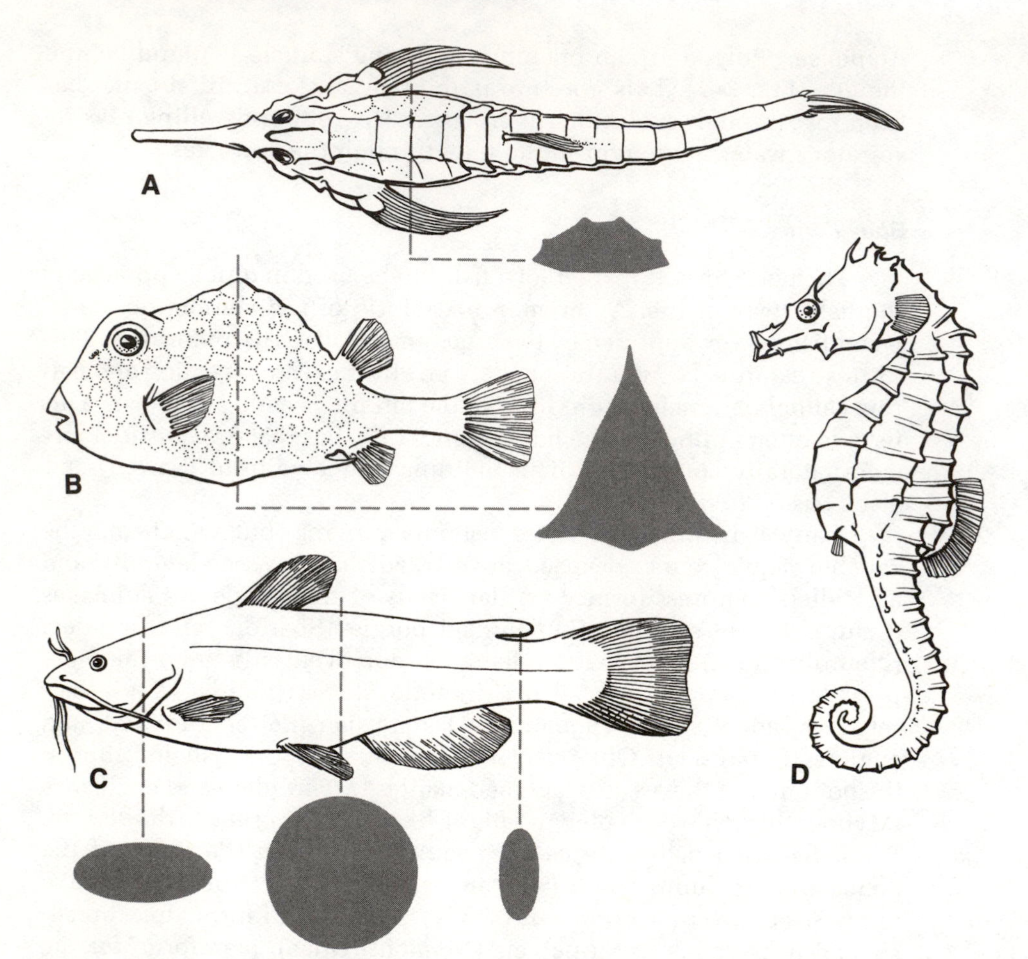

FIGURE 2–9. Examples of body shapes in fishes: *A,* Sea moth (Pegasidae); *B,* cowfish (Ostraciidae); *C, bullhead* (Ictaluridae); *D,* seahorse (Syngnathidae). (*B* and *C* based on Jordan and Evermann, 1900.)

Topography of the Body. Some regions of the body are described by terms that aid in locating identifying features (Fig. 2–10). The dorsal surface just behind the occiput is the nuchal region, often characterized by a hump. The most anterior ventral part of the body is usually the narrow isthmus that extends far forward below and between the gill openings. Directly behind this is the breast, and behind that is the belly. The narrow "handle" of the fish just forward of the tail fin is the caudal peduncle.

Conspicuous along each side of the typical fish is the lateral line, a continuation of the network of sensory canals on the head. The usually single line may be an open groove in the skin, as in some chimaeras, or a row of pores in the skin or scales. Lines may be multiple, as in the greenlings (Hexagrammidae), or variously reduced. Herrings (Clupeidae), for instance, lack the extended lateral line; it appears only on a few anterior scales.

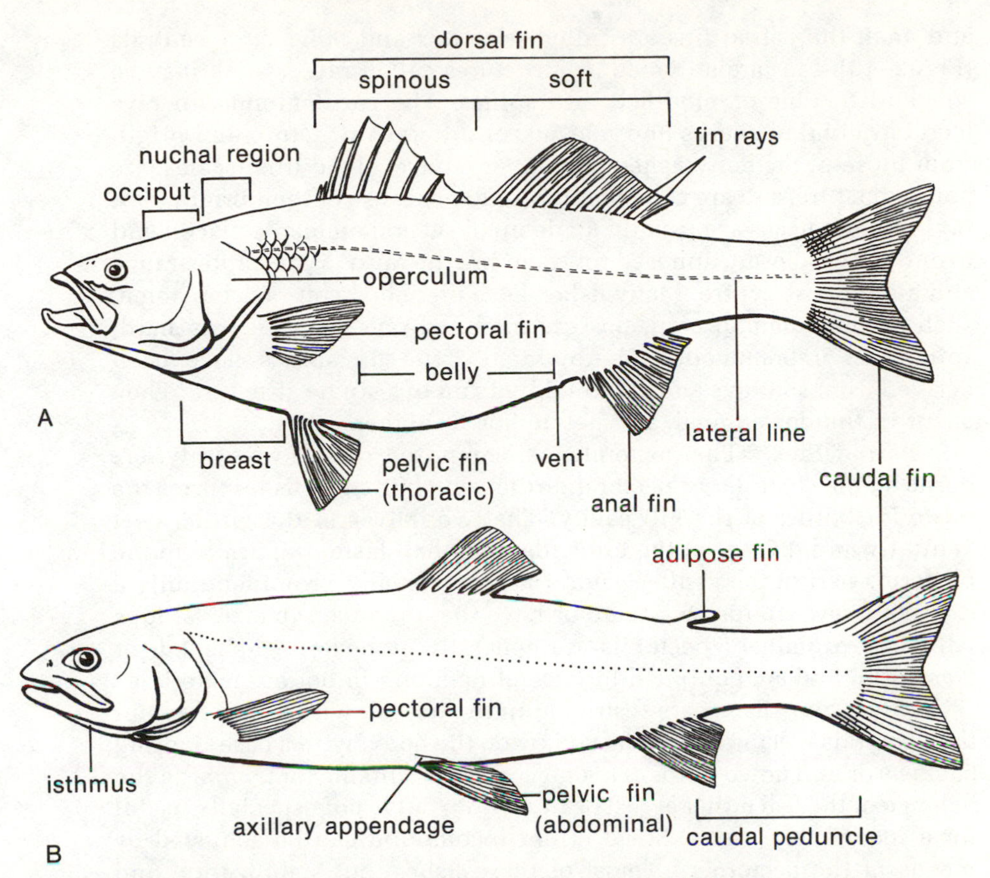

FIGURE 2–10. Body regions and fins. *A,* Spiny-rayed fish; *B,* soft-rayed fish.

The gill openings are typically in a lateral position covered by the operculum, but they may be placed farther back, as seen in eels, or in a ventral position in some groups. In sharks, rays, and lampreys, several individual gill openings occur in a series.

In most fishes the vent or anus represents just the opening of the gut for, other than rare exceptions, only the sharks, rays, and lungfishes have the cloaca, which receives the openings of the gut and the urinary and genital systems. Bony fishes may have, behind the anus, a combined urogenital opening, or the urinary opening may be separate from and behind the genital pore. Often a urogenital papilla is developed. A few families, such as Clupeidae, have an opening from the air bladder posterior to the anus. A peculiar situation obtains in female chimaeras — the two oviducts have separate external openings.

Fins

Fins, of course, are conspicuous features on the fish body. These are supported by the appendicular skeleton and are composed of two groups, unpaired and paired. The unpaired fins are the dorsal, caudal, and anal; the paired fins are called pectorals and pelvics, or ventrals (Fig. 2–10). Fins are stiffened by structures called rays, which may be soft and flexible or modified into spines. The cartilaginous fin rays (ceratotrichia) of sharks and rays are of different structure and origin from those of the bony fishes. The latter have evolved from scales, are bony in nature, and are called lepidotrichia because of their origin. Soft rays of bony fishes are usually made up of several elements placed end to end in two closely apposed rows, giving the ray a jointed appearance and a double structure. Many fishes have branched soft rays but some, such as the common carp, possess modified rays that are very hard, spine-like, unbranched, and unjointed. True fin spines have been derived from soft rays and are unjointed and of a single structure. They occur in the dorsal, anal, and pelvic fins of higher fishes.

Paired Fins. The pectoral fins, composed of soft rays only, are borne by the shoulder or pectoral girdle, which in most fishes forms the posterior border of the gill cavity. The two halves of the girdle meet ventrally and diverge at the top ends, each half fastening firmly to the posterior part of the skull. Pectoral fins are usually prominent; only a relatively few groups lack them or have them reduced in size. Groups with less prominent pectorals are generally elongate, eel-shaped, or taeniform, and are equipped for wriggling along in bottom materials.

Lower bony fishes are restricted in the uses to which they can put the pectorals. Pectorals are placed low on the body, with a base slanting backwards and downwards. This placement is suitable for trimming the balance of the fish either at rest or in motion, but is not especially useful for a locomotory organ. Some minor locomotion can be achieved by means of the pectorals in most of these fishes, but stabilization and changing direction of motion are probably more important functions.

The low position of the pectorals in primitive fishes allows the fins to touch bottom in species living close to the substrate. It may not be surprising to learn that these fins often bear numerous taste buds and touch receptors, or that the fish may rest upon them. The bichirs,

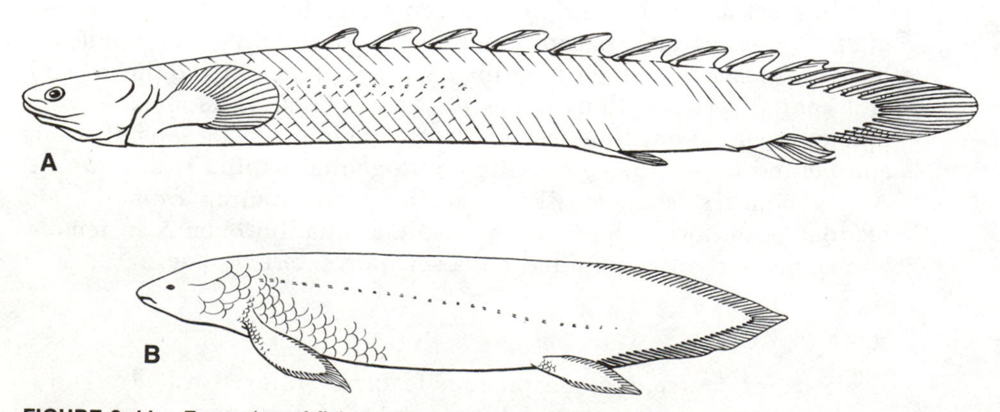

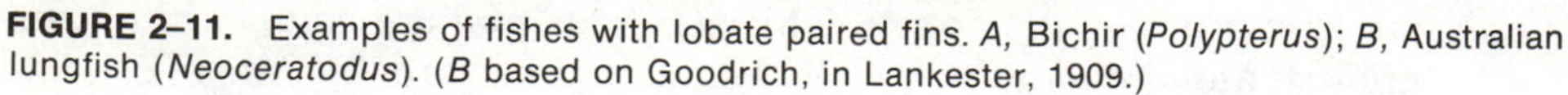
FIGURE 2–11. Examples of fishes with lobate paired fins. *A*, Bichir (*Polypterus*); *B*, Australian lungfish (*Neoceratodus*). (*B* based on Goodrich, in Lankester, 1909.)

Polypterus, bear the fins on a lobate base with a skeleton peculiar to the group, and use them somewhat as arms. This arm-like pectoral is called a brachiopterygium (Fig. 2–11). The crossopterygian, *Latimeria*, and the Australian lungfish, *Neoceratodus*, have lobate fins of a different structure. Highly modified pectorals supported by a central joined axis with no radials (fin ray supports) or fin rays are seen in the African and South American lungfishes, *Protopterus* and *Lepidosiren*.

Some species have pectorals that aid them in remaining at the bottom or even on a vertical surface. Sisorid catfishes apparently use the pectorals as well as a pad between the fins in clinging to surfaces (Fig. 2–12). The hill stream loaches (Gyrinocheilidae) have pectoral fins modified for use in maintaining suction.

An African genus, *Pantodon*, is called butterflyfish because of its large expanded pectorals. These 12-cm fishes make 2 meter leaps over the water surface. Unique among fishes is the pectoral apparatus of "freshwater flying fishes," the gastropelecine characins of South America. Bones of the pectoral girdle are expanded ventrally to give a somewhat hatchet-like appearance and to provide a wide area for attachment of massive pectoral musculature. The pectorals can be vibrated like wings and the fishes have a habit of jumping. One could imagine that we are observing an evolutionary process that might lead to true flight in fishes.

Catfishes have developed spine-like structures consisting of consolidated soft rays at the leading edge of the pectorals. Anyone who has made a bad grab while sorting a seineful of mixed warm water fish and has embedded the spine of a bullhead or madtom in a thumb knows that this is a potent defensive weapon. Use of these spines in terrestrial locomotion is encountered in a few catfishes, the most well-known of which is the genus *Clarias*, the walking catfishes. *Clarias batrachus* has been accidentally released in Florida, where conditions for life and reproduction are suitable enough so that it is spreading and is reported as becoming a pest.

The more derived bony fishes, those that might be justifiably called middle and higher fishes, have lost a bone called the mesocoracoid from the pectoral girdle. That bone aids in holding the pectoral fins of the lower fishes in the low, oblique position. The pectorals of higher fishes can then be moved upward to the lateral aspect of the body, and can have a vertical base. This positioning is more versatile than that seen in the more primitive fishes, and the fins can be used for locomotion, turning, braking to sudden stops, aggressive displays, and other purposes.

Many higher fishes are structurally specialized for certain habitats and ways of life, and the pectorals are often involved in the specialization (Fig. 2–12). Threadfins (Polynemidae) have pectorals that are each divided into two parts, the lower one consisting of several filaments that reach great lengths in some species. These filaments are thought to function as tactile organs and, when extended and fanned out, can apprise the foraging fish of edibles over a wide area. Others with detached pectoral rays are the flying gurnards (Dactylopteridae) and some

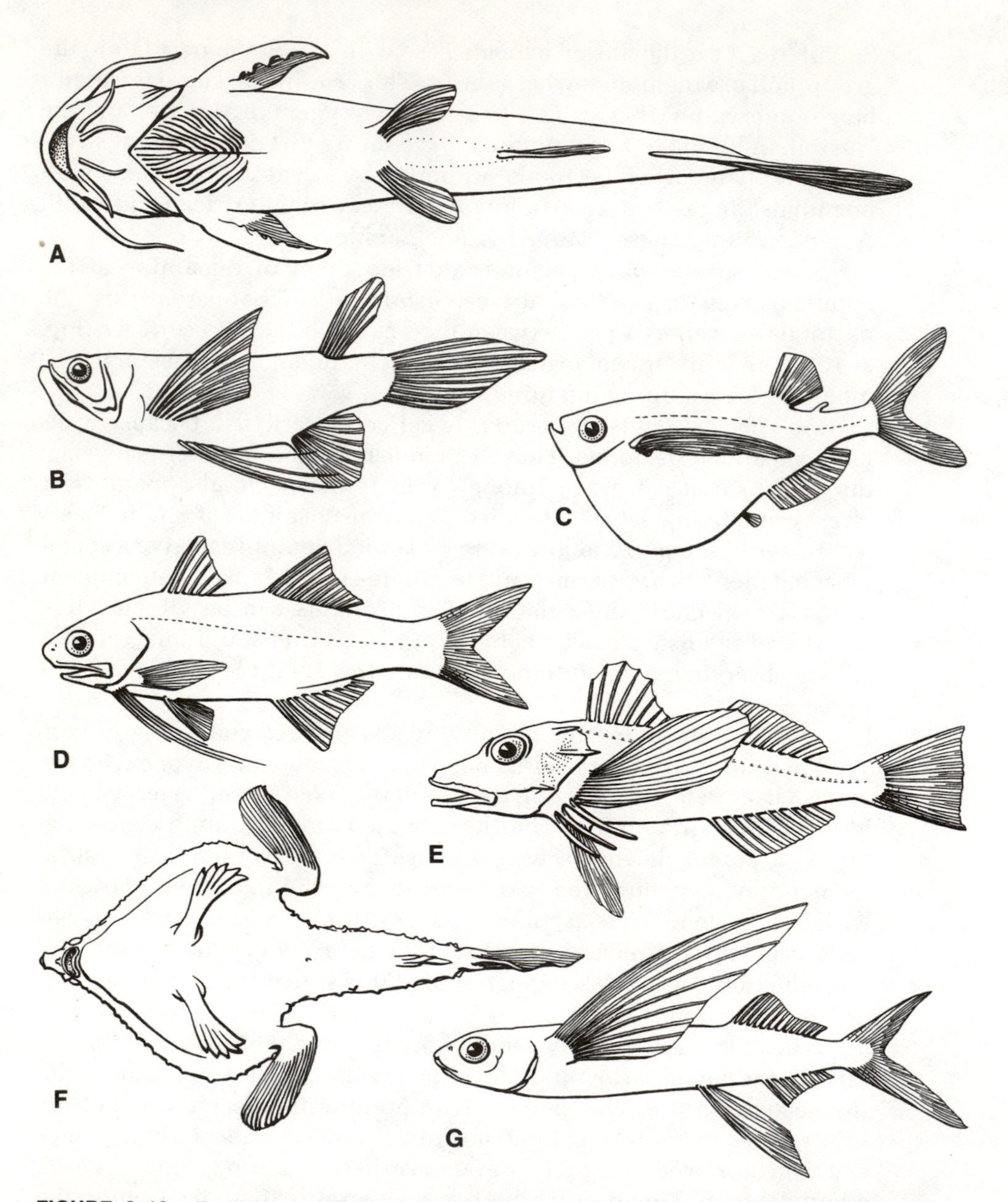

FIGURE 2–12. Examples of fishes with modified pectoral fins. *A*, Sisorid catfish (*Glyptothorax*); *B*, freshwater butterflyfish (*Pantodon*); *C*, hatchetfish (*Gastropelecus*); *D*, threadfin (Polynemidae); *E*, gurnard (Triglidae); *F*, batfish (Ogcocephalidae); *G*, flyingfish (Exocoetidae). (*B* based on Herald, 1961; *D*, *E*, and *G* based on Jordan and Evermann, 1900.)

of the stonefishes (Synanceidae); both families have finger-like rays that are probably tactile in function as well as useful in crawling along the bottom. Other kinds of fishes that walk over the bottom using the pectorals include the batfish, *Ogcocephalus*.

Development of the pectorals into "wings" for gliding is found in

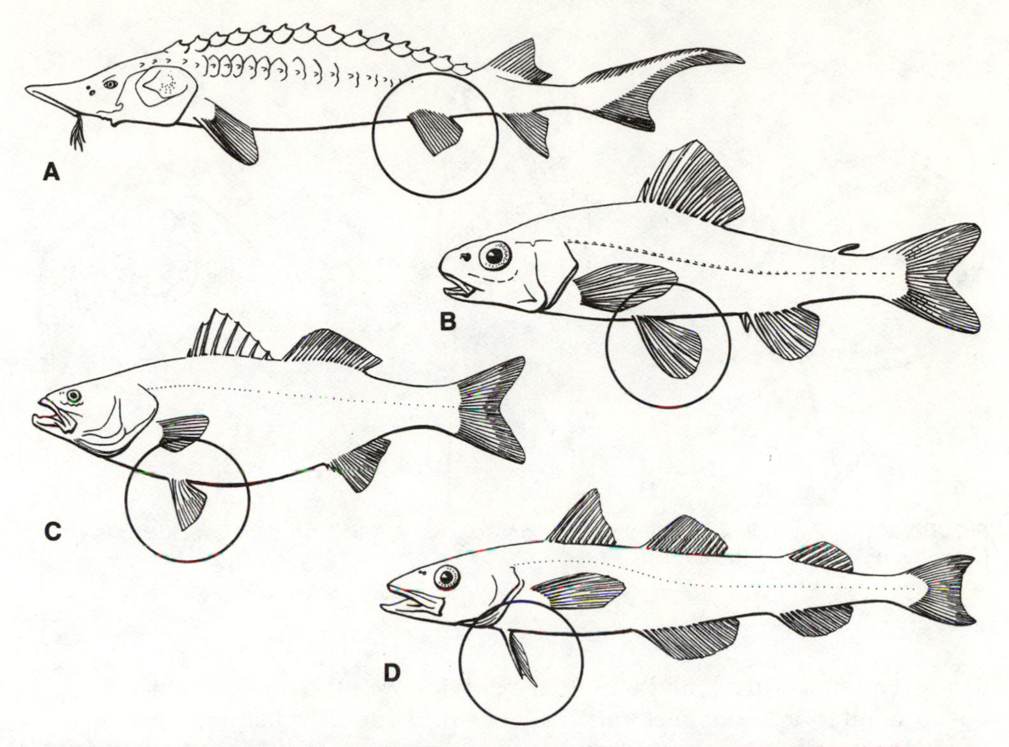

FIGURE 2–13. Examples of pelvic fin placement, pelvic fins circled. *A,* Abdominal (sturgeon, Acipenseridae); *B,* subabdominal (sand roller, Percopsidae); *C,* thoracic (bass, Percichthyidae); *D,* jugular (pollock, Gadidae). (Based on Jordan and Evermann, 1900.)

the flyingfishes (Exocoetidae). Members of this family remain airborne for as long as 20 seconds and can glide a distance of 150 meters or more.

Sexual dimorphism involving the pectorals (and other fins) is common among fishes. Even in fins without specialized reproductive function, coloration and size as well as other features may differ between the sexes. Fishes with sexually dimorphic paired fins include the dragonets (Callionymidae), topminnows (Cyprinodontidae), snappers (Lutjanidae), and suckers (Catostomidae).

Pelvic appendages of fishes are generally smaller than the pectorals, more restricted in function, and subject to greater variation in placement (Figs. 2–10, 2–11, 2–12, and 2–13). Lower bony fishes are characterized by abdominal pelvics, as in the sharks. Supporting structures are embedded in the flesh of the belly and have no internal connection with other skeletal elements. A few groups have the pelvics moved forward toward the pectorals, but still lack internal connections. Higher bony fishes usually have thoracic pelvics, placed below or a little behind the pectorals, with a more or less firm connection between the pelvic and pectoral girdles. Higher fishes usually have a spine and a few soft rays in the pelvic fin. Lower bony fishes tend to have many-rayed pelvics. In the cods and relatives (Gadidae), some of the blennies (Blenniodei), toad-

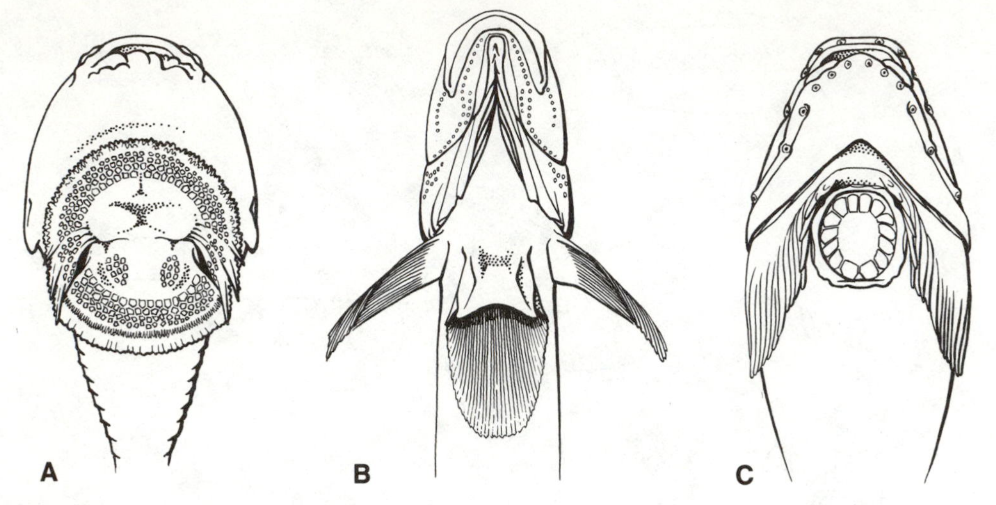

FIGURE 2–14. Pelvic fins modified as sucking devices. *A,* Clingfish (Gobiesocidae); *B,* goby (Gobiidae); *C,* Snailfish (Liparidae).

fishes (Batrachoididae), and others, the pelvics are placed in a jugular position anterior to the pectorals. Pelvics are reduced or lost in many forms, especially in elongate fishes that wriggle along the bottom.

Pelvic fins usually function in stabilizing and braking; they are of very little use in locomotion. Several groups of fishes show modification and specialized function of the pelvics. In the males of sharks, rays, and chimaeras, the pelvics are modified for use as intromittent organs in copulation. Many benthic species such as sculpins, which live on hard surfaces, use pelvics to create friction to help hold them in place. In gobies (Gobiidae), clingfishes (Gobiesocidae), lumpfishes (Cyclopteridae), and some hill stream loaches (Gyrinocheilidae), pelvic fins are involved in ventral sucking structures which aid the fish in holding to the substrate (Fig. 2–14). Use of pelvic fins as tactile devices may be exemplified by the gouramies, such as *Osphronemus.* Flyingfishes (Exocoetidae) may have pelvics modified into planing organs.

Median Fins. These are the dorsal fin, along the back, the caudal or tail fin, and the anal fin, located ventrally just behind the anus in most fishes (Fig. 2–10). The dorsal fin(s) may extend the length of the back, be divided into two or three separate fins, or be single and small. It is seldom lost, as in the Gymnotidae of South America. In some groups—trout (Salmonidae), various catfishes, and the lanternfishes (Myctophidae)—there can be a small fleshy, rayless adipose dorsal on the caudal peduncle. In the higher bony fishes, the anterior part of the dorsal—or the first dorsal if there are two—is supported by spines, which can be stiff and sharp or modified into flexible structures as in the freshwater sculpins, *Cottus.*

The usual functions of dorsal fins appear to be in stabilization and in helping to achieve quick changes in direction, but they can be used in

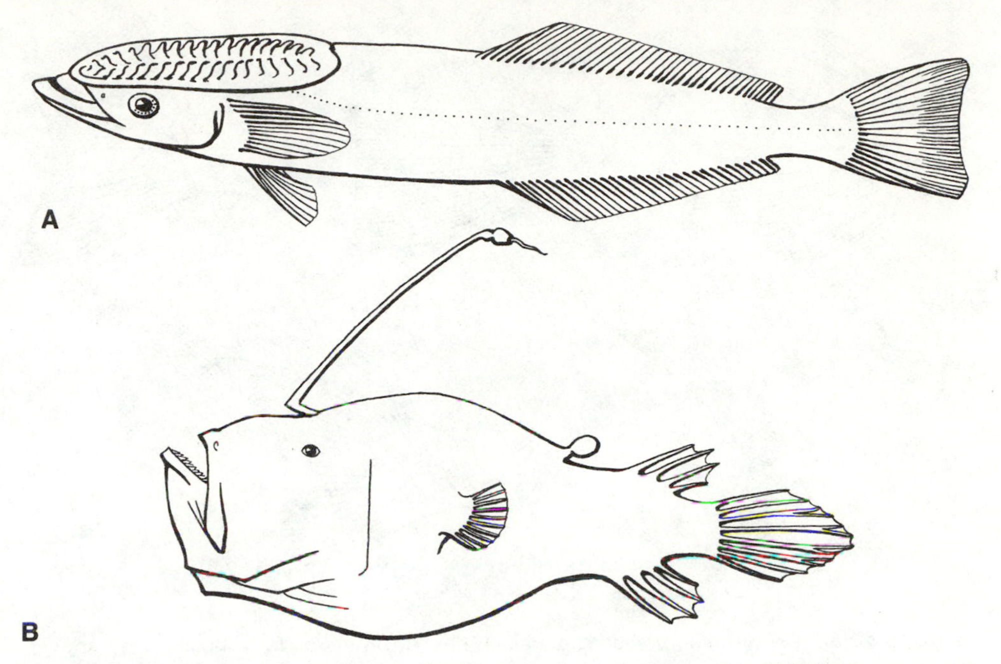

FIGURE 2–15. Modified dorsal fins. *A*, Sucking disc (remora, Echeneidae); *B*, fishing rod and lure (angler, Ceratiidae). (*B* based on Jordan and Evermann, 1900.)

conjunction with the caudal and anal fins in braking. Many species that have dorsal fins extending the length of the back can move by undulating the fin, but some species with short dorsals, such as pipefishes and seahorses, also use the dorsal for locomotion.

Modifications of the dorsal fin (Fig. 2–15) include the sucking disc atop the head of the remoras (Echeneidae) that allows them to cling to sharks or to other large fishes and be carried along as hitchhikers. The lures of anglerfishes (Lophiiformes) are modified dorsal spines. A few species involve a showy dorsal in displays for a variety of social purposes.

In the bichirs (Polypteridae) of Africa, the dorsal is divided into a series of unique finlets consisting of a spine-like structure and a few soft rays each. Mackerel-like fishes often have a series of finlets posterior to the dorsal fin. These consist of detached soft rays, usually branched and set in tough skin.

The anal fin is generally short-based but there are many species with anals exceeding the dorsal in length; some have elongate anals stretching from the anus to the caudal fin, even when the anus is located nearly under the chin, as in the knifefishes (Gymnotidae and Rhamphichthyidae) of South America. Flounders (Pleuronectidae) and gouramies (Osphronemidae) are compressiform fish with long-based anals. Only a relatively few fishes, such as cods (Gadidae), have more than one anal fin. Some, such as the jack mackerel (*Trachurus*), have the anal fin spines separate from the soft rays, forming a small spinous anal. Usually, however, if anal fin spines are present, they are located at the

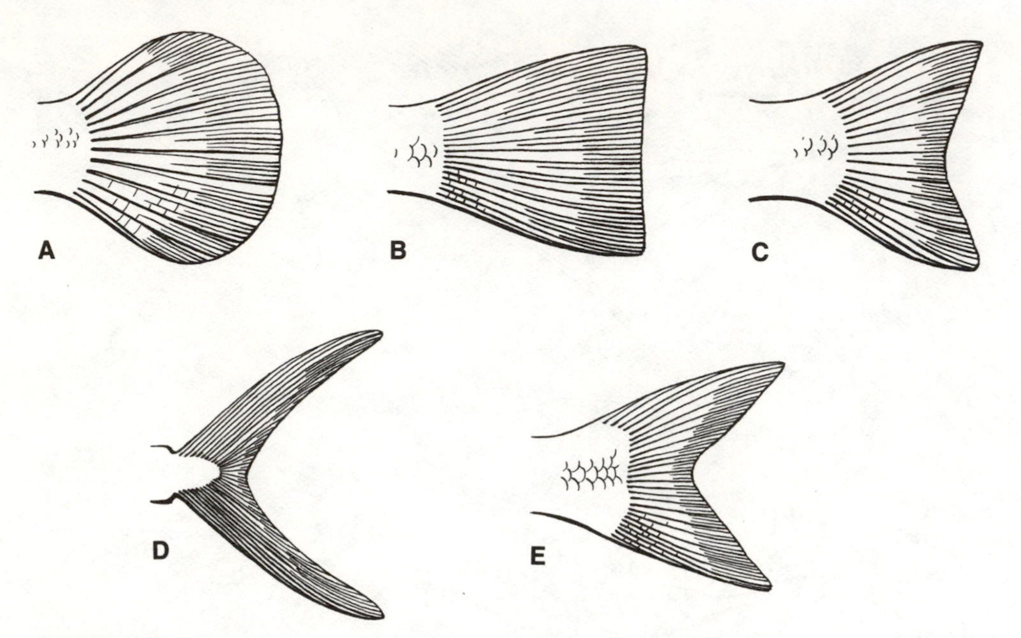

FIGURE 2–16. Representative shapes of caudal fins. *A,* Rounded; *B,* truncate; *C,* emarginate; *D,* lunate; *E,* forked.

front end of the single anal. Finlets behind the anal are present in the tunas, mackerels, and allied fishes (Scombroidei).

The anal fin is lacking in chimaeras, skates, and rays, and some sharks (*Squalus, Somniosus,* etc.), but is usually present in bony fishes; some exceptions are the male pipefish (*Syngnathus*) and king-of-the-salmon (*Trachipterus*). Males of the livebearers (Poeciliidae) have anals modified into an intromittent organ called a gonopodium. Some of the Opisthoproctidae (spookfishes) have the anal fin at the posterior terminus of the body, displacing the caudal upward.

Caudal fins appear in a variety of shapes, sizes, and kinds, and often reflect evolutionary levels and relationships more than the other fins. Swimming habits may be deduced to some extent from the caudal (Fig. 2–16). Those fishes having a crescent (lunate) caudal and a narrow caudal peduncle are generally among the speediest of fishes and are capable of rapid, sustained motion. Many pelagic species have forked tails and are constantly on the move. Species with truncate, rounded, or emarginate caudals may be strong swimmers but somewhat slower than those mentioned above. Fishes with small caudals or caudals continuous with the dorsal and anal tend to be weak swimmers, or may move by wriggling along the bottom.

Most familiar bony fishes have homocercal caudal fins—the type that externally appears symmetrical but that is actually asymmetrical internally, because the tip of the vertebral axis turns upward with most of the fin attached below it. Evidently this is an evolutionary stage

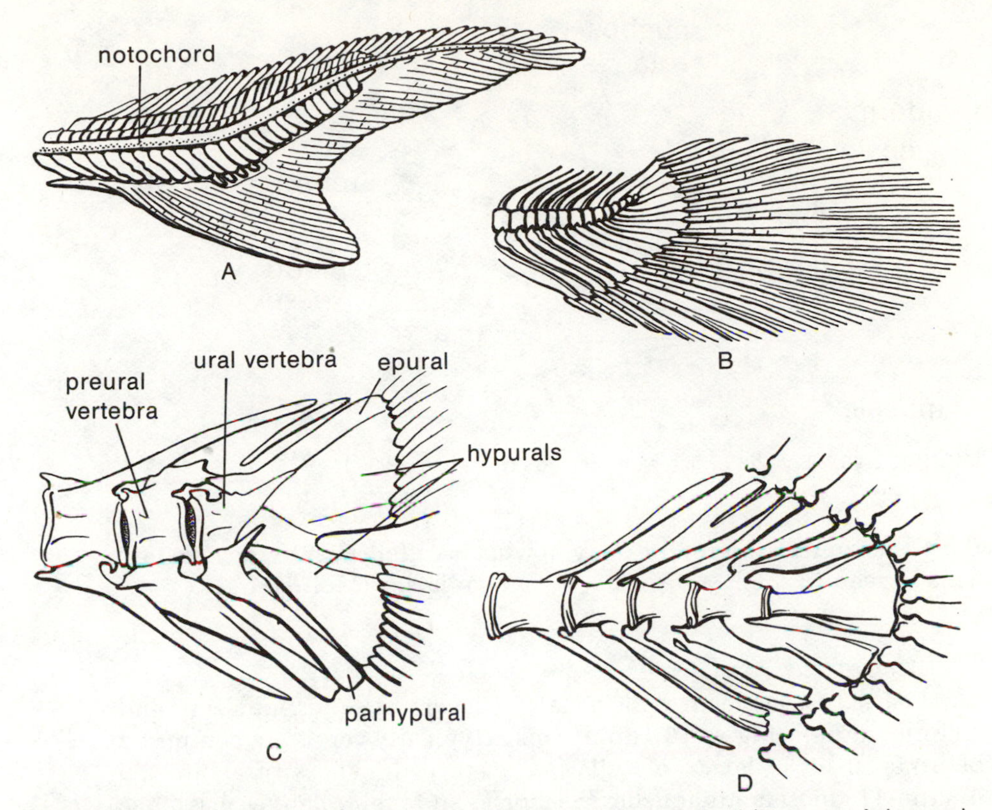

FIGURE 2–17. Types of caudal fins, showing structure. *A,* Heterocercal (sturgeon, Acipenseridae); *B,* abbreviate heterocercal (bowfin, Amiidae); *C,* homocercal (striped bass, Percichthyidae); *D,* isocercal (cod, Gadidae). (*A* based on Goodrich, 1930; *B* based on Jordan and Evermann, 1900.)

following the markedly asymmetrical heterocercal structure of the sharks, sturgeons, and several extinct groups in which the body axis obviously turns upwards and almost all of the caudal is actually borne on the lower side of the end of the tail. Intermediate stages can be seen in the gars (Lepisosteidae) and bowfins (*Amia*), in which the fins are called "abbreviate" heterocercal since they are only slightly asymmetrical externally. Actually, the homocercal tail of some of the lower bony fishes differs from that seen in the higher forms in that, like the gar and bowfin, more than one vertebra is involved in the upturned portion (Fig. 2–17).

The symmetrical tail of the cods and hakes (Gadidae) is called isocercal. The true caudal of cods is small and is borne on a symmetrical plate at the end of a tapering series of vertebrae. Most of the apparent caudal is actually composed of dorsal and anal fin elements (Fig. 2–17).

Long, tapering, or whip-like tails are called leptocercal. Other symmetrical tails that come to a more abrupt point, as in lungfishes, are called diphycercal. In a few fishes the caudal portion of the body is absorbed during development so that the dorsal and anal fins bridge over

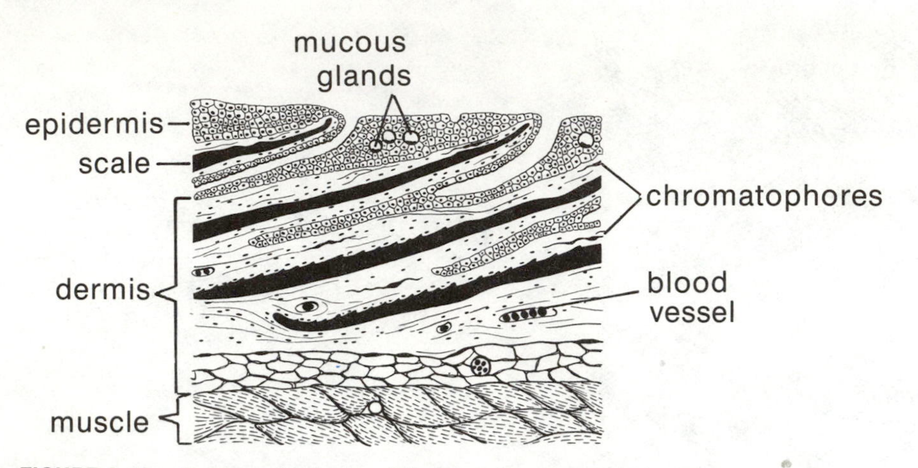

FIGURE 2–18. Section of fish skin. (Based on Wunder, 1936.)

the posterior terminus of the body in what is called a gephyrocercal tail. This is seen in the headfishes or ocean sunfishes (Molidae).

Skin

Fish skin is made up of the usual two layers, an outer epidermis and an inner dermis (Fig. 2–18). Epidermis is typically very thin, composed of from 10 to 30 layers of cells (an average thickness of about 250 microns) in most familiar fishes. Seahorses and their relatives may have only two or three layers of epidermal cells on the surface of their armor. The thickness of the epidermis on these fishes (about 20 microns) contrasts greatly with that found on the lips of sturgeons, which is up to 3 millimeters thick. At the exterior the epidermis consists of squamous cells produced in a columnar germinative layer next to the dermis; these move outward, where they eventually die and are sloughed off. Essentially, except for the mucous covering, live cells of the epidermis are in contact with the medium, for no cornified layer is present. However, in lampreys, a nonliving secretion of the epidermis called cuticle covers the cells. Certain bony fishes, notably minnows (Cyprinidae) and their relatives, secrete horny nuptial tubercles or pearl organs that cover part of the skin. These tubercles roughen the skin and provide friction that helps to keep breeding fishes in contact.

Some epidermal cells are unicellular mucous glands that discharge the mucus that forms the slimy outer covering of fishes. There appear to be two types of mucous cells, one that discharges abruptly and refills, and another that produces slime over a longer period. Mucus consists largely of glycoproteins that can absorb great amounts of water. The champions among mucus-producers are the hagfishes, which have cells that discharge mucoprotein threads of considerable length. There is a published report of a hagfish that turned the water of a display tank so viscous that an octopus could not pass it through its respiratory apparatus.

The slime of fishes is largely protective. In addition to protecting the epidermis and making fishes difficult to grasp, slime can tie up particulate irritants and some heavy metal salts and slough them off. Bacteria may be kept from the live epidermal cells by the mucus. Although mucus performs a lubricatory function by helping fishes slip through the water, it appears to give slight advantage to most fishes and may function mainly in fast starts. There is evidence that mucus aids in osmoregulation in some species by retarding the passage of chloride. Mucus also can precipitate certain suspended solids in muddy water. Special functions of mucus from the skin include use as nestbuilding material in gouramies and as cocoon material in the African lungfishes (Protopteridae), which require an airtight covering during estivation. Parrotfishes (Scaridae) form a mucous envelope around themselves when they rest at night. The mucus of snakeheads (Channidae = Ophiocephalidae) is used in western India to mix extra strong mortar.

Although thin epidermis of a few layers of cells is relatively simple in structure, thicker epidermis may contain nerve endings and pigment cells. Blood vessels usually are absent; nutritive materials diffuse through the intercellular matrix that holds the cells together. Protoplasmic bridges often connect epidermal cells. In some fishes epidermal cells specialized for venom production are found associated with fin or head spines. Chimaeras (Holocephali), stingrays (Dasyatidae), stonefishes (Synanceidae), weevers (Trachinidae), and certain catfishes are fishes with stinging spines. The thick epidermis of the sucking discs of clingfishes (Gobiesocidae) contains alveolar cells that form cushions that help shape the surface of the organ to the substrate.

The dermis is much thicker and much more complicated than the epidermis. It usually is made up of a stratum spongiosum, just beneath the epidermis, and a deeper stratum compactum. Generally, the dermis consists of connective tissue with a paucity of cells. A subcutis of connective tissue lies next to the musculature. The dermis contains pigment cells, blood vessels, nerves, lenses of light-producing organs, and the dermal skeleton, which may consist of plates or scales of various types.

In some bony fishes, especially those without scales, the skin may be thick and tough. Such fish may be skinned and the skin made into leather. Shark skin, with its small, rough placoid scales, has a variety of specialized uses.

Scales

Most fishes are provided with a covering of scales, which are dermal in origin. These may be lacking, as in catfishes, or modified into bony plates or scutes, as in sturgeons (Acipenseridae), sticklebacks (Gasterosteidae), the armored catfishes (Loricariidae, etc.), and seapoachers (Agonidae) (Fig. 2–19). Perhaps the most primitive type of scale among living fishes is the dermal denticle, or placoid scale, of the sharks and their relatives. This consists of a basal plate that is buried in the skin, with a raised portion exposed (Fig. 2–20). The overall structure is similar to that of a tooth, with which these scales are homologous, having a pulp cavity and tubules leading into the dentine. These denticles, with their

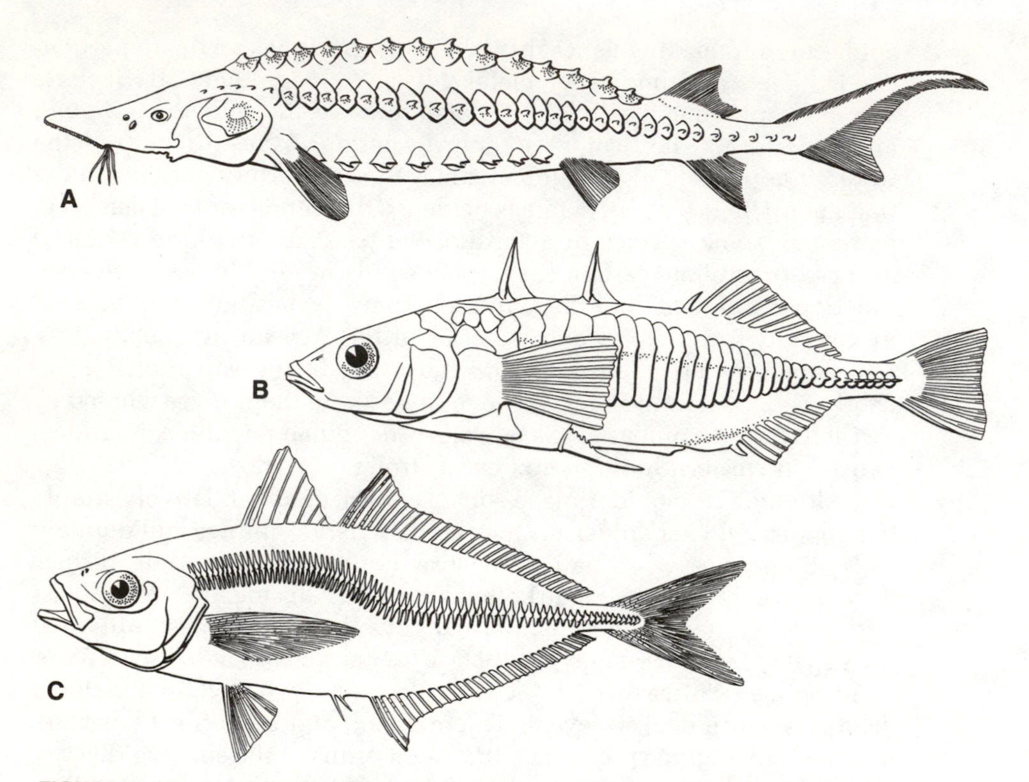

FIGURE 2–19. Examples of fishes with scutes. *A*, Sturgeon (Acipenseridae); *B*, stickleback (Gasterosteidae); *C*, jack (Carangidae). (*A* and *C* based on Jordan and Evermann, 1900.)

hard outer layer of vitrodentine, make possible the use of dried shark skin as an abrasive similar to fine sandpaper.

Many extinct lobefinned fishes and the living coelacanth (*Latimeria*) have cosmoid scales, so-called because of a layer of a hard substance termed cosmine that lies beneath the very thin outer layer of vitrodentine. Below the cosmine is a stratum of vascular bone, and the bottom layer is laminate bone. The scales of living lungfishes have been derived from cosmoid scales but are greatly simplified.

Ganoid scales are encountered on reedfishes (*Erpetoichthys*), bichirs (*Polypterus*), gars (*Lepisosteus*), and in modified form on the caudal fin of sturgeons (Acipenseridae) and paddlefishes (Polyodontidae). In these the ganoid scale has a typical rhomboid shape, with an anterior, peg-like extension of each overlapped by the scale in front. The outer layer of this scale is acellular ganoin, with a cosmine-like layer beneath it. In the extinct palaeonisciforms and the living bichirs and reedfish the cosmine layer is perforated by tubules similar to those present in cosmoid scales, and is underlaid by a vascular area of transverse canals. This type of scale is sometimes termed "palaeoniscoid." The cosmine and tubules are reduced in gars, sturgeons, and paddlefishes.

Scales of most bony fishes are variously called "bony," "bony-

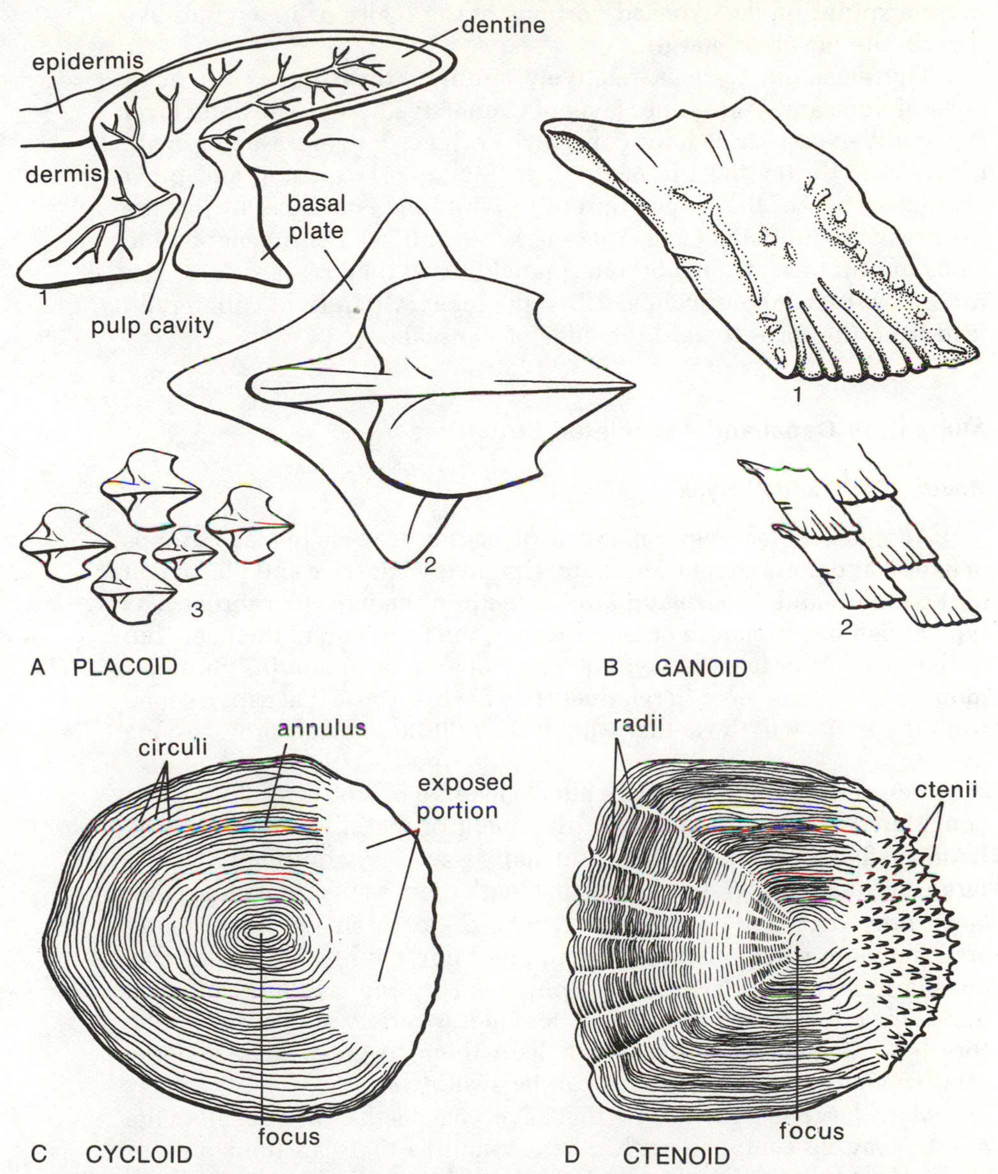

FIGURE 2–20. Examples of types of scales (anterior to left). *A*, Placoid– *1*, sagittal section, *2*, top view, *3*, disposition on skin; *B*, ganoid—*1*, single scale, *2*, disposition on fish; *C*, cycloid; *D*, ctenoid.

ridge," or "elasmoid." These are quite thin as compared to ganoid or cosmoid scales and lie in pockets of the dermis, usually overlapping the scale behind in an imbricated manner. Scales of lower bony fishes are generally ovoid to subcircular in shape and lack spines or projections on the surface or posterior margin; this type of smooth-rimmed scale is termed cycloid. Higher bony fishes tend to have ctenoid scales with minute spines on the exposed portions of the scales or in a comb-like row on the posterior margin.

The elasmoid scale is relatively simple, consisting of an outer layer of bone and a thin inner layer of connective tissue. The bony layer is usually characterized by concentric ridges that represent growth increments during the life of the fish (Fig. 2–20). Spacing and other characteristics of the ridges (circuli) give biologists clues to the life history of the individual fish. Year marks (annuli), spawning marks, and signs of other events may be interpreted by a skilled scale-reader. The innermost plate of the scale is called the focus. Often lines called radii lead from the focus toward the edge of the scale.

Alimentary Canal and Associated Structures

Mouth, Teeth, and Pharynx

Adaptations for diverse manners of feeding are seen in many groups of fishes, and these adaptations naturally involve the size and placement of the mouth and the size and kind of teeth in the mouth or throat. The typical fish has its mouth at, or very near, the front end of the head, but numerous bottom feeders have subterminal or inferior mouths. Superior mouths are encountered in relatively few fishes—those that capture food from the surface or those that wait at the bottom to catch prey passing over.

Size of the mouth can give a clue to feeding habits, especially when considered along with size and placement of teeth. Pikes (Esocidae), handsawfishes (Alepisauridae), and many sharks are equipped with the large mouths and big, sharp teeth that mark them as predators for rather large prey that may be swallowed whole. Some sharks have dental arrangements which enable them to bite large chunks out of animals too big to swallow. Barracudas (*Sphyraena*), piranhas (*Serrasalmus*), and some other characins may do the same. A variety of deepsea predators have dagger-like teeth which help them to grasp prey of large relative size and hold it until it can be swallowed.

Many large-mouthed fishes that have weak teeth or none at all in the mouth may be equipped with other structures that can hold prey or strain plankton out of the water. Pads of small conical or cardiform teeth are seen in many species that are opportunistic in capturing a variety of animal prey. The largemouth bass (*Micropterus salmoides*) is a good example.

Fish with specialized feeding habits may depart from the usual in remarkable fashion (Fig. 2–21). Wolf-eel (*Anarrhichthys*) habitually feed upon shelled animals of various sorts, and have strong canine teeth in the front of the jaws for grasping the prey. In the back of the jaws are blunt molars for crushing the shells. Parrotfishes (Scaridae) can bite off chunks of coral with a beak-like structure formed by the fusion of the front teeth. Slipmouths (Leiognathidae) are capable of extending the mouth half the resting length of the head to siphon in small prey. Many

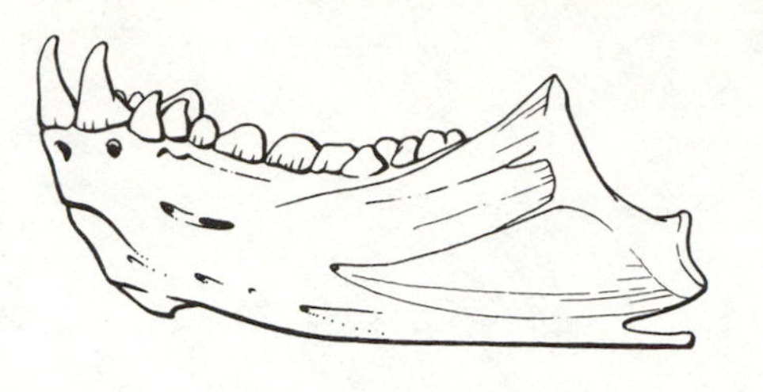

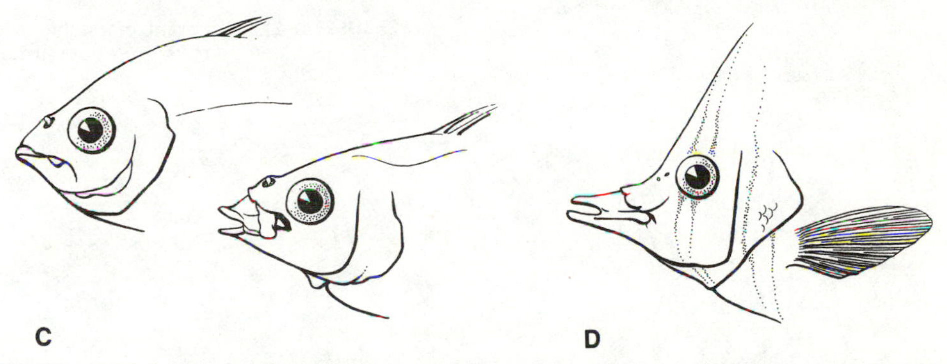

FIGURE 2–21. Examples of teeth and mouths. *A*, Dentary bone of wolf-eel (*Anarrhichthys*); *B*, beak-like teeth of parrotfish (Scaridae); *C*, protrusible mouth of slipmouth (Leiognathidae); *D*, small, specialized mouth of butterflyfish (Chaetodontidae).

butterflyfishes (Chaetodontidae) have small mouths at the end of a thin snout, an arrangement that must be useful in removing food items from crevices.

Teeth are borne on several of the head and face bones. Those in the upper jaw include the premaxillary, the maxillary in the lower bony fishes, and commonly the vomer and palatines. Many species bear teeth on the pterygoids and parasphenoid. In the lower jaw the dentaries are usually the main toothed bones, but teeth may be present on the tongue (glossohyal) and the basibranchials between the gills (Fig. 2–22). Most branchial elements may bear teeth, considering all fishes.

The gills lie just behind the oral cavity in the pharynx; there are commonly four pairs of gills in bony fishes, but sharks and rays may have gills on five to seven arches. The gill arches may be equipped with projections called gill rakers that aid in food gathering (Fig. 2–23). Gill rakers in fishes that consume large prey may be few in number and small, but may carry rough prominences or denticles that aid in holding and swallowing. Plankton feeders usually have an extensive straining sieve formed of long slender gill rakers. A peculiar epibranchial organ for collection and concentration of small food particles is present in certain groups of fishes that feed on plankton and similar materials. The plankton-eating herrings (Clupeidae) and the herbivorous milkfish (*Chanos*) are examples.

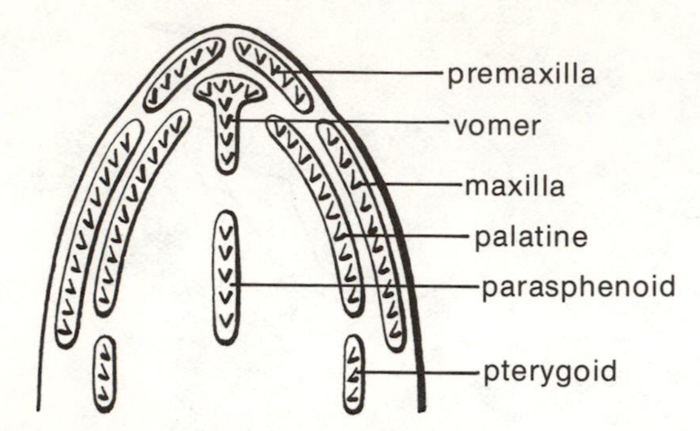

FIGURE 2–22. Diagram of positions of bones that can bear teeth in bony fishes.

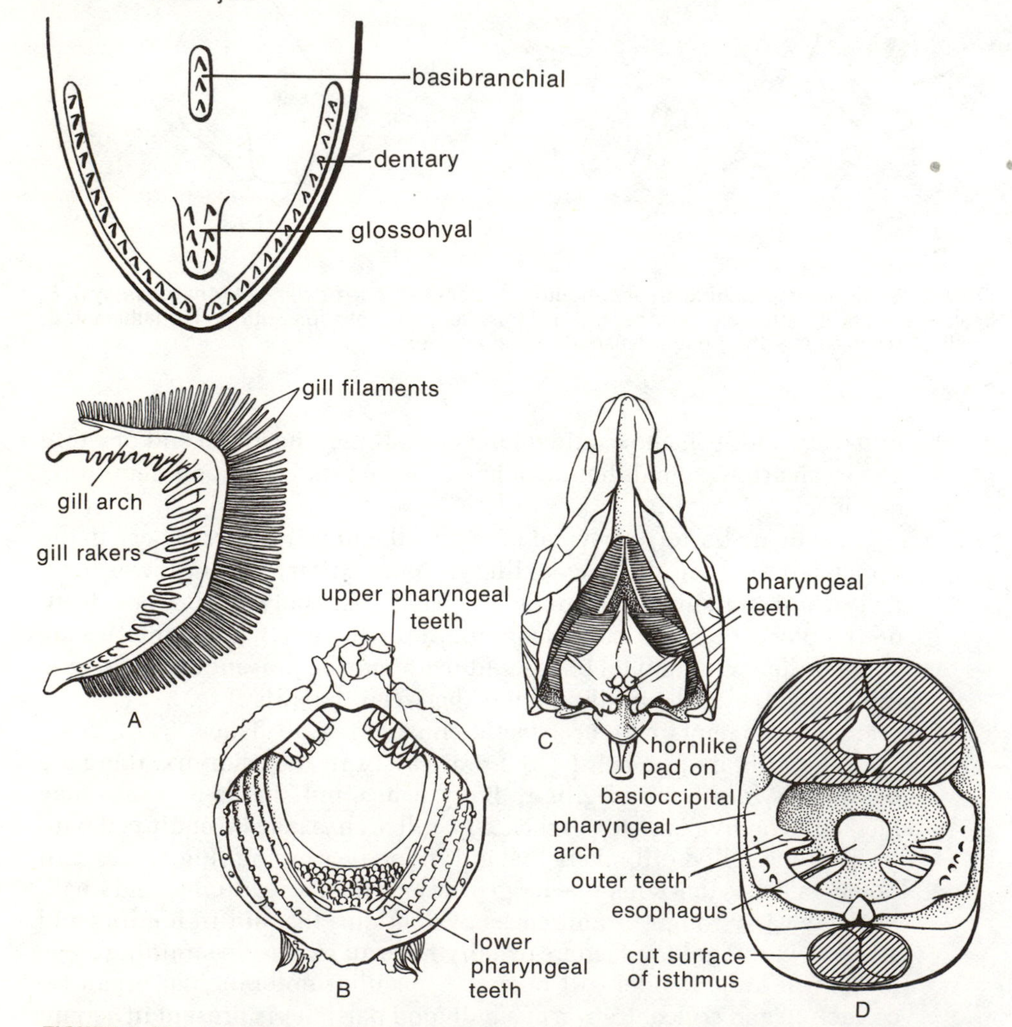

FIGURE 2–23. Examples of gill rakers and pharyngeal teeth. *A,* Diagram of gill arch with rakers and gills; *B,* anterior view of gill arches and pharyngeal teeth of surfperch (Embiotocidae); *C,* ventral view of the pharyngeal region of a carp (*Cyprinus*), with the pharyngeal arch displaced anteriorly to expose the basioccipital pad; *D,* anterior aspect of pharyngeal teeth of the squawfish (*Ptychocheilus*), cross section behind last gill arch with musculature and other soft tissue removed.

The fifth gill arches of bony fishes are usually reduced to a single lower element (the fifth ceratobranchial) on each side. This bone bears teeth which bite against teeth borne on the upper elements (pharyngobranchials) of one or all of the four branchial arches. In minnows (Cyprinidae) and suckers (Catostomidae) the lower pharyngeal bones bite against a pad borne on an extension of the basioccipital bone. Pharyngeal teeth are varied, ranging from small conical points to grinding plates (Figure 2–23).

Esophagus, Stomach, and Intestine

In general, the esophagus in fish is short and distensible so that relatively large objects can be swallowed, but microphagous fishes have less distensible tubes than those of predatory fishes. Esophageal walls are characteristically equipped with both circular and longitudinal layers of striated muscle. The lining of the esophagus consists of stratified epithelium and columnar epithelium with numerous mucous cells or glands. Taste buds are present in some species. Gastric glands appear in the posterior part of the esophagus in some mullets (Mugilidae) and sculpins (Cottidae). In lower bony fish, the esophagus is the site of the connection of the gas bladder with the alimentary canal via the pneumatic duct. (Fish with such a connection are said to be physostomous; those without the connection are called physoclistic.)

Several modifications of the esophagus are known. The butterfishes (Stromateidae) and their close relatives have muscular sacs connecting to the esophagus. In some of the genera (*Pampus, Nomeus*), the esophageal sacs are lined with teeth, which are attached to thin bone in the walls of the sacs. The sacs serve various functions in various species, such as production of mucus, storage of food, or preparation of food by trituration. In a few species the esophagus is modified for respiratory purposes. The rice eel (*Monopterus alba*) and the Alaskan blackfish (*Dallia pectoralis*) are examples of the latter.

The stomach is lacking in lampreys, hagfish, chimaeras and some bony fishes, including minnows (Cyprinidae), sauries (Scomberesocidae) and parrotfishes (Scaridae). In these, characteristic gastric glands are lacking and the esophagus empties directly into the intestine. In most fish, a stomach is present, and of course varies in shape and structure according to the diet of the various species. Usually the stomach is shaped like a "U" or "V", being essentially a bent muscular tube. Another common arrangement is a bag-shaped stomach with openings to the esophagus and gut on the anterior aspect. Heavy-walled, gizzard-like stomachs are found in mullets (*Mugil*), gizzard shad (*Dorosoma*) and a few others (Figure 2–24).

Herbivorous fishes often have intestines many times the length of the body. These long tubes are coiled or folded into distinct patterns in the body cavity (Figure 2–25). Carnivores generally have short guts, and omnivorous fishes have guts of intermediate lengths (Figures 2–24, 2–26). Sturgeons, lungfishes, polypterids, and some other primitive bony fishes, as well as the sharks and related cartilaginous fishes possess a spiral intestine (Fig. 2–27). The absorptive surface is increased by the

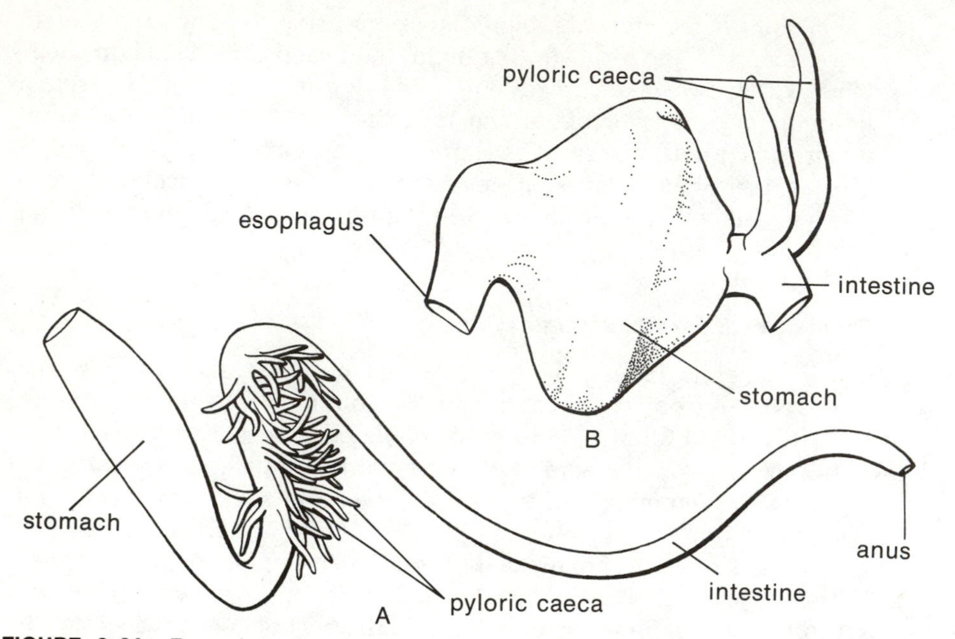

FIGURE 2–24. Examples of stomachs and pyloric caeca (anterior to left). *A,* Stomach, caeca, and intestine of trout (Salmonidae); *B,* stomach and pyloric caeca of mullet (Mugilidae).

corkscrew course of a fold of tissue down the length of the organ. A modification is the scroll-like rolled-valve seen in some sharks.

An interesting contrast is seen in the guts of jawless fishes. The intestine of the parasitic lamprey, which feeds on the blood and juices of its prey, is extremely thin-walled and may be greatly distended, but when empty it appears as a thin cord. The hagfish rasps away the flesh of its prey, and its intestine has a thick wall and an extensively folded lining.

A cloaca is present in sharks, rays, and lungfishes but, except for the male of *Latimeria* and possibly the female of *Nerophis,* a pipefish, the remainder of the fishes lack this structure.

Attached to or associated with the intestine, are the pyloric caeca, liver, pancreas, and air bladder.

Pyloric Caeca. On the intestine of most bony fishes, at the pyloric end of the stomach, there may be from one to many blind sacs, or pyloric caeca. A few groups such as catfishes (Ictaluridae), topminnows (Cyprinodontidae), and pikes (Esocidae) lack these structures. *Polypterus* has only one, the yellow perch has three, and in other groups such as flatfishes (Pleuronectiformes) the pyloric caeca are few, usually no more than five. In others such as mackerels (Scombridae), salmons (Salmonidae), and seasnails (Liparidae) the number of these caeca may range to 200 or more. Caeca of different species vary considerably in size, state of branching, and the connection with the gut. In the sturgeons (Acipenseridae) the many caeca form a large mass, but only a single duct leads

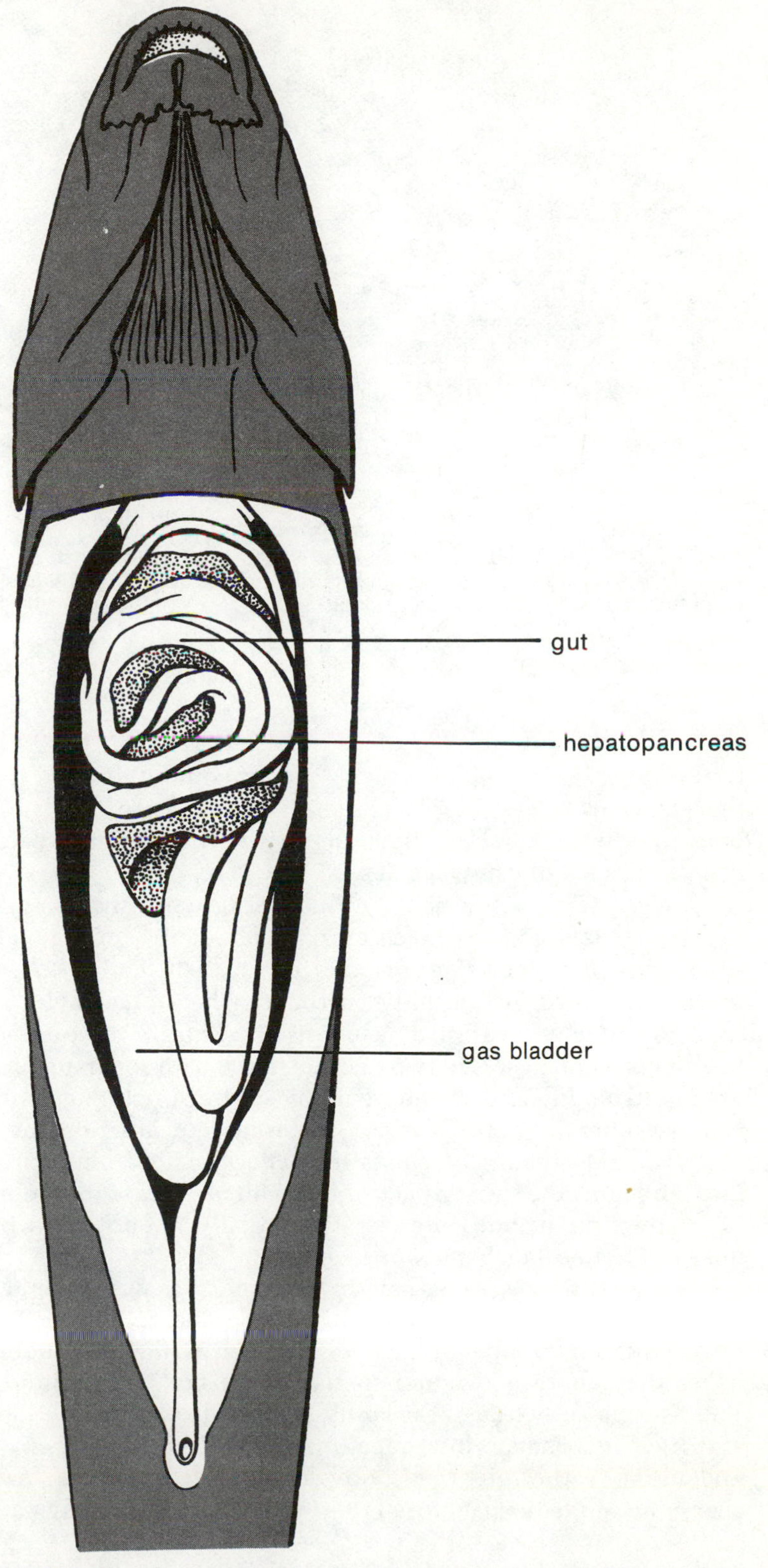

FIGURE 2–25. Elongate gut of sucker (*Catostomus macrocheilus*), a microphagous fish.

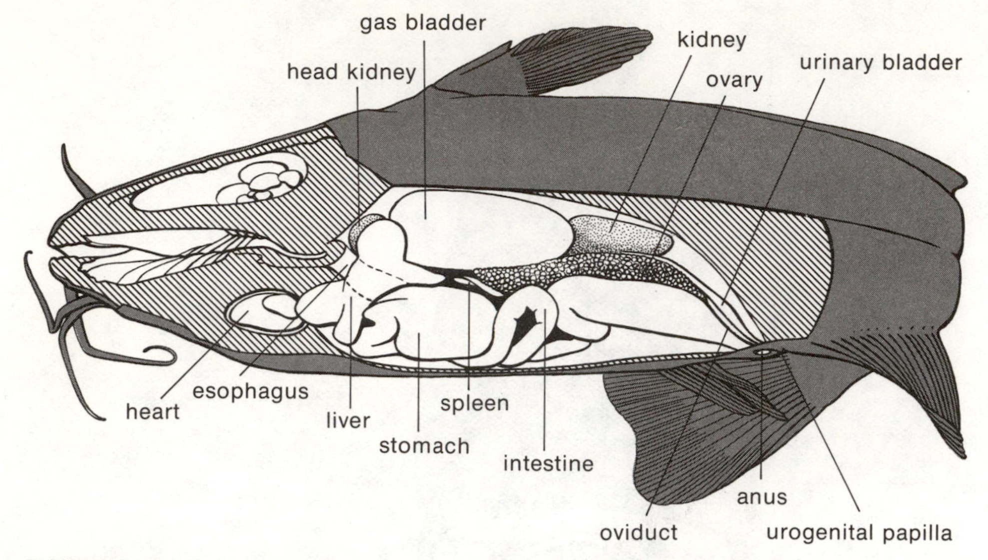

FIGURE 2–26. Bullhead (*Ictalurus*), head sectioned slightly to left of midline, body cavities opened to show relative positions of internal organs. (Note that the head kidney is separated from the renal kidney by the gas bladder.)

to the intestine. In salmon, each caecum communicates directly with the gut (Fig. 2–24). The functions of pyloric caeca probably involve both digestion and absorption. Digestive enzymes have been isolated from the caeca of many species.

Liver. This is a large gland in all fishes, but sharks and rays may have extremely large livers comprising about 20% of the body weight, especially in some pelagic sharks (Figs. 2–26 and 2–27). The liver usually lies over or partially surrounds the stomach. It is typically bilobed, but may have only one lobe as in salmon or three as in mackerel. In hagfishes the liver is in two distinct parts, with separate ducts leading to the gall bladder. Adult lampreys have no bile ducts or gall bladder, but in most other fishes the bladder is present and functions to store liver secretions. Ordinarily, one hepatic duct originates from each lobe of the liver and joins the cystic duct from the gall bladder to form the bile duct. Liver function includes bile secretion and glycogen storage, in addition to several other biochemical processes.

Pancreas. Hagfishes have a small pancreas with several ducts that empty into the bile duct. Lampreys have pancreatic tissue (endocrine only?) located throughout the liver and intestinal wall. Among the bony fishes the pancreatic tissue is usually diffused in or around the liver. This is especially true of the spiny-rayed fishes, in which the pancreas and liver incorporate in a hepatopancreas. The lungfish, *Protopterus*, and many of the soft-rayed bony fishes have a discrete pancreas. In sharks and rays the pancreas is a compact organ, usually consisting of

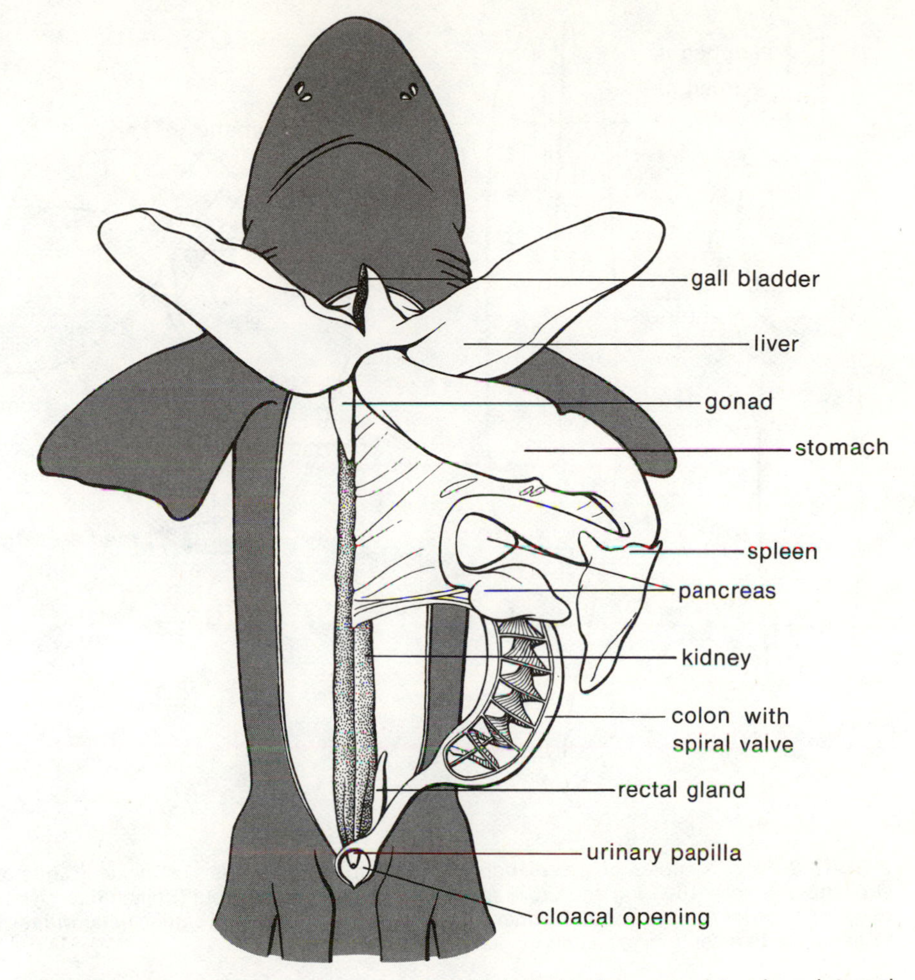

FIGURE 2–27. Diagram of viscera of shark (spiral valve opened to show internal structure). (Based on Daniels, 1934.)

two lobes. The pancreatic duct may reach the small intestine separately from the bile duct, as in the sharks, or may discharge into the bile duct, as in the gar (*Lepisosteus*) or lungfish (*Protopterus*). The pancreas secretes several enzymes that are active in digestion. In addition, the pancreatic islets have the endocrine function of producing insulin.

Spleen. The spleen is usually recognized as a dark red, often pryamidal, structure lying on or behind the stomach, to which it attaches by a band-like ligament. Although it is associated with the digestive organs it has no digestive function, but rather is instrumental in blood cell formation. The function of red blood cell destruction is ascribed to the spleen of higher bony fishes. In lampreys and hagfishes, which do not have a compact spleen, spleenlike tissue is diffused along the intestine. Lungfishes lack the spleen.

Gas Bladder. The gas bladder is a thin-walled sac typically found

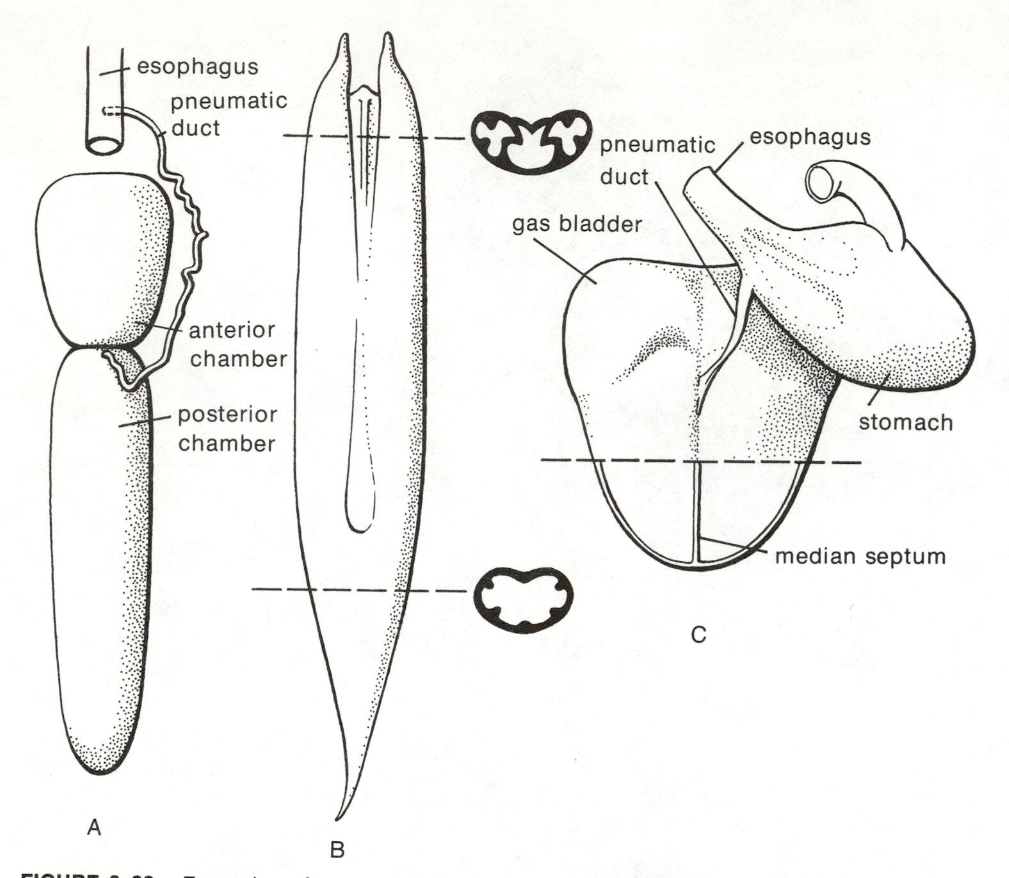

FIGURE 2–28. Examples of gas bladders, shown in ventral views. *A*, Sucker (Catostomidae) showing characteristic long and crooked pneumatic duct; *B*, seatrout (Sciaenidae, *Cynoscion*), showing anterior and posterior chambers in cross section; *C*, channel catfish (Ictaluridae) opened to show median septum (stomach displaced to side).

in the upper part of the body cavity immediately below the kidney (Fig. 2–26). In many fishes the shape is simple, usually somewhat torpedo-shaped, but there are many variations. Minnows and carps (Cyprinidae) have anterior and posterior chambers connected by an opening controlled by a sphincter. Featherbacks (Notopteridae) have the air bladder divided laterally, but the two chambers communicate anteriorly. Drums and croakers (Sciaenidae) have unusual air bladders in that variously shaped sacs or branching caeca may be arranged along each side of the organ (Fig. 2–28).

In the herrings (Clupeidae) the gas bladder has a posterior opening to the exterior near the anus. Some fishes have posterior extensions of the gas bladder reaching beyond the body cavity; viviparous perches (Embiotocidae) extend the gas bladder along the ventral surface of the vertebrae, whereas in the hairtails (*Trichiurus*) the posterior extension of the gas bladder runs along the concave anterior face of the first interhaemal (anal fin support).

In the embryology of bony fishes, the gas bladder originates as an outgrowth of the alimentary canal, and remains attached to the esophagus or the stomach via the pneumatic duct in physostomous fishes. Many physoclistous fishes retain the connection into the larval or juvenile stages, losing it only after they gulp some air into the gas bladder.

Some bottom-dwelling stream fishes such as darters (*Etheostoma*) and sculpins (*Cottus*) lack the gas bladder. Various bathypelagic fishes have lost the gas bladder, and it is also absent in agnaths and the cartilaginous fishes.

Functions of the gas bladder include hydrostatic balancing, sound production, aiding reception of sounds, and respiration. Those modified for respiration are usually compartmentalized and highly vascular, as in the lungfishes, bowfins, and gars. Sound production is often accomplished by special muscles attached to or near the gas bladder.

Gas bladders serve in sound reception by acting as a resonator connected with the ear, either by means of an extension that comes into contact with the ear or through a connecting chain of ossicles known as the Weberian apparatus, a common feature of the ostariophysine fishes including carps, catfishes, and characins (Fig. 11–7).

Gills

Agnatha

The gill tissue of lampreys and hagfishes is arranged as a series of radiating ridges inside expanded pouches that are internal to the branchial skeleton. This skeleton consists of an elaborate basket-like arrangement of considerable elasticity. The pouches are flattened anteroposteriorly, and are somewhat separated from each other. Lampreys have seven pouches on a side, each opening separately to the exterior via a short tube (Fig. 2–29). Internally they open into a special respiratory tube or "pharynx" beneath the esophagus. In larval lampreys, the pharynx is continuous with the oral cavity at the front and with the esophagus behind, similar to the pharynx of hagfishes, but during metamorphosis the pharynx disconnects from the esophagus posteriorly, with an ensuing separation of the two tubes. Anteriorly the entrance to the respiratory tube is guarded by a valvular velum. The special respiratory tube is unique among vertebrates.

Depending on the species involved, hagfishes may have five to fifteen pairs of gill pouches that open internally into the elongate pharynx. External openings are separate in some genera, but in others some or all of the excurrent tubes from the gills may connect with a collecting tube that conveys excurrent respiratory water to an external pore on each side. Hagfishes usually have the respiratory apparatus set well back behind the head. They are unique in possessing a duct that connects the esophagus with the exterior, opening on the left side behind the last branchial opening.

Gill irrigation in the cyclostomes is accomplished by contracting muscles around the branchial area that force water out of the several gill

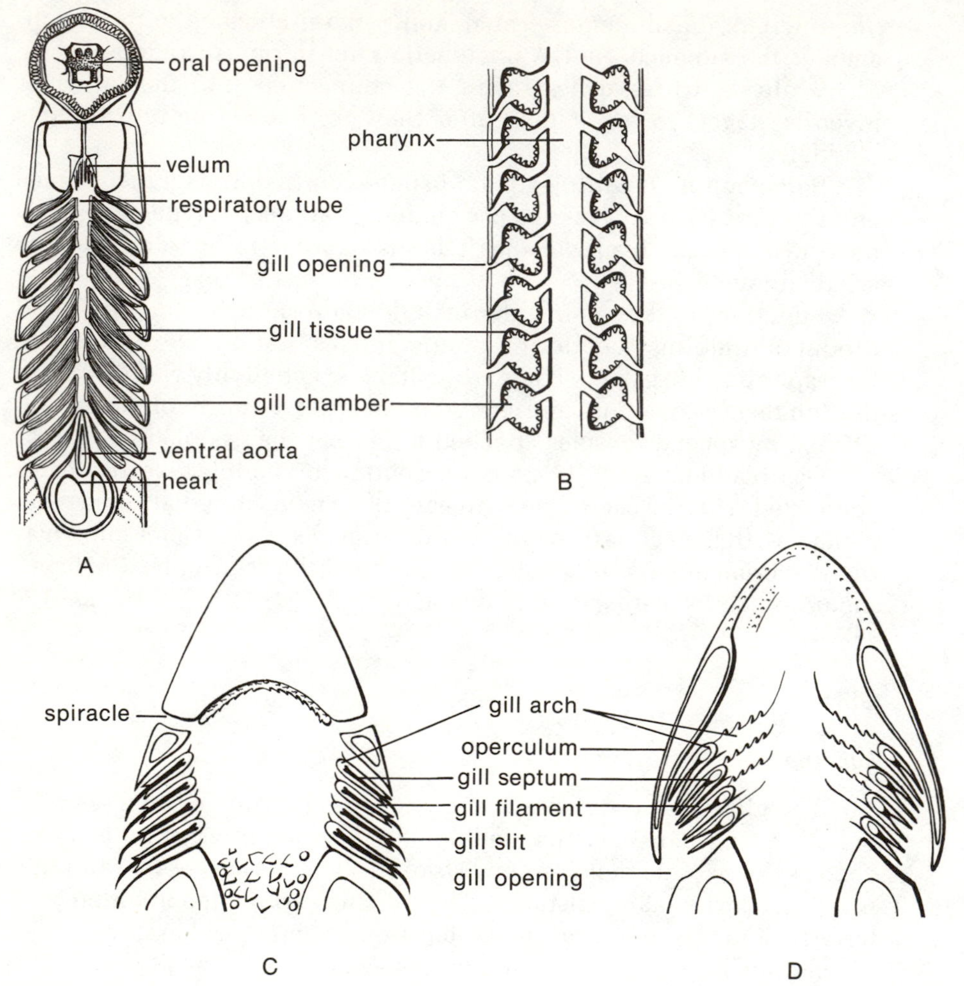

FIGURE 2–29. Diagrams showing arrangement of gills in frontal section. *A,* Lamprey (*Lampetra*); *B,* hagfish (*Eptatretus*); *C,* shark; *D,* bony fish. (*C* based on Weichert, 1951.)

pouches. Elasticity of the cartilaginous branchial basket aids in filling the pouches. Water can enter and leave the individual pouches through the separate external openings of lampreys and those hagfishes that possess separate openings. It can also enter through the mouth, or through the so-called esophageocutaneous duct of hagfishes. Gill irrigation in hagfishes is accomplished in part by a velar pump.

Gnathostomes

In gnathostomatous fishes the gill tissue appears in the form of filaments or ridges on interbranchial septa borne on the gill arches (Fig. 2–29). Both the septa and the gills are external to the branchial skeleton, contrasting with the agnaths. The gnathostomes have the branchial apparatus concentrated into a smaller proportion of the body than do the lampreys and hagfishes.

Cartilaginous Fishes. The interbranchial septa of sharks and rays extend to the body wall so that each branchial chamber is entirely separated from the others, and each has its own opening to the exterior. There are from five to seven of these openings, or gill slits. Cartilaginous branchial rays extend from the gill arch outward within the septum. Each arch and septum bears a series of gill lamellae on both the anterior and posterior face. The gill lamellae on one side of a septum constitute a hemibranch, or "half-gill." The two hemibranchs on a gill arch are called a holobranch. In most sharks the posterior hemibranch of the hyoid arch is present in the first gill pocket, so that an odd number of hemibranchs occurs on each side—nine, eleven, or thirteen, depending on whether the fish has five, six, or seven gill pouches. A remnant of the mandibular gill, called the mandibular pseudobranch, is associated with the spiracle, anterior to the functional gills.

Chimaeras have four branchial arches and four gill pouches covered by a fleshy operculum. Therefore, the gill septa do not extend to the body wall, and are only slightly longer than the gill filaments. Adults have no spiracle and the pseudobranch is absent. There are even numbers of hemibranchs on each side. The hyoid hemibranch is followed by holobranchs on the first, second, and third branchial arches and an anterior hemibranch on the fourth branchial arch. In general the branchial apparatus of chimaeras is more compact than that of the sharks.

Bony Fishes. Among bony fishes the gill septa are progressively reduced (Fig. 2–30). Some of the primitive types, such as sturgeons

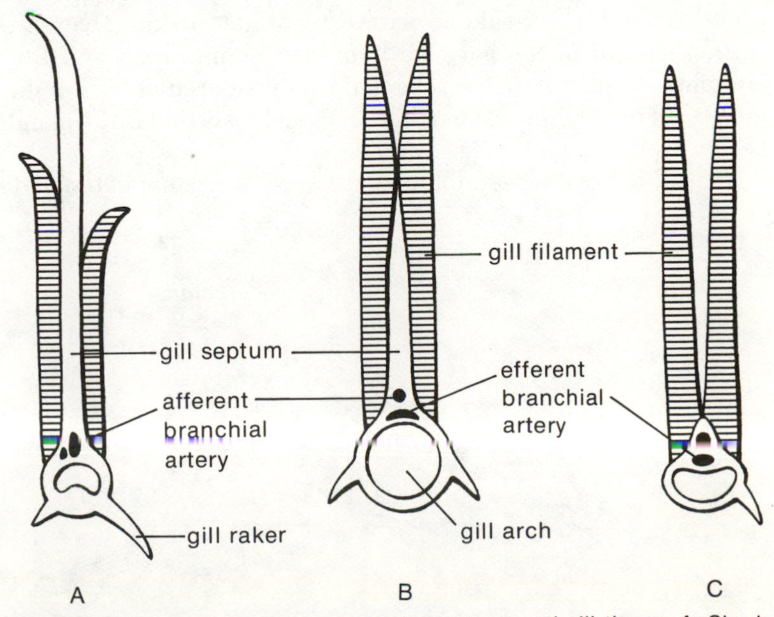

FIGURE 2–30. Relationship of branchial septum and gill tissue. *A*, Shark; *B*, sturgeon; *C*, teleost.

(Acipenseridae) and gars (Lepisosteidae), have slightly reduced septa with the tips of the gill lamellae extending beyond as free filaments. In higher bony fishes the septa become greatly reduced to no more than small ridges along each gill arch, from which the gill tissue extends as long filaments. The gill apparatus is thus more compact than in the sharks, rays, and chimaeras. The gill arches are closely apposed, and the entire chamber is covered on each side by the bony operculum. Loss of the septa results in greater respiratory efficiency because the flow of water through the secondary lamellae is facilitated.

Typical bony fishes retain a pseudobranch at the site of the hyoid arch and holobranchs on each of the four branchial arches. Exceptions to this can be seen in such primitive forms as sturgeons, gars, and coelacanths (Latimeriidae), which retain a hyoidean hemibranch as well as a pseudobranch. The deepsea *Eurypharynx* has five complete gills. Reduction of gills has occurred in some air-breathing teleosts. Members of the family Synbranchidae (swamp eels, etc.) may have only one well developed holobranch.

Among lungfishes the greatest modification is seen in the African lungfishes. *Protopterus*, which retain a hyoidean hemibranch but which have no gills on the first or second branchial arches. There are holobranchs on the third and fourth branchial arches and an anterior hemibranch on the fifth. The Australian lungfish, *Neoceratodus*, has four holobranchs plus a hyoidean hemibranch. None of the lungfishes retain a pseudobranch.

The pseudobranch is evidently the remnant of the primitive gill of the mandibular (or first visceral) arch. The mandibular pouch and slit may be retained as the spiracle in many sharks, all rays, and some of the primitive bony fishes such as sturgeons, paddlefishes, and bichirs. The pseudobranch is associated with the spiracle in the sharks, rays, and sturgeons, but in the gars the hyoidean hemibranch and the pseudobranch are adjacent on the inner side of the opercular base at the anterior portion of the branchial chamber. In the teleosts that is the usual location of the pseudobranch (Fig. 2–31).

In all fishes the pseudobranch receives oxygenated blood only, even

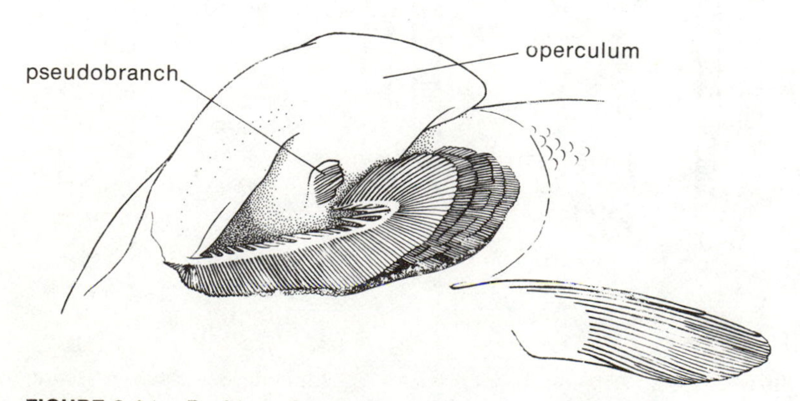

FIGURE 2–31. Position of pseudobranch in trout (Salmonidae).

though in some species it closely resembles a functional gill. Therefore, it cannot function in respiratory gas exchange. In fact, most bony fishes have the structure reduced to the appearance of a glandular organ that may be situated beneath the skin. The pseudobranch appears somehow to be involved in the function of the choroid rete, which secretes oxygen to the retina of the eye. Blood runs directly from the pseudobranch to the choroid gland, and there is speculation that it may receive some substance from the pseudobranch to aid in oxygen secretion. Some research has shown that secretion of gas into the gas bladders of certain species may be facilitated by secretions of the pseudobranch.

Typically, the nearly continuous flow of water across the gills of bony fishes depends upon the coordinated action of the mouth and the opercular apparatus, including the branchiostegal membrane and rays. As the mouth opens to take in water, the opercular flap is closed and the gill cavity is expanded by outward movement of the operculum and a flaring of the branchiostegal membrane, so that the pressure in the gill chamber is less than that in the buccal cavity. At this stage the water is flowing into the mouth and through the gills to the gill chamber. Next, the mouth closes, the opercular flaps open, and the operculum and branchiostegal apparatus narrow the gill chamber in a sequence that maintains water flow through the gill lamellae. Some fishes, such as pikes, have prominent oral valves in the anterior part of the mouth to prevent backflow.

Vascular System

The typical fish heart is said to have four chambers: the sinus venosus, the atrium, the ventricle and the conus arteriosus (Fig. 2–32). It is located in a pericardial cavity below or slightly behind the gills. Although the various groups of fishes have many features of the heart in common, there are some characteristic features that warrant considering the groups in separate paragraphs, at the cost of a little repetition.

The hagfish heart is the most primitive among the fish-like vertebrates. The organ is less compact than in other fishes; the sinus venosus is well developed, receives several veins (only the left duct of Cuvier is developed), and is partially divided into anterior and posterior portions by a fold of tissue. The atrium discharges through a narrow connection to the thick-walled ventricle, which pumps the blood into a non-contractile, but elastic, basal part of the ventral aorta called the bulbus arteriosus. A conus arteriosus appears to be lacking, as the semilunar valves preventing backflow from the bulbus are set into the walls of the ventricle itself. The expansible bulbus aids in maintaining a flow of blood to the gills. In addition to the "branchial" heart described above, the hagfishes have three additional hearts of differing design that aid in circulating the blood. A cardinal heart in the head pumps venous blood toward the branchial heart, as does a caudal heart, located

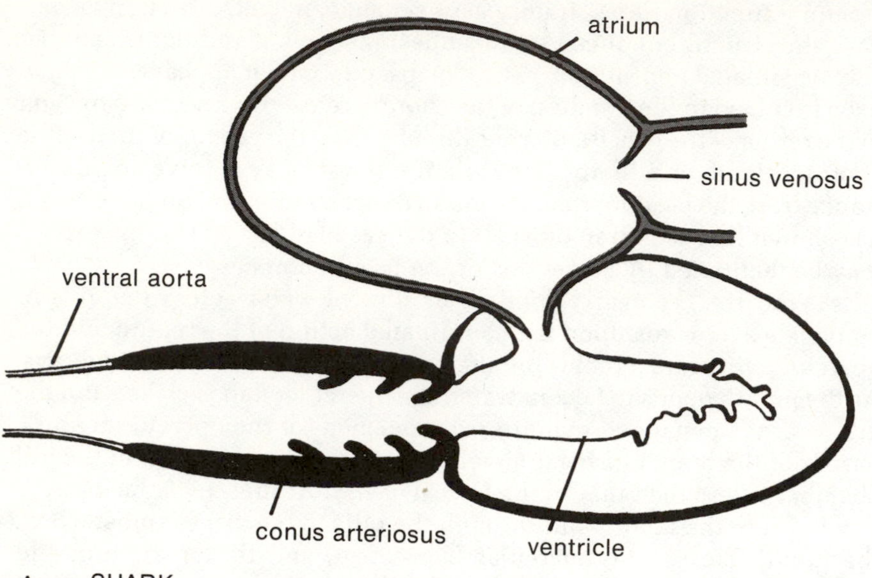

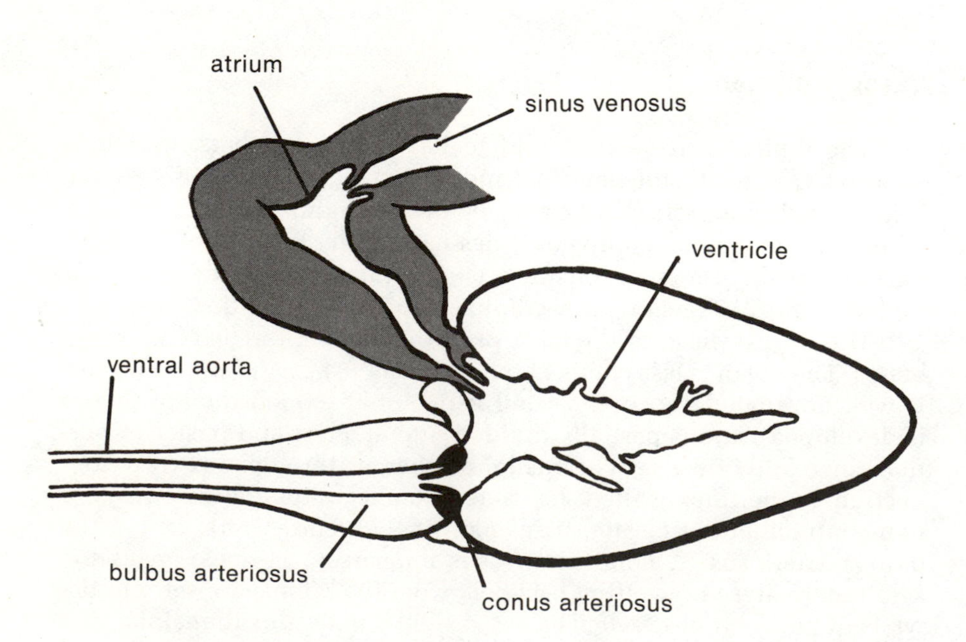

FIGURE 2–32. Diagrams of heart in shark and bony fish.

far back near the end of the tail. A portal heart pumps blood through the two lobes of the liver.

The heart of the lamprey is more compact than that of the hagfish and is relatively large in comparison to the hearts of most fish. The sinus venosus is a small, vertical, tubular structure, and the atrium overlies the ventricle. A bulbus arteriosus is present at the base of the ventral aorta. The conus is undeveloped. Only the right common cardinal (duct of Cuvier) is present.

In elasmobranchs the sinus venosus, atrium, ventricle and conus arteriosus are well developed. Unlike lampreys and hagfishes, both right and left common cardinals are developed and the bulbus arteriosus is absent.

Bony fishes are more variable in heart structure than the other groups because of the great evolutionary range seen in living forms. Lungfishes, chondrosteans (sturgeons, etc.), holosteans (bowfins, etc.) and some lower teleosteans (tarpons, etc.) retain a contractile conus arteriosus with two or more rows of valves. In most teleosts the conus is reduced to a very small structure bearing a set of valves between the ventricle and the nonmuscular, expansible bulbus arteriosus. This may appear as a small, white dilation in a dead fish but, when the heart is pumping, the bulbus expands to the size of the ventricle. The blood is prevented from flowing back into the ventricle by the valves on the reduced conus, so the elastic walls of the bulbus maintain a strong pressure on the blood flowing to the gills, maintaining an almost steady flow in contrast with the pulsed flow from a conus arteriosus.

The atrium of the lungfish heart is essentially divided into two parts by an incomplete septum. The right division receives unoxygenated blood via the sinus venosus, but the left side receives oxygenated blood via the pulmonary vein. Virtually complete separation of these two streams is thought to be maintained through the atrium, and mixing in the ventricle is minimized by another incomplete partition. The conus of lungfishes is flexed and provided with semilunar valves and, especially in *Protopterus,* a peculiar "spiral valve" that divides the separate streams coming from the ventricle. Oxygenated blood is mainly guided to the first and second gill arches (which lack gill tissue) and thence to the dorsal aorta.

The ventral aorta of fishes extends forward underneath the pharynx. In lampreys the aorta remains single to the fourth gill pouch and splits into right and left branches at the septum between this and the third pouch. Eight pairs of afferent branchial arteries branch from the aorta and enter the walls of the gill pouches (Fig. 2–33).

Afferent branchial arteries in elasmobranchs arise from the single ventral aorta and enter each of the branchial arches to supply blood to the holobranchs borne by these arches, and to the hyoidean hemibranch. There are five pairs of branchial arteries in most rays and sharks, but more in those having six or seven gill slits. Teleosts are similar in this respect to the elasmobranchs,. except that the afferent branchial artery leading to the hyoidean hemibranch is absent. The sturgeons more

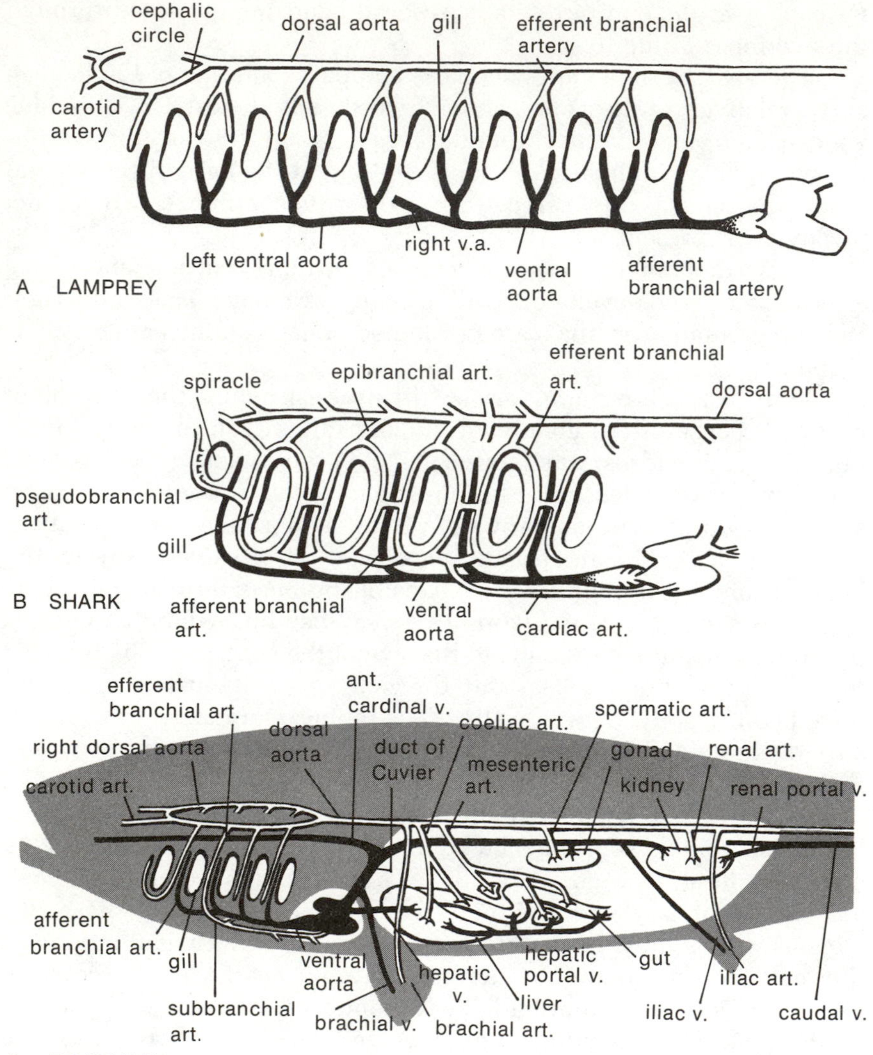

FIGURE 2–33. Diagrams of blood circulation. *A*, Lamprey, branchial arteries; *B*, shark, branchial arteries; *C*, bony fish, showing major blood vessels.

closely resemble the sharks, and some of the holosteans are intermediate between the sharks and bony fishes in pattern of branchial vessels. In lungfishes the ventral aorta is short; the afferent branchial arteries branch off from the conus arteriosus very close to the heart.

In the gill arch the afferent branchial arteries give rise to arterioles that terminate in capillaries or open spaces in the gill lamellae. After traversing these capillaries and lacunae the blood, now oxygenated, is collected by branches of the efferent branchial arteries and conveyed by those arteries to the dorsal aorta for distribution to the body.

The dorsal aorta of lampreys and hagfishes is single and median except for a peculiar section called the cephalic circle in the region of the first gill pouch or anterior to it. From this circle major arteries supply the head region with blood. In most other fishes, the dorsal aorta is paired anteriorly and continuations extend to the head as the internal carotid arteries. Posteriorly, the dorsal aorta is unpaired through the trunk region and continues into the tail as the caudal artery, running through the hemal arches of the vertebrae. Arteries supplying the viscera and musculature branch from the aorta and caudal artery.

The caudal vein also runs through the hemal arches and is ventral to the caudal artery. In lampreys and hagfishes, the caudal vein splits to form the paired posterior cardinals. In other fishes the caudal vein enters the kidneys through the renal portal system. Postcardinal veins receive blood from the kidneys and gonads and from the musculature as they run forward to join the common cardinal veins (ducts of Cuvier). Precardinal veins, subclavian veins, and the jugular veins enter the ducts of Cuvier. Blood from the ducts enters the sinus venosus, which receives blood from the hepatic portal system as well (Fig. 2–33).

Urogenital System

Kidneys

Kidney structure and function in fishes are extensive and complex subjects, considering the wide evolutionary span of the animals involved and their myriad adaptations of form and physiology. Various modes of life in fresh and salt water have required structural adjustment of the kidneys to accommodate changing function. Association of the kidneys with the genital system is close in some fishes, but the systems are virtually separate in most bony fishes. Utilization of the degenerate anterior portion of the kidney as a hemopoietic area varies from group to group. These and other considerations make it obvious that only a portion of this highly interesting and important subject can be treated here. Following a general orientation, a few examples of gross kidney structure and relationships will be given. Function of the kidneys will be discussed in Chapter 13.

Kidneys of most fishes are slender, elongate, dark red organs extending along the dorsal aspect of the body wall just ventral to the vertebrae (Figs. 2–26 and 2–27). When viscera are removed from the body cavity the kidneys can be seen through the peritoneum. Kidneys are paired, but are usually placed close together in most bony fishes; fusion along the midline is not uncommon. Excretory function is concentrated in the posterior section. The anterior part of the kidney is subject to modification in structure and function. In males of elasmobranchs, chimaeras, and primitive bony fishes such as sturgeons, gars, and bowfin, the anterior part is involved in the reproductive system. In

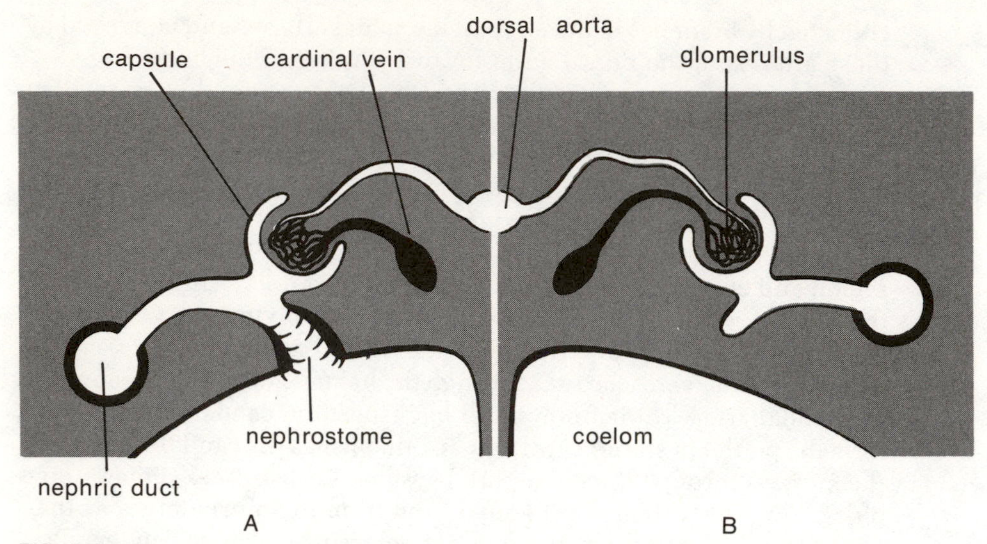

FIGURE 2–34. Diagrams of kidney types. *A,* Pronephros; *B,* opisthonephros. (Based on Goodrich, in Lankester, 1909.)

most teleosts, the anterior part of the kidney has a concentration of lymphoid and hemopoietic tissue, with chromaffin (suprarenal) and interrenal tissue distributed along the postcardinals.

The structural unit of the kidney is the nephron or kidney tubule, which consists of a renal corpuscle and a convoluted tubule, the latter leading to ducts that convey urine to the exterior. The renal corpuscle is made up of the double-walled Bowman's capsule and a glomerulus, a mass of capillaries coiled inside the capsule. The lumen between the walls of Bowman's capsule is continuous with the remainder of the tubule, which consists of several segments. Examples of tubule structure will be given later.

The excretory kidney of most fishes is an opisthonephros, which resembles the mesonephros that appears in the embryonic stages of amniotes (Fig. 2–34). In an opisthonephros, except in the hagfishes, in which the essential features of the mesonephros are retained, the tubules are not arranged segmentally and may be concentrated posteriorly. The renal corpuscle is present and the nephrostomes are absent. A more primitive pronephros, with segmentally arranged tubules opening by means of ciliated funnels (nephrostomes) to the abdominal and pericardial cavities, is present during embryonic development of fishes and persists in modified form in many species. It persists in functional form in hagfishes and in larval lampreys. In both of these the nephrostomes enter the pericardial coelom. A few bony fishes, such as certain lanternfishes (Myctophidae), eelpouts (Zoarcidae), and clingfishes (Gobiesocidae) retain the pronephros. In these instances the pronephros is always anterior to the opisthonephros, separated from it in hagfishes but intergrading in most others. A kidney retaining both pronephros and opisthonephros elements is sometimes called a holonephros, while others reserve the term for a kidney in which each trunk segment has a tubule.

Each kidney is provided with a duct draining posteriorly to a juncture with its fellow to form a median duct, or to a bladder or sinus. The major duct draining each kidney is called the archinephric (or nephric) duct in most fishes, although a new, special urinary duct draining the kidney is developed in the elasmobranchs and chimaeras.

Hagfishes. As previously mentioned, the hagfishes retain the pronephric kidneys. These are paired, small, and situated some distance anterior to the mesonephros and dorsal to the heart on the pericardial wall. Usually one large renal corpuscle, or glomus, consisting of three glomeruli occurs in each pronephros, at the posterior end. There appears to be no connection of the corpuscles and tubules to a functional archinephric duct or to any other excretory duct. The nephrostomes enter the pericardial coelom (which is in free connection with the body cavity) and the tubules connect with the pronephric vein. The function of the pronephros in the adult hagfish is problematical but the organ is thought to be periodically hemopoietic and may have a lymphoid function. The mesonephros of hagfishes is long and slender, extending nearly the length of the body cavity. The more than 30 renal corpuscles are irregularly arranged over the length of the kidney. Tubules are unciliated and quite simple, consisting of a short neck segment and a short proximal segment that empties into the archinephric duct. A urinary sinus receives the archinephric ducts and communicates to the exterior through a urogenital papilla.

Lampreys. The opisthonephric kidneys of adult lampreys are suspended from the dorsal body wall along a mass of adipose tissue. They are long and strap-like, somewhat comma-shaped in cross section with a dorsal portion tapering to a thin lower edge. The fine anatomy has been studied by Drs. J. H. Youson and D. B. McMillan (1970) on whose work the following discussion is based. The upper, thicker portion is dominated by lymphoid tissue, the archinephric duct courses along the lower, thinner edge, and the bulk of the organ consists of a single, elongate renal corpuscle having numerous tubules. This compound glomerulus or glomus has no typical Bowman's capsule; however, that structure is represented in the larvae, which have several renal corpuscles, with the posterior ones compounding into glomera.

The tubules of the lamprey kidney consist of a ciliated neck segment, a proximal segment divided into a convoluted and a straight portion, an intermediate segment, a distal segment with straight and convoluted parts, and a collecting segment. There are two types of intermediate segments, a short type and a longer type that may be involved in urine formation. The collecting tubules communicate with the archinephric duct, which has muscle tissue in the walls, possibly for the expulsion of urine. The archinephric ducts from each side merge and enter a urogenital sinus, which opens to the exterior via a urogenital papilla.

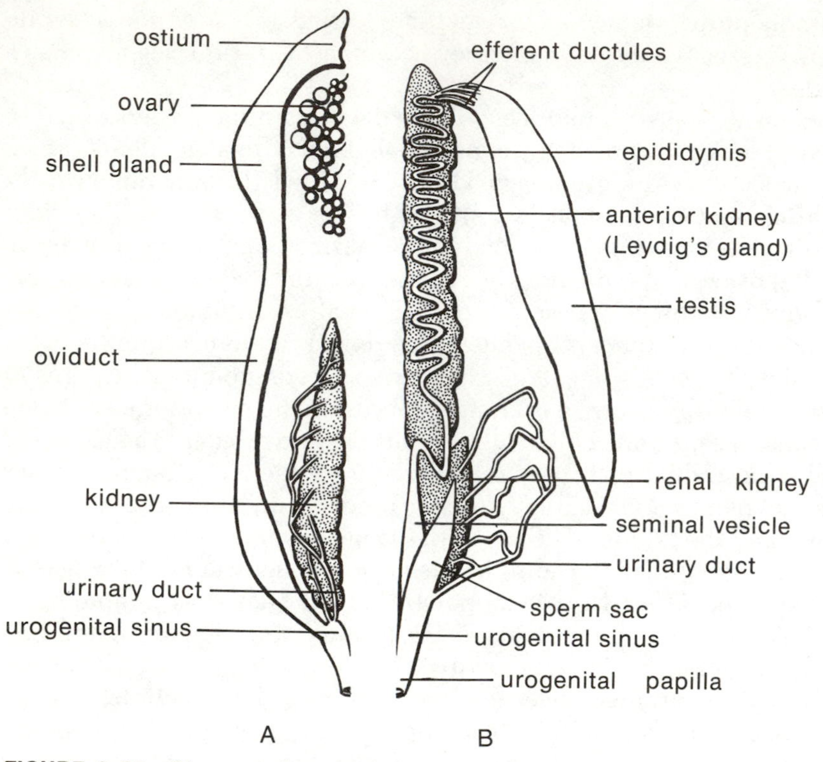

FIGURE 2–35. Diagram of urogenital organs of shark. *A*, Female; *B*, male. (Based on Goodrich, in Lankester, 1909.)

The kidney of larval lampreys is a holonephros, with the archinephric duct draining the pronephros and entering the opisthonephros dorsally. The larval kidney apparently does not contribute to the formation of the adult kidney. It degenerates and an entirely new structure forms during transformation.

Elasmobranchs. The opisthonephric kidneys of sharks and rays are usually flattened, band- or strap-shaped structures that are wider in the posterior sections (Fig. 2–35). Shapes may vary, with some species showing lobulations along the lateral edges. This lobate nature is especially prominent in rays, which in addition may have the kidney of the female confined to the posterior part of the body cavity. The anterior part of the kidney of females is usually reduced in sharks also. The anterior part of the kidney of male elasmobranchs is rather enlarged, and functions as part of the genital system. The most anterior part forms an epididymis, and between this and the functional part of the kidney is Leydig's gland (see p. 56).

Nephrons of the modified front part of the male kidney may become modified by losing the renal corpuscles and secreting seminal fluid. These tubules have enlarged lumens and empty into the archinephric duct which, depending upon the species involved, may have varying relationships to the functional kidney. It is usually converted into a sperm duct for the transport of sperm and seminal fluid, but in some species receives urine through separate urinary ducts. In others collecting tubules convey urine to urinary ducts that have no connection with the archinephric duct.

In males, the archinephric duct (sperm duct) usually enlarges into a seminal vesicle which, along with the urinary duct, empties into the urogenital sinus. In females the archinephric ducts typically drain into a urinary sinus.

Kidney tubules of sharks and rays consist of a long neck segment, an initial proximal segment composed of two cytologically distinct regions, a second proximal segment, and a distal segment followed by a collecting tubule. The kidneys of chimaeras (Holocephali) resemble the kidneys of elasmobranchs.

Bony Fishes. Sturgeons and paddlefishes have elongate kidneys extending the length of the body cavity. The pronephric portion, or head kidney, is included. The kidneys are fused posteriorly but taper forward separately. Anteriorly the kidney of the male is associated with the genital system, receiving many efferent ducts from the testis. Many of the tubules are modified for sperm transport. The nephric ducts from each side meet posteriorly in an expanded bladder-like section. In *Polypterus*, the bichir, it is the posterior section of the kidney that is most intimately associated with the testis, but there is no connection of the sperm duct with the archinephric duct short of the urogenital sinus.

In the bowfin (*Amia*), the gars (*Lepisosteus*), and lungfishes (*Neoceratodus, Protopterus, Lepidosiren*) the relationship between the testes and the kidneys resembles that seen in sturgeons and paddlefishes. In the lungfishes, the nephric ducts from the left and right kidneys run separately to the cloaca except for the ducts of male *Protopterus*. These unite before entering the cloaca.

Salmon, trout and other salmonoids have relatively large, thick kidneys of the holonephric type, with an expanded head kidney representing the pronephros at the anterior end.

Teleosts, being more numerous than other fishes, and having wider geographic and ecological distribution, show a variety of kidney structures. Collectively, the teleosts have the following nephron structures (Figure 2–36): glomerulus, neck segment, a two-part proximal segment, an intermediate segment, a distal segment and the collecting tubule. Typically, freshwater fishes have all the components mentioned; marine fishes usually lack at least the distal segment, and may lack others. Actually, the only nephron structures that all teleosts seem to have in common are the second proximal segment and the collecting tubule. Marine fishes, which lose water extra-renally, usually have smaller and fewer glomeruli than freshwater species.

Although there is probably insufficient information on the thousands of species to arrive at a definitive classification of teleost kidneys, Ogawa (1961) proposed a classification based on configuration or shape of marine teleost kidneys and pointed out general relationships to phylogeny and habitat, as some of the types are encountered in freshwater species.

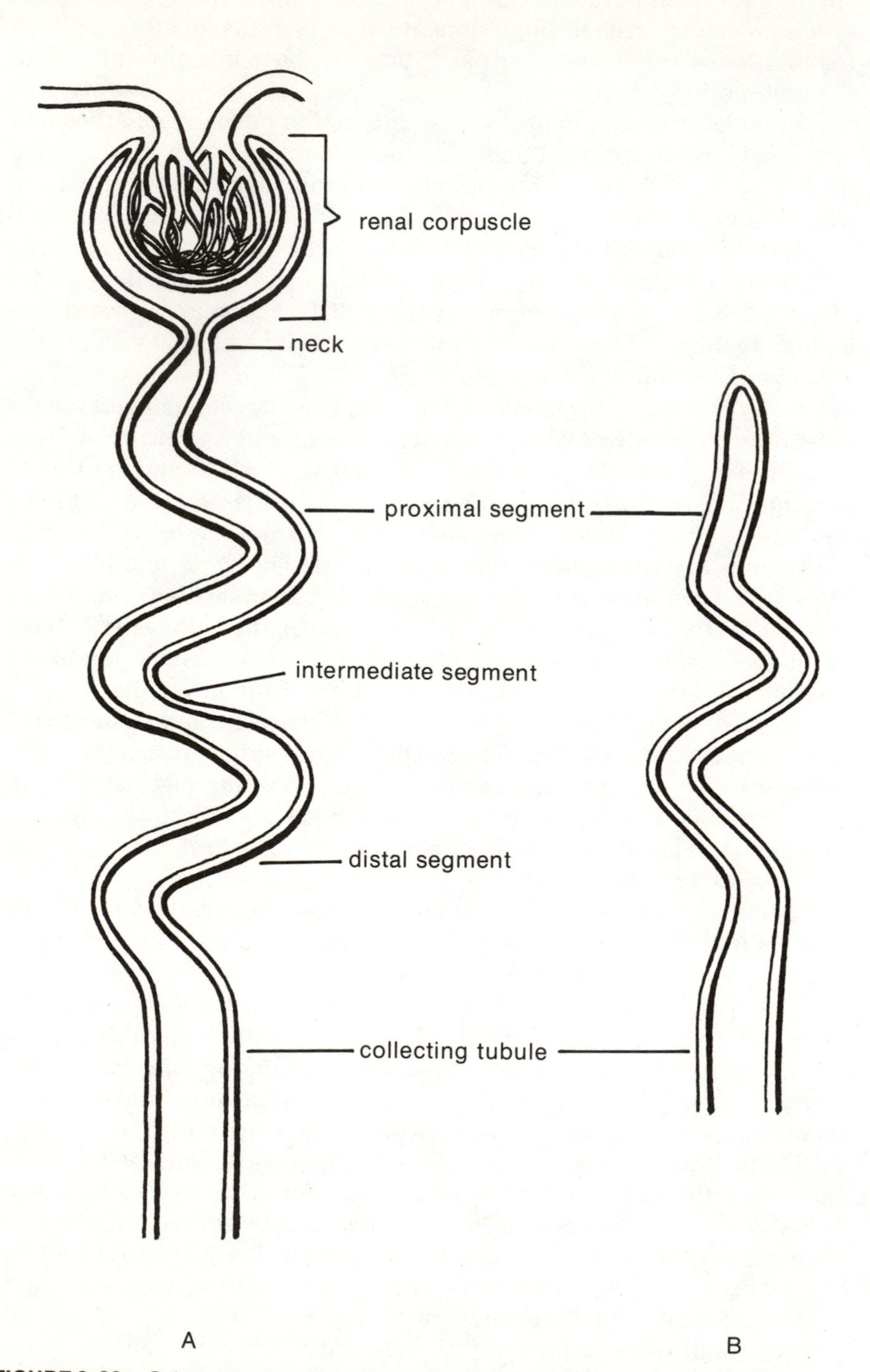

FIGURE 2–36. Schematic diagrams showing components of two types of nephrons found in bony fishes. *A*, Glomerular; *B*, aglomerular (in some marine fishes).

Salmon and trout (Salmonidae), herrings (Clupeidae) and the ayu (Plecoglossidae) have kidneys fused throughout their lengths and the opisthonephros continuous with the head kidney. The head kidney may be somewhat expanded laterally, especially in the salmonids. It consists of lymphoid tissue, with some suprarenal tissue included, and has lost the typical renal function. This corresponds to Ogawa's type I. The nephron is mostly typical of marine fish, although many of the clupeoids and salmonoids live in fresh water. The glomerular capsule is large (ca 85–105 μ) as in freshwater species, but the neck segment and distal segment are lacking according to Ogawa.

Many minnows (Cyprinidae), the loach (*Misgurnus*), some catfishes and eels have kidneys fused to each other anteriorly and posteriorly, but separate through the middle section. The head kidney is usually expanded laterally. This is Ogawa's type II. These have large glomerular capsules (ca. 60–95 μ) and all segments are present in the freshwater species and in the marine catfish genus *Plotosus*. Morays (*Gymnothorax*) lack the distal segment.

Ogawa recognized his type III in perch-like fish (Percoidei), gobies (Gobioidei), barracuda (*Sphyraena*), blennies (Blennioidei), the medaka (*Oryzias*), the snakehead (*Channa*), lantern fishes (Myctophidae), mackerels (Scombridae), sculpins (Cottidae), flounders (Pleuronectidae) and others. In these the kidneys are found posteriorly only and the head kidneys are well-differentiated from the opisthonephric kidney. The glomerular capsule is relatively small (ca. 40–70 μ) and the distal segment is absent in the marine forms. The freshwater *Channa* and *Oryzias* retain the distal segment.

Pipefishes and seahorses have narrow kidneys connected only at the most posterior portion. The head kidney is not developed. This represents type IV. The nephron is greatly reduced, there being only the second portion of the proximal segment and the collecting segment.

Type V kidneys are completely separated from each other. The head kidney is developed (and may even retain pronephric glomeruli in some angler fishes). In addition to angler fishes (Lophiiformes) this type occurs in puffers (*Fugu*, *Canthigaster*) and boxfishes (*Ostracion*) and other closely related fish. Some species are reported to lack glomeruli: the goosefish (*Lophius piscatorius*), the sargassumfish (*Histrio histrio*) and porcupine fish (*Diodon* spp.) are examples. These retain only the second proximal segment and the initial collecting tubule. Species with glomeruli lack the distal segment.

Gonads

The gonads of fishes are usually elongate structures suspended by mesenteries from the dorsal aspect of the abdominal cavity. Their relationship to the kidneys and their ducts differs widely among groups, so that presenting a meaningful general description is difficult. Examples will therefore be given for several groups, and in each notable exceptions from the general plan will be mentioned. Details of reproduction will be the subject of a later chapter.

Lampreys. The gonads are single, suspended from the midline, and reach most of the length of the body cavity. In immature specimens they appear as thin lobulated structures, but at maturity they may virtually fill the body cavity, crowding the other viscera. The gut, which may be greatly distended during the feeding stage, becomes small and nonfunctional in mature fish and may be almost completely hidden by the ripe ovary or testis. No sperm ducts or oviducts are present; at spawning both the eggs and sperm are shed into the body cavity, from which they exit through paired abdominal pores to the urogenital sinus. Eggs are somewhat over a millimeter in diameter at extrusion. A prominent urogenital papilla is developed in mature specimens.

Hagfishes. As in lampreys, a single elongate gonad is present in both sexes. The testis is irregular and lobate, without a sperm duct, and sperm are released into the body cavity. Unlike the lamprey, hagfishes produce large eggs with tough shells. There is no oviduct. Both eggs and sperm reach the exterior through an abdominal pore just behind the anus. In hagfishes the gonads are suspended by a mesentery from the gut.

Sharks. The testes are paired and usually placed anteriorly in the body cavity, suspended dorsally by means of a mesorchium. Often the right testis is larger than the left. Sperm discharges into a central canal network that communicates with the anterior part of the kidney through efferent ducts traversing the mesorchium. The front part of the kidney is modified into a glandular epididymis, where the archinephric duct receives the ductuli efferentes and runs posteriorly via a coiled and tortuous path. The time spent in traversing the epididymis may be involved in the maturation of sperm. Just behind the testis the kidney is modified into Leydig's gland, in which the tubules secrete a seminal fluid into the archinephric duct (Fig. 2–35).

As the archinephric duct runs posteriorly it courses along the functional kidney as the vas deferens, and then enlarges into a seminal vesicle from which a sperm sac opens dorsally. The vesicles and sperm sacs open into the urogenital sinus, which in turn empties into the cloaca. From the cloaca sperm enter the grooves of the claspers, through which they can be transferred to the female. Associated with the claspers, under the skin of the pelvic fin and abdomen, are glandular sacs called siphons that secrete a lubricatory fluid.

Ovaries are paired, but the left one may be greatly reduced in size in some species. Like the testes, they are placed well anteriorly in the body cavity; each is suspended by a mesovarium. The oviducts open anterior to the ovaries and usually have a common mouth or funnel. Eggs are released into the coelom, proceed into this funnel, and then travel down the oviduct to the region of the shell gland (nidamental gland) where fertilization occurs and a horny shell or membrane is secreted. In oviparous species the shell is tough and protects the developing embryo. In viviparous species, the shell is slight or vestigial, and the young develop in the posterior, uterine portion of the oviduct.

Chimaeras. In the male the testes are large, rather compact structures placed forward in the body cavity. The system differs from that of sharks in the greater number of efferent ducts traversing the mesorchium and in the expansion of the posterior part of the archinephric duct into a structure called an ampulla (or vesicula seminalis). This is glandular in nature, partially compartmentalized in some species, and has been described as a receptacle for maturation of the sperm in the spermatophores coming from the epididymis. The ampulla receives a number of ducts from the posterior part of the kidney.

The urogenital pore opens medially behind the anus; adjacent to the opening are pelvic claspers that convey the seminal fluid to the genital openings of the female. Anterior to the anus is a pair of abdominal claspers that apparently function during copulation. Females of chimaeras are unique in that the two oviducts open separately to the exterior, not connecting to each other or to the urinary system.

Bony Fishes. In most bony fishes the testes are whitish, lobulate organs lying along the gas bladder, although in some groups such as salmonids the organs are smooth and entire, without lobules. In most forms there is no connection at all between the reproductive and urinary systems, there being separate openings to the exterior for the two systems, with the urinary pore posterior to the genital pore. In some, the sperm ducts connect with the urinary system in a urogenital sinus located at the posterior end of the body cavity. Primitive bony fishes may differ from this plan (Fig. 2–37).

In *Polypterus* the testes are closely bound to the kidneys and, although the sperm ducts enter that organ, they do not join the urinary system until the urogenital sinus is reached.

In lungfishes the testes are elongate and may stretch the length of the body cavity. A longitudinal collecting duct or network of ducts lies

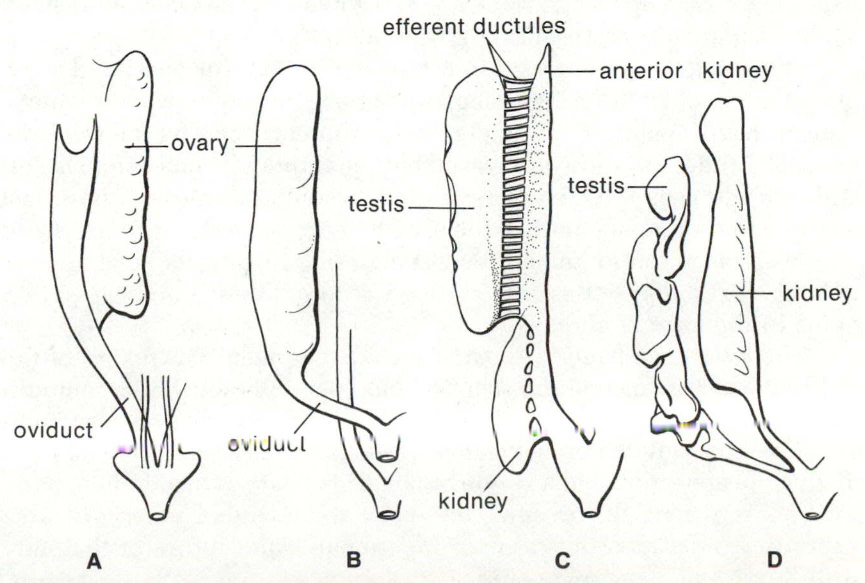

FIGURE 2–37. Diagrams of gonads and ducts in bony fishes. *A*, Bowfin female (Amiidae); *B*, representative teleost female; *C*, bowfin male (Amiidae); *D*, representative teleost male. (Based on Goodrich, in Lankester, 1909.)

along the medial edge of the testis, and in the posterior section efferent ducts connect with the kidney. In *Protopterus* the central ducts from each testis merge to form a median structure posterior to the testes, and ducts from this tube enter the kidneys.

The testes of *Amia* are in the anterior part of the body cavity and are closely associated with the kidneys. A longitudinal duct is situated along the medial edge of the testes and numerous efferent ducts extend from this to the anterior part of the kidney. *Acipenser* and *Lepisosteus* are similar, but the testes are more elongate and communicate with a greater length of the kidney.

Ovaries of bony fishes are typically saccular and continuous with the oviduct. This closed ovary-oviduct system arises in one of two ways during development. Some species show an endovarial condition in which two folds along the edge of the genital ridge meet and merge to enclose a hollow, which becomes the central cavity of the ovary and extends behind this as the oviduct. In others, the parovarial condition develops through the lateral growth of the edge of the genital fold, so that it curls upward to fuse with the body wall. This process also captures a bit of the coelom that becomes the cavity of the ovary and oviduct. Oviducts reach the exterior via a pore between the anus and the urinary pore.

Interesting exceptions to this system involve the more primitive forms, in which the structure of the female reproductive system is more typical of the general vertebrate plan. *Polypterus, Acipenser, Polyodon,* and *Amia* all shed eggs from the incompletely covered ovaries into the body cavity. Eggs are caught by a coelomic funnel opening partially back on the ovary and conveyed through the oviduct. (Homology of the funnels and ducts with the Müllerian duct is questionable.) In lungfishes the funnel opening is anterior to the ovary at the front of the body cavity, apparently representing a true Müllerian duct.

Other exceptions are seen in the smelts, salmonids, eels, and a few others, none of which have saccular ovaries continuous with oviducts. Smelts have coelomic oviducts with funnels opening behind the ovaries. Salmonids show parovarial but incompletely enclosed ovaries, and extrude eggs through a very short funnel leading to a pore just anterior to the urinary pore. Others with short oviducts or funnels and free ovaries are found among the Osteoglossiformes. Eels have no funnels; the eggs simply pass out through a pore. Morays possess paired pores in the form of slits.

The ovaries of fishes are usually well separated, but fusion of the right and left organs can be seen in some percoids. In the largemouth bass, the ovaries join posteriorly to produce a V-shaped structure. Ovaries of the yellow perch, *Perca flavescens,* are so completely fused as to give the appearance of a single organ. This ovary is fused to the body wall just posterior to the anus, and eggs are extruded when this area ruptures, so that oviducts are not functional. The rupture of the body wall heals soon after oviposition.

References

Ahsan-ul-Islam. 1949. The comparative histology of the alimentary tract of certain freshwater teleost fishes. Proc. Ind. Acad. Sci., B33:297–321.

Al-Hussaini, A. H. 1946. The anatomy and histology of the alimentary tract of the bottom-feeder, *Mulloides auriflamma* (Forsk.). J. Morphol., 28:121–154.

———. 1949. On the functional morphology of the alimentary tract of some fish in relation to differences in their feeding habits: Anatomy and histology. Q. J. Microsc., 90:109–139.

Anderson, B. G., and Loewen, R. D. 1975. Renal morphology of freshwater trout. Am. J. Anat., 143(1):93–114.

Antony, A. C. 1952. Use of fish slime in structural engineering. J. Bombay Nat. Hist. Soc., 50(3):682.

Barrington, E. J. W. 1957. The alimentary canal and digestion. *In*: Brown, M. E. (ed.), The Physiology of Fishes, Vol. 1. New York, Academic Press, pp. 109–161.

Bishop, C., and Odense, P. H. 1966. Morphology of the digestive tract of the cod, *Gadus morhua*. J. Fish. Res. Bd. Can., 23:1607–1615.

Branson, B. A., and Ulrikson, G. U. 1967. Morphology and histology of the branchial apparatus in percid fishes of the genera *Percina*, *Etheostoma* and *Ammocrypta* (Percidae: Percinae: Etheostomatini). Trans. Am. Microsc. Soc., 86(4):371–389.

Dobbs, G. H., III, and DeVries, A. L. 1975. The aglomerular nephron of Antarctic teleosts: a light and electron microscopic study. Tissue Coll., 7(1):159–170.

Hickman, C. P., Jr., and Trump, B. F. 1959. The kidney. *In*: Hoar, W. S., and Randall, D. J. (eds.), Fish Physiology, Vol. 1. New York, Academic Press, pp. 91–239.

Hoyt, J. W., 1975. Hydrodynamic drag reduction due to fish slimes. *In*: Yu, T. Y-T., Brokaw, C. J. and Brennen, C. (eds.), Swimming and Flying in Nature, Vol. 2. New York, Plenum Press, pp. 653–672.

Martin. W. R. 1949. The mechanism of environmental control of body form in fishes. Univ. Toronto Stud. Biol. Ser., 58:1–81.

Miller, R. V. 1964. The morphology and function of the pharyngeal organs in the clupeid, *Dorosoma petenense* (Gunther). Chesapeake Sci., 5:194–199.

Morgan, M. and Tovell, P. W. A. 1973. The structure of the gill of the trout, *Salmo gairdneri* (Richardson). Z. Zellforsch. Mikrosk. Anat., 142:147–162.

Nelson, G. T. 1970. Pharyngeal denticles (placoid scales) of sharks with notes on the dermal skeleton of vertebrates. Am. Mus. Nov., 2415:1–26.

Nursall, J. R. 1962. Swimming and the origin of paired appendages. Am. Zool., 2:127–141.

Ogawa, M. 1961. Comparative study of the external shape of the teleostean kidney with relation to phylogeny. Sci. Rep. Tokyo Kyoiku Daigaku, B10:61–68.

———1962. Comparative study on the internal structure of the teleostean kidney. Sci. Rep. Saitama Univ., B4(2):107–131.

Ørvig, T. 1968. The dermal skeleton; general considerations. *In* Ørvig, T. (ed.), Current Problems of Lower Vertebrate Phylogeny. Proc. 4th Nobel Symp., New York, John Wiley & Sons, pp. 373–397.

Pasha, S. M. K, 1964. The anatomy and histology of the alimentary canal of an herbivorous fish, *Tilapia mossambica* (Peters). Proc. Ind. Acad. Sci., Sec. B, 59(6):340–349.

Rosen, M. W. and Cornford, N. E. 1971. Fluid friction of fish slimes. Nature, 234:49–51.

Sharma, Shyama, 1971. Homology of the so-called "head-kidney" in certain Indian teleosts. Ann. Zool. (Agra), 7(2):20–40.
Thompson, K. S. 1976. On the heterocercal tail in sharks. Paleobiol., 2(1): 19–38.
Van Oosten, J. 1957. The skin and scales. *In:* Brown, M. E. (ed.), The Physiology of Fishes. New York, Academic Press, pp. 207–244.
Weisel, G. F. 1962. Comparative study of the digestive tract of a sucker, *Catostomus catostomus,* and a predaceous minnow, *Ptychocheilus oregonense.* Am. Midl. Nat., 68:334–346.
———. 1973. Anatomy and histology of the digestive system of the paddlefish *(Polyodon spathula).* J. Morphol., 140:243–256.
Wiley, M. L. and Collette, B. B. 1970. Breeding tubercles and contact organs in fishes: Their occurrence, structure and significance. Bull. Am. Mus. Nat. Hist., 143(3):147–216.
Youson, J. H. and McMillan, D. B. 1970. The opisthonephric kidney of the sea lamprey of the Great Lakes, *Petromyzon marinus* L. I. The renal corpuscle. Am. J. Anat., 127:207–232.
Youson, J. H., 1970. Observations on the opisthonephric kidney of the sea lamprey of the Great Lakes, *Petromyzon marinus* L. at various stages during the life cycle. Can. J. Zool., 48(6):1313–1316.

3
ENDOSKELETON AND MUSCULATURE

Axial Skeleton: Skull, Vertebral Column, Ribs

Development of the Chondrocranium

There is a general similarity among vertebrates in the embryonic development of the skull, although various groups show characteristic differences. Because there will be some referral to the development of fish skulls later in this chapter, a most generalized and abbreviated account of the process will be outlined here (Fig. 3–1).

The neurocranium forms at the anterior end of the notochord, beginning with the formation of two parachordal cartilages from somatic mesoderm. These cartilages, one on each side of the notochord, enlarge and form a structure called the basal plate by fusing around the notochord. The basal plate enlarges and fuses with paired occipital arch cartilages that develop over the hindbrain, thus forming the back wall of the neurocranium. Also uniting with the basal plate and occipital arch cartilages are paired otic capsules that form around the inner ears. The synotic tectum, a small cartilage that forms over the posterior part of the brain, joins with the occipital arch cartilages. Anteriorly, two prechordal cartilages called trabeculae form from the neural crest. As the trabeculae grow they fuse at their anterior ends to form the ethmoid plate, and unite with the developing basal plate posteriorly. In some fishes a pair of polar cartilages form between the trabeculae and the parachordals and join them, so that three paired elements enter into the formation of the cranial floor. Nasal capsules form anteriorly from their cartilages, and orbital cartilages extend forward from the otic capsules. Antorbital processes develop from prechordal somatic mesoderm. Enlargement and dorsal growth of the parts result in the cartilaginous cranium, which will be partially covered or replaced by bone in the bony fishes.

The Branchiocranium

A series of arches that form around the pharyngeal region are associated with the neurocranium. The primitive number of these visceral arches is not known but many anatomists believe there were eight, and that most or all of them functioned primitively in the sup-

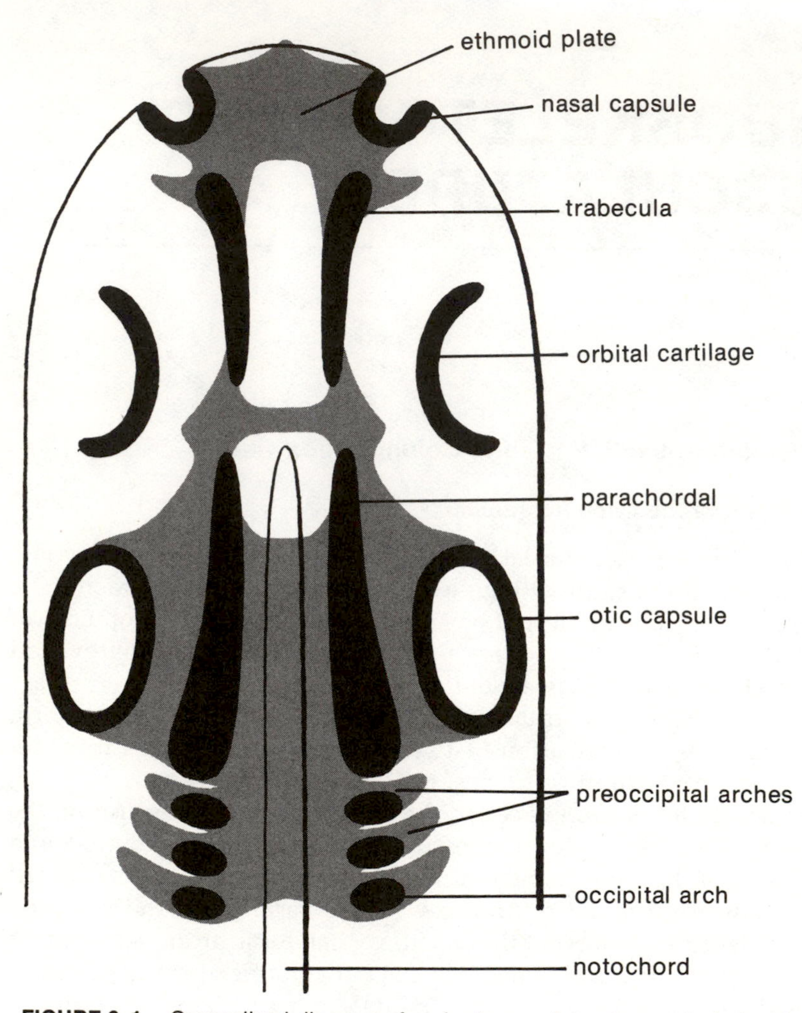

FIGURE 3–1. Generalized diagram of early stages of development of chondrocranium. *In black,* separate cartilages; *in gray,* the formation of ethmoid and basal plates.

port of gills. In fishes most of the branchiocranium is associated with the support of gills, but some arches have been modified and have other functions.

The anterior arch in living fishes is called the mandibular arch because of its contribution to the formation of the primary upper and lower jaws. As in the other arches, there are major upper and lower elements on each side. In the course of evolution, the upper elements (palatoquadrate cartilages) of the mandibular arch have slanted forward along the underside of the neurocranium and have joined anteriorly to complete the primitive upper jaw. Essentially, this remains as the functional tooth-bearing upper jaw of sharks and rays, but does not form the border of the mouth in bony fishes.

The lower elements of the mandibular arch are called the mandibular cartilages, or Meckel's cartilages. These have evolved into the primary lower jaw, remaining jointed to the palatoquadrate cartilage posteriorly and joining each other anteriorly. In sharks and rays the mandibular cartilage bears teeth, but in bony fishes it is hidden in the secondary lower jaw.

The second arch in the series is the hyoid arch. The upper element on each side, the hyomandibular cartilage, evolves into an important suspensory structure in fishes, aiding in the suspension of the jaws and tongue. The lower element of the hyoid arch is the ceratohyal cartilage, which becomes the supporting structure of the "tongue" apparatus. Remaining visceral arches typically support gills and are called branchial arches. The posterior arch of bony fishes is usually modified to bear pharyngeal teeth.

Skull of Living Agnathans

Hagfishes, generally considered the most primitive of living fishes, retain a cranium that appears to have been arrested in an early stage of

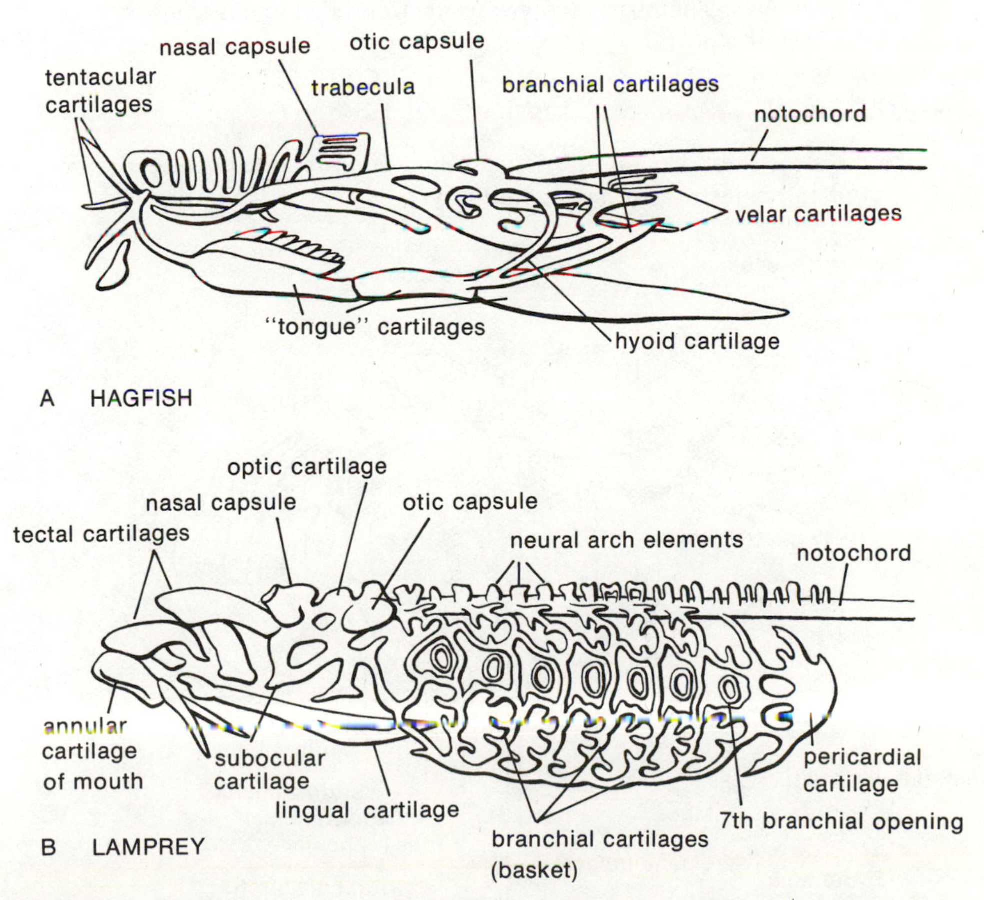

FIGURE 3–2. Diagrams of skulls of living agnaths. *A*, Hagfish; *B*, lamprey. (Based on Goodrich, in Lankester, 1909.)

development. The trabeculae are present and there is a cartilaginous floor consisting of the fused parachordals. The cartilaginous otic capsules are fused to the parachordals, but the remainder of the sides and top of the structure is membranous. Visceral arches are not well developed in hagfishes. Anatomists interpret three cartilaginous structures in the head far anterior to the gill pouches as the hyoid and third and fourth visceral arches. A rudimentary framework of cartilage external to the pharynx and gill pouches serves as a branchial skeleton, with a series of cartilages supporting the rasping organ and its muscles.

The cranium of lampreys is more complete than that of hagfishes (Fig. 3–2). The brain is roofed over anteriorly by extensions of the trabeculae. Posteriorly there are sidewalls, but only connective tissue covers the brain. This rudimentary chondrocranium, homologous with that of other vertebrates, constitutes a minor part of the total skull of lampreys. There are dorsal cartilages supporting the region anterior to the nasal opening, a series of cartilages around the circular mouth, and a long lingual cartilage ventrally located. Support of the gill region is provided by an intricate branchial basket of cartilage just under the skin. This basket is not definitely homologous with the branchial skeleton of gnathostomes. The heart receives protection from expansions of the posterior section of the complex.

Skull of Cartilaginous Fishes

In sharks the chondrocranium is usually a complete box, pierced by foramina and fenestrae for passage of nerves (Fig. 3–3). The olfacto-

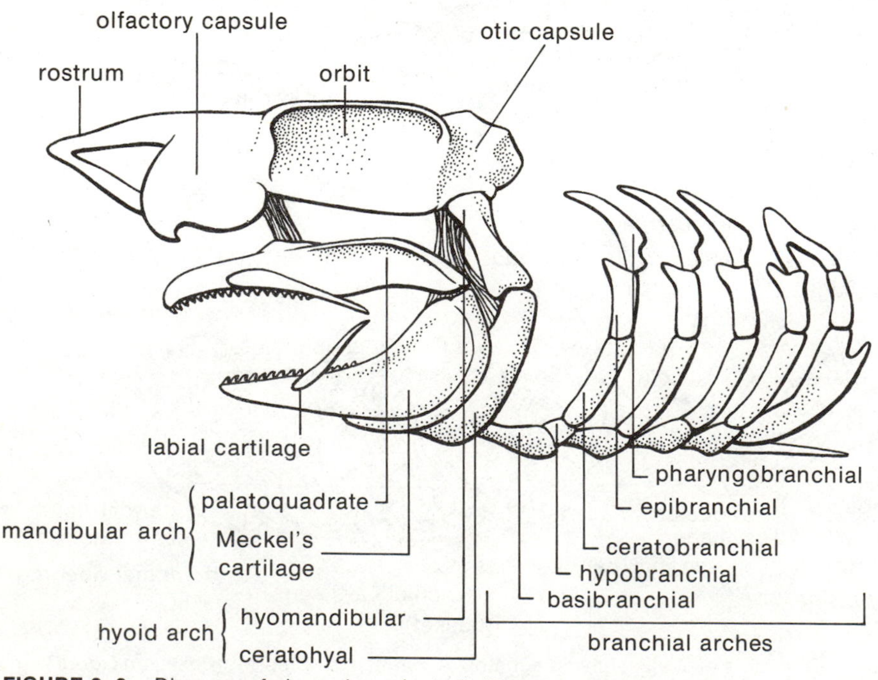

FIGURE 3–3. Diagram of elasmobranch skull. (Based on Bridge, 1904.)

ry and otic capsules are integral parts of this box. Anterior to the olfactory capsule is the rostrum, which may be extremely prominent, reaching a maximum in the sawshark, *Pristiophorus*.

Rays may have large dorsal fontanelles both in the rostral region and more posteriorly, so that the cranial roof appears to be incomplete. The rostrum in rays is often very long. The extremes of length are seen in the eagle ray, *Myliobatis*, in which the rostrum is undeveloped, and in the sawfish, *Pristis*, in which the rostrum may be nearly one third of the entire length of the fish.

In most elasmobranchs there is usually a prominent process both anterior and posterior to the orbit. Some species show various crests and other sculpturing on the roof of the cranium. Posteriorly, occipital condyles form a place of attachment for the vertebral column, and the foramen magnum allows passage of the spinal cord.

Although modern elasmobranchs have no bone, the cartilage of the skeleton may be variously calcified to some extent, even to the point of being as hard as bone. Prismatic calcifications may encrust major skeletal structures of adult elasmobranchs, so that a mosaic appearance is given to the surface of a dried chondrocranium. Vertebral centra are often heavily calcified.

The jaws of elasmobranchs consist of the modified first visceral arch, greatly enlarged in some species. The palatoquadrate cartilage extends forward along the underside of the neurocranium, remaining free and movable in most species. It is usually attached to the cranium by means of the upper part of the hyoid arch, the hyomandibular cartilage, which articulates with the otic region. The lower jaw, or Meckel's cartilage, articulates with the posterior part of the palatoquadrate. This type of jaw suspension is called hyostylic. In a few sharks the palatoquadrate is attached to the neurocranium as well as to the hyomandibular in what is termed amphistylic suspension (Fig. 3–4).

The lower part of the hyoid arch is called the ceratohyal cartilage, and both it and the hyomandibular may bear gill rays, slender cartilaginous rods that strengthen the interbranchial septa from which gill tissue projects as lamellae. A basihyal cartilage connects the ceratohyals ventrally.

There are five branchial arches in most sharks and rays, but a few species have six or more. These arches typically consist of pharyngobranchials, epibranchials, ceratobranchials, and hypobranchials. Pharyngobranchials may or may not be held by connective tissue to the roof of the pharynx, depending on the species. Basibranchial cartilages connect the hypobranchials on the ventral midline. Gill rays on the epibranchials and ceratobranchials form support for the gills on all but the last arch.

The chimaeras or holocephalans have a cartilaginous neurocranium with which the palatoquadrate cartilage is entirely fused, a type of autostylic jaw suspension generally referred to as holostylic (Fig. 3–4 C). The hyomandibular serves no suspensory purpose for either the upper

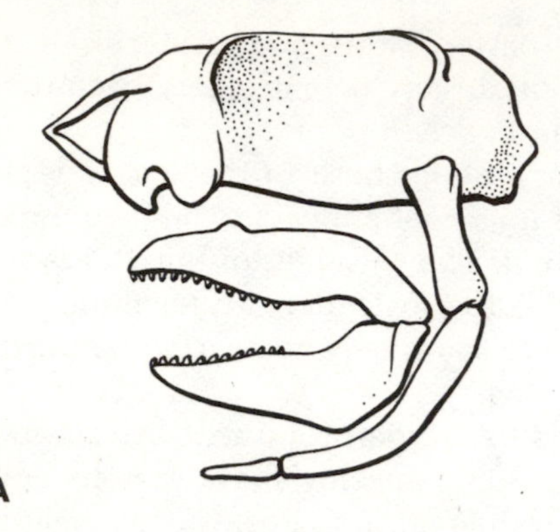

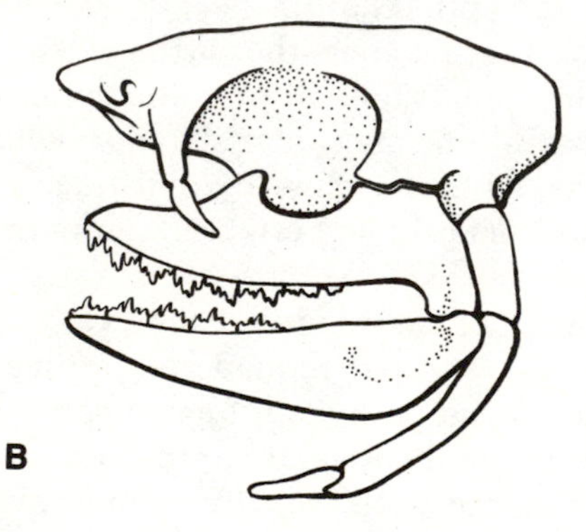

FIGURE 3–4. Schematic diagrams showing relationship of mandibular arch to neurocranium in three types of jaw suspensions. (The mandibular arch has been lowered in *A*; the hyoid arch has been flexed downward in *A* and *B*, and has been lowered in *C*.) *A*, Hyostylic (upper jaw not firmly attached to neurocranium, with ligamentous attachment to hyomandibular); *B*, amphistylic (upper jaw attaches anteriorly to basal angle of neurocranium and posteriorly to postorbital process); *C*, holostylic (upper jaw fused to neurocranium).

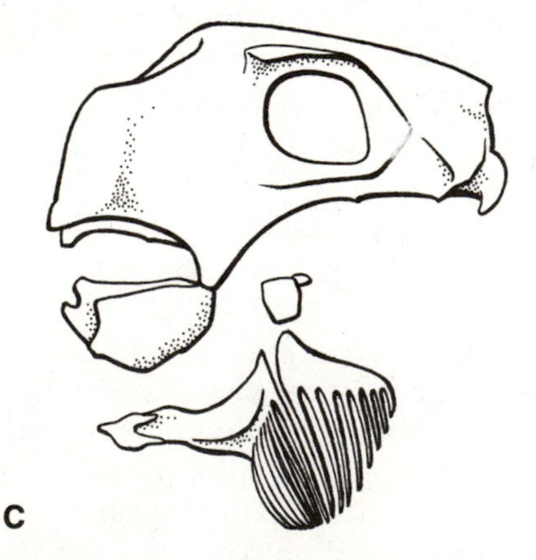

or lower jaw. The hyoid arch is slightly more modified than the branchial arches and bears gill rays.

Skull of Bony Fishes

Figures 3–5, 3–6, and 3–7 show the structures mentioned in this section. Frequent referral to these figures will be helpful.

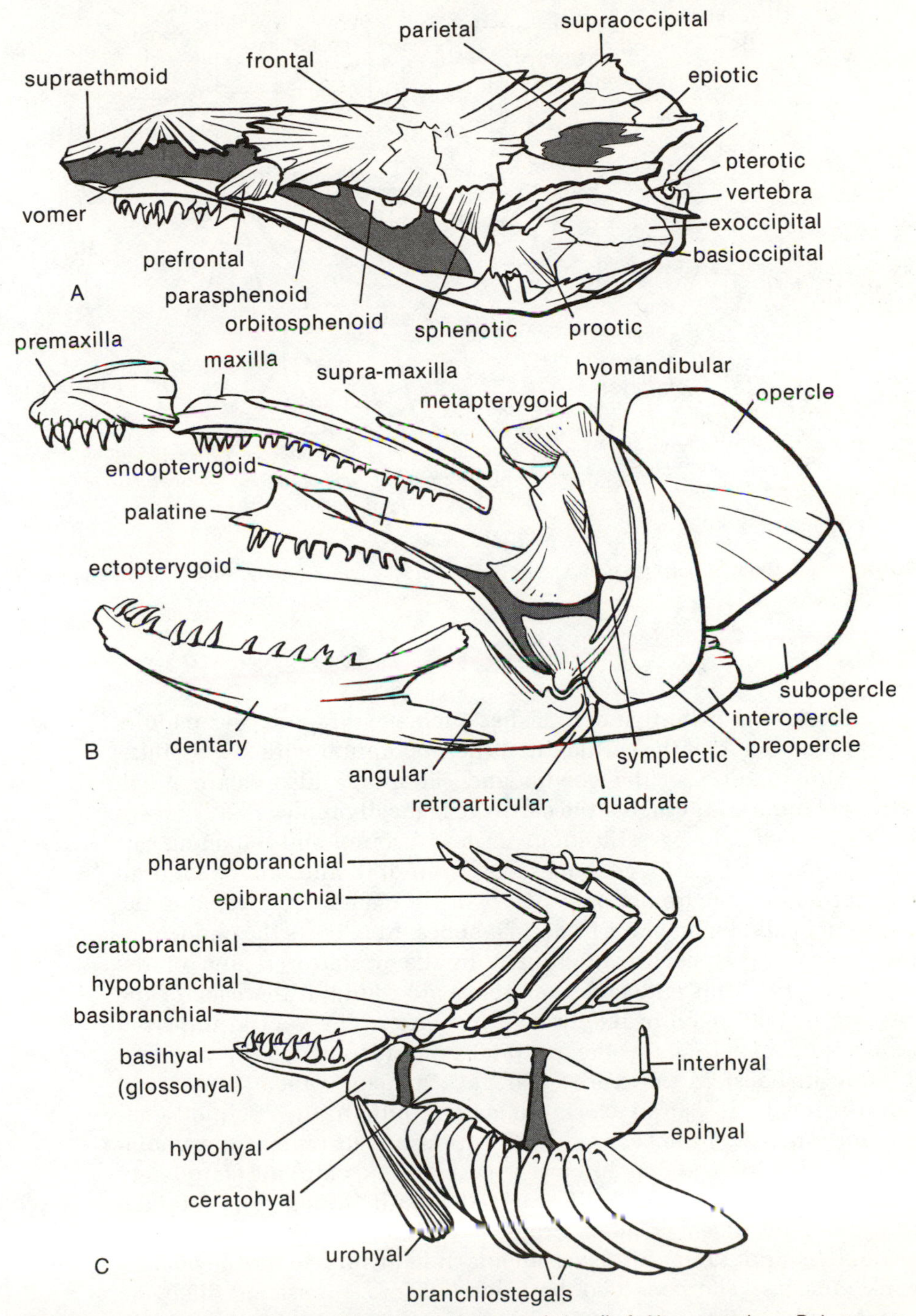

FIGURE 3–5. Skull bones of steelhead trout (*Salmo gairdneri*). *A*, Neurocranium; *B*, jaws, suspensorium, and operculum; *C*, branchiohyoid apparatus.

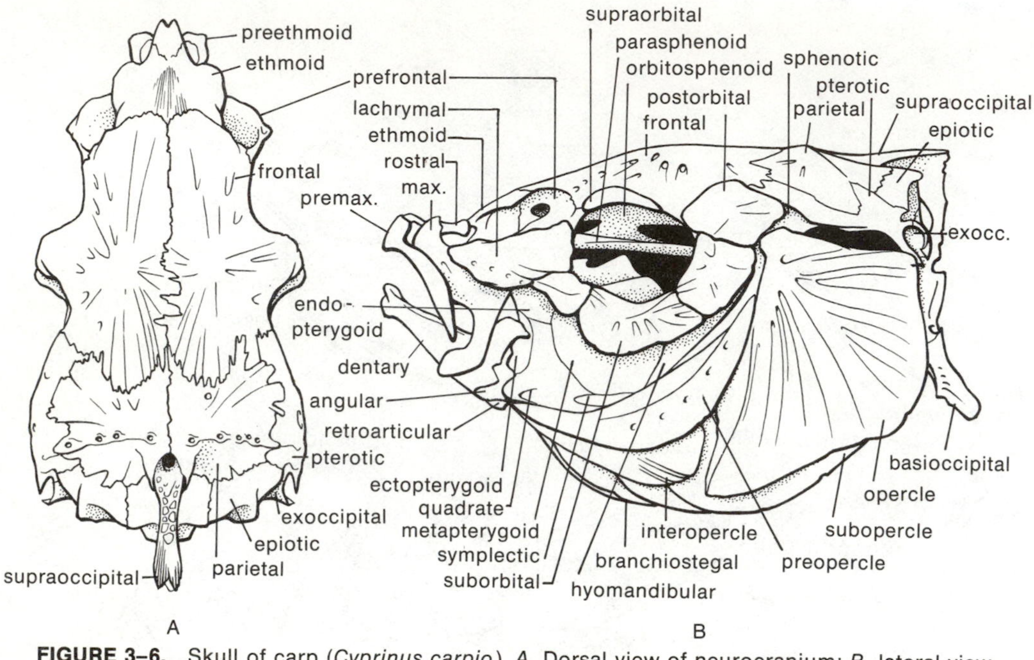

FIGURE 3–6. Skull of carp (*Cyprinus carpio*). *A*, Dorsal view of neurocranium; *B*, lateral view of skull.

Some of the primitive bony fishes such as sturgeons and paddlefishes retain much of the cartilaginous neurocranium with few ossifications. Others such as the bowfin and salmonids also retain much cartilage in the cranium, but the cartilage is mostly replaced by bone in the majority of fishes. Ossifications forming around and replacing cartilage are called cartilage bones, perichondral and endochondral, whereas bones that are not preceded by cartilage but are formed in the dermis are called membrane or dermal bones. In teleosts the endochondral bones are especially prominent in the posterior region of the neurocranium. Four endochondral bones are found at the back of the cranium in the region of the foramen magnum. The ventral unpaired basioccipital usually forms the occipital condyle to which the vertebral column attaches. In perch-like fish and in many other spiny-rayed groups the lateral, paired exoccipitals contribute to the occipital condyle forming, with the basioccipital, a tripartite structure for articulation with the first vertebra. In most teleosts the exoccipitals completely surround the spinal cord as it enters the skull through the foramen magnum. The dorsal median supraoccipital, in addition to forming part of the cranial roof, furnishes an attachment surface for the epaxial trunk muscles. The bone may be extended into a crest to which the muscles attach. The crest varies in size with the size of the muscle mass.

Much of the posterior part of the teleost skull consists of five endochondral bones that form in each otic capsule, protecting the membranous labyrinth. The largest of the complex is the prootic, which forms a considerable portion of the lateral floor of the cranium in many species. Each prootic meets its fellow just anterior to the basioccipital. Part of the posterior boundary of the orbit consists of the sphenotic, dorsal to the anterior part of the prootic. The sphenotic forms in part around the anterior semicircular canal. The hyomandibular, which supports the jaws, articulates with the sphenotic. The prominent ridges that usually mark the widest part of the cranium are formed by the pterotics, which ossify around the lateral semicircular canals and usually combine with a dermal element to produce a compound bone. The epiotic, which ossifies in part around the posterior semicircular canal, usually can be recognized as a process between the pterotic and the supraoccipital. The epiotic is the site of the attachment of the pectoral girdle to the cranium. An opisthotic appears in the back wall of the cranium between the pterotic and the exoccipitals. Although phylogenetically this was a cartilage bone, that remaining in most teleosts represents the ossified base of a ligament by which the post-temporal of the pectoral girdle attaches to the cranium. In recent literature the name intercalary is generally used, instead of opisthotic.

Cartilage bones of the trabecular section of the cranium include: the alisphenoid (paired), which makes up part of the posterior wall of the orbit, connecting with the prootic posteriorly; the orbitosphenoid

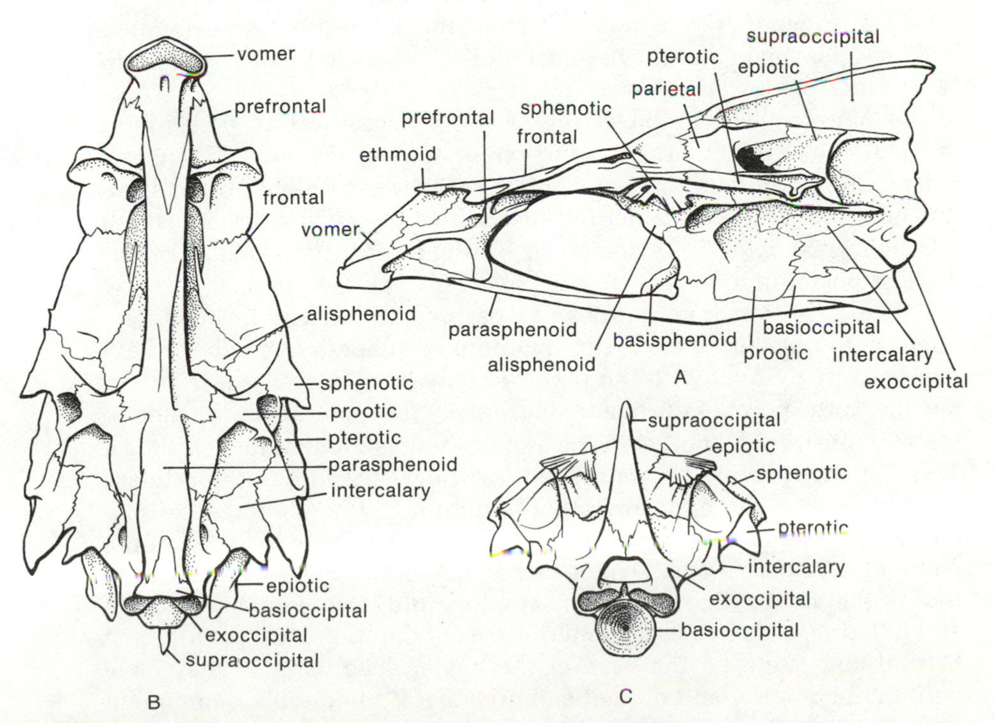

FIGURE 3–7. Neurocranium of percoid (based on *Morone saxatilis*). *A*, Lateral view; *B*, ventral view; *C*, posterior view.

(median), which forms a bony interorbital septum in clupeiforms, salmoniforms, cypriniforms, and other "lower" bony fishes, but is absent in higher teleosts (the olfactory nerves pass through this bone); and the basisphenoid (median), a Y-shaped bone in the posterior part of the orbit. The arms of the Y articulate with the prootics and the shaft connects ventrally to the parasphenoid. Anterior to the orbit are paired prefrontals (ectoethmoids). The prefrontal forms a complex with dermal elements in some fishes and it is the site of attachment of the lachrymal, a dermal bone.

The cartilage bones comprise the primary neurocranium, to which bones originating in the dermis are added dorsally and ventrally. The latter, called dermal bones, membrane bones, or investing bones may include plates originating from scales, plates formed from coalescing tooth bases, and bones that form directly from membranes. The most posterior dermal bones of the skull are the paired parietals, which usually flank the supraoccipital and constitute part of the roof of the cranium. Anterior to these are the large frontals, which make up most of the cranial roof. Both parietals and frontals are traversed by cephalic lateral line canals in some fishes. Typically, an unpaired ethmoid (supraethmoid, dermal mesethmoid) roofs the snout in front of the frontals. Paired nasals, which have developed around cephalic sensory canals, are on each side of the ethmoid. Ventrally the vomer usually forms the anterior point of the neurocranium; it is often attached to the ethmoid in higher teleosts but is separate in the lower bony fishes. The vomer forms part of the roof of the mouth and often bears teeth. The long parasphenoid forms the ventral midline of the cranium, extending between the vomer and the basioccipital. The parasphenoid bears teeth in some of the lower teleosts.

A more or less complete ring of bones (circumorbitals) surround the orbit, although the number and extent of these are reduced in many fishes. Circumorbitals enclose cephalic sensory canals and protect head musculature. Often the anterior bone of the group is enlarged, called the lachrymal. Others are named according to position, such as suborbital or postorbital.

The palatoquadrate cartilage is ossified in part as the quadrate bone at the posterior end. The quadrate is shaped somewhat like a quadrant of a circle, with the point downward. The lower jaw articulates with the point of the quadrate. Dorsally the quadrate is attached to the metapterygoid, another ossification of the palatoquadrate cartilage. This bone is instrumental in suspending the remainder of the primary upper jaw from the hyomandibular. Anterior to the quadrate and metapterygoid are two dermal bones that form along the lower edge of the palatoquadrate cartilage. One of these, the endopterygoid, stiffens the roof of the mouth. The other, the ectopterygoid, connects the quadrate and palatine, which is in the anterior part of the roof of the mouth, just behind and lateral to the head of the vomer. The palatine may have both endochondral and dermal components. If the dermal component is lacking, the bone is called the autopalatine.

In the upper jaw the vomer and palatines, and sometimes the ectopterygoids, endopterygoids, and the parasphenoid, may bear teeth. However, the so-called secondary upper jaw, composed of the premaxillary and maxillary, both dermal bones, constitutes the main dentigerous surface.

Within the lower jaw Meckel's cartilage remains largely unossified, except for an anterior mentomeckelian element and a posterior articular, both of which may form complexes with the dermal elements of the lower jaw. The major tooth-bearing bone of the lower jaw is the dentary. Between the dentary and the quadrate, from which the lower jaw is suspended, is the angular, called the "articular" in most older literature. The retroarticular, consisting of endochondral and dermal elements, is on the posterior lower corner of the angular, and has often been called the "angular" in older literature. A sensory canal runs through the angular and the dentary.

The hyoid arch is ossified in several parts. The uppermost is the hyomandibular, which articulates with the otic region of the cranium, and acts as a suspension for the primary upper jaw, the lower jaw, the hyoid apparatus, and the operculum. The metapterygoid articulates with the anterior face of the hyomandibular. A peg-like bone, the symplectic, extends from the bottom of the hyomandibular to the quadrate. The interhyal attaches to the hyomandibular just behind the symplectic and suspends the remainder of the hyoid arch, which consists of the epihyal, ceratohyal, upper hypohyal, lower hypohyal, and the unpaired basihyal (glossohyal). The latter bears teeth in many fishes.

An unpaired dermal bone, the urohyal, extends backward from the basihyals into the isthmus, constituting the firm ventral connection between the head and trunk. Important dermal bones that connect with the epihyal and ceratohyal are the branchiostegals, which in some fishes protect the gills ventrally. In others the branchiostegals stiffen a membrane that can be of greater importance than the operculum in pumping water over the gills. The operculum, which serves both as a shield for the gills and as part of the branchial pump, is composed of four pairs of dermal bones. These are usually plate-like and are associated with the hyoid apparatus. The largest bone in the operculum is usually the opercle, which attaches by its anterodorsal corner to a condyle on the posterior edge of the hyomandibular. The preopercle usually attaches along the hyomandibular for much of its length. The interopercle is below the preopercle, and the subopercle lies ventrally to the opercle.

The branchial arches of teleosts consist of a series of endochondral bones. The first three arches consist of a pharyngobranchial and an epibranchial in the upper section, and a ceratobranchial and a hypobranchial in the lower part. The pharyngobranchial of the fourth arch is typically fused to that of the third arch, or reduced so that only the epibranchial and ceratobranchial are evident. The fifth arch is reduced further to one bone that may represent a ceratobranchial, but is generally modified to bear pharyngeal teeth. A series of basibranchials is set between the left and right halves of the arches. These sometimes bear teeth, which appear just behind the teeth on the basihyal.

The Vertebral Column

In hagfishes the notochord persists without constriction, and the only rudiments of vertebrae are small cartilages resembling neural arches in the caudal region. Lampreys show neural arch elements the length of the notochord.

The notochord of elasmobranchs is constricted by cartilaginous vertebral centra so that if extracted, as many have noted, it would resemble a string of beads because the constricted portions would contrast with the unconstricted portions that fit into the concavities of the amphicelous centra. Some species have a single calcified cylinder formed within the centrum (cyclospondylous condition), while others may possess two or more concentric cylinders (tectospondylous). In some, calcified radiating lamellae extend from the calcified cylinder, giving a somewhat star-shaped pattern in cross section (asterospondylous). Each centrum in the trunk of the elasmobranch has ventrolateral transverse processes, or basapophyses, which bear the cartilaginous ribs. Dorsally there is a neural process surrounding the neural canal through which the spinal cord runs. Intercalary plates alternate with the neural processes. Caudal vertebrae bear ventral hemal arches and spines.

The typical teleost has ossified biconcave (amphicelous) centra, with the notochord filling the concavities (Fig. 3–8). Basapophyses (parapophyses) are present, but may not be fused to the centra. Neural arches and spines are present, and the caudal vertebrae have haemal

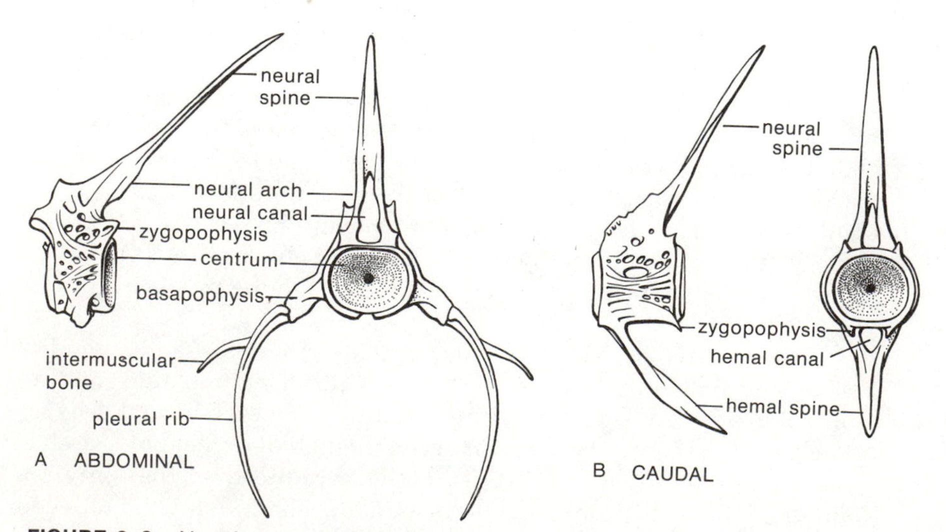

FIGURE 3–8. Vertebrae of teleost. *A,* lateral and posterior views of abdominal vertebrae; *B,* lateral and posterior views of caudal vertebrae.

arches and spines. Zygapophyses may occur both anteriorly and posteriorly on the centra. These are generally small in fishes and those of adjacent vertebrae do not make contact, but they may be large and interlock in powerful swimmers such as the tunas and their allies. The interlock prevents excessive rotation of vertebrae on each other.

Ventral ribs (pleural ribs) usually attach to the basapophyses. Intermuscular bones that extend into the horizontal skeletogenous septum are often called dorsal ribs. Other intermuscular bones may take their names from the structures that bear them; for example, those borne on the neural arch are called epineurals, and those on the ribs are called epipleurals.

There is much modification of the vertebral column in the region of the caudal fin (Fig. 2–15). In some of the lower teleosts the column may be upturned, with three or more progressively smaller centra involved. The upturned portion is called the urostyle by some anatomists. Below these there is a supporting structure made up of about six hypurals, which appear to be modified hemal spines. Above the column there are one or more uroneurals. The hypural complex supports the rays of the caudal fin. In higher teleosts the vertebral column ends in a urostyle, which is a turned-up portion of the last vertebral centrum. The hypurals of higher teleosts are usually fused into larger plates, sometimes with one supporting the upper lobe of the caudal fin and one supporting the lower.

The Appendicular Skeleton (Fins and Fin Girdles)

The median fins of elasmobranchs are supported by basal cartilages that are often segmented into proximal, middle, and distal elements (Fig. 3–9). In some sharks, the proximal elements may fuse into a single plate. In some flattened species the basals may join with the neural spines.

In teleosts each ray of the median fins is typically supported by two ossified and one cartilaginous pterygiophores. Proximal pterygiophores are elongate tapered bones set deeply into the median skeletogenous septum, usually between the neural or haemal spines. Because of this they are often called interspinous bones, or those supporting the dorsal fin may be called interneurals and those of the anal fin be called interhaemals. The middle pterygiophores are ossified and jointed flexibly to the proximal elements on one end and to the distal pterygiophores, if present, on the outer end. In soft-rayed fins the cartilaginous distal pterygiophores usually fit between the bases of the two halves of the rays. In spinous fins the distal pterygiophore may be lost or fused to the middle element. In extreme cases fusion of the three elements is complete and fin spines attach to single supporting structures.

The pectoral fins of elasmobranchs are supported by a cartilaginous girdle consisting of an upper, or scapular, section and a lower coracoid piece (Fig. 3–10). A small suprascapular cartilage may be

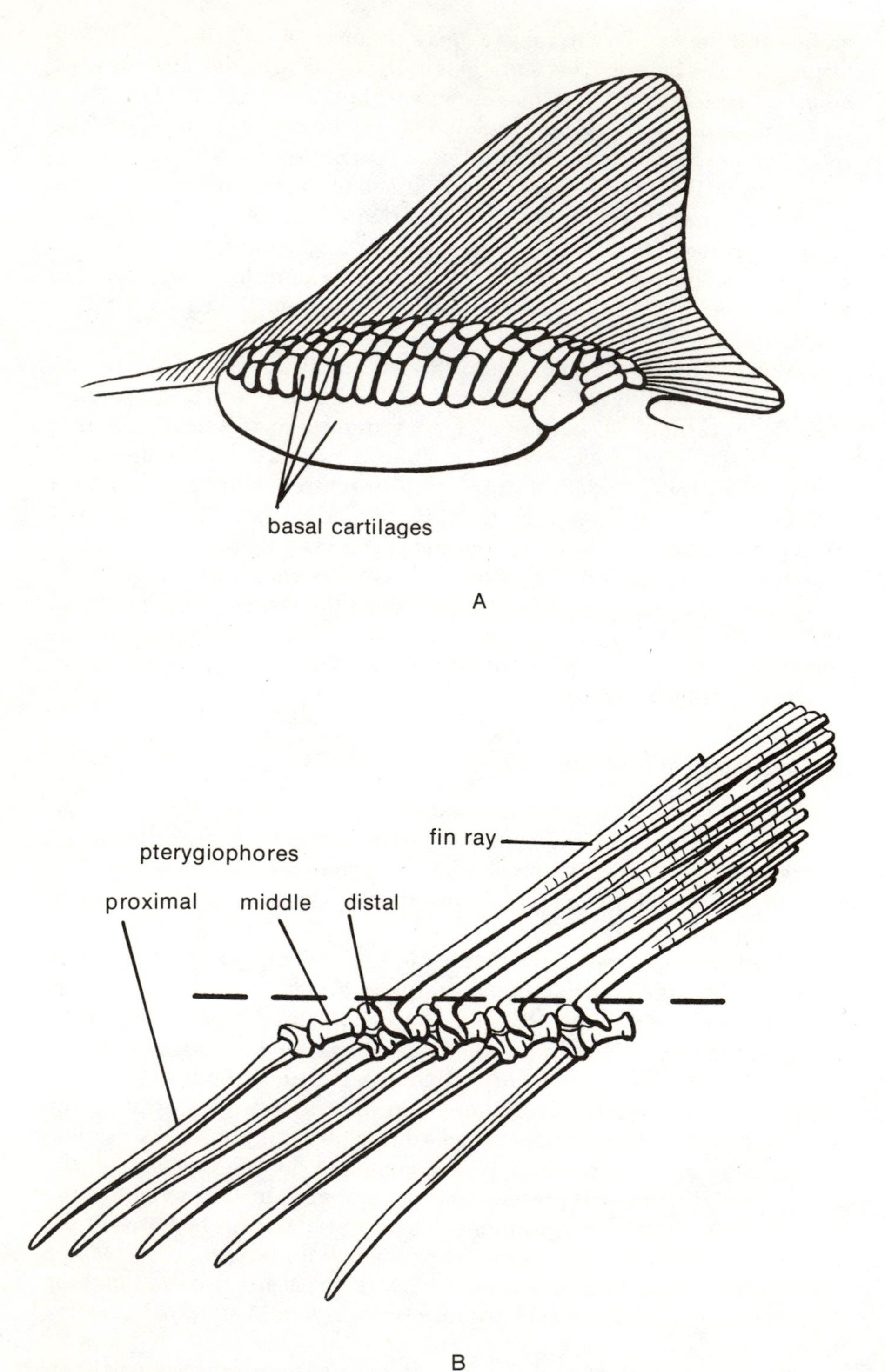

FIGURE 3–9. Skeletal supports of dorsal fin. *A*, Shark; *B*, bony fish (dashed line shows approximate body contour). (*A* based on Goodrich, 1930.)

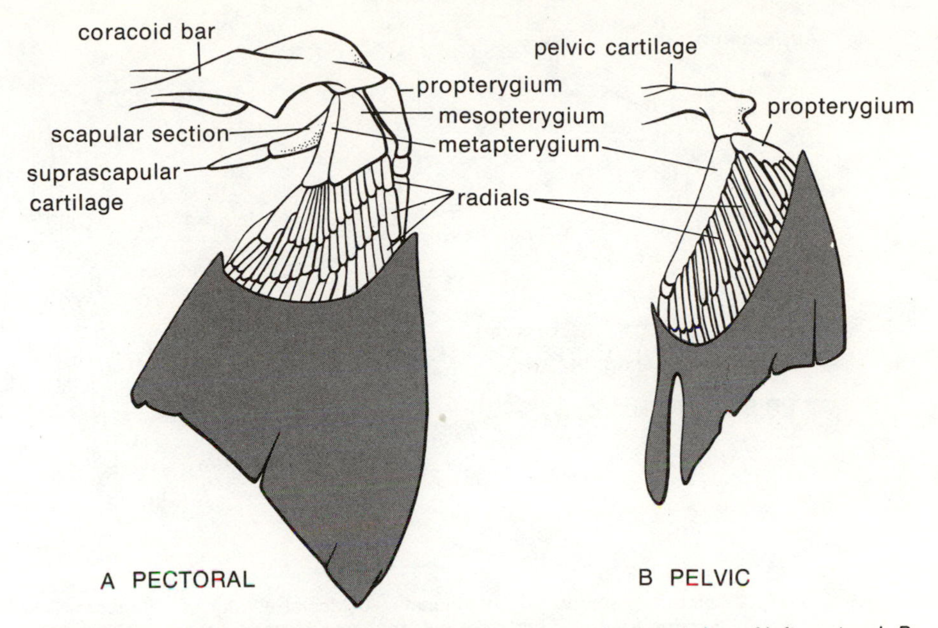

FIGURE 3–10. Skeletal supports of paired fins of shark. *A*, ventral view of left pectoral; *B*, ventral view of left pelvic. (Based on drawings by John McKern.)

present. In sharks the two halves of the girdle are separate from each other dorsally and do not attach to the vertebral column. In rays the two halves join each other or the vertebral column.

The pectoral fin articulates with the coracoid. There are three basal cartilages — the propterygium anteriorly, a middle mesopterygium, and a posterior metapterygium. In rays the articulation is horizontal and the propterygium and metapterygium extend far forward and backward, respectively. Series of jointed radials, which attach to the basal cartilages, bear the fin rays at their distal ends.

The pelvic girdle of elasmobranchs is rather simple, being a bar of cartilage crossing the ventral midline. Extending posteriorly from the ends of the cartilage are the elongate basipterygia (one in each fin in some species) that bear the jointed radials of the fin.

In the pectoral girdle of typical teleosts the scapula and coracoid are ossified as endochondral bones and part of their outer edges form the articular surface for the radials (actinosts) of the pectoral fin (Fig. 3–11). This complex is applied to the inner surface of a secondary pectoral girdle consisting of a series of dermal bones. Actual attachment is to the cleithrum, usually the largest of the series. Cleithra meet at the ventral midline and extend upward toward the cranium. A supracleithrum meets the cleithrum and extends forward, where it attaches to the posttemporal, a forked bone. The upper branch of the posttemporal attaches to the epiotic and the lower branch to the pterotic or opisthotic (intercalary). A series of postcleithra appears in most teleosts.

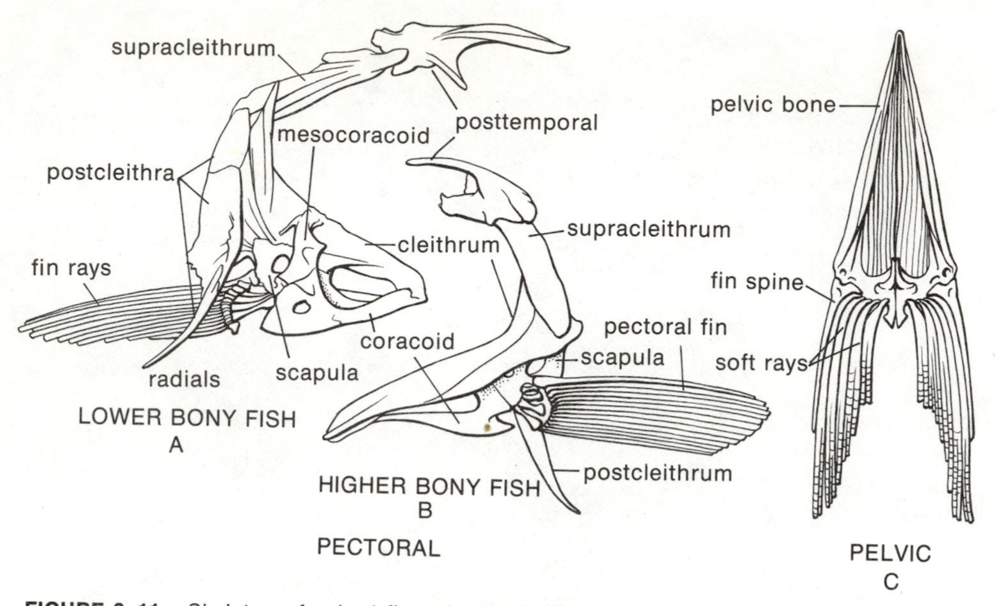

FIGURE 3–11. Skeleton of paired fins of teleosts. *A,* Mesial aspect of left pectoral bones of *Salmo gairdneri; B,* lateral view of left pectoral bones of *Morone saxatilis; C,* ventral view of pelvic skeleton of *M. saxatilis.*

Teleosts in which the pectorals are low and have oblique bases possess a mesocoracoid bone that braces between the coracoid and the cleithrum. This is typical of soft-rayed fishes such as herrings, salmons, carps, and catfishes, but is absent from higher teleosts — perches, basses, and relatives, among others — that have pectorals set higher on the sides with vertical bases.

The pelvic fin skeleton in teleosts is made up of plate-like basipterygia, one for each fin. These bones usually are joined to each other posteriorly and may meet anteriorly. Remnants of pterygiophores may be present where the fin rays join the basipterygium.

Fin rays of various groups of fishes differ in structure and origin. Those of lampreys and hagfishes are simply rods of cartilage, whereas elasmobranchs have fibrous, horny fin rays arising from the dermis. These structures, usually unbranched, are called ceratotrichia. The dermal fin rays of teleosts are generally called lepidotrichia because of their presumed origin from rows of scales. Soft rays of bony fishes are composed of two halves (Fig. 3–12). The dorsal fin of a trout, for instance, can be split rather easily into left and right parts with a dissecting needle. In relatively unmodified form these rays are jointed and branched. Some "soft" rays, such as those at the leading edges of the dorsal and anal fins of carp, are not really soft to the touch, but have been modified by loss of the branched and jointed characteristics and have hardened into spine-like structures.

True spines such as those in the fins of acanthopterygian fishes are undivided, typically hard, sharp structures, although these may become modified toward flexibility in some groups, such as sculpins.

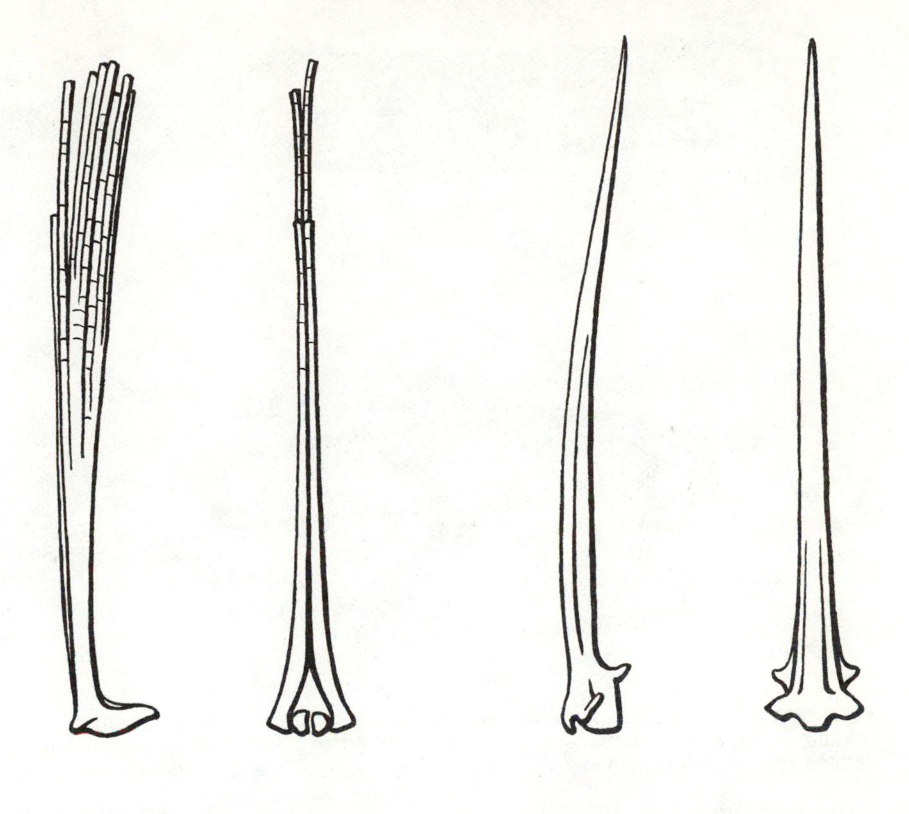

FIGURE 3–12. Comparison of soft and spinous rays. *A,* lateral and anterior views of soft ray; *B,* same views of fin spine. Note branching, segmentation, and double construction of soft ray.

Musculature

Skeletal musculature in fishes consists mainly of the large muscles of the trunk and tail. Other skeletal musculature is found associated with the jaws, the branchial arches, and the fins. The trunk musculature consists of a series of muscle blocks, myomeres or myotomes, separated by connective tissue sheets called myosepta or myocommata. These myomeres represent the segmentation seen in all higher animals. The fibers of these blocks are oriented more or less parallel with the body axis. The myotomes are not simply arranged, but are flexed so that, just under the skin, their outer edges resemble the letter W tipped at a 90° angle (Fig. 3–13). In lampreys the curvature of the myotome is slight, especially anteriad, but in bony fishes and sharks the bends are sharper and more evident. The modification and folding of the myotomes is so great that a short and simple description is difficult. Reference to illustrations and to specimens will help in understanding the trunk musculature.

The origin of each myotome on the axial skeleton is anterior to the

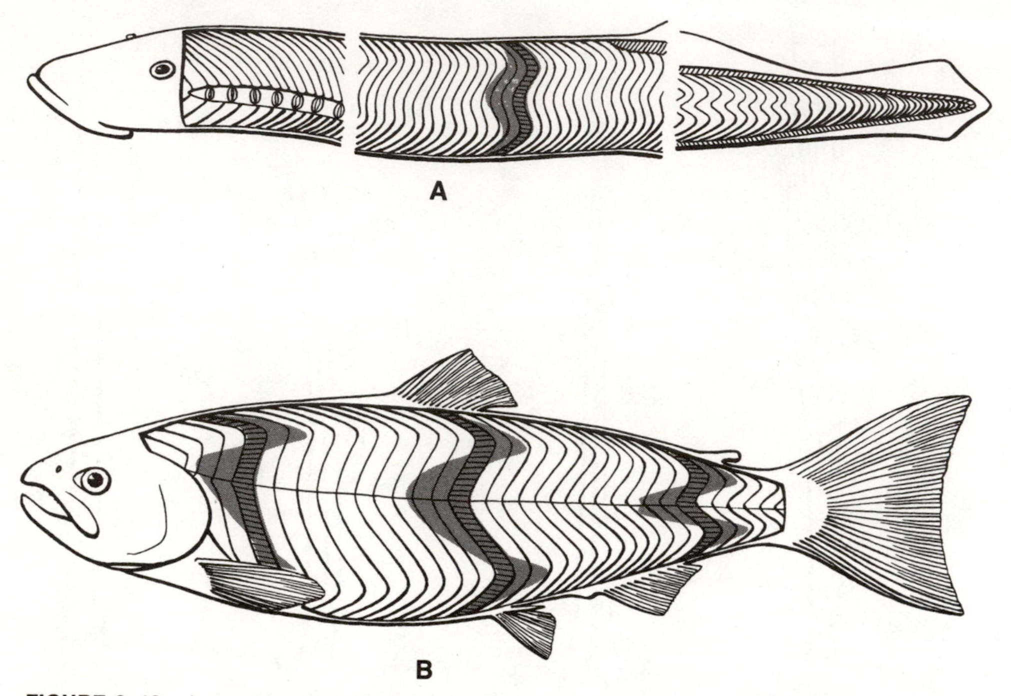

FIGURE 3–13. Lateral body musculature. *A,* lamprey, showing myotome patterns in anterior, middle, and posterior sections; *B,* diagram of myotome patterns in salmon, myotome number reduced. Full extent of selected myotomes shown in gray. (*B* based on Greene and Greene, 1914.)

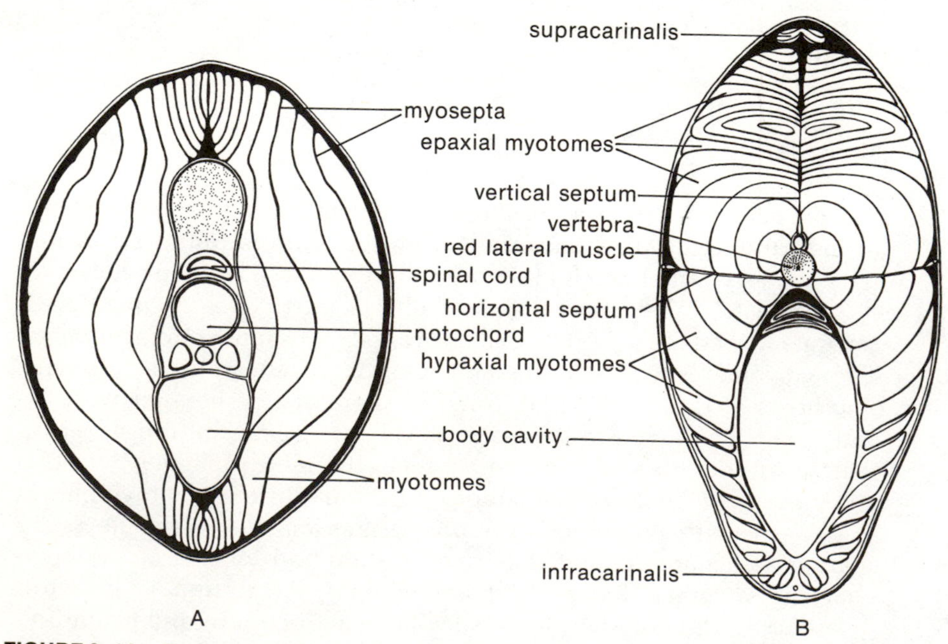

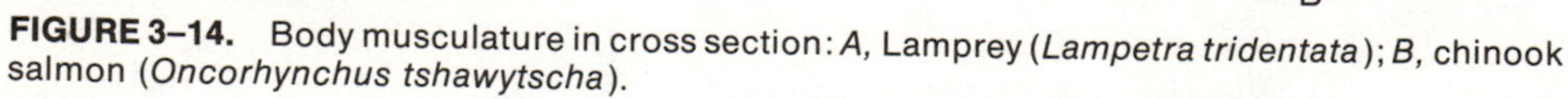

FIGURE 3–14. Body musculature in cross section: *A,* Lamprey (*Lampetra tridentata*); *B,* chinook salmon (*Oncorhynchus tshawytscha*).

visible edge in lampreys, but in bony fishes and sharks the myotomes extend posteriorly from the regions of the backward-pointing upper and lower flexures, and forward from the middle, forward-pointing flexure. This brings about an intricate system of cone-shaped structures, each fitting inside another so that a cross section of the trunk or tail cuts through several myotomes on each side, showing myosepta as concentric lines (Fig. 3–14).

A vertical septum separates the muscles into the left and right halves and, on each side, a horizontal septum divides them further into epaxial muscles dorsal to the septum and hypaxial muscles below it. The horizontal septum is missing in lampreys and hagfish. Along the sides just under the skin lie the lateral superficial muscles, which are usually dark in color, well-supplied with blood vessels, and of high fat content. These muscles are used in normal sustained swimming activity, and are fatigue-resistant at slow or cruising speed. These "red" muscles are larger in active than in sedentary fishes, being extensive and very bloody in tunas. The remainder of the lateral musculature is used for sudden bursts of swift or strong swimming, as in escape or the capture of prey.

A thin muscle called the anterior supracarinalis passes from the skull, usually from the supraoccipital, to the first pterygiophore of the dorsal fin. A posterior supracarinalis connects the last pterygiophore of the dorsal with the posterior neural spine or caudal fin supports. Paired anterior infracarinales stretch between the cleithrum and the pelvic bone, along the ventral midline. Another, infracarinalis medius, passes from the pelvic bone to the first basal pterygiophore of the anal fin, and the posterior infracarinalis connects the anal and caudal fins. Carinal muscles serve mainly as protractors and retractors for the dorsal and anal fins (Fig. 3–15). Muscles of the fins are derived from the myotomes but may no longer correspond with body segments in adults. Each ray

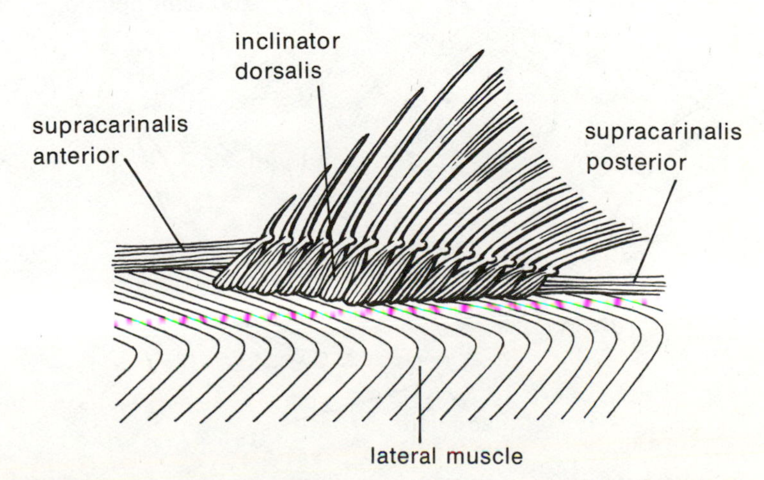

FIGURE 3–15. Diagram of inclinator muscles of dorsal fin. (Based on Greene and Greene, 1914.)

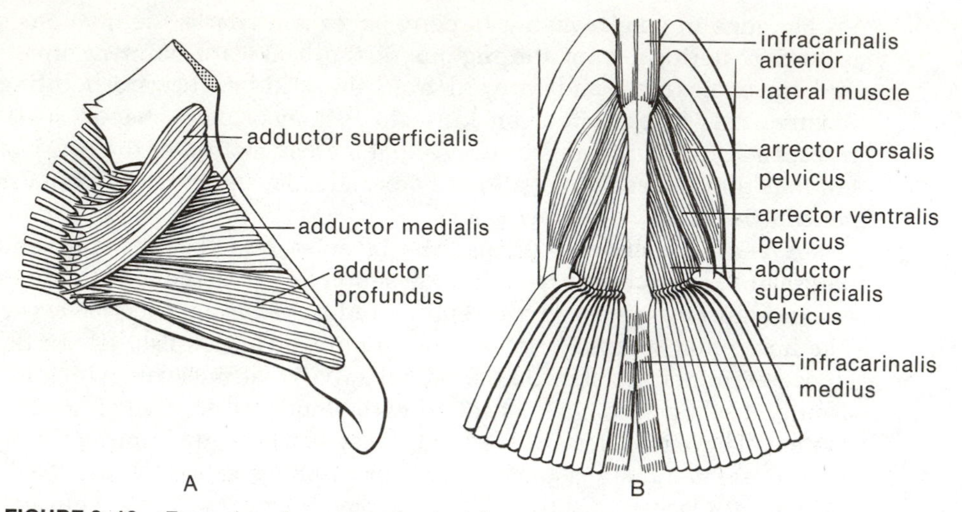

FIGURE 3–16. Examples of muscles of paired fins. *A*, Medial aspect of pectoral muscles, *Perca fluviatilis; B*, ventral view of pelvic muscles. (*A* redrawn from Winterbottom, 1974; *B* based on Greene and Greene, 1914.)

has a set of erectors and depressors. Soft fins have inclinators capable of bending the rays. Musculature of the paired fins (Fig. 3–16) are abductors, adductors, and arrectors, with some fibers attaching on individual fin rays or their basals to give great flexibility — especially of the pectorals — in some fishes.

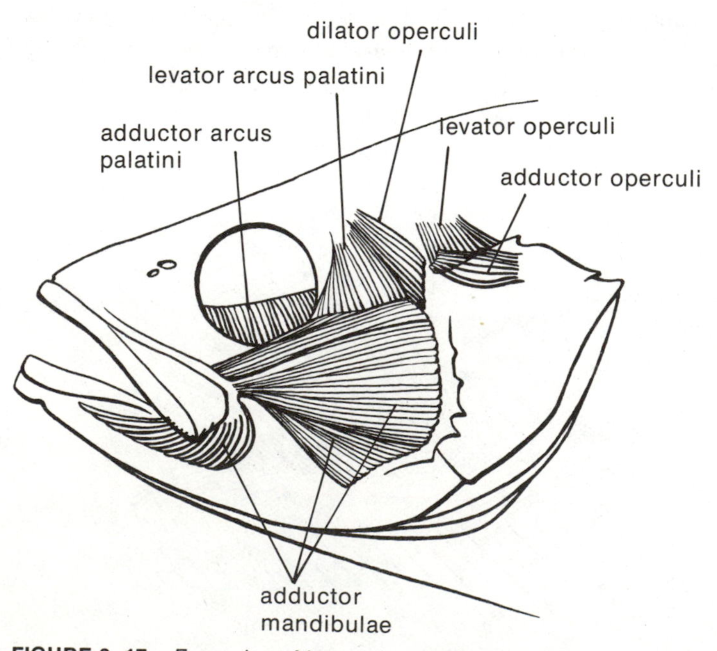

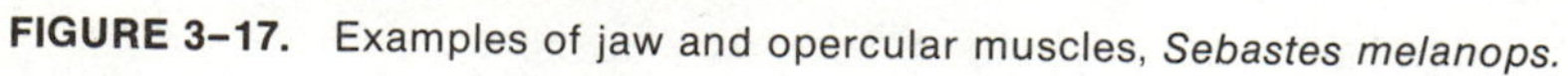

FIGURE 3–17. Examples of jaw and opercular muscles, *Sebastes melanops.*

The numerous moving parts of the jaws and branchiohyoid apparatus, plus the operculum, require complex sets of muscles that may differ among fishes of different lineages or habits. Examples of these muscles are shown in Figure 3–17. The adductor mandibularis, which closes the mouth, is one of the largest of the head muscles, and reaches a large proportionate size in those fishes with crushing teeth or those that bite chunks out of prey. The jaw muscles of the wolf-eel are remarkably large, and the posterior portion of the cranium is greatly compressed and smooth, giving a large place of attachment for these muscles. Opercular musculature is well developed in most fishes, especially those of sedentary habits. Bottom fishes generally have better developed branchiostegal muscles than do active species. Swift fishes that depend on continuous movement for gill irrigation tend to have small opercular and branchiostegal muscles.

References

Alexander, R. M. 1969. The orientation of muscle fibers in the myomeres of fishes. J. Mar. Biol. Assoc. United Kingdom, 49:263–290.

Bainbridge, R. 1963. Caudal fin and body movement in the propulsion of some fish. J. Exp. Biol., 40:23–56.

Boddeke, R., Slijper, E. J., and van der Stelt, A. 1959. Histological characteristics of the body musculature of fishes in connection with their mode of life. Proc. K. Ned. Akad. Wet., Ser. C, 62:576–588.

Bone, Q. 1966. On the function of the two types of myotomal muscle fibre in elasmobranch fishes. J. Mar. Biol. Assoc. United Kingdom, 46:321–349.

Branson, B. A. 1966. Guide to the muscles of bony fishes, excluding some special fibers in siluroids and a few others. Turtox News, 44(4):98–102.

Breder, C. M. Jr. 1926. The locomotion of fishes. Zoologica, 4:159–297.

DeBeer, G. R. 1937. The Development of the Vertebrate Skull. Oxford, The Clarendon Press.

Dineen, C. F., and Stokely, P. S. 1954. Osteology of the central mudminnow, *Umbra limi*. Copeia, 1954:169–179.

Greene, C. W., and Greene, C. H. 1914. The skeletal musculature of the king salmon. Bull. U. S. Bur. Fish., 33:21–60.

Greer-Walker, M., and Pull, G. A. 1975. A survey of red and white muscle in marine fish. J. Fish Biol., 7(3):295–300.

Gregory, W. K. 1933. Fish skulls. Trans. Am. Phil. Soc., 23:75–481. (Offset reprint edition, Laurel, Florida, Eric Lundberg Publ. Co., 1959.)

Harrington, R. W., Jr. 1955. The osteology of the American cyprinid fish, *Notropis bifrenatus*, with an annotated synonymy of teleost skull bones. Copeia, 1955:267–290.

Hollister, G. 1936. Caudal skeleton of Bermuda shallow water fishes: I. Order Isospondyli: Elopidae, Megalopidae, Albulidae, Clupeidae, Dussumieriidae, Engraulidae. Zoologica, 21:257–296.

Holstvoogd, C. 1965. The pharyngeal bones and muscles in Teleostei, a taxonomic study. Proc. K. Ned. Akad. Wet., Ser. C, 68:209–218.

Jessen, H. 1972. Schultergürtel und Pectoralflosse bei Actinopterygiern. Fossils and Strata, (1):1–101, Plates, 1–25.

Krumholz, L. A. 1953. A comparative study of the Weberian ossicles in North American Ostariophysine fishes. Copeia, 1953:33–40.

Laerm, J. 1976. The development, function, and design of amphicoelous vertebrae in teleost fishes. Zool. J. Linn. Soc., 58(3):237–254.

Le Danois, Y. 1958. Système musculaire. *In*: Grassé, P. P. (ed.), Traité de Zoologie, 13(1):783–813.
Leim, K. F. 1963. The comparative osteology and phylogeny of the Anabantoidei (Teleostei, Pisces). Ill. Biol. Monogr., 30:1–149. Urbana, Illinois, University of Illinois Press.
———. 1970. Comparative functional anatomy of the Nandidae (Pisces: Teleostei). Fieldiana: Zoology, 56:1–166.
Miller, P. J. 1973. The osteology and adaptive features of *Rhyacichthys aspro* (Teleostei: Gobioidei) and the classification of gobioid fishes. J. Zool. Lond., 171:397–434.
Mujib, K. A. 1967. The cranial osteology of the Gadidae. J. Fish. Res. Bd. Can., 24:1315–1375.
Nag, A. C. 1967. Functional morphology of the caudal region of certain clupeiform and perciform fishes with reference to taxonomy. J. Morphol., 123:529–558.
Nelson, E. M. 1948. The comparative morphology of the Weberian apparatus of the Catostomidae and its significance in systematics. J. Morphol., 83(2):225–251.
Norden, C. R. 1961. Comparative osteology of representative salmonid fishes, with particular reference to the grayling (*Thymallus arcticus*) and its phylogeny. J. Fish. Res. Bd. Can., 18:679–791.
Nursall, J. R. 1956. The lateral musculature and the swimming of fish. Proc. Zool. Soc. Lond., 126:127–143.
Osse, J. W. M. 1969. Functional morphology of the head of the perch (*Perca fluviatilis* L.): an electromyographic study. Netherl. J. Zool., 19(3):289–392.
Ridewood, W. G. 1904*a*. On the cranial osteology of the families Elopidae and Albulidae with remarks on the morphology of the skull in lower teleostean fishes generally. Proc. Zool. Soc. Lond., 2:35–81.
———.1904b. On the cranial osteology of the clupeoid fishes. Proc. Zool. Soc. Lond., 2:448–492.
Shufeldt, R. W. 1899. The skeleton of the black bass. Bull. U. S. Fish. Comm., 19:311–320.
Starks, E. C. 1926. Bones of the ethmoid region of the fish skull. Stanford Univ. Publ. Biol. Sci., 4:139–338.
———. 1930. Primary shoulder girdle of the bony fishes. Stanford Univ. Pub. Biol. Sci., 6:149–239.
Willemse, J. J. 1959. The way in which flexures of the body are caused by muscular contraction. Proc. K. Ned. Akad. Wet., Series C, 62:589–592.
Winterbottom, R. 1974. A descriptive synonymy of the striated muscles of the Teleostei. Proc. Acad. Nat. Sci. Phila., 125(12):225–317.

4

NERVOUS SYSTEM AND ENDOCRINE GLANDS

Nervous System

Brain

The brain lies in the cavity of the neurocranium, protected by cartilage or bone, by the surrounding meninges and cerebrospinal fluid, and by a fatty matrix that fills much of the cranial cavity. Cranial nerves extend from the brain through foramina of the skull. The spinal cord, with which the brain is continuous, leaves the cranium posteriorly through the foramen magnum. Interesting differences appear among the brains of fishes, even though they are all representative of the same basic plan (Fig. 4–1). Some differences are of phylogenetic significance

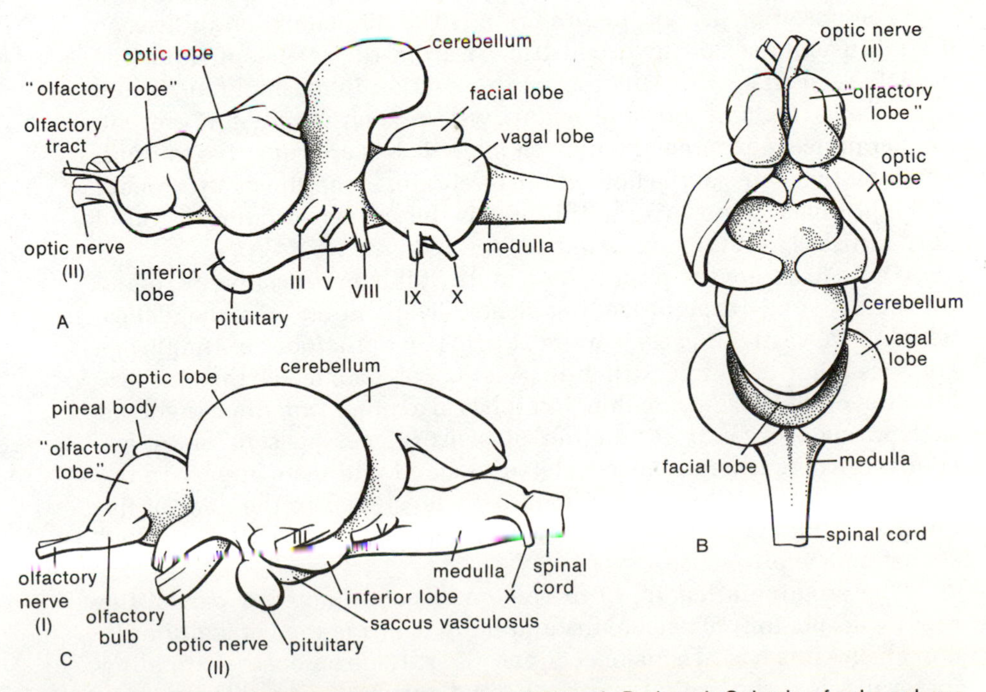

FIGURE 4–1. General topography of carp brain: *A*, lateral; *B*, dorsal; *C*, brain of coho salmon. (*A* and *B* based in part on Tuge, et al., 1968.)

and may be manifested in the complexity and compactness of the organ, whereas others are apparently due to the degree of development of certain sensory and motor functions. Differences in the relative sizes of fish brains are evident. The brains of sharks and some predatory fishes weigh considerably less than 0.1% of the total fish weight, while the relative weight of the brains of many minnows (Cyprinidae) may be twice that. The largest relative size among fish brains is that of the African elephant fishes or mormyrids (Mormyridae), in which the brain may be more than 1% of the fish weight.

The fish brain develops as the forebrain or prosencephalon, the midbrain or mesencephalon, and the hindbrain or rhombencephalon, as in other vertebrates and then differentiates further. The forebrain is the site of the olfactory sense, and its anterior part (telencephalon) is characterized by a pair of primary olfactory centers, the olfactory bulbs, from which olfactory nerves extend to the olfactory organ. Posterior to the bulbs, the telencephalon swells into what are often called "olfactory lobes." These contain a secondary olfactory center and are usually larger than the bulbs, but are mainly concerned with nonolfactory function. In most bony fishes the bulbs are situated just anterior to the lobes, but in elasmobranchs and in some bony fishes, such as certain catfishes, carp, and cods, the bulbs are adjacent to the organ of smell and a long olfactory tract separates them from the lobes. In these species, especially in sharks and rays, the bulbs may reach a relatively large size. Part of the telencephalon is developed as the cerebrum in elasmobranchs and bony fishes, although in the latter the cerebral hemispheres are prominent only among the more primitive members.

The posterior part of the forebrain is the diencephalon or "tween brain," usually set off by a constriction from the telencephalon. The pineal organ arises from the roof of this section in elasmobranchs and bony fishes. In lampreys and hagfishes two such organs are present, the parapineal and pineal, which in many fishes are sensitive to light. The hypothalamus, at the floor of the forebrain, is the site of attachment of the pituitary gland. Also in this area is the saccus vasculosus, found only in fishes; its function is not definitely known.

Optic lobes are the prominent feature of the mesencephalon, being especially large in sight-feeding fishes. The midbrain of hagfishes, which have vestigial eyes, is quite small. The optic lobes of lungfishes are fused into a single structure. Vision, of course, is the primary function of the mesencephalon, but it has also other functions involved in learning and with correlation of sensory messages to motor responses. Afferent nerve fibers related to smell and taste appear in the optic tectum. The torus semicircularis is developed in the floor of the midbrain, and is especially large in fishes with a well developed acoustico-lateralis sense.

The prominent feature of the metencephalon (anterior part of the rhombencephalon) of selachians and bony fishes is the cerebellum. In bony fishes this has two major sections: the valvula cerebelli, extending forward below the optic tectum; and the corpus cerebelli, extending forward dorsally. The valvula cerebelli is absent in elasmobranchs. At the area of attachment of the corpus cerebelli to the medulla there are lateral swellings called eminentia granularis, which may be enlarged in some species (certain catfishes and hairtails, Trichiuridae). The size of the corpus cerebelli is variable, being especially well developed in large sharks and in the family Mormyridae. In the latter group the cerebellum is relatively enormous and may overlie the forebrain. Catfishes, mackerels, and tunas usually have enlargement of the cerebellum. In lampreys the cerebellum is very small; and the structure is not recognized in hagfishes.

Function of the cerebellum involves coordination of movement, muscle tone, and posture or balance — most fishes with an enlarged cerebellum are fast swimmers. Catfishes and mormyrids are slower moving yet have this structure enlarged, so integration of certain sensory impulses appears to be an important function in some types of fishes. Mormyrids, of course, are noted for their "electrical" sense, and many catfishes are equipped with organs for electrical impulse reception.

The most posterior part of the brain, the myelencephalon, is comprised chiefly of the medulla oblongata, which includes somatic and visceral sensory and motor areas. Cranial nerves V through X arise from the medulla oblongata, and impulses to and from the spinal nerves are also relayed here. Various parts of the medulla are enlarged according to the sensory equipment and habits of fishes. For instance, certain suckers (Catostomidae) and minnows (Cyprinidae) have characteristic enlargements at the roots of the seventh, ninth, and tenth cranial nerves. The large restiform bodies of sharks and rays are related to their acute lateral line sense. These bodies are connected to the medulla but may actually be part of the metencephalon. A pair of giant neurons usually called the cells of Mauthner are found in the floor of the medulla of actinopterygians. These are believed to function in the coordination of locomotion, as the dendrites are associated with acoustico-lateralis centers and the axons travel the length of the spinal cord to caudal musculature.

Cranial Nerves

There is a series of nerves extending from the brain to certain sensory organs and muscles. Most of these cranial nerves serve the head region but some, especially the vagus, have both sensory and motor components serving parts of the body. Eleven cranial nerves are developed in fishes, including the terminal nerve (0) that is found only in lower vertebrates. The nerves and their major roles are:

0. The terminal nerve is a small nerve associated with the olfactory nerve, which connects with the forebrain, and its fibers are distributed around the olfactory bulb. Its function may include both somatic and special sensory roles.

I. The olfactory nerve runs from the olfactory organ to the olfactory center of the forebrain. Its function is sensory.

II. The optic nerve arises in the retina of the eye and connects to the optic tectum. It carries visual impulses.

III. The oculomotor nerve supplies the inferior oblique and the superior, inferior, and internal rectus eye muscles. Its connection with the brain is with the mesencephalic brainstem. It is a somatic motor nerve.

IV. The trochlear nerve innervates the superior oblique muscle of the eye. This somatic motor nerve connects with the mesencephalon.

V. The trigeminal nerve is divided into three branches. Two of these, the ophthalmic and maxillary, are somatic sensory nerves and the third, the mandibular, carries both somatic sensory and somatic motor fibers. The trigeminal connects portions of the head and jaws to the medulla oblongata. It carries impulses from tactile and thermal receptors.

VI. The abducens nerve is a somatic motor nerve coursing from the anterior part of the medulla oblongata to the external rectus muscle of the eye.

VII. The facial nerve is closely associated at the medulla with nerves V and VIII (acoustic). It is often considered together with nerve VIII as one nerve. The facial nerve is usually composed of three branches — the superficial ophthalmic, the buccal, and the hyomandibular. These branches supply the lateral line canals on the head, taste receptors on the head and body, touch receptors, and certain head muscles. The facial nerve has components involved with special and general somatic sensory, special and general visceral sensory, and visceral motor functions.

VIII. The acoustic nerve has a somatic sensory function, serving the inner ear. It is often considered a branch of the combined acousticofacialis nerve in fishes.

IX. The glossopharyngeal nerve is composed of both sensory and motor components serving mainly the region of the first gill slit. A dorsal group of branches serves the temporal lateral line canal and proprioceptors; branchial branches are involved with taste organs of the pharynx and with branchial muscles.

X. The vagus nerve is a large mixed complex with several branches, two of which, the supratemporal and body lateral line branches, serve the lateralis system. Branchial branches travel to the region of the posterior four gill slits. A dorsal recurrent branch innervates body taste receptors, and the visceral branch supplies internal organs.

Spinal Cord and Nerves

The spinal cord is continuous with the medulla oblongata and extends to the tail region. It is essentially a hollow tube, but the central canal is of small diameter compared to the walls. Around the central canal, making a pattern in cross section similar to a pair of butterfly wings, is gray matter composed of myelinated nerve fibers running longitudinally.

The paired spinal nerves are arranged segmentally and arise from the gray matter as dorsal and ventral roots which merge and then typically branch into three parts. The dorsal root has a ganglion outside the spinal cord. Dorsal and ventral branches (or rami) serve the axial muscles and skin, while a visceral branch (ramus) serves the internal organs. Lampreys differ from the other fishes in lacking the connection between the dorsal and ventral roots. In these forms the dorsal roots originate opposite the myosepta and the ventral roots opposite the myotomes.

The dorsal roots of spinal nerves in fishes carry somatic and visceral afferent fibers, and some visceral efferent fibers. Somatic and visceral efferent fibers enter the spinal cord through the ventral roots. The visceral efferent components of the spinal nerves contribute to the autonomic nervous system, which is involved in the control of smooth muscle and certain glands. Bony fishes have a chain of interconnected segmentally arranged ganglia. Sympathetic ganglia are found in an irregular series in the trunk region of elasmobranchs. Parasympathetic fibers are largely associated with cranial nerves, almost entirely with the vagus.

Sensory Organs

Eye. Although there are many modifications of eye shape and structure among fishes, the general plan is similar throughout. As with other vertebrates, the major features of the eye are an anterior chamber, an iris, a lens, and a vitreous chamber containing the vitreous humor and lined by the retina. The eye is flattened anteriorly so that the spherical lens is nearly in contact with the cornea, which is essentially a transparent section of the scleroid coat of the eyeball. A vascular choroid layer is situated between the retina and the sclera. The sclera of elasmobranchs and teleosts may be stiffened by cartilaginous structures or, in the case of the latter, by scleral ossicles (Fig. 4–2). A choroid body or "gland," actually a rete mirabile, is prominent in the choroid of many teleosts (see p. 368).

Lampreys differ from other fishes in that the conjunctiva and cornea are not fused to each other, and in the mechanism for accommodation to near and distant vision. The lens is not suspended from the interior of the eyeball but is held in place by the pressure of the fluid in the vitreous cavity. In accommodation for distant vision the lens is forced back by contraction of a muscle that flattens the cornea. The lamprey eye is rather solidly attached to the rim of the eye socket; the eyeball lacks any cartilaginous or bony reinforcement.

The elasmobranch lens is suspended dorsally, as is that of teleosts. In the elasmobranch the lens is pulled forward for accommodation, and in the bony fishes it is retracted. The retractor muscle of the teleost is attached to the anterior end of the falciform process, which is an outgrowth of the choroid extending through a fissure in the retina.

Fishes have no eyelids except for the nictitating membranes of certain elasmobranchs. The so-called "adipose eyelids" of some teleosts, herrings, and tunas, for example, are immobile and serve mainly

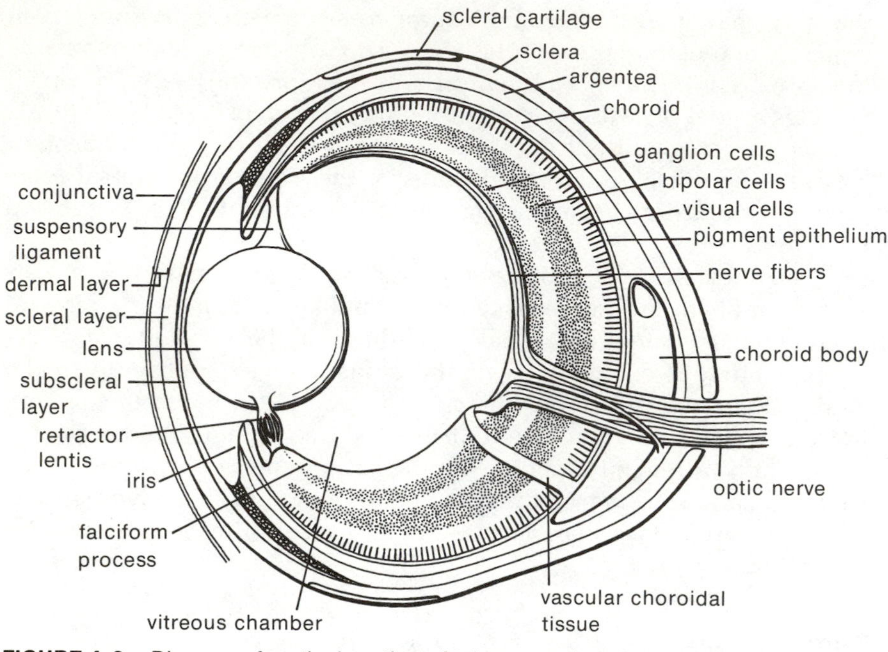

FIGURE 4–2. Diagram of vertical section of teleost eye (not drawn to scale), showing the relationships of its parts.

to streamline the slight bulge of the eye beyond the surface of the head. A few elasmobranchs can control the opening of the pupil because of the musculature associated with the iris, but most fishes lack this control or have it only feebly developed. The retina consists of a pigment epithelium, the visual cells (rods and cones), a layer of bipolar cells and, closest to the vitreous humor, the ganglion cells and the nerve fibers leading to the optic nerve.

The Acoustico-Lateralis System. This system is usually considered to include the inner ear (fish have no middle or outer ear), the grooves and tubes that make up the lateral line complex, and various related pits and follicles. These structures have common embryological origins. The functions of the ear appear to be mainly equilibration and reception of sound. The organs of the lateral line respond to displacement of water and pressure, with some of the related structures thought to be temperature receptors. Others are known as receivers of electrical impulses, while some are of unknown function.

The inner ear of bony fishes is composed of the osseous and membranous labyrinths. The membranous labyrinth (Fig. 4–3) typically consists of three more or less distinct chambers — utriculus, sacculus, and lagena — and three semicircular ducts or canals. The ducts and the utriculus constitute the pars superior of the organ, and the sacculus and lagena are the pars inferior. These sections are nearly separated in some minnows and completely separate in some gobies. In sharks, rays, and chimaeras, an endolymphatic duct communicates with the exterior. In bony fishes this duct is shortened or lacking

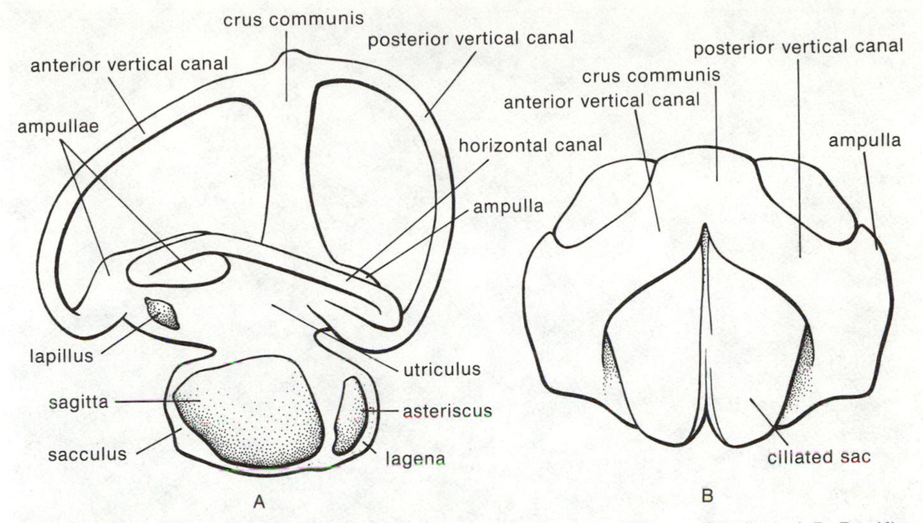

FIGURE 4–3. Membranous labyrinth of *A*, cutthroat trout (*Salmo clarki*) and *B*, Pacific lamprey (*Lampetra tridentata*), both showing left labyrinth from outer aspect.

altogether. Considering the utriculus as a reference point for descriptive purposes, it can be seen as the place of attachment of the semicircular ducts, each of which is oriented in a different plane in relation to the others. One is horizontal, situated on the outer aspect of the organ. The other two are vertical, one posterior and one anterior, at right angles to each other, placed so that each is at an approximately 45° angle to the axis of the body as viewed from above. The vertical component of the labyrinth to which they are attached dorsally is continuous with the utriculus and is termed the crus communis.

Each semicircular duct has an ampulla at its junction with the utriculus (the ampulla of the horizontal canal is at the anterior connection). Rising from the floor of each ampulla is an eminence called a crista, upon which is a gelatinous cupula, surrounding hair-like extensions of neuromast cells. Neuromasts are the basic receptors of the acoustico-lateralis system. These consist of a clump of supporting cells surrounding pyriform sensory cells, each of which bears the sensory filament, or "hair." Movement of the head in any plane will cause the endolymph of the canals to deform one or more of the cupulae, thus stimulating the neuromasts.

The sacculus is attached to the utriculus ventrally and the lagena, which attaches to the posterior part of the sacculus, may be well delineated but is not distinct in many species. There are beds of neuromasts called maculae in all three of these sections, upon which the otoliths lie. Otoliths of elasmobranchs consist of calcareous granules (otoconia) in a soft matrix, and have been reported to include mineral particles such as sand grains that enter through the endolymphatic duct. These exogenous particles are called otarena. In most bony fishes the otoliths (Fig. 4–4) are hard structures, with those of most species having a characteristic shape and size. These are generally held in

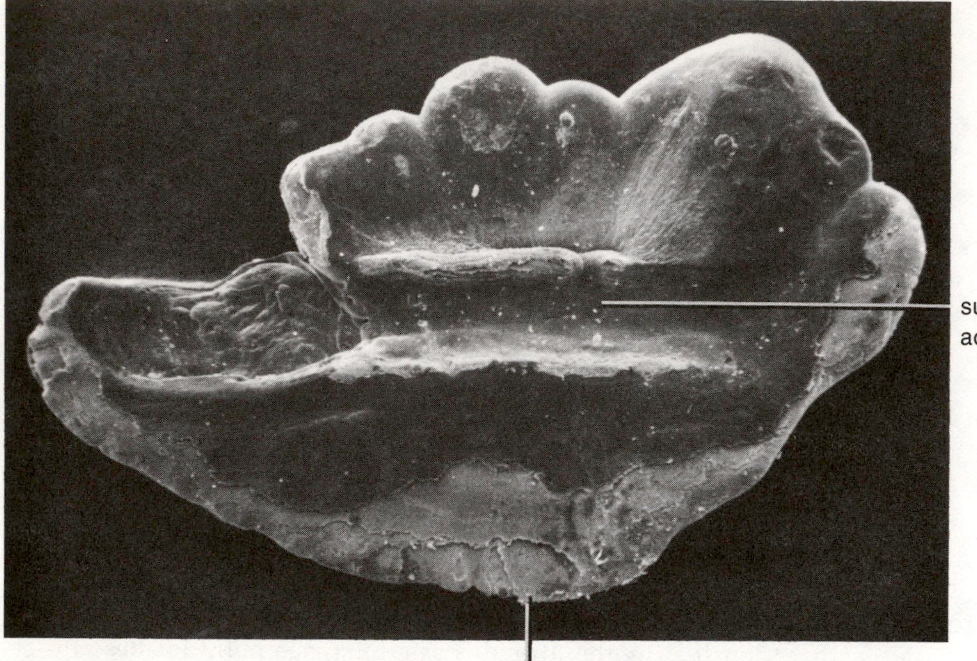

FIGURE 4–4. Scanning electron micrograph of medial aspect of right otolith from pink salmon (*Oncorhynchus gorbuscha*). (Photograph courtesy of Dr. Richard Casteel.)

place by connective tissue. The otoliths of the utriculus, sacculus, and lagena are called respectively lapillus, sagitta, and asteriscus. In the labyrinth of lampreys and elasmobranchs there is a macula (macula neglecta) without an otolith.

The membranous labyrinth of the hagfish consists of a lower chamber having a single macula associated with a membrane that carries calcium carbonate inclusions, which serve the function of an otolith. Above the lower section a continuous canal forms an arch. This has been called a single semicircular canal but has also been considered to represent the anterior and posterior vertical canals, because ampullar swellings with annular cristae appear near the junctions with the pars inferior. In lampreys the labyrinth is more complicated. The common macula is functionally divided into parts that appear to correspond to the sacculus, utriculus, and lagena, but is still covered by an "otolith" made up of calcareous crystals. There are two vertical semicircular canals with divided cavities. Two ciliated sacs that make up a large portion of the labyrinth are not seen in any other vertebrate. The function of these structures is not known.

Lateral line organs may be free neuromasts on the skin or in pits, or may consist of a series of the receptors in the canals or grooves on the head and body (Fig. 4–5). The canals are typically filled with mucus and, depending on the species, may run through the skin or through scales. The cephalic portion of the canal courses through many of the

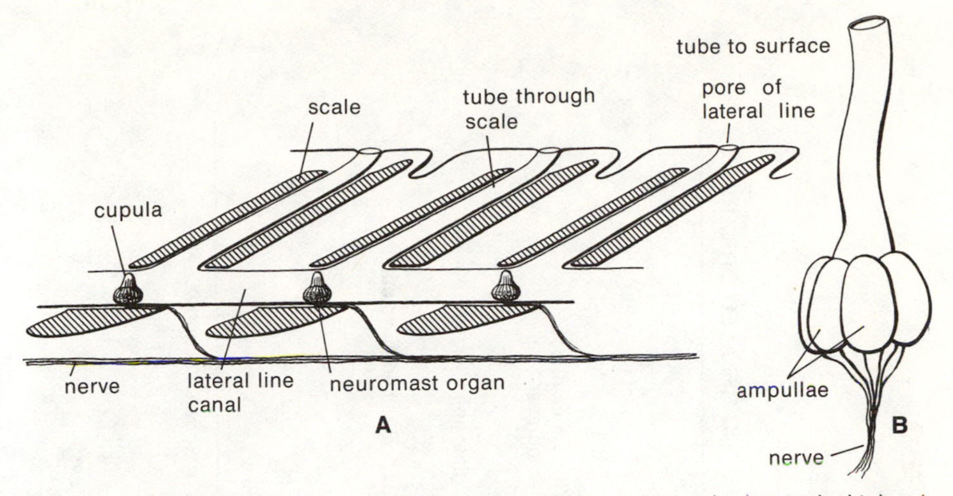

FIGURE 4–5. *A*, Diagram showing relationship of lateral line canal to scales in a typical teleost. The drawing represents a horizontal section with the thickness of the scales and size of the lateral line exaggerated; *B*, diagram of ampullae of Lorenzini.

skull bones. Usually these tunnels through the bones are of small diameter but in some groups, such as the drum family (Sciaenidae), they are wide and cavernous.

Herrings and their close relatives are peculiar in possessing an intercranial space called the recessus lateralis, which connects with all the head canals of the lateralis system. It is situated between the pterotic and frontal bones above and the prootic below. A membrane separates the recessus from the labyrinth of the auditory organ. It may function in the reception of vibrations.

Associated with the lateral line system in elasmobranchs and in the marine catfish *Plotosus* are organs called ampullae of Lorenzini (Fig. 4–5). These usually are disposed in groups following a definite pattern on the head, and also on the pectoral fins of skates. These groups take their names from their respective placement — supraorbital, buccal, hyoid, and mandibular. Their positions are marked by pores that open into the canal or tubule of each ampulla; a gelatinous material fills the canal. The ampullae of various species may be of specific types. Some are globular, some consist of a clump of rounded structures (alveolar), some have a central plate set into the base, and those of *Hexanchus*, the sixgill shark, have several diverticula, each with its own separate division of the canal. The sensory cells of the ampullae are globular, in contrast to the pyriform cells of the lateral line neuromasts. They are set into the epithelium of the walls of the ampullae, with each sending a cilium into the lumen of the ampulla. The ampullae function as receptors of electrical stimuli.

Similar in function to these ampullae are the cutaneous electroreceptors of mormyroids, gymnotoids, and various other freshwater fishes. One type of electroreceptor is ampullary in nature, with a canal leading to the surface, while another type has no opening to the exteri-

Table 4–1. PRIMARY INNERVATION OF THE ACOUSTICO-LATERALIS SYSTEM AND RELATED ORGANS

Structure	Nerve
Inner ear	Acoustic (VIII)
Lateral line system	
Lateral canal	
Anterior section sharks	Vagus (X), ramus lateralis dorsalis, facial (VII), ramus oticus, Glossopharyngeal (IX), ramus supratemporalis
Other	Vagus (X), ramus lateralis
Supratemporal commissure	Vagus (X), ramus lateralis
Supraorbital canal	Facial (VII), ramus ophthalmicus superioris, glossopharyngeal (IX)
Infraorbital canal	Facial (VII), ramus buccalis
Preoperculomandibular canal	Facial (VII), truncus hyomandibularis
Ampullae of Lorenzini	Facial (VII), rami ophthalmicus, superioris, buccalis, hyomandibularis
Vesicles of Savi	Trigeminal (V)
Electroreceptors of freshwater fishes	Facial (VII), vagus (X)
Spiracular organs	Facial (VII), ramus oticus
Pit organs of elasmobranchs	
Mandibular	Facial (VII), ramus mandibularis externus
Remainder of head and body	Vagus (X), mostly ramus lateralis*

*The rami supratemporalis, dorsalis, and postrematicus are known to innervate a few pit organs in *Squalus acanthias.*

or. The latter, referred to as tuberous organs, appears to be more sensitive to higher electrical frequencies than the ampullary type.

Other organs found in some electric fishes are the vesicles of Savi. These are closed sacs filled with a gelatinous material and enclose three discs, a median one flanked by two smaller ones. They are found on the snout region of electric rays (Torpedinidae) and some other rays, mostly on the ventral surface. Usually there are about 200 vesicles, with only 30 to 40 on the dorsal surface of the snout.

Additional organs of doubtful function are the spiracular organs of sharks, rays, and such primitive bony fishes as sturgeons and the polypterids, and the organs of Fahrenholz, found only on the head of larval lungfishes. In each instance, the organs are quite similar to the typical lateral line sensory organ in structure.

The physical stimuli of touch and temperature are apparently received by fishes through free nerve endings in the skin but the ampullae of Lorenzini, now better known as receptors of electrical stimuli, have been shown to react markedly to changes in temperature. In addition, some sharks, sea robins, and eels have corpuscle-like structures that may be specialized touch receptors.

Table 4–1 outlines the primary innervation of the acousticolateralis system and related organs.

Olfactory organs. The organs of smell in fishes are located in sacs on the snout, usually directly in front of the eye. These nasal sacs do not communicate with the pharynx in the majority of fishes, as there are external nares only. Typically there are anterior and posterior nares on each side. Fishes with internal nares are lungfishes, hagfishes, and stargazers (Uranoscopidae).

In order for chemical substances to come into contact with the olfactory epithelium in the nasal sacs, these structures must be irrigated with a considerable volume of water. Respiratory movements can facilitate the irrigation of the olfactory rosettes of those fishes with internal nares, but maintaining a flow of water in most fishes requires some special hydraulic engineering. Circulation of water in the nasal sac is accomplished in three ways: by forward movement of the fish in relation to the water; by the action of cilia in the sac or extensions of it; and by pumping effected by constriction of the nasal sac and accessory sacs.

Fishes that depend on their forward motion or on water currents for movement of water through the olfactory organ usually have flaps or ridges behind the anterior nares to guide the water over the olfactory epithelium. These cutaneous structures are easily seen in common freshwater fishes such as trout, minnows, and suckers. Sharks and rays are notable for their occasionally elaborate complement of flaps and grooves associated with the nares. In eels, some catfishes, and probably in other fishes with a long nasal sac and widely separated anterior and posterior nostrils, cilia are important in moving water through the system, even though contraction of facial muscles may be involved. Many such fishes have tubular extensions of the nares.

Modifications for pumping water in and out of the olfactory chamber in a tidal sequence are common among bony fishes, but can also be seen in the more primitive lampreys. Lampreys have a single olfactory opening leading to an olfactory rosette that probably represents a fusion of two organs of smell. However, the nasal portion of the canal is continuous with a blind tube called the nasohypophyseal canal or pouch that runs posteriorly under the brain and ends between the brain and the pharynx. Contractions of the branchial muscles apply pressure to the pouch and cause rhythmic emptying and refilling. As water flows into the pouch, some is shunted into the olfactory organ by a valve at its entrance.

In bony fishes accessory sacs continuous with the nasal sac very commonly appear in the region lateral to the nostrils, and some species have such sacs under the skin between the nasal organs. Respiratory and other movements involving the jaws and other face bones affect the volume of the nasal and accessory sacs, and cause water to flow in and out over the olfactory epithelium. The front nostril is usually incurrent and the back excurrent. A few bony fishes, such as cichlids, sticklebacks (Gasterosteidae), and damselfishes (Pomacentridae) have a single nostril on each side.

The olfactory epithelium is arranged so that a considerable area is presented to the water that is being tested for odors, either by folding into the typical "rosette" or by disposition as a series of finger-like projections (Fig. 4–6). The epithelium is columnar and consists of supporting (sustentacular) cells, in addition to numerous interspersed olfactory

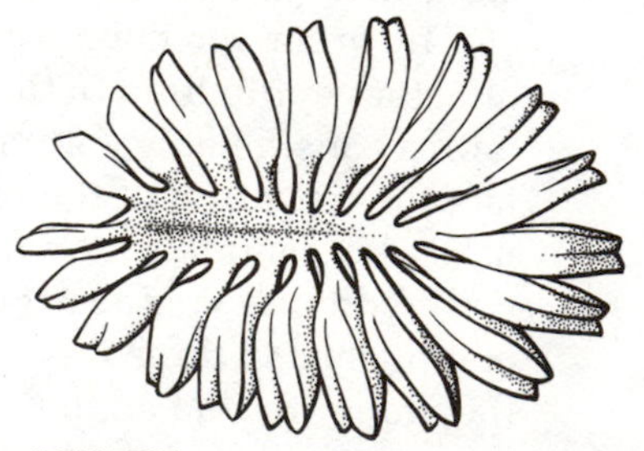

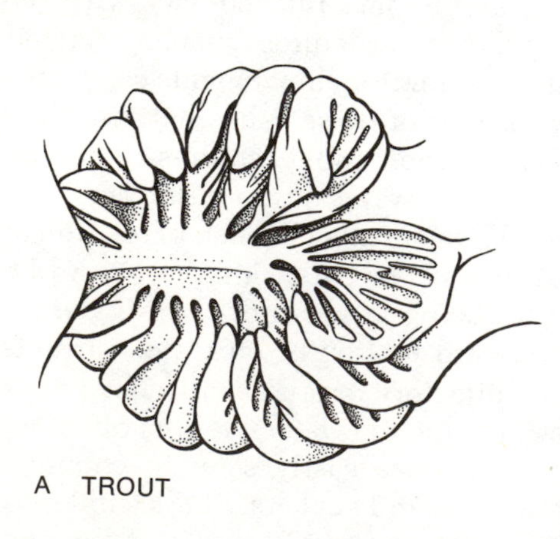

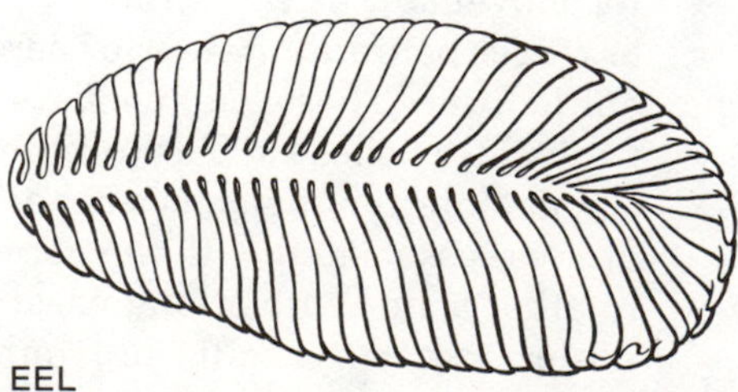

FIGURE 4–6. Dorsal views of olfactory rosettes. *A,* Trout (*Salmo gairdneri*), showing concave rosette with finger-like processes; *B,* staghorn sculpin (*Leptocottus armatus*), convex rosette; *C,* American eel (*Anguilla rostrata*), flat rosette. In eel, incurrent water moves from left to right; in others, incurrent water impinges on middle of rosette.

sensory cells. A basal membrane underlies the remainder of the epithelium. Mucus-secreting goblet cells have been identified in *Anguilla*, *Ictalurus*, and *Mustelus*. The olfactory cells are somewhat elongate, a vesicle at their outer ends reaching the lumen of the olfactory sac, where a series of olfactory cilia or "sensory hairs" extends from them. Inwardly, thin axons extend from the cells to the olfactory bulb. The identity of the olfactory cells as ganglionic cells and the direct connection with the brain has led many anatomists to conclude that the olfactory receptor is essentially a primitive structure that has remained relatively unchanged.

There are about 20 olfactory cilia per cell in the burbot, *Lota*, and these appear to be 20 to 30 microns long. Ciliar movements observed in vitro have shown no regular pattern; bending and straightening occur in unorganized sequences. The relationship of the cilia and their movements to the olfactory sense has not yet been fully described, but they seem to be the site where chemical substances come into contact with the appropriate nervous tissue so that the smell impulse can be generated.

The number of olfactory cells per square mm of olfactory epithelium in freshwater fish is usually in the range of 50,000 to 100,000. The number of folds or lamellae in the olfactory rosette varies greatly among species. The sand lance, *Ammodytes*, has an unfolded epithelium; the stickleback, *Gasterosteus*, has but two lamellae; many familiar fishes such as trout, grayling, minnows, and pike have fewer than 20; and the eel, *Anguilla*, may have up to 80 or 90. The number of lamellae apparently increases with the size of the individual. The relationship between the number of lamellae and olfactory acuity is obscure, but there seems to be some correlation between the size and shape of the rosette and acuity. Species such as the pike with round rosettes are said to have weak powers of olfaction (microsmatic) and those such as the eels, with long rosettes, appear to have the greatest acuity (macrosmatic). Most fishes have oval rosettes and intermediate powers of olfaction. The area of the olfactory epithelium in relation to the surface area of the body is smallest in those fishes with round rosettes but greatest in those having an oval shape, so the disposition of receptors in the olfactory epithelium may be more important in governing acuity than the size of the organ.

Taste Receptors. The receptors of the gustatory sense are called taste buds. These are somewhat football-shaped or pear-shaped structures made up of several elongate cells, including basal and supporting cells as well as the receptor cells (Fig. 4–7). Taste buds are set in the skin, with the outer end at the surface of the epithelium. On the outer ends of the receptor cells are tiny taste hairs or microvilli that protrude from the surface or into a slight depression. Innervation of the taste buds is usually by the vagus, the glossopharyngeal, and the facial nerves.

Taste buds are generally concentrated in the mouth and pharyngeal region, appearing even on the gill arches of bony fishes. However,

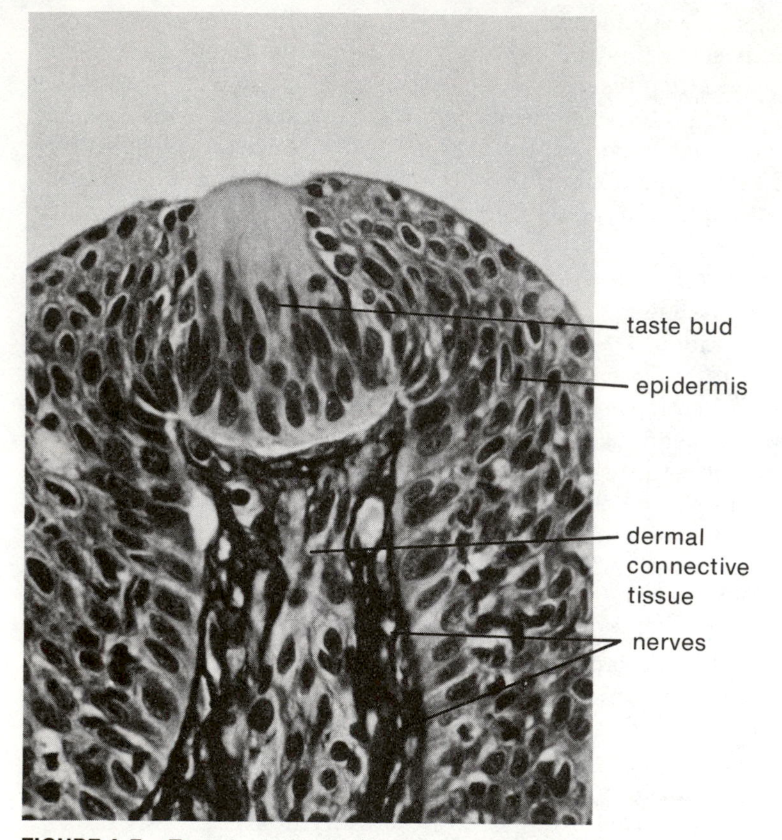

FIGURE 4–7. Taste bud of juvenile rainbow trout (*Salmo gairdneri*). (Photomicrograph courtesy of Prof. Joseph H. Wales.)

many species have an abundance of taste buds on specialized structures such as barbels and elongate fin rays. Some catfishes have them over much of the body. Taste buds commonly occur on the lips and head of bottom-feeding fishes.

Somewhat related to the taste sense is the common chemical sense of the skin of fishes. Free nerve endings apparently respond to various chemical irritants and to other biologically significant substances, but much research needs to be done to understand this phenomenon more fully.

Some additional cells with a presumed chemosensory function have been described from the epidermis of fishes. These are called spindle cells and are structurally similar to the receptor cells in taste buds.

Endocrine Glands

The endocrine glands of fishes comprise a system comparable to that of higher vertebrates, but various endocrine tissues do not form

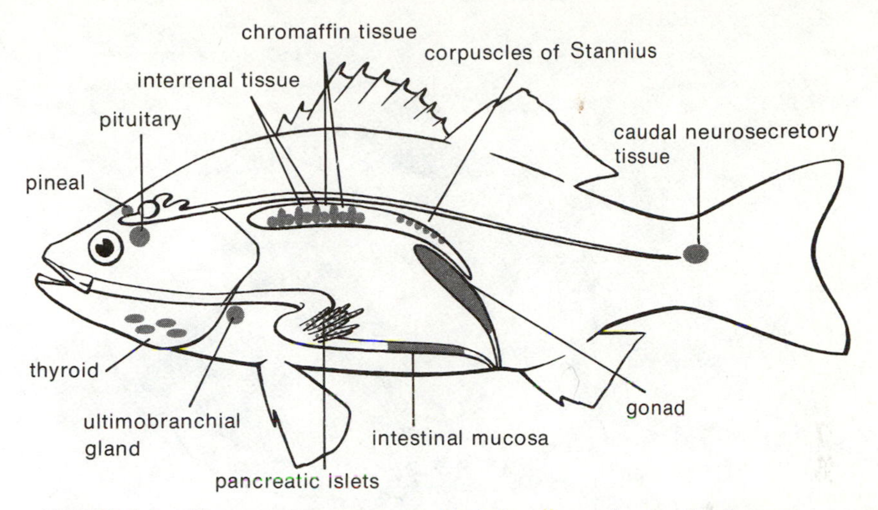

FIGURE 4–8. Diagram showing approximate locations of endocrine tissues.

discrete glands in fishes, and the sites of the tissues may be different from the sites of glands in higher forms. However, for the most part, the same or very similar hormones are produced. In addition, fishes possess some endocrine tissues, such as the caudal neurosecretory system and the Stannius corpuscles, that do not have counterparts in higher vertebrates. Following is a description of the endocrine glands of fishes with the general location of the tissue and a brief mention of the function of the secretions. In later chapters the role of endocrine activity in various aspects of the life of fishes will be pointed out where appropriate. General locations of the endocrine glands are shown in Figure 4–8.

Pituitary Gland. This gland, also referred to as the hypophysis or hypothalamohypophyseal system, is located under the diencephalon (Fig. 4–8). Its embryonic origin involves a neural downgrowth from the diencephalon and epithelial material (Rathke's pouch) growing up from the dorsal part of the embryonic mouth cavity. The neural component of the gland, called the neurohypophysis (posterior lobe), is closely associated with the hypothalamus, which apparently is the site of the neurosecretions associated with the pituitary. The remainder of the gland is the adenohypophysis, consisting of histologically distinct sections that are functionally equivalent to parts of the mammalian pituitary. The anterior section is called the pars distalis, and is composed of a rostral part (pro-adenohypophysis) and a distal part (meso-adenohypophysis). The posterior part of the gland is the pars intermedia (meta-adenohypophysis; Fig. 4–9).

The pituitary glands of lampreys and hagfishes are considered to be more primitive than those of elasmobranchs and bony fishes, and may not secrete as full a range of hormones. The pituitaries of sharks and rays are peculiar in having a small ventral lobe attached to the pars distalis by a short stalk. A similar structure in the chimaeras is detached from the remainder of the gland.

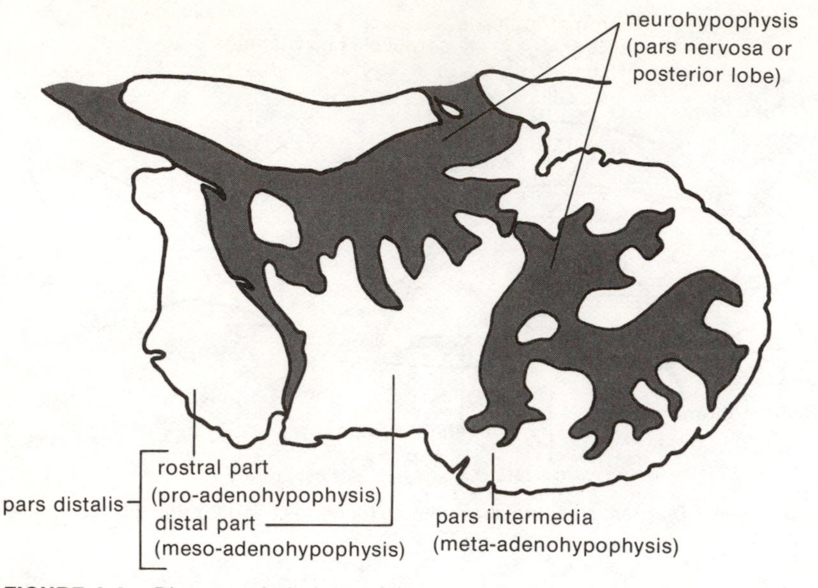

FIGURE 4–9. Diagram of pituitary of juvenile chinook salmon (*Oncorhynchus tshawytscha*).

There are considerable differences among bony fishes in the shape and internal structure of the gland. The pituitary of *Latimeria* is quite elongate and has a rostral division of the pars distalis connected to the remainder of the pars distalis by a thin tube, and is directly supplied with blood from the carotids. It somewhat resembles the ventral lobe of elasmobranchs. Usually the glands are somewhat elongate, but in some salmonids and cyprinids they may be roughly globular. In the rice eel, *Monopterus albus*, the adenohypophysis completely surrounds the neurohypophysis except at the connection with the infundibulum. In the Polypteridae the vestige of Rathke's pouch persists as an orohypophysial duct. This is encountered in teleosts only in an Indian shad, *Hilsa ilisha*.

In most fishes the pituitary lies close to the hypothalamus, but in some it is on a short stalk. A notable case is the goosefish, *Lophius*, in which the gland is on a long stalk. In a goby, *Lepidogobius*, the pituitary is surrounded by the hypothalamus.

The neurosecretions of the hypothalamus, oxytocin and vasotocin, appear to be stored and released by the neurohypophysis. In general, these secretions are involved in osmoregulation and reproduction. The adenohypophysis contains a variety of hormone-producing cells. Those secretions associated with the pars distalis are fish prolactin, important in sodium regulation in fresh water, growth hormone, corticotropin (ACTH), gonadotropins, and thyrotropin. The pars intermedia secretes at least one melanotropic hormone, intermedin (MSH), which is involved in control of the melanophores and possibly also in melanogenesis. Although the pituitary is usually called the "master gland"

and has been shown to govern many activities of the other endocrine glands, a strange cobalt blue mutant of the rainbow trout, *Salmo gairdneri*, lives to an age of at least five years with almost no pituitary. This appears in broods in Japanese hatcheries, but cannot reproduce because of abnormal oogenesis and spermatogenesis. It has some metabolic disorders, but the thyroid gland and the interrenal tissue appear normal.

Thyroid Gland. The thyroid of most bony fishes consists of follicles associated with the surface of the heart, ventral aorta, and the lower parts of the branchial arteries. In elasmobranchs and a few bony fishes the thyroid is a discrete gland surrounded by connective tissue. In sharks the gland is anterior to the ventral aorta beneath the basihyal cartilage. The compact type of thyroid has been described in a variety of bony fishes, such as the skipping goby, *Periophthalmus*, swordfish, *Xiphias*, parrotfishes, Scaridae, and a few others. Exceptionally, as in the goldfish, thyroid follicles may concentrate in the head kidney as well as in the usual subpharyngeal location.

Thyroid hormones appear to have a variety of effects upon the physiology of fish. Many of these actions are yet not well understood, but there is evidence that they affect the rate of oxygen consumption, promote the deposition of guanine in the skin, and alter carbohydrate and nitrogen metabolism. In addition, effects upon motor activity, skeletal growth, and function of the central nervous system have been noted. Involvement in osmoregulation is suspected, and administration of thyroxine heightens the "preference" of young salmon for salt water.

Interrenal Tissue. In elasmobranchs this tissue, which is homologous with the adrenal cortex of higher vertebrates, is organized into glands situated between the posterior region of the kidneys. The interrenal tissue of bony fishes is usually associated with the head kidney, appearing as cells or groups of cells scattered there, especially along the cardinal veins. Cells similar to adrenocortical cells are found in the walls of the cardinal veins of lampreys.

The secretions of the interrenal tissue are steroids, most notably cortisol, corticosterone, and cortisone, the latter being much more prominent in bony fishes than in elasmobranchs and cyclostomes. The adrenocorticosteroids appear to exert some control over osmoregulatory processes, acting upon the kidney, gills, and gastrointestinal tract. Metabolism of proteins and carbohydrates is affected by the corticosteroids, especially in such fishes as the Pacific salmons, *Oncorhynchus*, which make lengthy migrations while fasting and must utilize muscle protein in order to gain sufficient energy to complete its travels and the ensuing spawning process.

Chromaffin Tissue. This tissue is homologous with the adrenal medullary tissue of higher vertebrates, but there appears to be only one family of fishes (Cottidae) in which the interrenal and chromaffin cells are organized into a compact gland. Usually chromaffin cells of bony fishes are distributed along the postcardinal veins, and may intermingle to some extent with interrenal cells. Chromaffin tissue in elasmobranchs is associated with the sympathetic ganglia and the dorsal aorta anterior to the interrenal tissue. A separation of the two tissues is seen also in the cyclostomes, in which the chromaffin cells appear as strands along the dorsal aorta.

This tissue secretes adrenalin, which has strong effects upon heart rate, blood pressure, dilation of pupils, and concentration of melanin in the melanophores. The action of the hormone is similar to that of the sympathetic nervous system, with which it is closely associated in both origin and location.

Ultimobranchial Gland. In bony fishes this gland is located below the esophagus near the sinus venosus, often on, or closely associated with, the pericardium. In elasmobranchs the gland is on the left side of the midline beneath the pharynx. It secretes the hormone calcitonin, which is involved in the inhibition of bone resorption in mammals, and is thought to be involved in calcium metabolism in fishes. The gland does not occur in cyclostomes.

Islets of Langerhans (Pancreatic Islets). In bony fishes the islet tissue is usually distributed around the pyloric caeca, small intestine, spleen, and gallbladder. A few species have a compact mass of this tissue on or near the latter organ. Elasmobranchs have a discrete pancreas that includes the islets. The islet tissue is found in the walls of the intestine in lampreys. The islets of Langerhans produce insulin, which is important in carbohydrate metabolism and in the conversion of glucose into glycogen, and is involved in oxidation of glucose and fat production.

Intestinal Mucosa. Hormones that regulate certain secretions of the pancreas are produced in the lining of the small intestine of fish. Pancreozymin helps to promote the production of zymogens, and secretin stimulates secretion of water and inorganic salts.

Gonads. The sex glands of both sexes are involved in the secretion of steroids that are important in the manifestation of courtship, nest-building, and other aspects of reproductive behavior, as well as in the development and maintenance of secondary sex characteristics and the production of gametes. The ovary produces estrogens but these have not as yet been well studied in fishes. Investigations have shown positive relationships of ovarian secretions to receptivity to males and to development of secondary sex characteristics. Estrogen-like secretions appear to act as attractants to the males.

The testis produces androgens, especially testosterone. Other hormones identified from the testes include dehydroepiandrosterone and androstenedione. Many studies have indicated that androgens are of great importance in the sexual behavior and spawning activity of male fishes.

Caudal Neurosecretory System. Near the termination of the spinal cord in sharks, rays, teleosts, and some other bony fishes such as *Lepisosteus* and *Polypterus* there are found enlarged neurons of a secretory nature known as Dahlgren cells. The axons of these neurosecretory cells terminate in a capillary bed that appears to function in the storage and release of secretions. In teleosts the capillary network is contained in a neurohemal structure called the urophysis. This complex, which includes the terminal filament of the spinal column, obviously is the site of production and release of some endocrine substance, but the exact biological activity of the hormone(s) is not well known. Experimentation has produced contradictory evidence on the role of the caudal neurosecretory system, but it appears to affect water balance and sodium regulation. In a year-round study of this gland in an Indian catfish, investigators noted that stored material disappeared during the breeding season, leading them to surmise that the caudal secretory system is involved in the reproductive cycle.

Corpuscles of Stannius. The corpuscles of Stannius are found in the opisthonephric kidney of holosteans and teleosts. They vary in position among species, being found dorsally, dorsolaterally, or ventrolaterally; they are seldom arranged symmetrically. The featherbacks (*Notopterus*) have only a single corpuscle near the head kidney. The corpuscles may be highly vascularized and lobulated.

The secretion(s) of the corpuscles appears to be involved in osmoregulation, since their removal brings about changes in plasma composition. A decline in sodium and a rise in potassium and calcium concentrations are typical reactions. Exposure of certain freshwater fishes to saline water or the injection of potassium salts effects histological changes in the corpuscles. There appears to be a relationship between the adrenal cortex and the activity of the corpuscles, as injections of corticosteroids bring about nuclear hypertrophy and other evidence of stimulation in the cells of the corpuscles.

The Pineal Organ. This body, attached to the roof of the diencephalon, is the remnant of the "third eye," and continues to have a light sensory function in lampreys and some elasmobranchs and teleosts. In most bony fishes a thickening of the pineal epithelium and great vascularization of the structure indicates a glandular function, but there is great variation among fishes, even those within certain families, in the state of transition from the light-sensitive stage to the glandular situation. The secretion of the pineal is melatonin, which aggregates melanin in amphibians and has been shown to have a similar effect in some fishes. Removal of the pineal from fishes can bring about changes in growth and can result in the stimulation of the pituitary and thyroid glands.

Kidney. In teleosts the hormone renin appears to be secreted by certain granular cells associated with the renal blood vessels.

Thymus. The thymus gland has its origin in the branchial pouches of fishes and is generally found above the branchial chamber or pockets in lampreys, sharks, and bony fishes. Little is known of its function in fishes, but it is probably not an endocrine gland.

References

American Society of Zoologists. 1973. Symposium: The current status of fish endocrine systems. Am. Zool., 13:710–936.
Ball, J. N., and Baker, B. I. 1969. The pituitary gland: anatomy and histophysiology. *In*: Hoar, W. S., and Randall, D. J. (eds.), Fish Physiology, Vol. 2. New York, Academic Press, pp. 1–110.
Belsare, D. K. 1973. Comparative anatomy and histology of the corpuscles of Stannius in teleosts. Z. Mikrosk. Anat. Forsch., 87(4):445–456.
———. 1974. Morphology of the pineal organ in some carps. Zool. Beitr., 20(1): 47–54.
Berlind, A. 1973. Caudal neurosecretory system: A physiologist's view. Am. Zool., 13:759–770.
Bern, H. A. 1967. Hormones and endocrine glands of fishes. Science, 158:455–462.
Bernstein, J. J. 1970. Anatomy and physiology of the central nervous system. *In*: Hoar, W. S., and Randall, D. J. (eds.), Fish Physiology. Vol. 4. New York, Academic Press, pp. 1–90.
Butler, D. G. 1973. Structure and function of the adrenal gland of fishes. Am. Zool., 13:839–879.
Colombo, L., Bern, H., and Pieprzyk, J. 1971. Steroid transformations by the corpuscles of Stannius and the body kidney of *Salmo gairdneri* (Teleostei). Gen. Comp. Endocrinol., 16(1):74–84.
———, and Johnson, D. W. 1972. Corticosteroidogenesis in vitro by the head kidney of *Tilapia mossambica* (Cichlidae, Teleostei). Endocrinology, 91(2):450–462.
Donaldson, E. M., and McBride, J. R. 1974. Effect of ACTH and salmon gonadotropin on interrenal and thyroid activity of the gonadectomized adult sockeye salmon, *Oncorhynchus nerka*. J. Fish. Res. Bd. Can., 31:1211–1214.
Henderson, I. W., and Jones, I. C. 1973. Hormones and osmoregulation in fishes. Ann. Inst. Michel Pacha (1972), 5(2):69–235.
Hoar, W. S., and Randall, D. J. (eds.). 1969. The endocrine system. Vol. 2, Fish Physiology. New York, Academic Press, 446 pp.
Igarashi, S., and Kamiya, T. 1972. Atlas of the Vertebrate Brain. Baltimore, London, Tokyo, University Park Press.
Jaiswal, A. G., and Belsare, D. K. 1973. Comparative anatomy and histology of the caudal neurosecretory system in teleosts. Z. Mikrosk. Anat. Forsch., 87(5/6):589–609.
Krishnamurthy, V. G., and Bern, H. A. 1971. Innervation of the corpuscles of Stannius. Gen. Comp. Endocrinol., 16(1):162–165.
Lagios, M. D. 1975. The pituitary gland of the coelacanth *Latimeria chalumnae* Smith. Gen. Comp. Endocrinol., 25(2):126–146.
Munford, J., and Greenwald, L. 1974. The hypoglycemic effects of external insulin on fish and frogs. J. Exp. Zool., 190(3):341–345.
Nieuwenhuys, R. 1959. The structure of the telencephalon of the teleost *Gasterosteus aculeatus*. I. Proc. K. Ned. Akad. Wet., Series C, 62:341–362.
———. 1962. Trends in the evolution of the actinopterygian forebrain. J. Morphol., 111:69–88.
O, Wai-sum, and Chan, T. H. 1974. A cytological study on the structure of the pituitary gland of *Monopterus albus* (Zuiew). Gen. Comp. Endocrinol., 24(2):208–222.
Parks, A. M. 1969. The neurohypophysis. *In*: Hoar, W. S., and Randall, D. J. (eds.), Fish Physiology, Vol. 2. New York, Academic Press, pp. 111–205.
Pickford, G., and Atz, J. W. 1957. The Physiology of the Pituitary Gland of Fishes. New York, New York Zoological Society.

Ray, D. L. 1950. The peripheral nervous system of *Lampanyctus leucopsaurus*. J. Morphol., 87:61–178.

Sharma, S., and Sharma, A. 1975. A note on the caudal neurosecretory system and seasonal changes observed in the urophysis of *Rita rita* (Bleeker). Can. J. Zool., 53(3):357–360.

Subdehar, N., and Prasado Ras, P. D. 1974. Effects of some corticosteroids and metopirone on the corpuscles of Stannius and interrenal gland of the catfish, *Heteropneustes fossilis* (Bloch). Gen. Comp. Endocrinol., 23(4):403–414.

Tester, A. L., Kendall, J. I., and Milisen, W. B. 1972. Morphology of the ear of the shark genus *Carcharinus* with particular reference to the macula neglecta. Pacific Sci., 26:264–274.

Tuge, H., Uchihashi, K., and Shimamura, H. 1968. An atlas of the brains of fishes of Japan. Tokyo, Tsukiji Shokan.

Yamazaki, F. 1974. On the so-called "cobalt" variant of rainbow trout. Bull. Jap. Soc. Sci. Fish., 40(1):17–25.

Section Two

RELATIONSHIPS AND DIVERSIFICATION OF FISHES

The great antiquity of fishes, their adaptive radiation and consequent diversity of form, and the incompleteness of the fossil record all combine to make the delineation of the many phylogenetic lines difficult. Major groups, of course, can be recognized and their boundaries described, but many of these boundaries may have transitional zones where relationships with other major groups may be more or less evident. One of the great difficulties is in learning whether the apparent relationships among groups are real or are the results of convergence. Selection of criteria for assessment of phylogeny may be relatively easy on a broad scale, but ichthyologists do not always agree on the usefulness of these criteria at various taxonomic levels. A class erected and bounded by one researcher may be only a subclass in the classification scheme of another, even though both appreciate the relationships of the fishes involved.

Furthermore, there seem to be great differences of opinion among ichthyologists as to the morphological features best suited for use in phylogenetic studies in fish. Some may emphasize cranial osteology, or osteology of the caudal region, or scales, or fin structure, or various soft body parts. All these factors are pertinent and important, and all taxonomists are seeking the truths that eventually will provide knowledge of fish phylogeny complete enough to bring general agreement on fish classification. Until this is accomplished, especially during this period of active taxonomic research, the classes and orders of fishes are found to be arranged differently in most available publications on the subject.

If an orderly presentation must be made of the fishes, as must be done here in order to show their relationships and diversity, some framework must be adopted that is both intelligible to students and scientifically acceptable. The groupings used in this section will parallel those of most systems, but may differ as to taxonomic level, or may be somewhat simplified. The list of orders and families of teleosts is not presented as an attempt to demonstrate a new classification system, but simply to arrange them in a reasonable evolutionary sequence that will not diverge drastically from the major classifications now in use. Coverage will not be equal or parallel. Groups of great phylogenetic or general interest will receive more complete treatment than others.

5
THE JAWLESS FISHES

The superclass Agnatha is comprised of primitive fish-like craniate vertebrates without jaws. These are the oldest of the fish-like vertebrates in terms of their fossil history; the agnathans developed along several lines that may have had a common origin sometime before the Ordovician, but this is only speculation because the fossil record is lacking. Record of the processes that led to the development of the relatively complex armored forms that appeared in the Cambrian over 500 million years ago may never be found, but the known forms, both living and fossil, inherited certain common characteristics.

First and most important, jaws are not present, so that food must have been taken in by a sucking action, unless some rasping mechanism was present. Vertebrae do not replace the notochord, there are only two semicircular canals in each ear, and pelvic fins are absent, although some forms have pectoral fins of a primitive nature.

Perhaps the second most important characteristic, indicating that development was independent from the jaw-bearing vertebrates, is the entire arrangement of the gills and branchial skeleton. The gill arches are considerably different from those of jawed fishes and are fused to the neurocranium. All the gill tissue and the branchial arteries and nerves are internal to the branchial skeleton.

Agnaths seem to have been most abundant in the Silurian and Lower Devonian, but their numbers decreased through the Middle and Upper Devonian. Only two groups, the lampreys and hagfishes, have survived to modern times. An excellent coverage of the group is found in the 1971 revision by R. S. Miles of J. A. Moy-Thomas's *Palaeozoic Fishes.** The classification of Agnatha used here is from that book. The Russian ichthyologist, Leo S. Berg, has considered that the agnaths comprised six classes. Here we will accept but two. (Some authors present the entire group as the class Ostracodermi.)

Class Cephalaspidomorphi

This group is called Monorhina by some authors because of the single opening leading to the nasal sac and the hypophyseal area

*2nd ed., Philadelphia, W. B. Saunders.

beneath the brain. Another important characteristic is the great number of gill openings, as many as 15. This group flourished in the Silurian and the Lower Devonian, but only the lampreys and hagfishes are its descendants.

The class is subdivided as follows:*

Subclass Hyperoartii — agnaths with no connection between the nasohypophyseal cavity and the pharynx

†Infraclass Osteostraci

†Infraclass Anaspida

Infraclass Petromyzonida — lampreys

Subclass Hyperotreti — agnaths with the nasohypophyseal cavity communicating with the pharynx

Infraclass Myxinoidea — hagfishes

*Extinct groups in this chapter and throughout the text are designated by a dagger, †.

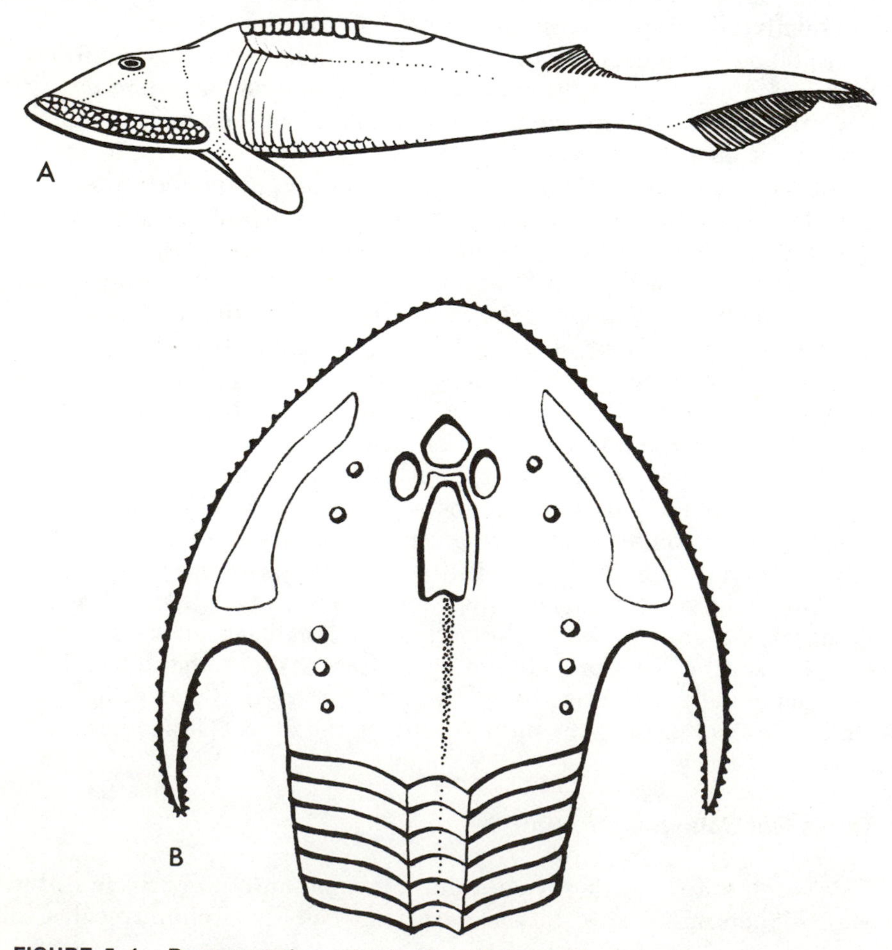

FIGURE 5–1. Representatives of infraclass Osteostraci. *A,* Lateral view; *B,* dorsal view of head.

Osteostraci

†Infraclass Osteostraci contains five extinct orders of small fishes (600 mm in length) characterized by a shield of bone that covers the head and some of the anterior portion of the body (Fig. 5–1). The head is flat and the mouth ventral. Eyes, nasohypophyseal opening, and pineal opening are dorsal. Bony overlapping scales are present, as are pectoral fins; the caudal fin is heterocercal. The head shield of many of the fossil specimens has preserved traces of gill chambers, blood vessels, and nerves that furnish an unusual opportunity for studying these structures. Some palaeontologists have described resemblances between the nerve and vascular patterns of the osteostracans and those of the lampreys.

Anaspida

†Infraclass Anaspida includes four extinct orders of very small fishes, usually less than 150 mm long (Fig. 5–2). If armor is present it consists of overlapping plates. Some specimens have a ridge of scutes down the back, some show lateral fin folds, and all have hypocercal tails. The body is rounded, the mouth terminal, the eyes lateral, and the nasohypophyseal and pineal openings dorsal.

Petromyzonida

Infraclass Petromyzonida has only one fossil representative, *Mayomyzon*, from the Carboniferous. The remainder of the known species are modern. The complete absence of hard parts apparently has reduced the chances of fossilization of the ancient representatives.

Lampreys, placed in the order Petromyzontiformes by most ichthyologists, are characterized by the lack of paired fins and scales (Fig. 1–1). They are eel-like, with lateral eyes and a ventral mouth consisting of a circular disc set with horny teeth. The nasohypophyseal opening is between the eyes. The skeleton is cartilaginous and not well developed except for the skull and branchial region. No vertebral centra are developed and the neural arches are rudimentary. Dorsal and caudal fins are present. Myotomes are not divided horizontally as in more advanced fishes.

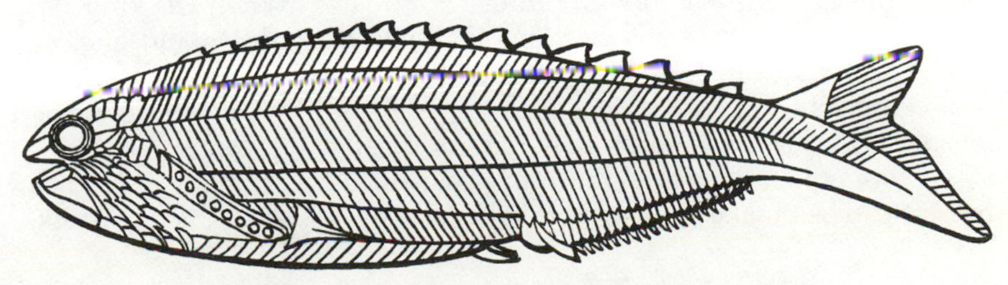

FIGURE 5–2. Representative of infraclass Anaspida.

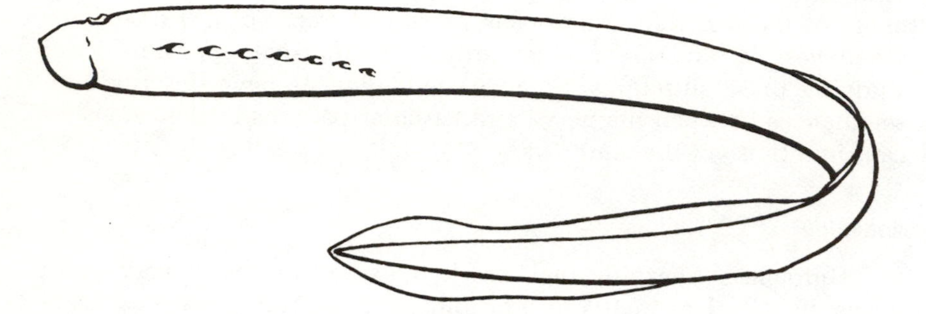

FIGURE 5–3. Larva of lamprey, infraclass Petromyzonida.

Lampreys possess some peculiar internal features. The internal labyrinth has a ciliated epithelium, the left duct of Cuvier is absent, and the dorsal and ventral roots of the spinal nerves are not connected with each other. In adults the gills open into a respiratory tube that begins at the mouth, extends under the esophagus, and ends at the seventh pair of gills. In the eyeless, toothless larvae (called ammocoetes, Fig. 5–3) the gills open to the long pharynx, but this is cut off posteriorly during metamorphosis and a new esophagus forms. The gallbladder and bile ducts disappear in adults.

There might be some relationship to bony fishes indicated by the fact that the embryo forms a medullary keel instead of a groove as in sharks and rays. In addition, a bulbus arteriosus is present instead of a conus arteriosus.

All lampreys share a long larval life. Very small eggs are deposited in nests made in gravel bottoms of streams. When the tiny larvae hatch they drift to soft bottoms in pools and eddies and begin a life of straining out organic matter at the mud-water interface. This period may be up to five years or so in length, after which metamorphosis takes place and a new type of existence is begun.

There are two types of lampreys, parasitic and nonparasitic. The parasitic types make their living by attacking fishes with their sucking mouths, rasping holes in the skin with their piston-like tongues, and pumping out blood and body fluids, a process aided by secretion of an anticoagulant from paired buccal or "salivary" glands. Some lampreys of this type are anadromous, spending their adult growth period in salt water before returning to streams to spawn and die. These may reach a meter in length. Others remain in fresh water and may grow a half meter or more, as does the landlocked sea lamprey (*Petromyzon marinus*) in the Great Lakes, or some strictly freshwater species may reach adult size at 15 cm or less.

Nonparasitic lampreys are usually called "brook" lampreys. These confine feeding to the larval stage, and after metamorphosis they spend a few months in hiding while the gonads mature. They then spawn and die.

Lampreys are creatures of temperate seas and streams, and occur in both the North and South Hemispheres. All are generally considered to be in the family Petromyzontidae, with the forms in South America and Australia in a separate subfamily from those in other continents. Their economic value is slight even though they are used as food in some areas, and they have been used as a source of a light oil. Their economic impact can be great locally if they attack and scar or kill fishes of greater value to man than themselves. The chronicle of the invasion of the upper Great Lakes by the sea lamprey is a sad one, for the commercially valuable lake trout (*Salvelinus namaycush*) and whitefish (*Coregonus clupeaformis*) virtually disappeared as commercial species when the mortality due to the lamprey was superimposed on fishing mortality. In many other instances the effect has not been as severe, but smaller lampreys such as the chestnut lamprey (*Ichthyomyzon castaneus*) or landlocked Pacific lampreys (*Lampetra tridentata*) are known to attack and injure, if not kill, various freshwater game fishes.

Larval lampreys are eaten by a variety of fishes, and are sometimes used as bait. Adult lampreys appear to be excellent bait for sturgeon and are found in the stomachs of other fishes, including sharks. Seasonally they may form a great portion of the food of the California sea lions that live near the mouths of rivers that sustain runs of the Pacific lamprey.

Myxinoidea

In the subclass Hyperotreti, the single infraclass Myxinoidea has no fossil representatives, and is known from a few recent genera (Fig. 1–1). In addition to the main characteristics of the subclass, it differs in several ways from the lampreys. It is marine and has no larval stage; instead, it develops directly, beginning with a large meroblastic egg. These eggs (Fig. 5–4) may be over 25 mm in length and equipped with anchor-like hooks that can hold them together in a string or clump.

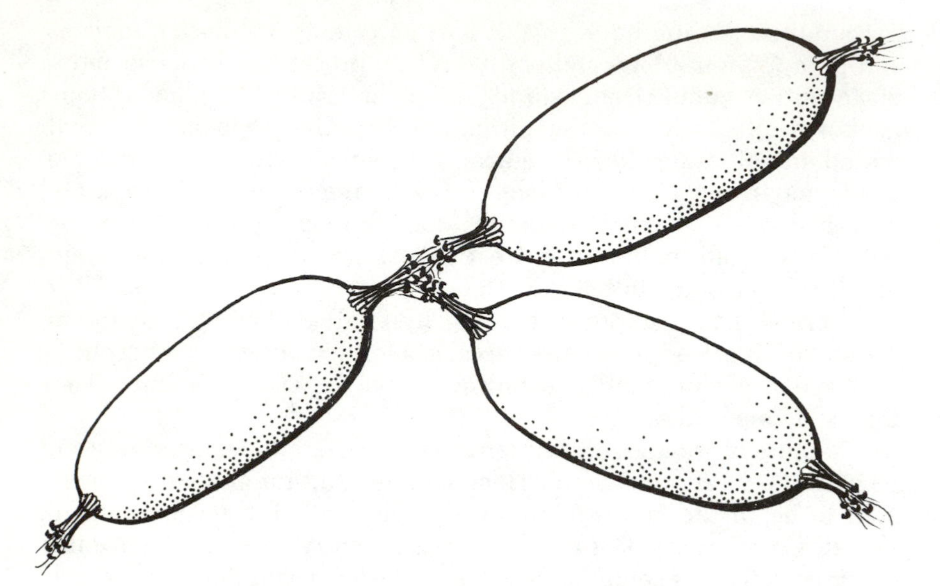

FIGURE 5–4. Eggs of hagfish, infraclass Myxinoidea.

The eyes are degenerate and completely covered by skin. Large tentacles surround the terminal nasal opening and the mouth, which is not developed as a sucking disc. The feeding action consists of rapid eversion and retraction of teeth situated on each side of the mouth, so that the effect is that of jaws operating laterally as opposed to the up-and-down motion of true jaws.

As few as five and as many as fifteen pairs of gills may open from the pharynx. In some genera the efferent branchial ducts may be collected into a tube with a single external opening on each side. On the left side a peculiar duct, apparently homologous with the gills, originates at the posterior portion of the pharynx and communicates with the exterior.

The skeleton is not well developed, with no rudiments of vertebrae, and only a membranous roof in the skull. The dorsal roots of the spinal nerves are united with the ventral roots. The two semicircular canals are connected in such a manner that they appear as one.

Hagfishes have been given considerable notice by medical researchers because of the array of contractile structures in their vascular system. In addition to the usual heart, considered to be a primitive one, there are a caudal heart, a cardinal vein heart, and a portal vein heart that pumps blood from an intestinal vein and the right jugular vein to the liver.

Hagfishes live on soft bottoms of mud, silt, or clay at depths of 25 to 600 m although a Japanese species may be found at 5 m and others have been taken at 1000 m. They are known to burrow in the bottom and perhaps to feed on worms and other soft-bodied animals while so doing. They act both as scavengers and predators, and may attack fishes in nets close to the bottom.

Several genera are known, all from temperate waters. Two of the best known species are *Myxine glutinosa* of the Atlantic and *Eptatretus stouti* of the eastern Pacific. Economic importance is slight. There is some use as food in Japan, and in many areas there have been instances of fish tangled in gill nets or hooked on long lines having been mutilated by hagfish.

†Class Pteraspidomorphi

This class is comprised of extinct agnathans that appear to have had paired nasal sacs with separate openings and no nasohypophyseal canal. Because of this they have been called Diplorhina by some authors. Usually there is only one pair of branchial openings, although there are several gill pouches. The class is divided as follows:

†Subclass Heterostraci — seven orders
†Subclass Thelodonti — two orders

Heterostraci

The heterostracans (Fig. 5–5) are the earliest known vertebrates. Fossils are known from the Upper Cambrian, and reached their

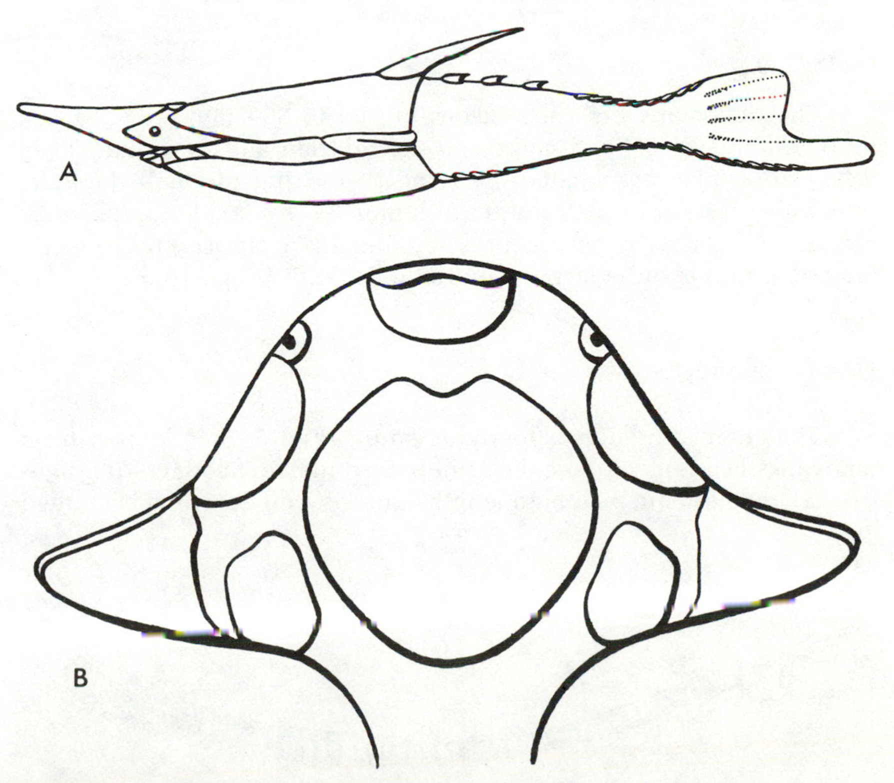

FIGURE 5–5. Representatives of subclass Heterostraci. *A*, Lateral view; *B*, dorsal view of head, showing laterally produced plates.

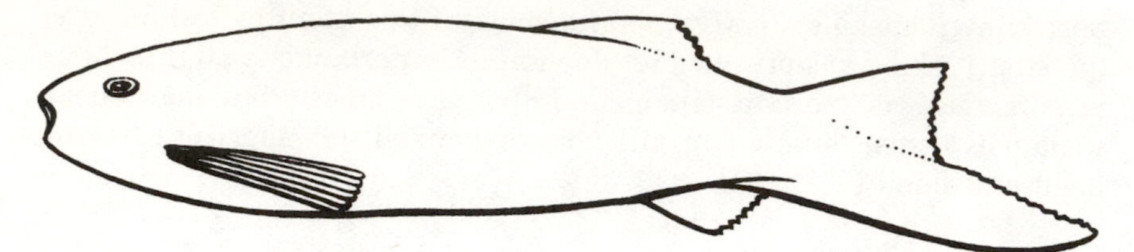

FIGURE 5–6. Representative of sublcass Thelodonti.

greatest development in the Upper Silurian and the Lower Devonian. They are characterized by bony plates forming a shield on the head and forebody. Typically the head is flat, the eyes lateral, and the mouth subterminal to slightly superior. The caudal is hypocercal. Paired fins are not present. Length is usually less than 300 mm, but some reach 1.5 m.

Some genera have part of the shield produced laterally to form what appear to be "gliding" surfaces (Fig. 5–5). One has an anterior extension or pseudorostrum similar in appearance to the rostrum of the saw shark *(Pristiophorus).* Specializations such as tubular mouths, dorsally placed mouths, and stabilizing keels has led palaeontologists to believe that the heterostracans, in their adaptive radiation, took advantage of more varying habitats than did the cephalaspidomorphs, which were primarily bottom-feeders.

Thelodonti

The thelodonts are small fishes of 100 to 200 mm (at most 400 mm) in length that lived mainly in the Silurian and Devonian. They differ from other agnathans in having a covering of small denticle-like scales instead of plates or solid armor. Dorsal, anal, a hypocercal caudal, and lateral fins are present, and there appear to be eight branchial sacs opening separately to the exterior (Fig. 5–6).

†Palaeospondylus

The enigmatic *Palaeospondylus gunni* (Fig. 5–7) is known from many fossil specimens from the Middle Devonian of Scotland. It ranges from 12 mm to about 50 mm in length, and has well developed calcified

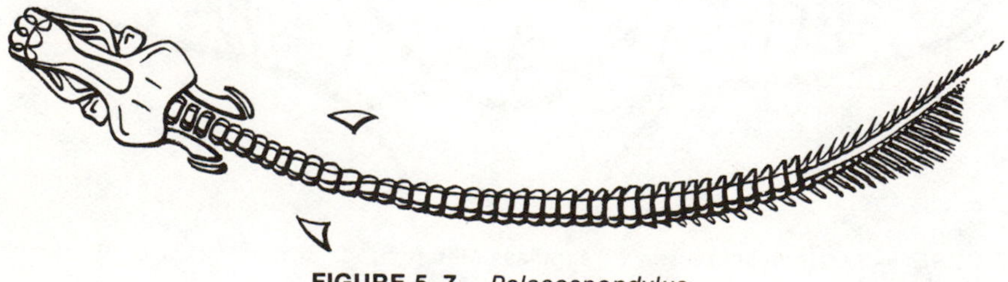

FIGURE 5–7. *Palaeospondylus.*

vertebrae, paired fins, and a rather intricate skull. The caudal is heterocercal.

The small size and the state of preservation have not allowed conclusive work to be done on the genus, and consequently many different interpretations have been presented. Some investigators consider it to be an agnathan related to the hagfish, others think that it is an adult gnathostome, and still others believe it represents the larval form of some gnathostome.

Certainly the possession of vertebrae seems to be an advancement over the agnathans, but the head structure does not appear to lead to a firm interpretation of true jaws, so the animal remains a mystery.

References

All references for this chapter are included in the complete listing for Section Two at the end of Chapter 7, pp. 206–210.

6

EARLY GNATHOSTOMES, ELASMOBRANCHS, AND RELATIVES

The remainder of the vertebrates bear jaws and constitute the great superclass Gnathostomata. In the fossil record, gnathostomes are known from the Upper Silurian, more than 400 million years ago. They flourished in the Devonian and several lineages survived into the Carboniferous and beyond, while the main lines of agnaths dwindled. Eventually the primitive gnathostome lines produced evolutionary sequences resulting not only in the modern fishes but in amphibians, reptiles, and their derivatives. The following discussion will be limited to the fishes.

Early gnathostomes share characteristics that contrast with those of the agnaths. Jaws are present; gill tissue, branchial arteries, and branchial nerves are external to the gill arches; pectoral and pelvic fins are present; three semicircular canals are present; and the branchial skeleton is not fused to the neurocranium. The notochord, persistent in the early forms, is partially or completely replaced by vertebrae in later gnathostomes.

Gnathostomatous fishes are considered by some recent authors to comprise two main groups, the shark-like Elasmobranchiomorphi or chondrichthyan fishes, and the "bony" Teleostomi or osteichthyan fishes. Among living fishes, there is no difficulty in assigning the chimaeras and the sharks and rays to the elasmobranchiomorph group and the lungfishes, coelacanths, and ray-finned (actinopterygian) fishes to the teleostome group. However, there can be some reservations in aligning the two major extinct lines of gnathostomes, the acanthodians and placoderms, with either of the groups. Some recent works suggest that the acanthodians may be aligned with the ancestors of teleostomes, but others have considered them as being related to chondrichthyans. Placoderms show strong trends leading toward the Elasmobranchiomorphi and, although they differ in the possession of bone and other important features, should probably be aligned with them. Acanthodians and placoderms will be presented as classes in the following treatment without placement in either of the lineages. Extinct groups are marked with the dagger,†.

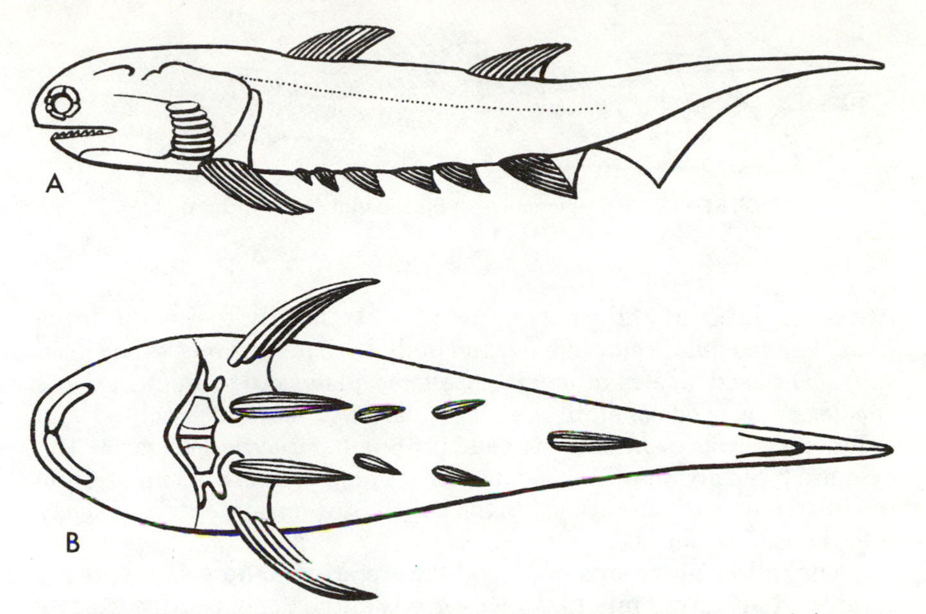

FIGURE 6–1. Representatives of Acanthodii. *A,* lateral view, *B,* ventral view.

†Class? Acanthodii

These are elongate fishes with heterocercal tails, large eyes set far forward in the head, and prominent spines at the leading edges of all fins except the caudal. In some there are spines in a row between the pectorals and pelvics. Gill clefts are covered by gill covers borne on the hyoid and branchial arches but, in advanced forms, the hyoid gill cover is enlarged and appears to cover all the slits. The endocranium is ossified and dermal bone is present on the head. Some show ossifications in the vertebral column. Small scales, with bony bases covered by dentine, are present. They are known from the Upper Silurian to the Lower Permian.

Acanthodians (Fig. 6–1) were apparently fairly active swimmers, not adapted to life on the bottom as are the contemporaneous agnaths and placoderms. Some had the teeth and gill rakers of predators; others lacked teeth and possessed long gill rakers that must have been effective in collecting small invertebrates.

Acanthodians are sometimes divided into as many as seven orders. In recent literature Moy-Thomas and Miles have recognized three orders. These are †Climatiida (= Climatiiformes), †Ischnacanthida (= Ischnacanthiformes), and †Acanthodida (= Acanthodiformes).

†Class? Placodermi

Placoderms are rather diverse in body form, some being shark-like and others flattened like rays, but with all typified by armor of bony

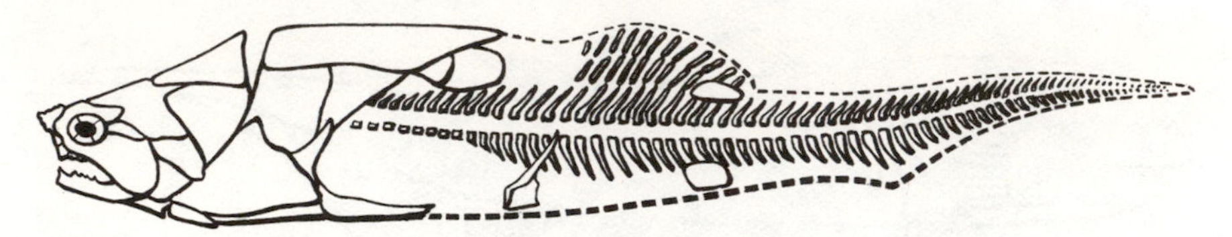

FIGURE 6–2. Representative of Placodermi, order Arthrodira.

plates on the head and forebody (Fig. 6–2). Usually there is a "neck joint" between the armor of head and body so that, apparently, the head could be raised. Scales or small tessellated plates are present in many species. The endoskeleton is at least partially ossified. Pectoral and pelvic fins are present and the caudal fin is heterocercal in most. The eyes are typically rather far forward. Most placoderms are known from the Devonian, but some appear in the Upper Silurian and some persisted into the Carboniferous.

The following groups of placoderms referred to here as orders are given various taxonomic ranks by various authors, some appearing as subclasses or classes.

†Order Arthrodira, also referred to as Coccostei, contains most of the known placoderms. These are characterized by a heavily armored head and forebody, with some having impressive tusk-like plates forming biting surfaces. The gills open between the head and body armor, the slit usually covered by a bony plate. A few genera contain species that reached more than 6 meters in length, but most are considerably smaller.

†Order Ptyctodontida contains small placoderms usually less than 200 mm long. Head and body armor is not as extensive or as heavy as in the arthrodires, there being armor only on the very anterior part of the body and none on the snout. A plate on the cheek apparently covers the gill opening. In body form and in several other characters including the tooth plates they resemble holocephalans. Pectoral fins are large, as are the pelvics. Associated with the latter are claspers tipped with bony plates. As in holocephalans prepelvic claspers are present.

†Order Phyllolepidida is represented by one depressiform genus with a reduced number of armor plates. The nuchal plate on the head and the median dorsal body plate are enlarged. The neck joint does not appear to have been movable. The snout region seems to be unarmored.

†Order Petalichthyida is characterized by numerous head plates, scales on the body, on the pectoral fins and in the snout region, and large lateral spines in front of the pectorals. The eyes are dorsal but anterior. The caudal fin is thought to be diphycercal.

†Order Rhenanida includes two types of small fishes. One, the Palaeacanthaspidoidei, is unusual in that the head armor may be represented within a species by large plates or by smaller plates resembling scales. The other, the Gemuendinoidei, are strongly depressiform and thus resemble rays. The eyes are dorsal, the mouth terminal, and the

gill openings lateral. There are a few large plates on the head, but much of the head and all the body and fins are covered by small scale-like plates arranged in mosaic fashion. In some the mouth was apparently protrusible. Vertebral centra are present.

†Order Antiarchi (Pterichthyes) has such peculiar pectoral appendages that some systematists have considered them to constitute a separate class of vertebrates. The pectorals are large and covered with plates of bone. They articulate with the large body shield, are jointed in the middle, and have an ossified or calcified endoskeleton. The head is relatively small. The body of some species is covered by overlapping scales. Some may have possessed lungs.

Class Holocephali

These are elasmobranchiomorph fishes with the palatoquadrate completely fused to the cranium (holostylic jaw suspension). The hyomandibular is little modified and plays no part in suspension of the jaws. The skeleton is cartilaginous and the notochord persistent. Branchial arches are all placed below the neurocranium; the gill openings are covered by fleshy opercula. Except in a few fossil lines, the teeth are grinding plates with no enamel. In living forms there is no spiracle and no cloaca, and the oviducts open to the exterior separately.

Holocephali are known from the Upper Devonian to the present. Extinct forms known as bradyodonts have been treated variously by modern authors, depending upon evaluation of certain aspects of structure; some are placed closer to sharks than to chimaeras. The following classification is based on that of Colin Patterson.

†Order Iniopterygiformes (Iniopterygia) is comprised of palaeozoic (Carboniferous) holocephalans with pectoral fins attached in the nuchal region (Fig. 6–3). These fins are characterized by prominent spines, often armed with hooks. Unlike modern forms, the teeth are rather shark-like, the dentition consisting of denticles arranged in rows. This is considered a primitive character for the class.

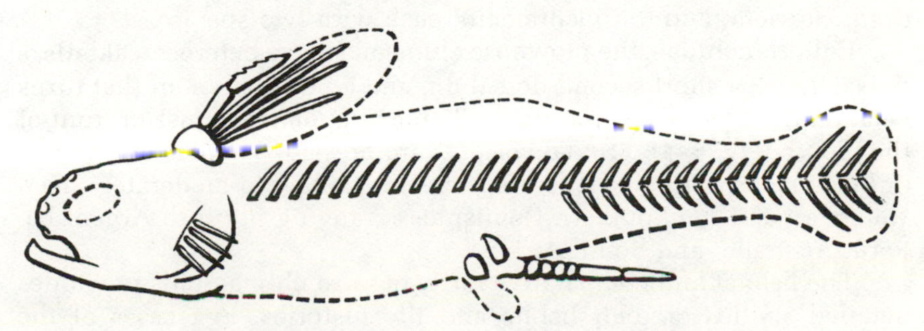

FIGURE 6–3. Representative of Iniopterygiformes.

Order Chimaeriformes (Chimaerida) contains the modern chimaeras and several extinct groups, here presented as suborders.

Suborder Chimaeroidei — modern chimaeras, Lower Jurassic to Recent.

†Suborder Squalorajoidei — Jurassic.

†Suborder Myriacanthoidei — Jurassic.

†Suborder Menaspoidei — Carboniferous to Permian; a bradyodont group possibly more closely related to sharks.

†Suborder Cochliodontoidei — Upper Devonian to Middle Carboniferous; a bradyodont group.

†Suborder Helodontoidei — Middle Carboniferous to Lower Permian; a bradyodont group.

†Order Chondrenchelyiformes — Carboniferous; bradyodont.

Living chimaeroids are among the most bizarre of fishes, as indicated by their various common names. These names are based on anatomical features or general appearance of the species and include spookfish, ghost shark, chimaera, ratfish, rabbitfish, and elephantfish. Modern species are usually placed in three families, all marine. Species are usually between 60 cm and 2 meters long.

Chimaeridae (ratfishes or shortnose chimaeras) contains about 15 species with a short snout, diphycercal tail, and a long second dorsal fin (Fig. 1–1). The first dorsal has a long, sharp venomous spine at the leading edge. Males are equipped with a frontal clasper on the top of the head. It is somewhat finger-like, with a patch of denticles on the ventral aspect of the tip. In addition, males have a set of abdominal claspers situated in pockets just in front of the pelvic fins, plus pelvic claspers of bifid or trifid construction.

Members of this widespread family are found in middepths or in shallow water. A north Pacific species, the ratfish, *Hydrolagus colliei,* is common in Puget Sound and is sometimes found intertidally. *H. colliei* is of little commercial significance, although fishermen at one time extracted a fine oil from the livers. *Hydrolagus* has nine species; *Chimaera* has six.

Rhinochimaeridae shares the sharp dorsal spine and diphycercal tail of the chimaerids, but has a long, depressed and pointed snout from which they take the name "longnose chimaeras." There are well developed frontal claspers and reduced paired claspers. This family contains deepwater forms of wide distribution with three genera, *Neoharriota, Harriota,* and *Rhinochimaera,* each with two species.

Callorhinchidae, the plownose chimaeras,have heterocercal tails, a dorsal spine, a short second dorsal fin, and a peculiar snout that turns back on itself ventrally, forming a flattened appendage just in front of the mouth (Fig. 6–4). Frontal claspers are present. There is one genus, *Callorhinchus,* with about four species in shallow to moderately deep waters around the Southern Hemisphere, ranging north to Argentina, Peru, Australia, and South Africa.

The habitat and comparative rarity of most chimaeroids precludes detailed studies of their habits and life histories. Egg cases of the

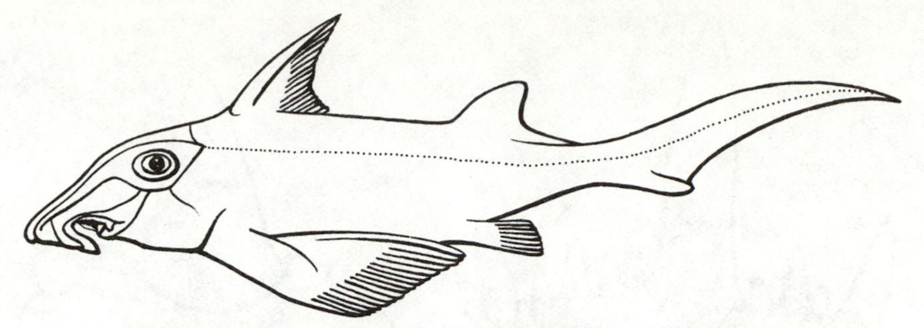

FIGURE 6–4. Representative of Chimaeriformes, *Callorhinchus.*

deepsea species are sometimes obtained, so that some details of the embryology are known. The shallow-water *Hydrolagus colliei* is somewhat better known. This species appears to have wide tastes in food, eating fishes, crustaceans, and molluscs. Young emerge from the spindle-like egg cases in the fall at about 140 mm and appear to grow to around 300 mm in the first year, much of this length being made up of the tail and caudal filament. Sexual maturity seems to be reached in the fourth year of life.

Class Elasmobranchii

Under this name the modern sharks and rays will be treated, along with their fossil relatives. These fishes have cartilaginous endoskeletons that are often hardened by calcifications, which may appear superficially on the endocranium in prismatic form or within the cartilage, as in the vertebrae. Dermal denticles in the form of placoid scales appear in the skin of most forms, but some have no scales. There is no operculum; the five to seven gill openings on each side are separate. Males are equipped with pelvic claspers. Fins are stiffened by horny rays called ceratotrichia. Jaw suspension is hyostylic or amphistylic. There is no gas bladder. A cloaca is present.

Elasmobranchs have existed since the Upper Devonian and constitute a diverse group that has been classified in several different ways. In 1967 Dr. Bobb Schaeffer pointed out that there are three general levels of organization evident in the evolution of the elasmobranchs: a primitive cladodontoid level; an intermediate and related hybodontoid level; and a modern level, with a few living forms occupying transitional positions between the latter two. Cladodonts are characterized by such features as teeth bearing an enlarged central cusp flanked by smaller cusps of the same conical shape (Fig. 6–5), a persistent notochord, a short rostral section of the brain case, palatoquadrate articulation with the enlarged postorbital processes of the cranium, and long jaws reaching from the snout to behind the skull.

The teeth of hybodonts are variable, with some genera being similar to cladodonts in dentition and others having more flattened teeth

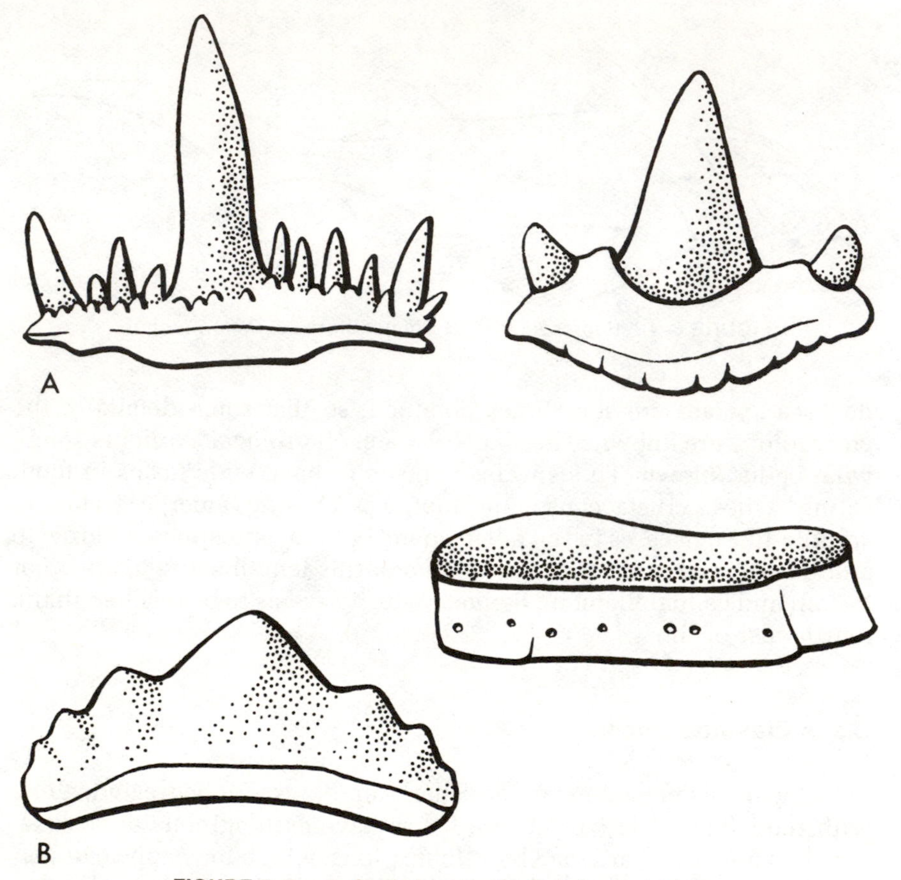

FIGURE 6–5. *A,* Cladodont teeth; *B,* hybodont teeth.

suitable for grinding (Fig. 6–5). Hybodonts share most of the characteristics mentioned for the cladodonts, but differ in the structure of the pectoral fin skeleton, possession of an anal fin, and reduction of caudal fin radials. In some the rostral portion of the skull is enlarged. Modern elasmobranchs have vertebral centra replacing the notochord, shorter and protrusible jaws with hyostylic suspension, and larger neural and haemal elements of the vertebrae, plus other features. Schaeffer has indicated that the living sharks, except for the transitional Chlamydoselachidae, Heterodontidae, and Hexanchidae, represent two main phyletic lines, the galeoids and the squaloids. The rays, or batoids, represent another line of modern elasmobranchs.

As more is learned about palaeozoic sharks and rays because of new fossil discoveries and subsequent research, there will be further alterations in our views of relationships among elasmobranchs. The recent discovery of several hitherto unknown sharks in Montana deposits should add to our understanding. At the present, some investigators consider that palaeozoic sharks represent a single premodern level of evolutionary organization, although there may be several types or designs.

As a matter of interest, two groups of sharks mentioned in L. S. Berg's 1947 classification are presented below. These were considered subclasses of Elasmobranchii.†Subclass Xenacanthi (Pleuracanthodii) contains extinct freshwater sharks with a prominent occipital spine. The paddle-like pectoral fins have a segmented axis that bears radials on both the anterior and posterior aspects. Two anal fins and claspers are present (Fig. 6–6a). †Subclass Cladoselachii (Fig. 6–6b) embraces sharks with a contrasting pectoral fin structure. The fins appear as wide horizontal flaps, stiffened by radials that extend almost to the fin margins. The pelvics are similar. Some authors have called these "fin-fold" sharks, contending that the broad-based fins could be remnants of lateral fin folds.

Dr. L. J. V. Compagno in a system published in 1973 considers holocephalans, sharks and rays all to belong to the Class Chondrichthys, placing the latter two into the subclass Elasmobranchii along with their palaeozoic relatives. Modern sharks and rays and their extinct near relatives are included in the cohort Euselachii, but Compagno does not follow the practice of splitting Euselachii into sharks and rays as lines of equal rank. Instead he recognizes four evolutionary lines, presented as three superorders of sharks and one superorder of rays.

In the following treatment of modern superorders and orders (Table 6–1), Compagno's arrangement will be followed, but these groups are presented as belonging to the class Elasmobranchii and subclass Selachii.

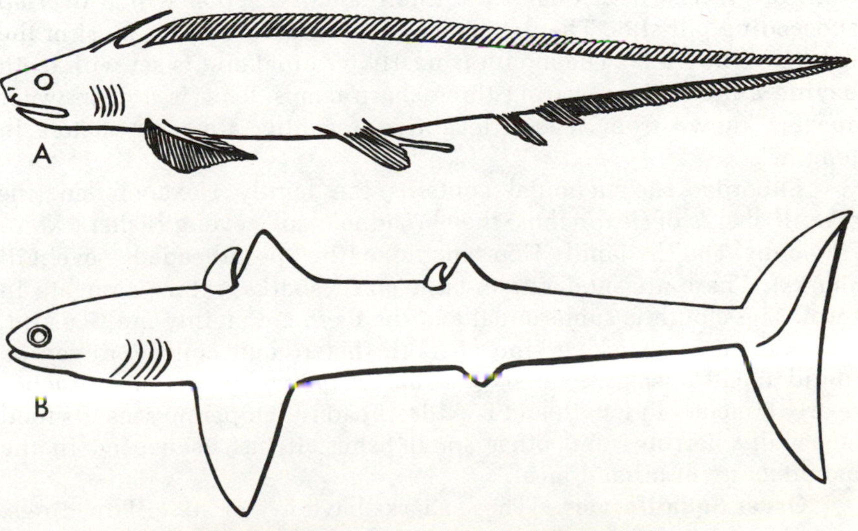

FIGURE 6–6. *A*, Xenacanthi; *B*, Cladoselachii.

Table 6–1. CLASSIFICATION OF MODERN SHARKS AND RAYS

Superorder Squalomorphii
  Order Hexanchiformes
    Suborder Chlamydoselachoidei, frill shark
    Suborder Hexanchoidei, sixgill and sevengill sharks
  Order Squaliformes, dogfish sharks, bramble sharks, etc.
  Order Pristiophoriformes, saw sharks
Superorder Batoidea
  Order Rajiformes
    Suborder Rhinobatoidei, guitarfishes
    Suborder Rajoidei, skates
  Order Pristiformes, sawfishes
  Order Torpediniformes, electric rays, torpedoes
  Order Myliobatiformes, stingrays, eagle rays, etc.
Superorder Squatinomorphii
  Order Squatiniformes, angel sharks, monkfishes
Superorder Galeomorphi
  Order Heterodontiformes, bullhead sharks, horn sharks
  Order Orectolobiformes, carpet sharks, whale shark
  Order Lamniformes, mackerel sharks, sand sharks, etc.
  Order Carcharhiniformes, requiem sharks, smoothhounds, etc.

Superorder Squalomorphii

Because of similarities of cranial and pectoral anatomy, this superorder consolidates three orders of diverse appearance and dentition, one of which, the Hexanchiformes, has not usually been considered as being closely related to the others (Fig. 6–7).

Order Hexanchiformes. In this order are sharks with six or seven gill arches and slits. There is an anal fin and a single dorsal that does not have a spine. The suborder Chlamydoselachoidei includes only the family Chlamydoselachidae with one species, *Chlamydoselachus anguineus,* the frill shark. This almost eel-shaped shark takes its name from the frilly extensions of the interbranchial septa, which overlap succeeding gill slits. The notochord is unconstricted over most of the length of the trunk. The mouth is nearly terminal and is set with teeth having a broad base bearing three sharp cusps. This is a deep-water species known from several localities, reaching almost 2 meters in length.

Suborder Hexanchoidei contains the family Hexanchidae (the six-gill sharks of *Hexanchus,* the "broadheaded" sevengill shark *Notorhynchus*) and the family Heptranchidae (the "pointheaded" sevengill sharks). These are moderate-to-large-sized sharks, rather elongate in form. The mouth is subterminal and the teeth are mainly multicuspid, although more than one kind of tooth (heterodont condition) can be found in all species. The sixgill shark, *Hexanchus griseus,* reaches nearly 8 meters in length and is widespread in temperate seas. Its food is usually herrings and other small fishes. It has been used in the manufacture of oil and meal.

Order Squaliformes. These sharks have five or six gill openings, two dorsal fins, often with a spine, and usually no anal fin. Families included are Echinorhinidae, the bramble sharks, and Squalidae, the

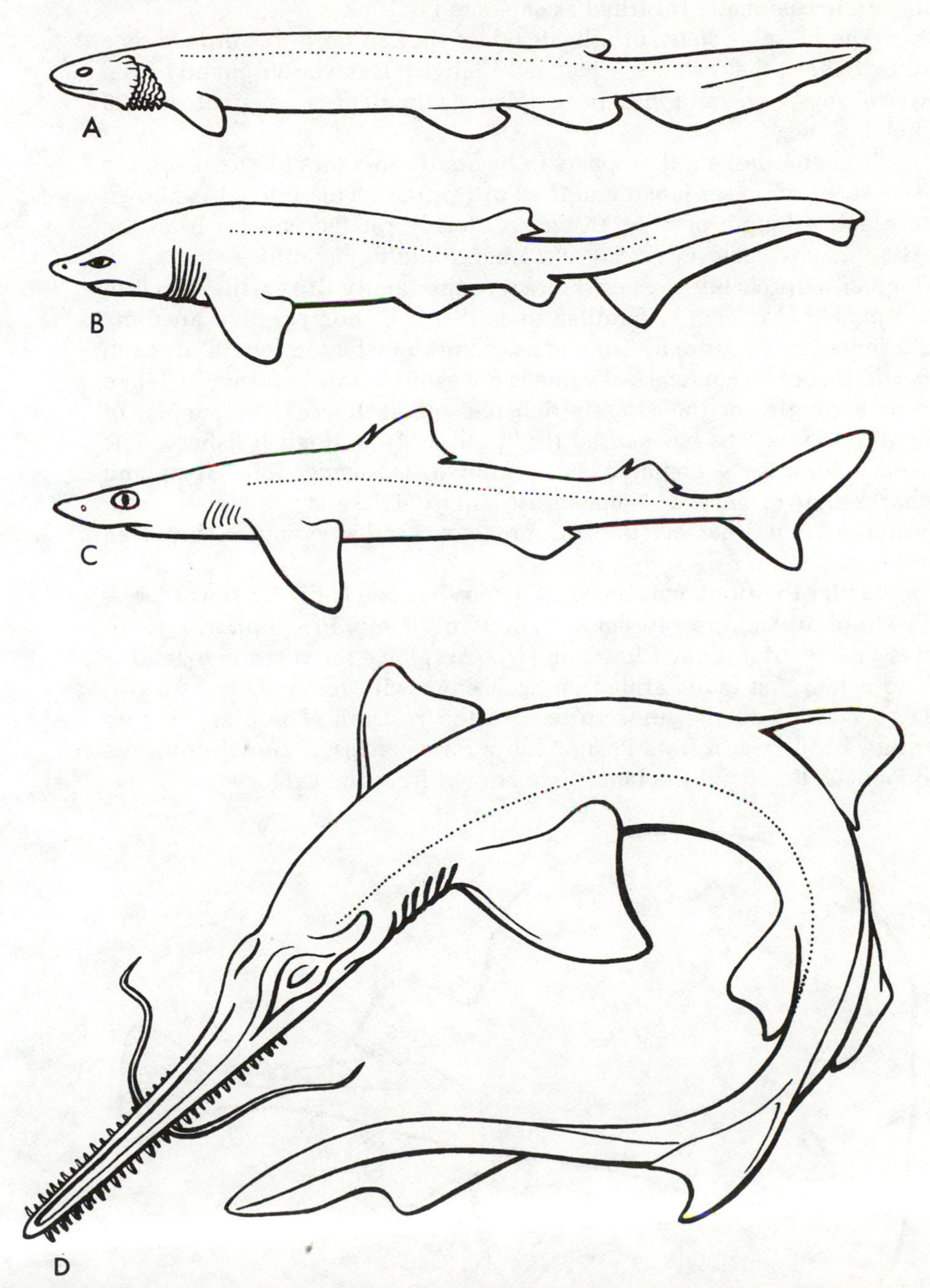

FIGURE 6–7. Representatives of superorder Squalomorphii. *A,* Frill shark (*Chlamydoselachus*); *B*, sixgill shark (*Hexanchus*); *C*, spiny dogfish (*Squalus*); *D*, saw shark (*Pliotrema*).

dogfish sharks and allies. The latter is usually considered to consist of several subfamilies, some of which, such as Dalatiinae and Somniosinae, are occasionally regarded as separate families.

The bramble shark or alligator dogfish, *Echinorhinus brucus,* is a robust shark that reaches 3 meters in length. It is known mainly from warm seas, where it has been taken from depths as great as 900 meters.

The smallest shark appears to be *Squaliolus laticaudus,* from the eastern Pacific near Japan and the Phillippines. This midget is known to reach a length of only 15 cm. A closely related species from the Atlantic has been measured at 22 cm. *Squalus acanthias,* the spiny dogfish, is probably the best known of the family. It is widespread in temperate seas and is familiar to millions of comparative anatomy students. This abundant fish reaches about 2 meters in length. Its flesh is edible, but its commercial value is not as high now as formerly, when it was sought for the vitamin-rich oils of the liver. Development of synthetic vitamins has caused the decline of the dogfish fishery. The genus *Somniosus* contains fishes known as sleeper and Greenland sharks that reach over 7 meters in length. These are sluggish, cold-water animals that act both as predators and scavengers, but feed primarily upon fishes.

Order Pristiophoriformes. There is but one family in the order — Pristiophoridae, the sawsharks. The two genera, *Pristiophorus* (with five gill openings) and *Pliotrema* (with six) have the rostrum extended into a long flat blade armed on each edge with teeth. There are two large barbels on the undersurface of the rostrum. These are mainly found in the warm Indo-Pacific, but a rare species is known from the Bahamas. Fossils of the family are known from the Cretaceous.

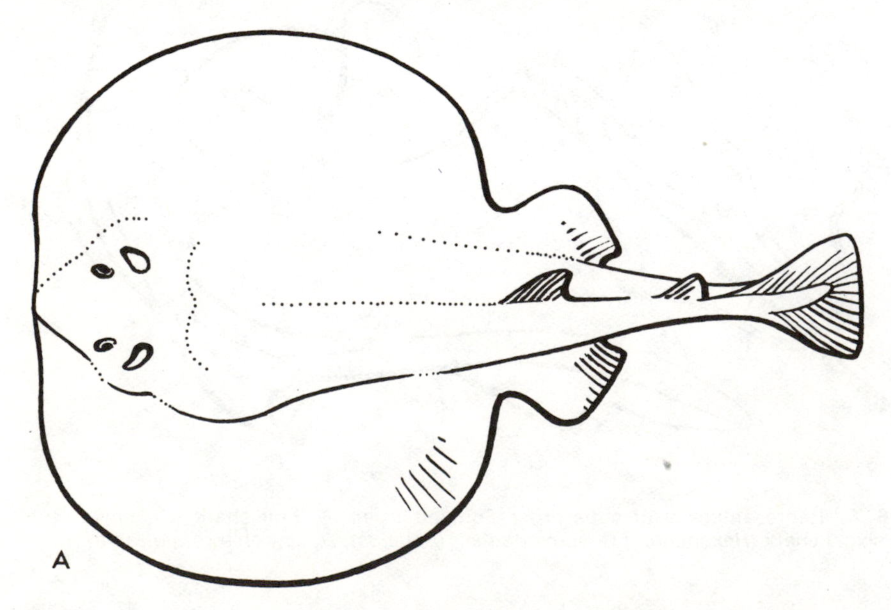

FIGURE 6–8. Representatives of superorder Batoidea. *A*, Electric ray (*Torpedo*).
(Illustration continued on opposite page.)

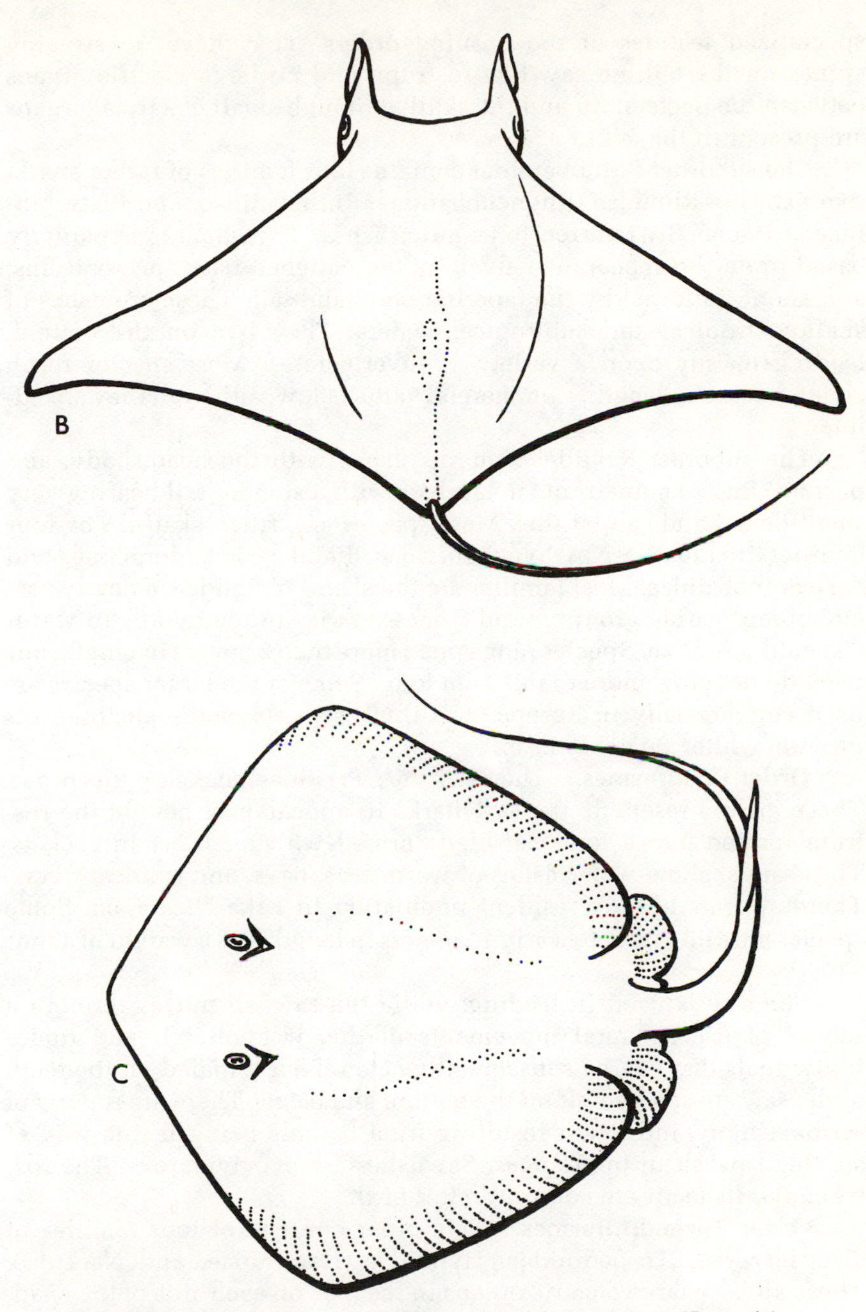

FIGURE 6–8 *Continued.* *B,* manta ray (*Manta*); *C,* stingray (*Dasyatis*).

Superorder Batoidea

The skates and rays of the superorder Batoidea are recognized by a depressiform body, with the pectoral fins extending forward and fusing to the head so that the five pairs of gill openings are ventral (Fig. 6–8). There are, in addition, several skeletal characteristics that distinguish them from sharks.

Order Rajiformes. This order contains batoids that lack some specialized features of the ensuing orders. They have no stinging spines on the tail, no saw-like rostrum, and no large electric organs between the pectoral fin and the skull, although small electrical organs are present in the tail of skates.

The suborder Rhinobatoidei contains four families of rather shark-like fishes — Rhinidae, Rhynchobatidae, Rhinobatidae, and Platyrhinidae, all generally referred to as guitarfishes. The name is apparently based upon the appearance given by the flattened head, pectoral fins, and snout, followed by the tapering body and tail. These are fishes of shallow tropical and subtropical waters. They live on the bottom, feeding mainly upon a variety of invertebrates. Most species reach about 1 meter in length. Commercial value is low, although they are edible.

The suborder Rajoidei contains fishes with the head, body, and pectoral fins combined into a flat disc, with a slender tail bearing very small dorsal and caudal fins. Most species are called skates. The four families included are Arhynchobatidae, Rajidae, Pseudorajidae, and Anacanthobatidae. Most familiar are the skates of Rajidae, a nearly cosmopolitan marine group, found from estuaries to the depths in warm and cold seas alike. Species range up to more than 2 meters in length, but most do not grow more than 75 cm long. Some of the larger species are used commercially in Europe. The thickest parts of the pectoral fins provide white, palatable flesh.

Order Pristiformes. The sawfishes, Pristidae, make up this order. These greatly resemble the sawsharks in appearance, having the rostrum formed into a long flat blade armed with teeth set in sockets. These are shallow-water fishes of warm seas, bays, and tropical rivers. There appears to be a resident population in Lake Nicaragua. Some species are said to reach nearly 11 meters in length and a weight of about 2400 kg.

The saw is used in feeding; while the sawfish moves through a school of fishes lateral movements of this weapon kill and injure individuals that can be subsequently eaten. Fish impaled on the teeth of the saw are scraped off on the bottom and eaten. There are reports of serious injury and death resulting from bathers being in the way of startled sawfish in the Ganges. Sawfishes are ovoviviparous. The rostrum and its teeth are soft until after birth.

Order Torpediniformes. This order consists of four families of electric rays — Torpedinidae, Hypnidae, Narcinidae, and Narkidae. These all have large electric organs in the disc on each side of the head. This organ is composed of columns of modified muscle tissue and allows the fish to deliver powerful shocks that can stun prey or possibly discourage predators, although they are known to be eaten by sharks. Up to 200 volts have been recorded from large specimens. The shock may be powerful enough to put humans to sleep, as evidenced by the word roots of the family names. Mediterranean species were used by the ancients as a form of electrotherapy for ailments such as arthritis and gout. Electric rays are found in tropical and temperate waters over a considerable depth range. Some of the deepwater forms, such as

Typhlonarke, are blind. The largest species is thought to be *Torpedo nobiliana*, an Atlantic species that reaches 1.8 meters in length.

Order Myliobatiformes. Rays of this order have large pectoral fins that combine with the head to form a broad disc, with a slender tail usually having strong stinging spines. The caudal and dorsal fins are reduced or absent. Included are seven families — Dasyatidae, the stingrays, Potamotrygonidae, river stingrays, Urolophidae, round stingrays; Gymnuridae, butterfly rays, Myliobatidae, eagle and bat rays, Rhinopteridae, cownose rays, and Mobulidae, mantas or devil rays.

These are all warm-water fishes, seldom entering cold temperate waters, and are usually found close to shore. Potamotrygonidae is found in rivers of South America. The various stingrays and the butterfly rays live on the bottom, often concealing themselves in sand or other fine materials. Their food is shellfishes and bottom-living fishes. The tail spines are typically barbed and grooved along the edges. The venom produced in the groove can make a wound caused by the spine to be both painful and dangerous. The largest stingrays may reach a width of 2 meters.

The eagle rays, bat rays, and cownose rays feed on the bottom, often dislodging bottom materials through the hydraulic action of powerful movements of their large pectorals. Clams, oysters, and other invertebrates make up most of their food. The teeth of these rays are in the form of broad grinding plates. Locomotion is by flying movements of the wing-like pectorals. The long whiplike tail is usually held straight behind. Some species reach 1.2 m in width.

Mantas or devil rays have turned from bottom feeding to seeking plankton and small schooling fishes. They swim through the water by means of the wide, slender-tipped pectoral "wings," holding the mouth open. The peculiar cephalic fins, positioned on either side of the mouth, are used to guide food into the mouth. These fins, when curled into the spiral resting position, give the impression of horns, hence the name "devil" rays. Although some species reach less than 1 meter in width, others may reach several meters. A specimen of *Manta birostris* was measured at 6.6 meters and is thought to have weighed over 1600 kg. Many of the mobulids, rhinopterids, and myliobatids have a habit of leaping clear of the water and landing with a loud noise. Cartwheeling is another behavioral trait. Cownose rays are occasionally seen in more or less regularly ordered schools of up to 6000 individuals.

Superorder Squatinomorphii

This group contains one genus of depressiform fishes, having the pectorals expanded forward but not fused with the head. The gill openings are mainly lateral, the spiracles large as in most batoids. They have two spineless dorsal fins set on the tail, no anal, and an essentially hypocercal caudal.

Order Squatiniformes. The order contains the single family Squatinidae and the genus *Squatina*, the monkfishes or angel sharks.

These are tropical to temperate in distribution, usually being found in shallow water. Despite their ray-like appearance their locomotion is shark-like, accomplished by movements of the tail. The largest species reaches about 2.4 meters in length and a weight of 72 kg.

Superorder Galeomorphi

Four orders of sharks are brought together under this name because of similarities in the cranial skeleton and in the structure of the pectoral fins. All have an anal fin and five gill openings. Most have dorsal fins, but only the Heterodontiformes have dorsal fin spines. Most of the familiar genera of sharks belong to this superorder (Fig. 6–9).

Order Heterodontiformes. Only one genus, *Heterodontus,* of the family Heterodontidae is included. Species are referred to as horn sharks because of the strong spines at the front of each dorsal fin. The generic name alludes to the condition of the teeth in the short and modified mouth. Anterior teeth are small and sharp whereas those in the back of the jaws are molariform. The horn sharks are found in the Indo-Pacific in tropical to warm temperate waters. They may reach 1.5 meters in length.

Heterodontus has often been considered to be closely related to such primitive sharks as hybodonts and ctenacanthids. Some evidence for this involves the suspension of the upper jaw, which might be said

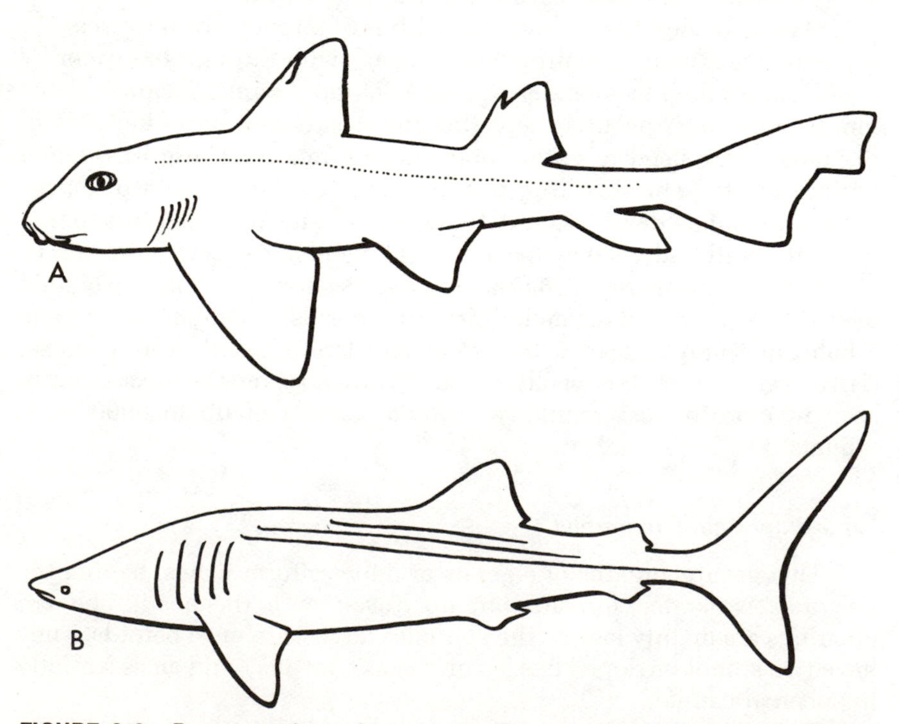

FIGURE 6–9. Representatives of superorder Galeomorphi. *A*, Horn shark (*Heterodontus*); *B*, whale shark (*Rhiniodon*).

to be structurally hyostylic but functionally amphistylic, according to Schaeffer. Heterodontid teeth are known from the Jurassic.

Order Orectolobiformes. The sharks of this group are usually placed into two families — the Orectolobidae, containing the carpet sharks, nurse sharks, zebra sharks, and wobbegongs, and the Rhiniodontidae, or whale sharks. In Compagno's arrangement the Orectolobidae is restricted to a few genera and the remaining fishes are placed in the following families: Parascyllidae, Brachaeluridae, Hemiscyllidae, Stegostomatidae, and Ginglymostomatidae. These sharks are found mainly in the tropical parts of the Indo-Pacific, with most of the genera being present in the Australian waters. One genus, *Ginglymostoma*, reaches the Atlantic. For the most part these are small sharks, reaching less than a meter long, but the Atlantic nurse shark has a maximum length of about 4.2 meters, and others such as the zebra shark, *Stegostoma*, and some species of *Orectolobus* may exceed 3 meters in length. The latter genus contains the strikingly marked wobbegongs of Australia. Some of the carpet sharks in the western Pacific are known as cat sharks, a name usually reserved for fishes of another order and family.

Rhiniodon (Rhincodon) typus is the largest living fish, attaining a length of 15 meters. One specimen of 11.5 meters was estimated to weigh about 12,000 kg. This sluggish giant is present in all tropical seas. Its mouth is terminal and broad but cannot be opened wide. The fish feeds upon great quantities of small schooling fishes such as herring, and upon squid and planktonic crustacea. The mouth is equipped with numerous rows of small teeth, but most food items appear to be captured by straining through the fine gill rakers. The gill slits are very long and set rather high on the sides, partially above the pectoral fin. The body has a humpbacked appearance and the caudal fin is very large. The color pattern is striking, consisting of yellowish or white spots on a gray to brown background. These huge animals are oviparous; the eggs are surprisingly small (300 × 90 mm) considering the size of the sharks.

Order Lamniformes. Usually these are large and active sharks of shallow waters, with a few exceptions. Families included are: Odontaspididae (= Carchariidae), the sand sharks; Pseudocarchariidae; Mitsukurinidae (= Scapanorhynchidae), the goblin shark; Alopiidae, the thresher sharks; Cetorhinidae, the basking sharks; and Lamnidae (= Isuridae), the mackerel sharks. The lamnids are among the best known sharks because of the large size and great appetite of some. One, the great white shark, *Carcharodon carcharias*, is known as a man-eater, and has been implicated in many fatal attacks on human beings. This giant has been measured to 11 meters in length and estimated to over 12. Specimens of 4.5 meters are reported to weigh 1350 kg, and one of 7 meters was weighed at 3200 kg. The white shark is found in all warm seas, ranging out of the subtropics in the summer. In the eastern Pacific it is known as far north as Petersburg, Alaska. Other large lamnids are the mako sharks of the genus *Isurus* and the salmon and mackerel sharks of *Lamna*. The mako is often sought as a big game fish;

L. nasus, porbeagle, has some importance as a commercial fish in Norway.

The goblin shark, *Mitsukurina owstoni*, is a deepwater fish that has been found in the Indian Ocean, the western Pacific off Japan, and the Atlantic near Portugal. It is characterized by a flat elongate snout and protrusible jaws set with slender, sharp teeth. Some of these teeth were found in a malfunctioning communications cable brought up from 1300 meters deep in the Indian Ocean. Whether the shark actually attacked the cable or was feeding on animals growing on it is an interesting point to ponder. The thresher sharks, *Alopias*, have the upper caudal lobe elongated so that it comprises about one half of the total length of the fish. This great tail is used to herd the small schooling fishes upon which the sharks prey. When in shallow water, the tail slaps and splashes the surface. The several species of the genus are found in warm seas.

Unlike its swift and ferocious relatives, the gigantic basking shark is a slow-moving plankton feeder. It is close to the whale shark in reported maximum length, reaching 13.5 meters. Its teeth, although numerous, are very small and would appear to have limited function. The gill rakers, however, are long and slender, constituting an excellent sieve for the small crustaceans upon which the shark feeds. Basking sharks are distributed in cold and temperate seas and have often been the object of harpoon fisheries for their liver oil.

Order Carcharhiniformes. This is a large assemblage that includes eight families and about 40 genera, encompassing many of the more familiar species of sharks. The families included are: Scyliorhinidae, the cat sharks and spotted dogfishes; Proscyllidae, also called cat sharks; Pseudotriakidae, false cat sharks; Leptochariidae; Triakidae, the smooth hounds or smooth dogfishes; Hemigaleidae; Carcharhinidae, the requiem or "typical" sharks; and Sphyrnidae, the hammerhead sharks.

Cat sharks are small sharks, often with striking color patterns, found in warm seas in many parts of the globe. Some are found at considerable depths and, like the brown shark of the eastern North Pacific, are of drab coloration. The swell sharks of the genus *Cephaloscyllium* are capable of swallowing air and inflating their stomachs when brought out of the water. The inflated specimens then float until they can deflate themselves, a task that appears easy for some but harder for others, taking hours or days to accomplish. Smooth dogfishes are widespread shallow-water forms, some of which reach 2 meters in length. Familiar North American species are *Mustelus canis* of the Atlantic coast and *Rhinotriacis henlei* of the Pacific.

The requiem shark family contains several well known medium to large species from tropical and temperate waters. Blacktip and whitetip sharks, so named because of their fin coloration, are of the genus *Carcharhinus* and are found in warm seas. One member of the group, the bull shark of the western Atlantic, *C. leucas*, is found in the fresh waters of Lake Nicaragua, where it is known to make fatal attacks on bathers. Similar species of Africa and India also enter fresh water. Another man-eater of the family is the tiger shark, *Galeocerdo cuvieri*. This is a circumtropical species that reaches about 5.5 meters in length and has some fame as a sport fish. The blue shark *Prionace glauca* (Fig. 1–1C) is found in most warm and temperate waters. It has a slender body, a remarkable blue coloration, and is an active feeder, often attacking hooked salmon to the dismay of the angler. The topes, *Galeorhinus* sp., are found in the Indo-Pacific and the eastern Atlantic. The soupfin shark, *G. zyopterus*, was once the target of a valuable fishery on the west coast of North America because of the vitamin content of its liver oil. Production of synthetic vitamin A lowered the price so that the fishery was abandoned.

The hammerhead sharks were also sought for the vitamin-rich liver oil. These medium to large sharks are characterized by flat lateral expansions of the head, so that from above or below the outline is that of the letter T. The eyes are borne on the outer edge of the structure. These are confined to warm waters and the family is circumtropical. *Sphyrna mokarran* has been measured at 5.4 meters in length, and *S. zygaena* to over 4. They are reported to have hearty appetites and to feed on a variety of animals, including other hammerheads and the formidable stingrays.

Several hypotheses have been advanced to explain the function of their expanded head. Some researchers have suggested that the flattened surface aids the swift animals to make tighter turns. Others have postulated that because the nostrils are out on the forward corners of the expansion, ranging in on sources of odors is facilitated.

References

All references for this chapter are included in the complete listing for Section Two at the end of Chapter 7, pp. 206–210.

7 CLASS OSTEICHTHYES — BONY FISHES

The endoskeleton of bony fishes is typically at least partly ossified, with dermal bones in the head region. There is usually a gas bladder, but it is secondarily lost in some specialized species. Scales are ganoid or bony. An operculum covers the gills, and the gill septa are reduced.

Primitive members of the class may share certain characteristics with Elasmobranchs and Placoderms in that the caudal fin may be heterocercal, and various members may have the spiracle, spiral valve, valvular conus arteriosus, cloaca, or the primitive position of the vent between the bases of the pelvic fins. In most members of the class, however, the caudal is homocercal, the spiracle, spiral valve, conus arteriosus, and cloaca are absent, and the vent is variously placed, usually just anterior to the anal fin.

As mentioned earlier, there are strong resemblances between the Acanthodii and primitive osteichthyans, especially in the structure of the neurocranium. Several opinions exist on the closeness of the evolutionary relationships of the bony fishes and the acanthodians, and many modern investigators postulate a common ancestor for the two groups.

Osteichthyes, or Teleostomi, is commonly divided into three subclasses; one of which, the Dipnoi, is considered to represent a separate class in some classification schemes. Dipnoi, the lungfishes, and the subclass Crossopterygii, the coelacanths, share some features of the cranial and appendicular skeleton and may be more closely related to each other than to the third subclass, the Actinopterygii, which contains most of our living bony fishes.

Table 7–1 shows the order in which the bony fishes will be considered. Ordinal names end in "iformes," and subordinal names take the suffix "oidei."

Subclass Dipnoi

The lungfishes (Figs. 1–1*F* and 2–11*B*) are often separated into their own class because of their skull structure, but they will be considered here as part of the Osteichthyes. Characteristics that distinguish them are: the suspension of the upper jaw is autostylic — i.e., the palatoquadrate is fused to the neurocranium; internal nares are present (in later forms), as is a cloaca; the teeth are fused into grinding plates and the maxillae and premaxillae are lacking; the connection of the lung or gas bladder to the esophagus is ventral.

The paired fins of lungfishes consist of a long central axis with, in the Australian species, the fin rays disposed along it. In the South American and African lungfishes the fin rays are lacking and the fins are produced as filaments. A special pulmonary circulation is developed in living forms, and the atrium of the heart is divided into left and right chambers by an incomplete septum. The spiral valve is present in the intestine.

As many as five extinct orders are recognized by some authorities, but only the two living orders will be considered here.

Order Ceratodiformes. This order contains the living Australian *Neoceratodus forsteri* and an extinct genus †*Ceratodus* that was distributed more generally over the continents. Both are placed in the family Ceratodidae, which is known from the Lower Triassic.

Neoceratodus reaches nearly 2 meters in length and is a heavy-bodied fish, with large scales and paddle-like paired fins. The caudal fin is diphycercal. It differs structurally from other living lungfishes by having an unpaired lung and a cartilaginous endocranium, as well as four pairs of gills. There are many differences in life history and habit between this and other species. The Australian species can depend upon the oxygen in the water unless conditions become stagnant. No special nest is constructed, the eggs are laid among vegetation, and the young have no special external gills. *Neoceratodus* apparently feeds on vegetation and the many small forms of animal life that live among the plants. It frequents permanent water bodies and is incapable of aestivation.

Order Lepidosireniformes. This order contains two families, Lepidosirenidae, the South American lungfish, and Protopteridae, the African lungfishes. Both have paired lungs, filamentous paired fins, and a membranous endocranium.

Both families contain elongate fishes with fairly small scales. The habitat of these species is generally swampy, often containing low concentrations of dissolved oxygen, so that they have evolved the ability to utilize atmospheric oxygen. The gills are reduced and relatively ineffective as compared to those of *Neoceratodus*. If the swamps dry up both the African and South American lungfishes can burrow into the muddy bottom and form a cocoon of mucus in which they can remain for several months in an inactive state, waiting for the next rainy season. This aestivation has been prolonged for as long as four years in the laboratory, using an African species.

When these fishes return to full activity after aestivation, nests are constructed and breeding begins. The African species of *Protopterus* make simple holes near the edge of the swamp, whereas the South American *Lepidosiren* constructs a burrow. In both the eggs and larvae

Table 7–1. SUBCLASSES, ORDERS, AND SUBORDERS OF CLASS OSTEICHTHYES, BONY FISHES*

Subclass Dipnoi
- Order Lepidosireniformes – African and South American lungfishes
- Order Dipteriformes – Australian lungfishes
- †(Five extinct orders)

Subclass Crossopterygii
- †Order Osteolepiformes
- Order Coelacanthiformes – coelacanths

Subclass Actinopterygii
- "Chondrostei"
  - †Order Palaeonisciformes
  - Order Polypteriformes – bichirs and reedfishes
  - Order Acipenseriformes – sturgeons and paddlefishes
- "Holostei"
  - Order Lepisosteiformes – gars
  - †Order Pycnodontiformes
  - Order Amiiformes – bowfins
  - †Order Aspidorhynchiformes
  - †Order Pholidophoriformes
- "Teleostei"
  - †Order Leptolepiformes
  - Order Elopiformes
    - Suborder Elopoidei – tarpons
    - Suborder Albuloidei – bonefishes
  - Order Anguilliformes
    - Suborder Anguilloidei – eels
    - Suborder Saccopharyngoidei – gulpers
  - Order Notacanthiformes—spiny eels
  - Order Clupeiformes
    - Suborder Denticipitoidei – toothheads
    - Suborder Clupeoidei – herrings and allies
  - Order Osteoglossiformes
    - Suborder Osteoglossoidei – bonytongues, arapaima
    - Suborder Notopteroidei – featherbacks, mooneye
  - Order Mormyriformes—elephantfishes
  - Order Salmoniformes
    - Suborder Argentinoidei—deepsea smelts, slickheads, etc.
    - Suborder Stomiatoidei – anglemouths, viperfish, etc.
    - Suborder Salmonoidei – salmon, trout
    - Suborder Galaxoidei – galaxiids, icefish, etc.
    - Suborder Esocoidei – pikes, mudminnows
  - Order Myctophiformes
    - Suborder Myctophoidei – lanternfishes
    - Suborder Aleposauroidei – lancetfishes
  - †Order Ctenothrissiformes
  - Order Gonorhynchiformes
    - Suborder Gonorhynchoidei – sandfishes
    - Suborder Chanoidei – milkfishes
  - Order Cypriniformes
    - Suborder Characoidei – characins
    - Suborder Gymnotoidei – knifefishes, electric eels
    - Suborder Cyprinoidei – minnows, carps, suckers
  - Order Siluriformes—catfishes
  - Order Polymixiiformes—beardfishes, barbudos
  - Order Percopsiformes
    - Suborder Percopsoidei – trout-perches
    - Suborder Aphredoderoidei – pirate perches
    - Suborder Amblyopsoidei – blind cavefishes
  - Order Gadiformes
    - Suborder Muraenolepoidei
    - Suborder Gadoidei – cods
    - Suborder Ophidioidei – cusk-eels
    - Suborder Zoarcoidei – eelpouts
    - Suborder Macrouroidei – rat-tails

Table 7–1. SUBCLASSES, ORDERS, AND SUBORDERS OF CLASS OSTEICHTHYES, BONY FISHES* (*Continued*)

Order Beloniformes—flyingfishes, halfbeaks
Order Cyprinodontiformes—topminnows
Order Atheriniformes—silversides
Order Lampridiformes
  Suborder Lampridoidei—opahs
  Suborder Veliferoidei—sailbearers
  Suborder Trachipteroidei—crestfishes, oarfishes
  Suborder Stylephoroidei—tube-eyes
  Suborder Ateleopoidei—tadpole fishes
  Suborder Mirapinnatoidei—hairy fishes, tapetails
Order Beryciformes
  Suborder Stephanoberycoidei—pricklefishes
  Suborder Berycoidei—soldierfishes, pinecone fishes
  Suborder Cetomimoidei—whalefishes
  Suborder Giganturoidei—giganturids
Order Zeiformes—dories
Order Gasterosteiformes
  Suborder Gasterosteoidei—sticklebacks
  Suborder Aulostomoidei—tube-snouts
  Suborder Syngnathoidei—pipefishes, seahorses
Order Perciformes
  Suborder Mugiloidei—mullets and barracudas
  Suborder Anabantoidei—gouramies, climbing perches, pikeheads
  Suborder Percoidei—perches, basses, wrasses, grunts, snappers, jacks, etc.
  Suborder Stromateoidei—butterfishes
  Suborder Icosteoidei—ragfishes
  Suborder Blennioidei—blennies
  Suborder Trachinoidei—weevers, etc.
  Suborder Ammodytoidei—sand lances
  Suborder Callionymoidei—dragonets
  Suborder Gobiodei—gobies
  Suborder Acanthuroidei—surgeonfishes
  Suborder Kurtoidei—forehead brooders
  Suborder Scombroidei—tunas, mackerels
  Suborder Mastacembeloidei—spiny eels
  Suborder Schindleroidei—"Schindler's fishes"
  Suborder Echeneoidei—remoras
Order Scorpaeniformes
  Suborder Scorpaenoidei—rockfishes, scorpionfishes, waspfishes, etc.
  Suborder Hexagrammoidei—greenlings
  Suborder Platycephaloidei—flatheads, spiny flatheads
  Suborder Cottoidei—sculpins
Order Pleuronectiformes
  Suborder Psettodoidei—toothed flounders
  Suborder Pleuronectoidei—flounders
  Suborder Soleoidei—soles
Order Tetraodontiformes
  Suborder Balistoidei—triggerfishes, trunkfishes
  Suborder Tetraodontoidei—puffers, molas
Order Pegasiformes—sea moths
Order Batrachoidiformes—toadfishes
Order Gobiesociformes—clingfishes
Order Lophiiformes
  Suborder Lophioidei—anglerfishes
  Suborder Antennarioidei—frogfishes
  Suborder Ceratioidei—deepsea anglers
Order Synbranchiformes
  Suborder Synbranchoidei—swamp eels

*Selected extinct orders are included and are indicated by a dagger, †.

are guarded by the male. Larvae are held in place by a secretion from a cement organ on the breast region, and have feathery external gills similar to those of salamanders.

There is apparently only one species in Lepidosirenidae, *Lepidosiren paradoxa.* It is characterized by reduced paired fins, five gill arches, and the development in breeding males of feathery gill-like structures on the pelvic fins. These seem to act as gills in reverse, releasing oxygen in the vicinity of the young. *Lepidosiren* feeds on animals, especially snails, but also takes algae. Maximum size is about 1.25 meters in length.

Protopteridae contains several species, all in the genus *Protopterus.* Some of these may reach more than 2 meters in length. The paired fins of these species have radials along the central axis. They have six gill arches. The males do not develop gill-like structures on the pelvic fins, and the young retain vestiges of the external gills for some time after metamorphosis. *Protopterus* is carnivorous and said to be very destructive to other fishes. They often attack others of their own species if held together. One that arrived at Oregon State University with several inches of the tail missing regenerated the section almost perfectly within two years.

Subclass Crossopterygii

In these fishes the palatoquadrate is separate from the endocranium and the jaw suspension is hyostylic. The paired fins are paddle-like, some with a median axis in which a proximal bone articulates with two distal ones with radials on each side. The skull is divided into front (ethmosphenoid) and rear (otico-occipital) sections at the juncture between the frontals and parietals (Fig. 7–1). Premaxillae and maxillae may be present. A cloaca is absent.

†Order Osteolepiformes. This extinct order, known from the Lower Devonian to the Upper Carboniferous, is also known as Rhipidistia, and is viewed by some systematists as being a superorder containing at least two orders. It includes many large predatory fishes (Fig. 7–1) with bony heads and a rather lizard-like appearance. They have cosmoid scales, two dorsal fins, and internal nares. These are the forms that apparently gave rise to tetrapods, and some of their osteological characteristics appear to be easier to relate to primitive amphibia than to other bony fishes. An example is the pectoral skeleton of †*Eusthenopteron,* which shows elements resembling the humerus, radius, and ulna.

Order Coelacanthiformes. The only living member of this order is *Latimeria chalumnae* of the family Latimeriidae (Fig. 7–1). Until a specimen was captured off southeast Africa in 1938, the order was thought to have become extinct in the Cretaceous. The South African ichthyologist Dr. J. L. B. Smith described the first recent coelacanth from the museum mount to which it had been converted, and initiated

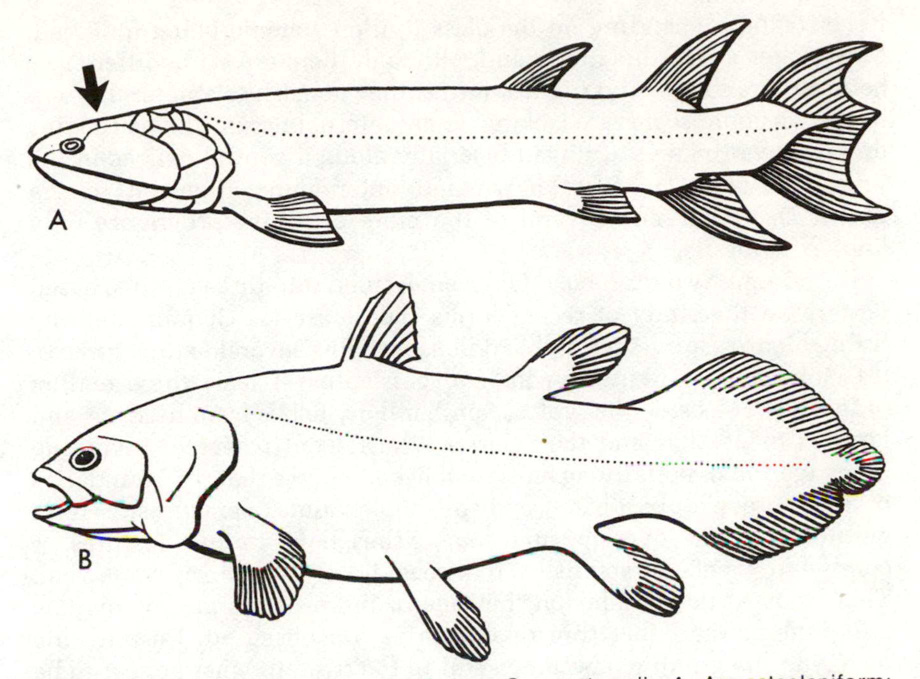

FIGURE 7–1. Representatives of subclass Crossopterygii. *A,* An osteolepiform; *B,* a coelacanth (*Latimeria*).

a search for further specimens. Not until 1952 did an additional specimen come to his attention, from the Comoro Islands northwest of Madagascar. Since 1952 several additional specimens have been obtained in the Comoro region at depths of from 150 to 400 meters. The finding and subsequent study of *Latimeria* were of great significance to palaeontologists, who had a rare opportunity to check their interpretations of fossil coelacanths against a living species.

Latimeria is large, with a maximum length of about 2.75 meters and weight that can be as much as 85 kg. Its color is dark blue and its large cosmoid scales are so rough that the Comoro Islands people use them like sandpaper to roughen inner tubes being repaired. The pectoral fins are large and strong and can rotate 180 degrees. Unlike extinct species the gas bladder of *Latimeria* is not ossified; instead, it is small and filled with fat. It maintains a connection to the lower part of the esophagus. Internal nares are lacking.

The reproductive biology of *Latimeria* is not fully known, but the species is ovoviviparous and the eggs are among the largest of bony fish eggs, being about 8 cm in diameter. *Latimeria* is a predator, and most specimens have been taken on hooks baited with fish.

Subclass Actinopterygii

The "ray-finned fishes" are usually placed together in the group called Actinopterygii, which may rank as a subclass or as a somewhat lesser taxon, depending on the classification scheme being followed. Most fishes are in this group and, although there are wide differences between the lowest and highest forms, they constitute a natural group in that cosmoid scales are lacking, as are internal nares. The paired fins do not have the rays arranged biserially along a central axis as in the Crossopterygii. The first known actinopterygians appeared in the Lower Devonian, and several of the more primitive orders are only known as fossils.

Actinopterygians have often been divided into three groups, based largely on the study of recent forms. These are (1) Chondrostei, the living bichirs, sturgeons, and paddlefishes, plus several extinct groups; (2) Holostei, the living bowfin and gars, plus at least three extinct orders; (3) Teleostei, the typical familiar bony fishes from herrings and tarpons to perches and their derivatives. Recent research, including investigation of both living and fossil fishes, shows that the boundaries of these three groups are not easily definable. Some characteristics form gradients, others overlap, and some fishes have unique features or combinations of characteristics that seem to set them apart somewhat. This leads to the conclusion that one or more of the groups may be polyphyletic, and that true relationships are obscured. Despite this objection the group names are useful in referring to what appear to be three levels of organization among the actinopterygians. Here the three traditional names will be used to designate groups of orders.

Chondrostei

This more or less artificial group or grade contains the extinct palaeoniscoids and allies, sturgeons, paddlefishes, bichirs, and reedfish.

†Order Palaeonisciformes. This is a large extinct group of about 38 families and 200 known genera sometimes interpreted as consisting of up to five orders related to some primitive living fishes. In the long existence of various lines from the Devonian to the Cretaceous considerable evolution occurred. Body forms range from the typical fusiform, as in palaeoniscoids (Fig. 7–2), to the compressiform platysomoids such as the Bobasatraniidae, which has a superficially symmetrical caudal fin and lacks pelvic fins. Tarrasiidae differs from the others in having a diphycercal tail and long dorsal and anal fins, continuous with the caudal.

Order Polypteriformes (Cladistia). This order contains the curious bichirs and reedfish of Africa (*Polypterus* and *Erpetoichthys*), and comprises a group well differentiated from other actinopterygians. The group is known from the Eocene of Egypt. It is sometimes placed by itself in the superorder or subclass Brachiopterygii or Polypteri, and sometimes placed with Crossopterygii. The dorsal fin of these fishes is completely unlike those of any other group. It consists of a series of separate small fins with one large spine-like ray and one or more soft

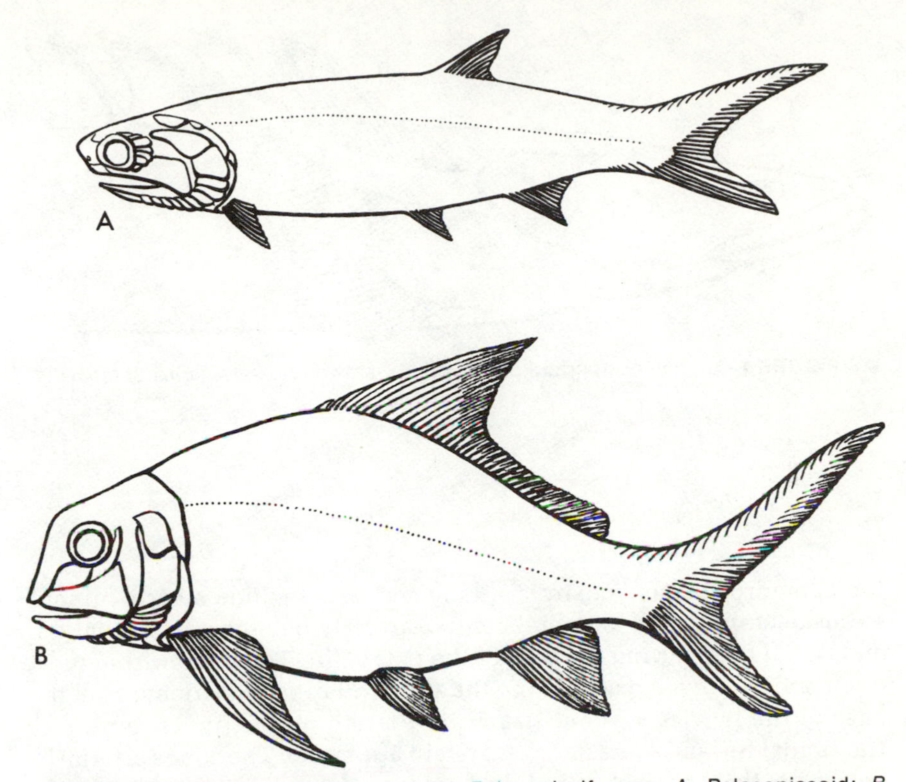

FIGURE 7–2. Representatives of order Palaeonisciformes. *A*, Palaeoniscoid; *B*, platysomoid.

branches from the posterior edge of the spine (Fig. 2–11). The spines are said to make handling these fishes dangerous.

The pectoral fin has a unique structure as well, being lobate with a skeletal support consisting of two bony elements, the propterygium and metapterygium with a cartilaginous plate-like mesopterygium between. Scales of polypteriforms are ganoid, but with three layers — ganoin, cosmine, and isopedine — as in the palaeonisciforms. Despite the hard scales their bodies are very flexible. The air bladder is lunglike and functions in respiration. A spiracle and spiral valve are present, and the heart has a conus arteriosus.

Polypterus contains about 10 species, all of African fresh waters. The species are medium- to large-sized and are predators. *P. bichir*, which reaches a length of about 1 meter, is sought as food, and is generally roasted in coals with the scales left on. The larvae have a rather amphibian-like appearance due to the large feathery external gills (Fig. 7–3). *Erpetoichthys (Calamoichthys)* of West Africa is elongate and lacks pelvic fins.

Order Acipenseriformes. The sturgeons and paddlefishes are closely related and are usually placed together in this order. One family,

FIGURE 7–3. Anterior portion of *Polypterus*, showing external gills of larva.

the Chondrosteidae, is extinct. Characteristics include a cartilaginous endoskeleton, lack of vertebral centra, strongly heterocercal caudal fin, and radials supporting the rays of the pelvic fin. They differ from most other actinopterygians in having the anus and urogenital opening at the base of the pelvics. Ganoid scales are present on the upper portion of the caudal fin, and some members retain a spiracle. The conus arteriosus has more than one set of valves, and a spiral valve is present in the intestine. A cellular air bladder is present. Dermal bone is prominent on the head of sturgeons.

Family Acipenseridae — Sturgeons. Sturgeons (Fig. 2–19*A*) are superficially distinguishable from paddlefishes by having bony scutes along the sides and back and four barbels on the underside of the rostrum, which is shorter than in the paddlefishes. The family contains four genera and over 20 species. They are holarctic in distribution and have marine, freshwater, and anadromous members. The waters of the Soviet Union are especially noted for the numbers of species of sturgeon found there. Sturgeons are sought by fishermen for their flesh and their roe, from which caviar is made. North American sturgeon fisheries are now relatively minor, but considerable tonnages are taken in the USSR and Iran.

The largest sturgeon is the beluga *(Huso huso)* of the Caspian and Black seas. It is reported to reach a length of about 9 meters and a weight of 1500 kg. Large specimens may be 100 years old and carry over 7 million eggs. One of the most important commercial sturgeons is the Russian sturgeon *(Acipenser guldenstadti)* of the Caspian and the Sea of Azov, reaching about 2.3 meters in length. The largest North American sturgeon is the white sturgeon *(Acipenser transmontanus)* of the Pacific coast. Lengths of 6 meters and weights of about 850 kg have been reported. This giant is found in both fresh and salt water from

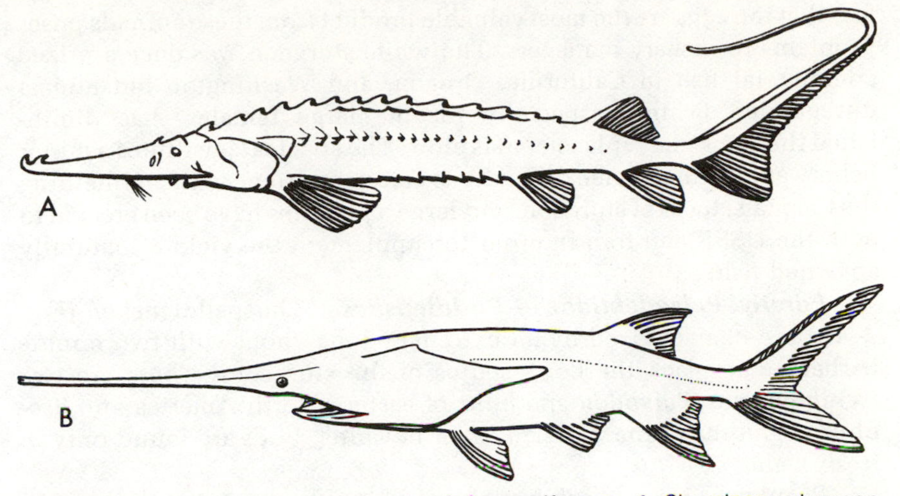

FIGURE 7–4. Representatives of order Acipenseriformes. *A,* Shovelnose sturgeon (*Pseudoscaphirhynchus*); *B,* paddlefish (*Polyodon*).

southern California to Cook Inlet of Alaska. It appears to be at home in large rivers such as the Snake, Columbia, Sacramento, and Fraser.

The smallest members of the family appear to be the curious shovel nose sturgeons (Fig. 7–4*A*). *Scaphirhynchus platorhynchus* of the Mississippi reaches about 1 meter in length, whereas the small shovelnose of the Amu Darya River of the USSR (*Pseudoscaphirhynchus hermanni*) is reported to reach a maximum length of only 27 cm. Other sturgeons of North America are: *Acipenser oxyrhynchus,* found on the Atlantic coast; *A. fulvescens,* found in the Mississippi drainage and north through the Great Lakes, the Red River of the North, the St. Lawrence River, and Hudson Bay; and *A. medirostris,* the green sturgeon, primarily a marine species that ranges from southern California to Alaska and westward to Asia, entering bays and rivers to spawn.

Food of sturgeons includes worms, crustaceans, and small fishes that can be sucked up by the greatly protrusible mouth. Beluga are reported to eat larger fishes and occasionally the young of the Caspian seal. A typical sturgeon life history includes a migration from feeding grounds to breeding grounds, usually to a river. Spawning takes place over gravel in fairly swift water. The demersal, adhesive eggs hatch after 3 to 5 days and the larvae—about 1 cm long—drift downstream to suitable rearing areas in the river or the sea. Growth is slow, with many species reaching only about 1.2 meters in length after 10 years. Males reach sexual maturity earlier than the females. Medium-sized species may become mature at 8 to 12 years for the males and 10 to 15 years for the females.

The requirement of clean water, slow growth, late maturity, and the fact that the eggs are the most valuable product from these animals poses problems for fishery managers. The white sturgeon was once a prized commercial fish in California, Oregon, and Washington but almost unregulated fishing—especially for the large females—has diminished the stock and replacement is slow. The species constitutes a minor fishery at this time. Some difficulty is being experienced in maintaining the Caspian stocks of sturgeon, and large hatcheries have been erected in both the USSR and Iran in order to supplement the yield of naturally spawned fish.

Family Polyodontidae — Paddlefishes. The paddlefishes (Fig. 7–4*B*) are characterized by an extremely long snout with two minute barbels. They lack the bony scutes of the sturgeons. There are two living species, *Polyodon spathula* of eastern North America and *Psephurus gladius* of the Yangtze River of China. Both are found only in fresh water.

Psephurus has a sword-like snout, relatively short gill rakers, and a protrusible mouth. Their food appears to be other fishes. Some specimens of *Psephurus* have been reported to 7 meters in length, but authentic records go only to 4 meters.

Polyodon has a broad, paddle-shaped snout, long gill rakers, and a nonprotrusible mouth. Crustaceans and other plankton are strained from the water by means of the long gill rakers. The paddle does not seem to function in food gathering, unless its role is sensory. *Polyodon* reaches a length of about 2 meters and weight of 76 kg, and spawns in swift rivers over gravel bars in the spring. Growth is more rapid than that of sturgeons. Although the range of the paddlefishes has been reduced by man's activities, the species has become numerous in several reservoirs in the Mississippi drainage and forms the basis of a popular fishery based on snagging with treble hooks.

Holostei

This diverse group of about five orders (some extinct) differs from the Chondrostei in that each supporting element of the fins bears a single fin ray. Most fishes of the group, including the extant gars, have rhomboid-shaped ganoid scales, but advanced forms such as the living bowfin have cycloid scales that lack ganoin. Holosteans are not greatly different from the primitive teleosts, into which they appear to grade through the extinct order Pholidiphoriformes, but have a more complicated lower jaw and different caudal fin support. The caudal fin is abbreviate heterocercal with a tendency to evolve toward homocercal. In living forms a spiral valve is present in the intestine and the gas bladder is cellular. Most of the hundred or so known genera of holosteans flourished in the Jurassic, after replacing chondrosteans; they then declined during the Cretaceous evolution of teleosts, leaving only two living orders.

Order Amiiformes. The bowfin (*Amia calva*) of the family Amiidae is the only living representative of this once widespread diverse order (Fig. 7–5*A*). *Amia* is found in eastern North America from the

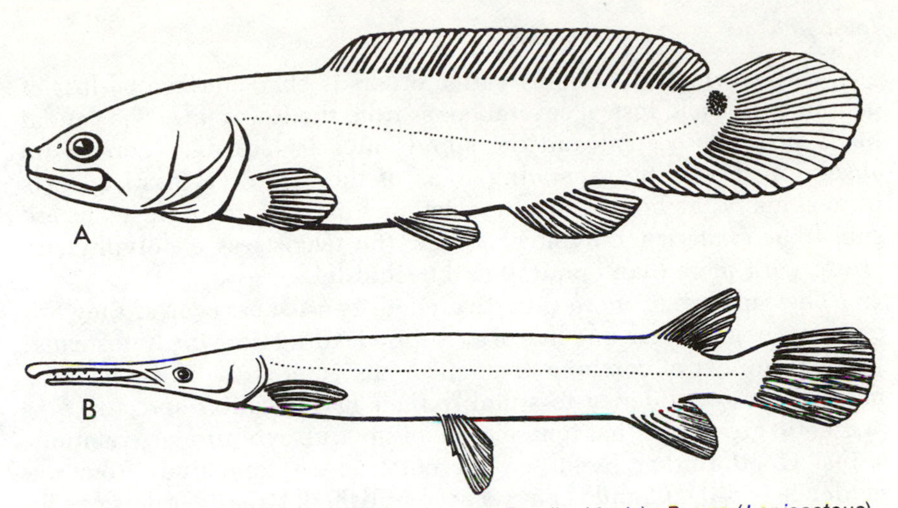

FIGURE 7–5. Representatives of Holostei. *A,* Bowfin (*Amia*); *B,* gar (*Lepisosteus*).

Great Lakes south, and appears to prefer warm shallow water. Males guard the nest — a depression made in aquatic vegetation — and the young, sometimes harboring the young in the mouth. The gas bladder is septate and can function in aerial respiration, allowing the species to inhabit stagnant waters having a low oxygen concentration. The bowfin is found only in fresh water. Females may reach 1 meter in length, but 60 cm is a more common length, with males being somewhat smaller than the females. Despite its interesting habits and the scientific interest in the species as a "living fossil," the mediocre flesh and predatory habits of the bowfin have led many conservation agencies to regard them as unwanted.

Order Lepisosteiformes. The gars form one living family, the Lepisosteidae, found in eastern North America from the Great Lakes region to Costa Rica with one species reaching Cuba. These are elongate fishes with the body covered by heavy ganoid scales and the head with equally hard bone (Fig. 7–5*B*). Both jaws are elongate and are armed with several rows of strong sharp teeth. The dorsal and anal fins are set far back and the caudal is abbreviate heterocercal. Gars differ from all other living fishes in having opisthocoelous vertebrae; these vertebrae are concave posteriorly and convex anteriorly.

Gars live in quiet, often weedy, waters and can usually be observed lying almost motionless near the surface. They utilize atmospheric oxygen in addition to that obtained through the gills. Prey is captured by means of a rapid lateral strike with the jaws, often after waiting for it to come into range. Most of the seven or eight species of gars inhabit fresh water but the alligator gar, *Lepisosteus spatula* and its close relatives may enter salt water. These are the largest of the gars, reaching 3 meters in length. Gars have little economic value — they are used as food to some extent in the southern United States and Mexico, and the ganoid scales are sometimes used as souvenirs and as ornaments.

Teleostei

This assemblage of many varied orders is characterized by loss of ganoid scales and loss of several bones from the lower jaw. The caudal fin is not heterocercal and the spiral valve is lacking. There is no absolutely clear line separating some of the primitive fossil teleosts from some of the holosteans; the relationships with the holosteans are such that modern ichthyologists view the teleosts as a polyphyletic group with more than one link to the Holostei.

The "lower" or more primitive teleostean fishes, even if they are products of more than one line of evolution leading from the holosteans, share a number of common anatomical characteristics. These features may lead to a tendency to simplify their classification and arrive at synthetic groupings that ignore some of the true evolutionary relationships. On the other hand, sorting out true phyletic lines from the evidence is difficult and is not yet accomplished. Usually orders can be well defined, but the affinities among orders and their placement along real or hypothetical phyletic lineages remains open to question.

The problems are much the same with the "higher" or more derived bony fishes as well as with the intermediate groups. Although many features are held in common, anatomical specialization may obscure true relationships. This is evidenced by the placement of some groups into the intermediate fishes by some scientists, and into the specialized derivatives of the perch-like fishes by others. The following discussion is based upon a rather conservative view of teleost relationships, and outlines the changes in certain characteristics progressing from the lower to the higher forms.

Most lower bony fishes have only soft rays in the fins and are sometimes placed into an evolutionary grade called Malacopterygii. This contrasts with the higher fishes, or Acanthopterygii, with both soft rays and spines in the fins. Lower fishes tend to have only one dorsal fin, but may have a small fleshy "adipose" fin set on the caudal peduncle. The dorsal is usually set about midway along the back, or even behind the midpoint. Higher fishes may have one or more dorsal fins, often originating rather far forward.

Pectoral fins of lower fishes are typically set low on the body, with the base slanting downward and backward, being nearly horizontal in some. There are certain limitations of movement inherent in this positioning. These fins must be useful in guiding and braking, but perhaps are not as versatile as the pectorals of higher fishes that are set higher on the sides with a base tending more to the vertical. The latter position might be more suitable for locomotion as well as for maneuvering and braking. Several modifications of the pectoral girdle are involved in the positioning of the fins, including changes in relative sizes of parts and twisting of the axis of the girdle, but the consistent difference between the lower and higher groups is the presence of the mesocoracoid bone in the lower orders. This bone forms an arch with the scapula and coracoid and braces the fin base at an angle. The loss of the bone in higher groups allows the base of the pectoral to be aligned with the vertical axis of the girdle.

Pelvic fins in lower bony fishes are placed rather far back on the belly and are said to be abdominal. The supporting bones, the reduced pelvic girdle, are not firmly connected to any other bony structure, being situated in the musculature of the body wall. Higher fishes have pelvics placed far forward on the breast region below the pectorals in what is called a thoracic position. The bony supports are attached to the lower portion of the pectoral girdle. A few groups of fishes have pelvics in an intermediate position with no connection to the pectoral girdle, and some maintain the intermediate position and a ligamentous connection. These are variously called subabdominal or subthoracic pelvic fins. In a few instances the pelvics are moved forward of the pectoral base into a jugular position, usually considered to be an advanced characteristic.

There is an additional difference in the pelvic fins of lower and higher bony fishes. The pelvics of the former may have numerous rays and those of the latter a reduced number, typically one spine and five soft rays in the perch-like fishes. There may be reductions in numbers of rays in specialized offshoots of lower, middle, and higher groups — or even loss of pelvics — but the trend of reduction of rays is evidently a valid evolutionary trend.

Lower fishes differ from the higher forms in the structure of the upper jaw. The outer edge of the jaw consists of both the premaxilla and the maxilla in the primitive fishes. The premaxillae are seldom protractile, and both the bones may bear teeth. On the other hand, the derived forms have only the premaxillae forming the border of the mouth. The maxillae are excluded from the gape, are situated dorsally to the premaxillae, and never bear teeth. Higher fishes may have a protrusible upper jaw with long ascending processes of the premaxillae sliding along the anterior part of the skull.

Another important characteristic of the more primitive teleosts is the retention of an open pneumatic duct from the alimentary canal to the gas bladder. This condition is known as physostomous and contrasts with the physoclistous condition of higher teleosts, in which the duct is absent or closed. An additional internal contrast is in the character of the pancreas, which is often a discrete gland in the lower forms but is most usually combined with the liver in higher fishes.

The orbitosphenoid bone, which forms a large part of the interorbital septum in many lower teleosts, is lacking in most of the middle and higher fishes. Scales of lower teleosts are usually cycloid, while those of the higher are usually ctenoid.

The foregoing are some of the generally recognized differences between lower and higher groups of teleosts. (There are others that will be mentioned later in the discussions of orders.) Generally, these differences are those that place the higher fishes farther from the generalized vertebrate body plan than the lower ones. These characteristics will occur in various combinations in the following material, and in some instances the difficulty in following well defined lines of relationships will be evident.

Layering the teleosts into lower, middle, and higher strata is an oversimplification that obscures the apparently separate lines leading from the Holostei. Therefore, mention is made here of a widely accepted scheme of classification proposed by P. H. Greenwood et al. in 1966 and refined considerably since that time. In that interpretation, the teleosts are seen to fall into four cohorts and to represent at least two phylogenetic lines. The cohort Archaeophylaces contains the bony tongues, mooneyes, featherbacks, and mormyroids, comprising the superorder Osteoglossomorpha. These are all soft-rayed "lower" fishes and are related to the cohort Clupeocephala, represented by the superorder Clupeomorpha, which contains the herrings and close relatives. The cohort Taeniopaedia, considered to be comprised of the single superorder Elopomorpha, contains the tarpons and allies, and the eels and allied groups, including the spiny notacanths. Therefore soft-rayed fishes and some that are sometimes thought of as "middle" fishes are included. The related cohort Euteleostei includes the remainder of the teleosts — lower, middle, and higher — in the following superorders: Protacanthopterygii, the salmons, smelts, and relatives; Ostariophysi, the carps, catfishes, characins, and allied groups; Scopelomorpha, the lanternfishes and relatives; and Paracanthopterygii, a controversial grouping of middle and higher fishes forming a line somewhat parallel to the superorder Acanthopterygii, which contains the bulk of the spiny-rayed fishes and close relatives. In the following list of orders, their relationships to the cohorts and superorders of the Greenwood et al. classification will be pointed out.

†Order Leptolepiformes. This is an extinct order known from the Upper Triassic to the Middle Cretaceous. It bears a strong resemblance to Holosteans in some features such as the cephalic lateral line arrangement and the ganoid-like covering of some dermal bones and scales. However, the caudal skeleton and the essentially teleostean preopercle, among other features, has caused students of the group to place it among the Teleostei. The order has structures that resemble both the clupeiform and elopiform lineages.

Order Elopiformes. This order embraces two suborders — the Elopoidei, which includes the tarpons and close relatives, and the Albuloidei, including the bonefishes and allied forms. They are known from Jurassic to the present, with special abundance in the Cretaceous. These fishes show some relationships to the herring-like fishes and are sometimes placed with them in an order called Isospondyli. On the other hand, they share the larval stage known as the leptocephalus with the eel-like fishes and are considered by some ichthyologists to be part of an evolutionary line (Taenopaedia, superorder Elopomorpha) that includes the eels, notacanths, and halosaurs. The leptocephalus larvae of various fishes range in shape from ribbon-like to that of a willow leaf. The flesh of the larvae is almost colorless and translucent and the teeth are prominent.

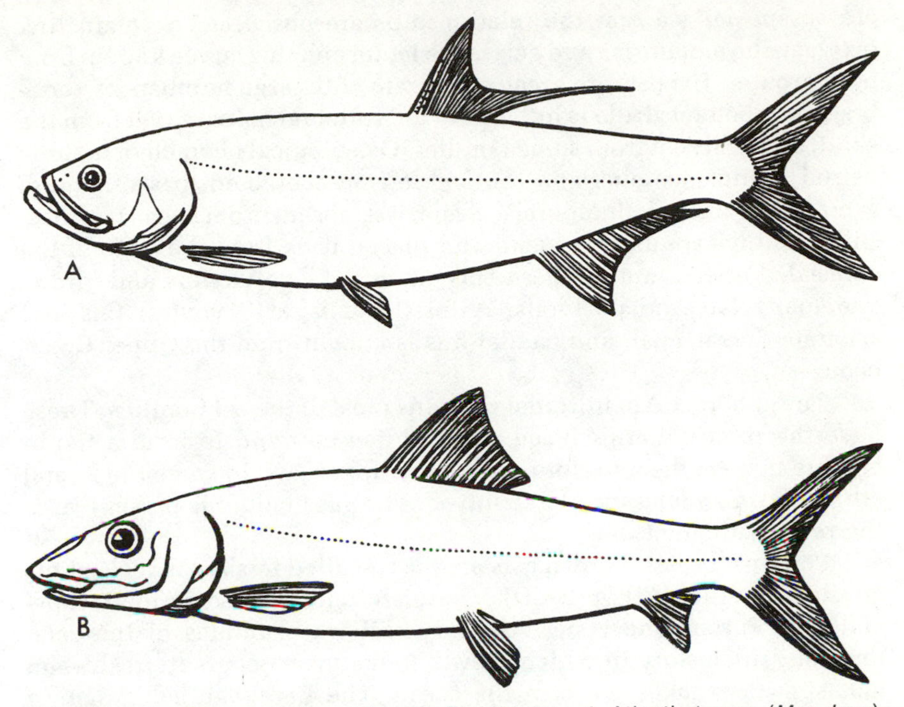

FIGURE 7–6. Representatives of order Elopiformes. *A,* Atlantic tarpon (*Megalops*); *B,* bonefish (*Albula*).

In addition to the general primitive features of a lower teleost, the elopiforms have a conus arteriosus with two rows of valves, a gular plate, and a commissure of the cephalic sensory canal system in the ethmoidal region. The number of branchiostegals ranges from 6 to 36, and the pelvics may have from 10 to 18 rays.

The suborder Elopoidei contains the tarpons, ladyfish and machete of the family Elopidae (Fig. 7–6*A*). These are large, big-scaled fishes of warm seas. Not prized as food, they are sought mainly for sport. The ladyfish or tenpounder, *Elops saurus,* reaches somewhat less than 1 meter in length, a weight of 14 kg, and is circumtropical. The Atlantic tarpon, *Megalops atlanticus,* may reach 2.6 meters in length and 150 kg, but the Pacific form, *M. cyprinoides,* usually does not exceed 1 meter in length. The tarpons contrast somewhat with *Elops* because they possess a conus arteriosus, lack pseudobranchiae, and have a connection of the gas bladder to the otic region of the skull

The suborder Albuloidei contains the family Albulidae, bonefishes, and the deepsea bonefish. The bonefish, *Albula vulpes* (Fig. 7–6*B*), is a prized game fish of tropical and subtropical waters around the world. *Dixonina nemoptera* is restricted to the West Indies. *Pterothrissus* lives in deep water off Japan.

Order Anguilliformes (Apodes). This is a large order of chiefly marine fishes comprised of up to 26 families. These are the true eels and are thought to have had their origin in an elopiform ancestor, but which are so specialized that the relationships are obscure. The main link between the elopiforms and eels is the leptocephalus larvae known from both groups. The eels are greatly elongate with large numbers of vertebrae, the pectoral girdle is hung from the vertebral column, not from the skull, and is absent from some families. Osteological characters include paired orbitosphenoids and the lack of mesocoracoid, basisphenoid, symplectic, and posttemporals. Scales are absent from most families. Gill openings are usually small and placed back from the edge of the concealed operculum. Modern eels all lack the pelvic fins and girdle, and many have no pectorals. A fossil genus with ventral fins and separate dorsal, anal, and caudal fins is known from the Upper Cretaceous.

The suborder Anguilloidei contains most of the eel families. These have the premaxillaries fused with the mesethmoid to form a tooth-bearing bone at the anterior point of the upper jaw. In some the lateral ethmoids and vomer may be involved. The gas bladder is present, as is the pneumatic duct.

Perhaps the best known eels are the so-called freshwater eels of the family Anguillidae (Fig. 2–8*D*). These are widely used as food, especially in Europe and Asia, and are excellent examples of the catadromous life history in which growth to maturity occurs in freshwater and spawning takes place in the ocean. The European eel, *Anguilla anguilla,* and the American eel, *A. rostrata,* spawn in adjacent or overlapping areas of the Sargasso Sea. The fragile leptocephalus larvae then drift for months or years before becoming glass eels, and then become elvers in fresh water where they feed and grow. In the Orient, especially Taiwan and Japan, the glass eels of the Japanese eel, *A. japonica,* are collected by the millions and transferred to culture stations where they are fed heavily until they reach marketable size.

Most of the remaining typical eel families live in warm marine waters, but some are found in the deep sea. Some families of interest include: Muraenidae, the morays; Congridae, the congers; Ophichthyidae, the snake-eels, and Simenchelidae, the snubnosed or parasitic eels, which live in deep water and feed by burrowing into, and feeding upon, larger fishes. This habit appears to be matched only by the hagfishes.

The snipe eels (Fig. 2–8*C*) (Nemichthyidae, Serrivomeridae, and Cyemidae) are characterized by elongate jaws that curve away from each other toward the tips. They lack the supraoccipital bone and the supracleithrum, and are sometimes given subordinal rank. These all live in deep water.

Suborder Saccopharyngoidei, the gulper eels, has usually been considered as constituting a separate order, called either Saccopharyngiformes or Lyomeri, but recently ichthyologists have combined it with the remainder of the eels. The three families that make up this suborder are all deepsea in habitat and are greatly specialized, in part by degeneration. They lack gas bladder, scales, opercular bones, branchiostegals, and ribs. The caudal and pelvic fins are reduced or

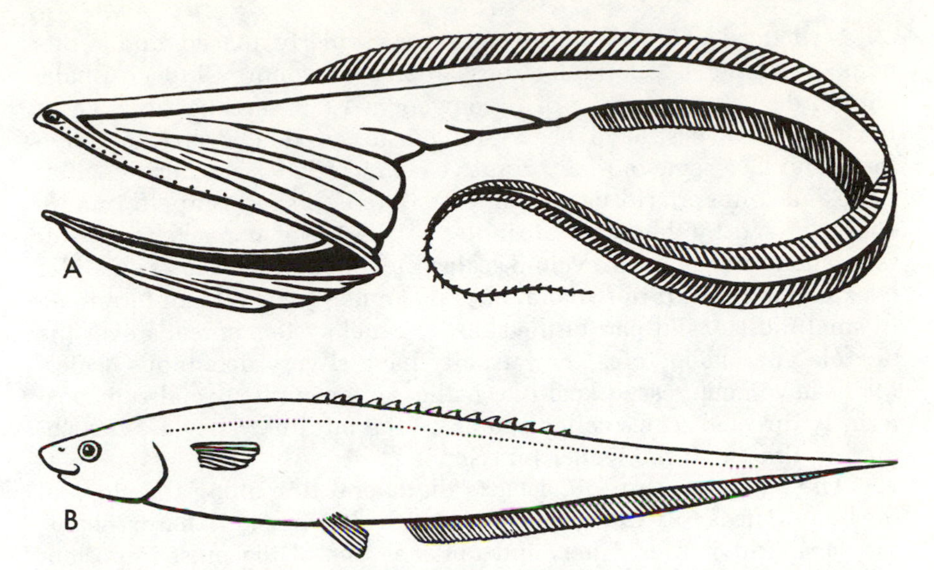

FIGURE 7–7. Representatives of: *A*, order Anguilliformes, gulper (*Eurypharynx*); *B*, order Notacanthiformes, spiny-backed eel (*Notacanthus*).

absent and the gill arches are degenerate. These are deepwater eels that may reach nearly 2 meters in length (Fig. 7–7*A*).

The Eurypharyngidae contains species with tremendous mouths and small teeth. The stomach is nondistensible. These fishes apparently feed on bathypelagic crustacea. This family is unique in having five complete gills. The Saccopharyngidae contains fishes with a slightly smaller mouth than the former family, but they have larger teeth, a more capacious stomach, and feed upon fishes. The remaining family, Monognathidae, lacks the typical upper jaw bones — maxillaries and premaxillaries — but its members nonetheless are predators on other fishes.

Order Notacanthiformes (Heteromi and Halosauriformes or Lyopomi). This order contains the spiny or spiny-backed eels (Fig. 7–7*B*) and halosaurs, both deepsea groups known from as far back as the Upper Cretaceous. They resemble the true eels in being elongate and lacking the mesocoracoid and basisphenoid. The pectoral girdle is not connected strongly to the skull. Some members at least have leptocephalus larvae. Although the pelvic fins are abdominal and the scales cycloid, these are physoclistic fishes that lack the orbitosphenoid. The tail tapers to a point, without a caudal fin.

The Halosauridae are soft-rayed and have the upper border of the mouth made up of the premaxillaries and maxillaries. Photophores are present in some. The spiny-backed eels of Notacanthidae and Lipogenyidae seem more advanced in that they have spines in the fins and the upper jaw bordered by the premaxillaries only.

Order Clupeiformes. Following several recent classifications, this order is here considered to embrace only the herrings and close allies, instead of the large assemblages formerly placed under this name or under Isospondyli. These represent cohort Clupeocephala, superorder Clupeomorpha of Greenwood's 1975 classification. These herring-like fishes have been thought to be related to the elopiforms but lack some of their primitive characteristics such as the gular plate and conus arteriosus encountered in those fishes. Clupeiforms are characterized by the usual attributes of the lower bony fishes — soft rays, abdominal pelvics, cycloid scales, etc. In addition, the gas bladder is extended forward in two branches that enter the skull and terminate in small bullae, and part of the sensory canal system spreads over the opercle and subopercle. They usually have silvery deciduous scales. Many have compressed keel-like bellies, often with specialized, posteriorly directed scales called scutes on the midline. They are distributed in time from the Upper Jurassic.

The suborder Clupeoidei lacks the lateral line along the sides of the body. The family Clupeidae includes the herrings, pilchards, shads, sardines, and similar fishes, and ranks as one of the most important groups of fishes. Many species occur in dense schools so that they are subject to mass capture, and their oily flesh makes them the object of fisheries all over the world. The uses to which they are put are numerous — from food for man and his domestic animals to fertilizers and oils. The annual world harvest of clupeids has been estimated at 20 million metric tons.

Some important members of the family are the herring, *Clupea harengus*, which has subspecies in both the north Atlantic and north Pacific, the menhaden, *Brevoortia* sp., of the western Atlantic, and the American shad, *Alosa sapidissima*, which is sought for both food and sport. Others are *Sardinops sagax*, the Pacific sardine, once extremely important on the California coast but now depleted, and the sprat, *Sprattus sprattus*, of the North and Baltic seas. In all there are over 170 species of herrings in about 50 genera. They are all small fishes, seldom exceeding 0.5 meter in length (Fig. 7–8).

Ecologically herrings are important as converters of plankton into fish flesh, so that they then form a great food resource for the large pelagic predators. Their life histories vary. Some spawn pelagically, while others spawn on seaweeds or other substrate. Some are anadromous, and some live in fresh water.

The family Engraulidae contains the anchovies, small silvery fishes with a rather rounded body cross section and a long snout and maxillary. The snout projects well beyond the lower jaw. Like herrings, anchovies are the objects of great fisheries. One of the greatest is the anchovetta fishery off Peru in the region of the Humboldt current, where tremendous biological production occurs. The anchovetta *Engraulis ringens* forms the basis for a fishery that produces a harvest of up to 13 million metric tons in some years, in addition to the tremendous tonnage eaten by the guano birds. Production and availability of the anchovetta depends on the ocean currents. In years when the wind patterns are wrong the fishermen, and the birds, go nearly fishless.

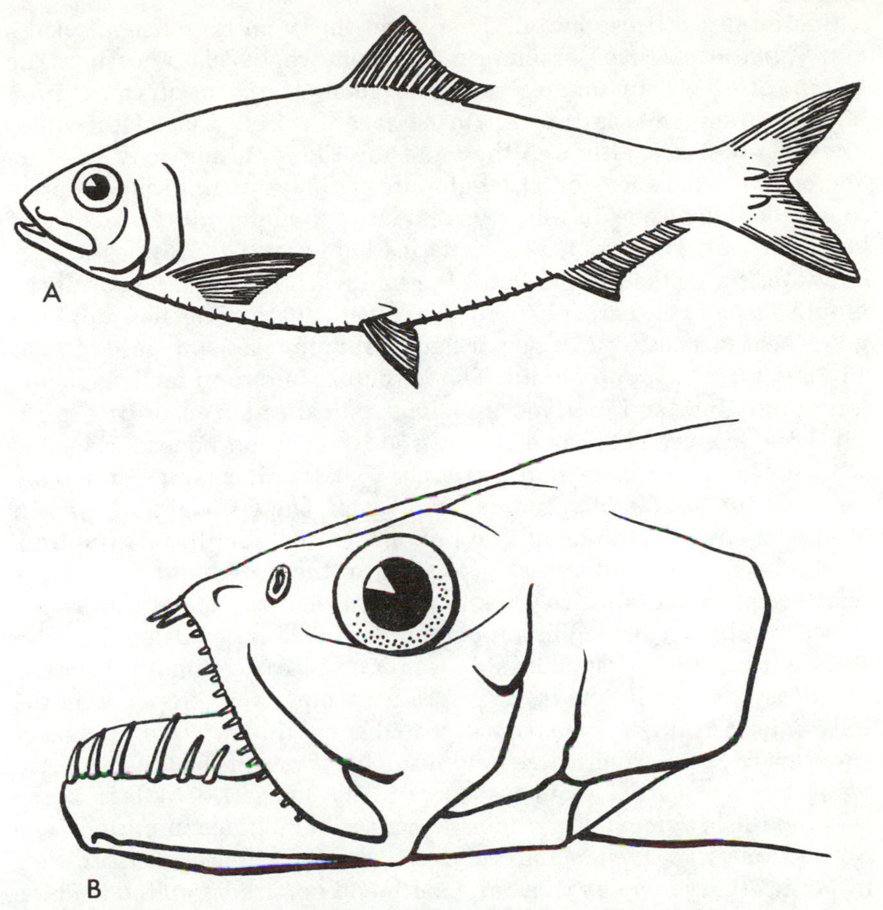

FIGURE 7–8. Representatives of order Clupeiformes. *A*, Shad (*Alosa*); *B*, head of wolf herring (*Chirocentrus*).

Anchovies are usually less than 25 cm long, and most species live in tropical or subtropical waters, There are 15 to 20 genera and over 100 species.

The family Chirocentridae contains the wolf herring or dorab, *Chirocentrus dorab*, which differs in several respects from typical herrings. It may reach a length of 3.5 meters and has numerous large canine teeth (Fig. 7–8*B*). The gas bladder is septate and there is a bony appendage in the axil of the pectoral fin. Wolf herrings are found in the Indo-Pacific and are used as food in some areas. Large specimens are said to be dangerous to fishermen.

The suborder Denticipitoidei contains one family, Denticipitidae, the tooth-head or denticle herring, in which a complete lateral line is present, and the skull bones bear dermal denticles. The only species, *Denticeps clupeoides*, is confined to West Africa.

Order Osteoglossiformes. Basically, the anatomy of the osteoglossiforms or bonytongues is similar to that of the other lower bony fishes, and they are often placed in a more inclusive interpretation of

Clupeiformes or Isospondyli. They form the cohort Archaeophylaces and Superorder Osteoglossomorpha of Greenwood's classification. The origin of the bonytongues and the related Mormyriformes from holostean ancestors is thought to be separate from that of the other lower bony fishes, although they and the Clupeiformes may have an ancient common ancestor. Osteoglossiforms show a similarity to many Mesozoic bony fishes in the arrangement of the bite, which, instead of being primarily between the dentaries below and the upper jaw, is between the toothed tongue and the toothed bones of the roof of the mouth. Usually the parasphenoid is involved, but in some the entopterygoids bear teeth as well. A peg or flange from the parasphenoid forms a support for the entopterygoid. That structure is rare in teleosts, being found only here and in alepocephaloids (slickheads), according to Dr. William Gosline. There is a pair of rod-like tendon bones associated with the base of the second gill arch. Usually some soft connection between the gas bladder and ear is present. The Osteoglossiformes is thought to be a very ancient group. Some extinct families dating from the Jurassic are considered as belonging in the order, and some close relatives of extant families are known from the Cretaceous.

The suborder Osteoglossoidei includes the families Osteoglossidae and Pantodontidae. Osteoglossids are found in Africa, South America, southeast Asia, and Australia, where a species of *Scleropages* is the only strictly freshwater teleost native to that continent. The arapaima or pirarucu of South America, *Arapaima gigas*, is one of the longest freshwater fishes, reaching about 3 meters in length (Fig. 7–9*A*). It is prized as a food fish. *Osteoglossum bicirrhosum* is sometimes imported from South America as an aquarium fish. The African *Heterotis* differs from the other genera by its small mouth and nonpredatory food habits. *Heterotis* and *Arapaima* have cellular air bladders that apparently make the use of atmospheric oxygen for respiration possible. *Heterotis* has a special suprabranchial respiratory organ that aids in that function.

Heterotis and *Arapaima* build large nests in shallow water and give protection to their young following hatching. The young of *Heterotis* are equipped with external gills. *Scleropages formosus* of Thailand and the Malay region is apparently a mouth brooder.

The family Pantodontidae contains one species, *Pantodon buchholzi* of Africa, which reaches a maximum size of less than 15 cm (Fig. 2–11*D*). This peculiar little fish has subthoracic pelvic fins that are made up of long, separate, rather filamentous rays. The pectorals are relatively large and expanded, and are thought to function as gliding planes during the long leaps made by this species over the surface of the water. The common names butterflyfish and freshwater flyingfish are derived from this habit. The male has a modified anal fin that is thought to function in internal fertilization. The eggs float at the surface as do the fry after hatching.

The suborder Notopteroidei includes the mooneyes, Hiodontidae, of North America, and the featherbacks, Notopteridae, of southeast Asia, the Indo-Malayan Archipelago, and Africa.

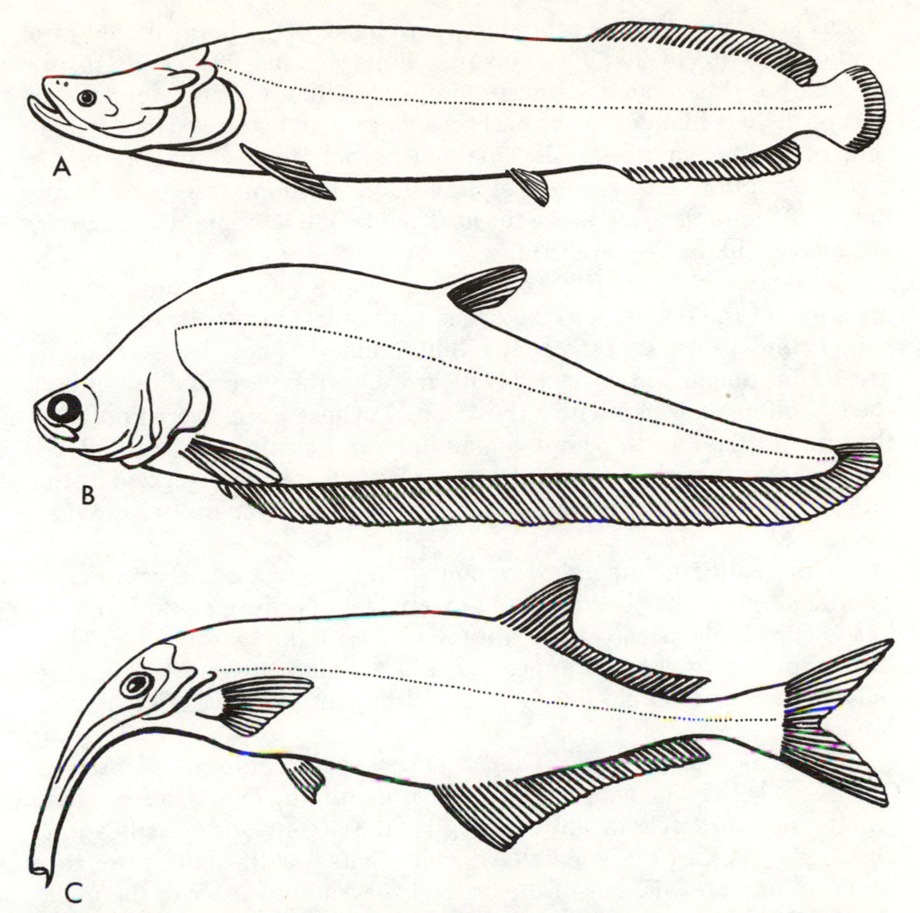

FIGURE 7–9. Representatives of orders Osteoglossiformes–*A*, arapaima (*Arapaima*); *B*, featherback (*Notopterus*)–and Mormyriformes; *C*, elephant-snout fish (*Mormyrus*).

Hiodontids, including the mooneye, *Hiodon tergisius* and the goldeye, *H. alosoides*, are silvery, herring-like fishes of the northern section of central and eastern North America. They rarely reach as much as 45 cm long and have limited use as food and sport fishes.

Featherbacks, mostly of the genus *Notopterus*, are rather strange-appearing fishes with a long anal fin beginning in the anterior third of the body and extending to the caudal, with which it is confluent (Fig. 7–9*B*). A very small dorsal fin is placed about midway down the back, except in the African genus, *Xenomystis*. Some have a rather humped back and a concave dorsal head profile. Several species are used as food.

Order Mormyriformes. This is the order of the elephantfishes or mormyrs of the family Mormyridae and their close relatives, the Gymnarchidae. All members of the order are from fresh waters of Africa. Mormyriforms are closely related to the osteoglossiforms and are sometimes placed with them in a suborder.

In this order the operculum is small and covered by skin. Electric organs are present in the caudal region. The skull is characterized by a lateral foramen covered by a flat supratemporal bone. The cavity inside the skull at this foramen is occupied by a vesicle which originates as part of the gas bladder in the early developmental stages. The cerebellum of the mormyriforms is enlarged as part of the adaptation for electroreception and is, relatively, the largest among the lower vertebrates. The premaxillaries are fused. They lack the opisthotic, angular, entopterygoid, and symplectic.

The family Mormyridae contains about 10 genera, among which are some of the oddest-looking fishes known from fresh water, the so-called elephantfishes of the genus *Gnathonemus*. These have the snout greatly produced and curved downward with the very small mouth at the tip, often equipped with a thick barbel. These fishes were known to the ancient Egyptians, who depicted them in paintings and sculpture. Many species have a more normal shape, but all are characterized by the narrow caudal peduncle (Fig. 7–9C). The family is generally sought as food.

The family Gymnarchidae contains one species, *Gymnarchus niloticus*, which has been studied extensively because of its ability to locate objects electrically, an ability shared with the Mormyridae. They differ from mormyrids in appearance, being elongate and lacking the pelvic, anal, and caudal fins. The dorsal fin runs most of the length of the back.

Order Salmoniformes. This order begins the cohort Euteleostei of Greenwood et al. The fishes considered as belonging to this order represent the superorder Protacanthopterygii of the Greenwood classification. This order typifies the generalized bony fishes, with many primitive structural features. Fishes of the order appear to be closer to the main stem of fish phylogeny than to the foregoing orders. Salmoniforms are soft-rayed and mostly physostomous, although the air bladder may be absent in some. There are no connections of the air bladder with the ear either through a chain of ossicles or otherwise. An adipose fin is often present. The order is often considered a basal one from which several of the higher groups could have evolved. Salmoniforms are known from the Cretaceous.

Suborder Argentinoidei includes over 100 species of deepsea fishes, arranged in from five to eight families. The caudal fin is forked, the gas bladder is physoclistic or absent, and an epibranchial organ for consolidation of small prey is present. Photophores are present in some species. Three families, Argentinidae or argentines, Bathylagidae or deep-sea smelts, and Opisthoproctidae or barreleyes, share the following characteristics: silvery color; no teeth on maxilla; adipose fin usually present; and dorsal fin close to middle of body. The Alepocephalidae, slickheads, and the Searsiidae are sometimes placed in a separate suborder. These dark colored fishes have the dorsal set over the anal and lack the adipose fin and gas bladder.

The suborder Stomiatoidei is characterized by rows of photophores along the body. Premaxillaries are present and functional. These are dark-colored deepsea fishes. Families are: Gonostomatidae, bristlemouths; Sternoptychidae, hatchetfishes; Chauliodontidae, viperfishes (Fig. 7–10*A*); Stomiatidae, scaly dragonfishes; Astronesthidae, snaggletooths; Melanostomiatidae, scaleless dragonfishes; Idiacanthidae, stalkeyes; and Malacosteidae, loosejaws.

As the common names of the families indicate, these fishes are often of bizarre shape and predatory habit. Some have extremely large teeth for their size. Most of the families, except for the first two, have barbels developed on the chin. Most are quite small fishes, being less than 100 mm long, although some reach about 150 mm.

The suborder Salmonoidei includes the salmons, trouts, and their relatives. All have an adipose fin. Oviducts are reduced or absent, and all retain a large proportion of cartilage in the cranium. Many of these are prized as food or sport fishes.

The family Salmonidae includes the trouts, salmons, whitefishes, and graylings, all native to the Northern Hemisphere. Trout and Atlantic salmon are in the genus *Salmo.* Such species as *S. gairdneri,* the rainbow or steelhead trout of North America, and *S. trutta,* the brown trout of Europe, are currently sought primarily for sport. Both have been transplanted to temperate parts of the Southern Hemisphere. The Atlantic salmon *S. salar* is still sought for commercial purposes in some parts of the North Atlantic. The first two species as well as other trout usually live in fresh water, but may be anadromous, feeding and growing in the ocean but spawning and spending a portion of their early life in fresh water. Members of this genus do not usually die following spawning.

The genus *Oncorhynchus* contains the Pacific salmons. Six anadromous species are found in the North Pacific, five of these along the coast of North America and all six in Asia. All normally die following spawning.

The Chinook salmon *O. tshawytscha* is the largest of the genus, reaching a maximum weight of about 57 kg (Fig. 7–10*B*). Its value as a commercial fish is great, and it supports a tremendous sport fishery in addition. Once it ascended the Columbia River and its tributaries for a thousand miles or more, but now hydroelectric and irrigation dams have cut off its spawning areas in the upper reaches of the river system. Hatcheries have been used with moderate success to maintain certain runs.

The coho salmon *O. kisutch* is also an excellent commercial fish, bringing slightly less per pound than the chinook. It is more numerous than the chinook in most areas, so it supports a larger sport fishery. The coho is well adapted to short coastal streams and lower tributaries of large river systems for spawning purposes, and thus has not been affected as severely as the chinook by dam construction. In addition the ease with which it can be reared in modern hatcheries makes possible the support of large runs in depleted streams within the natural range as well as in suitable streams outside the native range.

The introduction of the coho into the Great Lakes has provided a remarkable sport fishery, especially in Michigan. The salmon find ample food in the large stocks of alewife, *Alosa pseudoharengus,* which in recent years have become quite abundant in the lakes.

The Pacific salmon with the richest flesh is the sockeye salmon,

Oncorhynchus nerka. It forms the basis for a short intensive commercial fishery in Alaskan waters, especially in Bristol Bay, and good stocks are found in various Canadian localities. Because this salmon normally spends its early life in lakes, the best runs are into rivers with large lake systems. The great runs that entered the Fraser of Canada were once endangered and diminished by a man-caused rock slide at Hells Gate. Construction of passage facilities and excellent management have done much to restore the runs. The landlocked form of the sockeye, known as kokanee, is valuable as a game fish in lakes of the Pacific Northwest. The sockeye, unlike the piscivorous coho and chinook, depends heavily on pelagic crustaceans for food.

Chum salmon, *O. keta*, and pink salmon, *O. gorbuscha*, are sought as commercial fishes, and have only minor importance as game fishes. In most parts of their ranges these two species ascend rivers only a short distance to spawn, and the young drift to the ocean immediately after emerging from the gravel nests or redds. In the other species the young spend from a few months to two years in fresh water. The pink salmon is remarkable in that the life cycle is almost invariably two years, and some streams have heavy runs in odd-numbered years and others experience an abundance in even-numbered years. The odd-year and even-year populations have evolved some distinctions during thousands of years of chronological isolation.

The chars are placed in the genus *Salvelinus*, which differs in coloration from the trout and salmon in having light spots on a darker background instead of dark spots on a light background. These are freshwater and anadromous fishes found in cold waters of the Northern Hemisphere. In common usage they are usually called trout. Examples of the genus are: the Dolly Varden, *S. malma;* the brook trout, *S. fontinalis;* and the lake trout, *S. namaycush*, which supported a large commercial fishery in the Great Lakes prior to the spread of the sea lamprey. Some other salmonid genera are: *Coregonus*, the whitefishes; *Stenodus*, the inconnu; *Thymallus*, the graylings; and *Hucho*, the huchens of Europe and Asia. The latter are large voracious fishes of cold rivers.

The ayu of Japan, China, and Korea is placed in the family Plecoglossidae. This species *(Plecoglossus altivelis)* is quite trout-like, but has a row of large, rather square-cut, chisel-like teeth on the maxillaries and dentaries. The ayu's food is diatoms and associated organisms growing on the rocks in river bottoms. The ayu is known also as "sweetfish" and is sought as a delicacy in Japan, where it formerly was the object of the cormorant fishery. Now it is usually captured by snagging. One of the fishing methods is called "Tomozuri," involving the use of a previously captured live ayu to bring others of the territorial species near the snagging hooks. The ayu is an annual fish, without overlapping generations. They spawn in lower parts of rivers in fall and early winter and the young are carried to sea, returning to the streams in late winter and early spring. Their growth is rapid, so that they are large enough to sustain a fishery by early summer.

The family Osmeridae, sometimes placed in its own suborder, contains the smelts, small fishes found in both fresh and salt water of temperate and cold parts of the Northern Hemisphere. There are six genera and 10 species of smelts. They are slender silversided fishes of delicate flavor and are well accepted as food. Most species prefer to spawn on sand or small gravel. Some are anadromous, while others spawn on ocean beaches at high tide. Large congregations of spawners make it easy to capture them with various dip nets.

Salangidae, the icefishes or glassfishes, are considered to be closely related to Osmeridae by some ichthyologists, but some align them with the following suborder. Small, slender, transparent fishes with a flattened head, they are distributed in marine and fresh waters along the Asian coast from southern China to the Amur. Despite their small size (up to 10 cm) they are taken in commercial quantities on the spawning runs. As their general appearance is that of larval smelt, some ichthyologists believe them to be neotenical. Like the ayu, they may have only a one-year life span.

Suborder Galaxoidei, as presented here, includes three families of small fishes from the temperate waters of the southern hemisphere. The relationships of the galaxiids have been subject to several interpretations in recent years. The family Galaxiidae has been regarded as comprising its own order because, among other characteristics, the olfactory bulbs are at the olfactory organs and attached to the brain by long olfactory tracts. Otherwise, they have been set up in a superfamily or suborder with Aplochitonidae and Retropinnidae, and with or without Salangidae. Galaxoidei lack the mesocoracoid and supramaxillae; most lack pyloric caeca; an adipose fin is present in some members.

Retropinnidae, the "smelts" of Australia and New Zealand, are transparent little fishes usually less than 12 cm long. Some species are apparently confined to fresh water but others may be diadromous, descending streams to brackish water in the southern spring for spawning. Newly-hatched fish drift to sea and return upriver when 5 to 6 cm long. On the upstream migration these young are often captured, along with other species, as "whitebait" and used as food. Only the left ovary of the females is developed, and the spawning males have numerous pearl organs or nuptial tubercles. Some species reach maturity in one year. There are two genera—*Retropinna*, with five species, and *Stokellia*, confined to New Zealand, with one species.

Aplochitonidae, the southern smelts, contains the "grayling," *Prototroctes*, and whitebait, *Lovettia*, of Australia and New Zealand (Fig. 7–10C). Another species lives in the southern tip of South America. Formerly the Australian and New Zealand species were quite abundant, but now are scarce. Breeding takes place in fresh water; the young are carried to sea and return in a year to spawn. In mature males the anus and the urogenital opening are placed in the anterior part of the abdomen. The maximum size is about 30 cm in length. *Prototroctes* is sometimes placed in its own family, Prototroctidae.

Galaxiidae contains four or five genera (Fig. 7–11*A*). These are

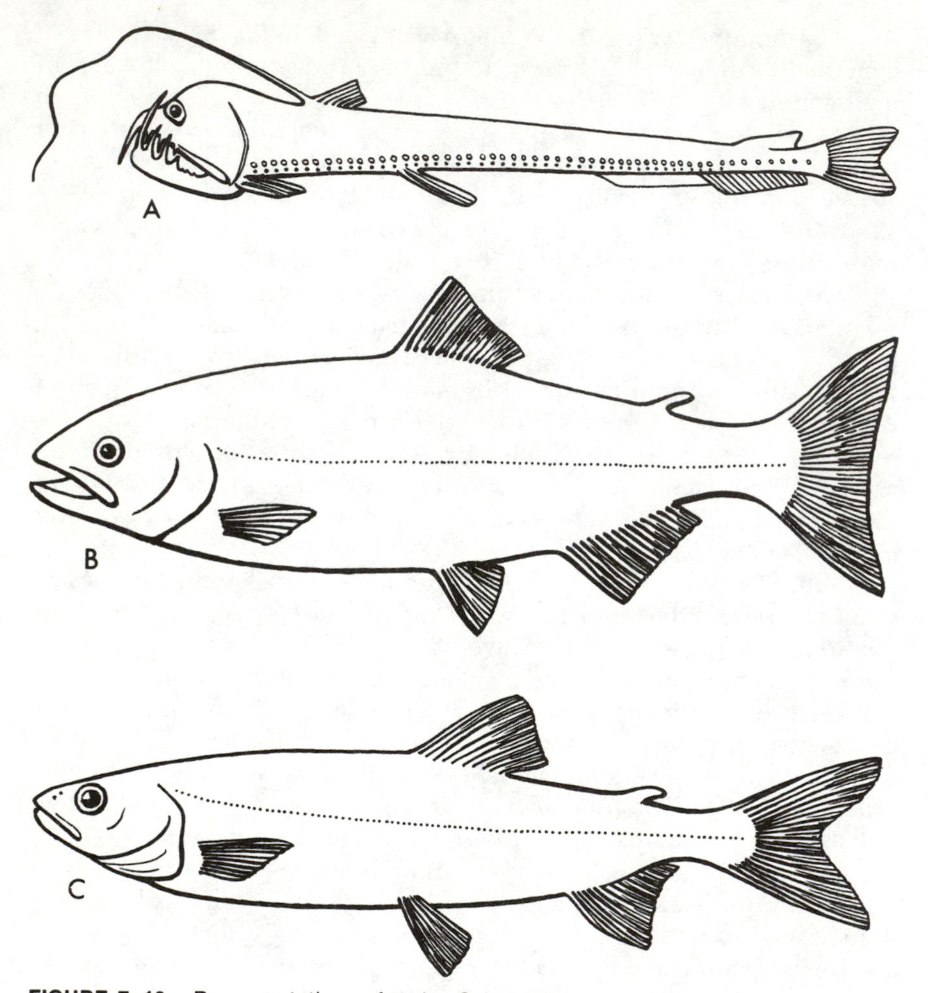

FIGURE 7–10. Representatives of order Salmoniformes. *A*, Viperfish (*Chauliodus*); *B*, Pacific salmon (*Oncorhynchus*); *C*, "New Zealand trout" (*Aplochiton*).

rather small freshwater or catadromous fishes of Australia, New Zealand, Tasmania, and the southern tips of Africa and South America.

Galaxias maculatus spawns in vegetation along the shore at high tide and leaves the eggs to incubate above the level of the sea until the next extreme high tides 2 weeks later. Eggs hatch when they are again covered by the water, and the larvae swim into the ocean, later to ascend the rivers. Landlocked populations are reported to spawn in tributaries to lakes on freshets that subside and leave the eggs on shore. The eggs hatch during a subsequent freshet. *Galaxias zebratus* of South Africa is a translucent scaleless fish of about 75 mm maximum length.

The suborder Esocoidei includes pikes (Fig. 2–8*F*), mudminnows (Fig. 7–11*B*), and the Alaska blackfish, all freshwater fishes of the Northern Hemisphere. They lack the orbitosphenoid and mesethmoid,

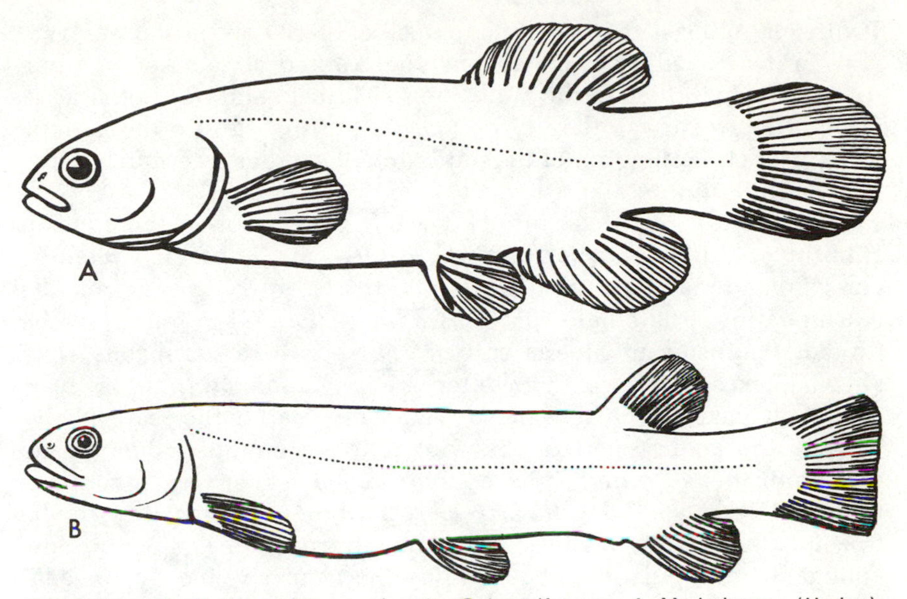

FIGURE 7–11. Representatives of order Salmoniformes. *A*, Mudminnow (*Umbra*); *B*, jollytail (*Galaxias*).

and there is no adipose fin. Although both the maxillaries and premaxillaries border the mouth, only the premaxillaries bear teeth.

The family Esocidae contains only the pikes (*Esox* spp.) which are medium-sized to large carnivorous fishes of lakes and slow rivers. Pikes are rather elongate fishes with snout and jaws produced into a long flattened mouth set with sharp teeth. The dorsal and anal fins are situated far back on the body. The northern pike, *E. lucius*, found in both North America and Eurasia, is a large fish, reaching over 25 kg in weight, and is noted for its voracious habits. Even larger, but not as widespread, is the muskellunge, *E. masquinongy*, of the upper Mississippi drainage, the Great Lakes, and some contiguous drainages. The "musky" is one of North America's greatest trophies for the game fishermen. There are some smaller species called pickerel in North America. *E. niger*, the chain pickerel, may reach about 2.25 kg while *E. americanus* is smaller. The Amur pike, *E. reicherti*, is found in Siberia.

Pikes usually spawn in the spring on flooded vegetation. Hatching occurs after about 2 weeks. The young initially feed on small organisms such as water fleas and insects, but turn to a fish diet after a few weeks. Pike are generally sought as game fishes, but in Europe and Asia they sustain commercial fisheries.

Mudminnows of the family Umbridae are found in both North America and Europe. Unlike the pike, these are very small fishes, usually less than 15 cm long. There are four species, *Umbra krameri* of central Europe, *U. limi* of central North America, *U. pygmaea* of the Atlantic Coast of North America, and *Novumbra hubbsi* of the Olympic

Peninsula in the state of Washington. All of these appear to prefer very slow water, usually bogs, stagnant ditches, and streams of low gradient. They will hide in the mud when disturbed, and *U. limi*, at least, is reported to survive the dry periods by burrowing into the bottom. The disjunct distribution of the mudminnows indicates a much wider distribution prior to glaciation.

The family Dalliidae, the Alaska blackfish, is often included in Umbridae, but its peculiar pectoral girdle, with unossified scapula, coracoid and radials, plus other peculiarities, seem to warrant separate consideration. The single species, *Dallia pectoralis*, is found on the Chukot Peninsula of Siberia and in Alaska, where it inhabits slow streams, lakes, and bogs. The winters are long and cold in these areas, so that the blackfish must be inactive for a great part of the year, usually passing the coldest part of the winter in the bottom. Some of the sphagnum bogs freeze to the bottom, and blackfish are sometimes immobilized in ice. They are extremely hardy. Apparently they can be frozen externally, but as long as their temperature does not drop low enough to crystallize the body fluids they can survive. In oxygen-deficient periods in summer they can seemingly utilize atmospheric oxygen. A peculiar edema has been noted by Mr. John Doyle of the University of Alaska in examples hibernating in water of low dissolved oxygen content. I once kept several for a few years in a refrigerated pan of water and used them in several different experiments. Although blackfish do not usually exceed 20 cm, they are useful as food for sled dogs, and are occasionally used as food by humans.

Order Myctophiformes (Scopeliformes, Iniomi). The fishes generally placed in this order are all marine and bear a resemblance to the Salmoniformes. They differ strongly from that order, however, in that they lack the mesocoracoid, the premaxillaries exclude the maxillaries from the border of the mouth, and those that have gas bladder are physoclistic. Unlike most of the salmoniforms, these fishes have well developed oviducts. Many of them possess photophores. There are several interpretations of the taxonomic position and relationships of myctophiforms. They have recently been placed in the superorder Scopelomorpha, usually following the Siluriformes. In the arrangement presented here, two suborders are accepted. Fossil myctophiforms are diverse from the Cretaceous.

The suborder Myctophoidei contains the lanternfishes and allies. The Synodontidae includes the lizardfishes, most of which are shore fishes, but which also contains some pelagic forms (Fig. 7–12*A*). Many of these have a strong odor. The family Harpadontidae contains the schooling *Harpadon nehereus*, the Bombay duck, a nearly colorless fish of the Indian Ocean; it is harvested in India where it is dried and used in the making of sauces. Aulopidae are shore fishes, usually highly colored. The Sergeant Baker of Australia, *Aulopus purpurissatus*, is a food fish of minor importance. Chlorophthalmidae, the greeneyes, Bathypteroidae, tripod fishes, and Ipnopidae are fishes of the deep ocean bottom. Neoscopelidae and Myctophidae are pelagic fishes.

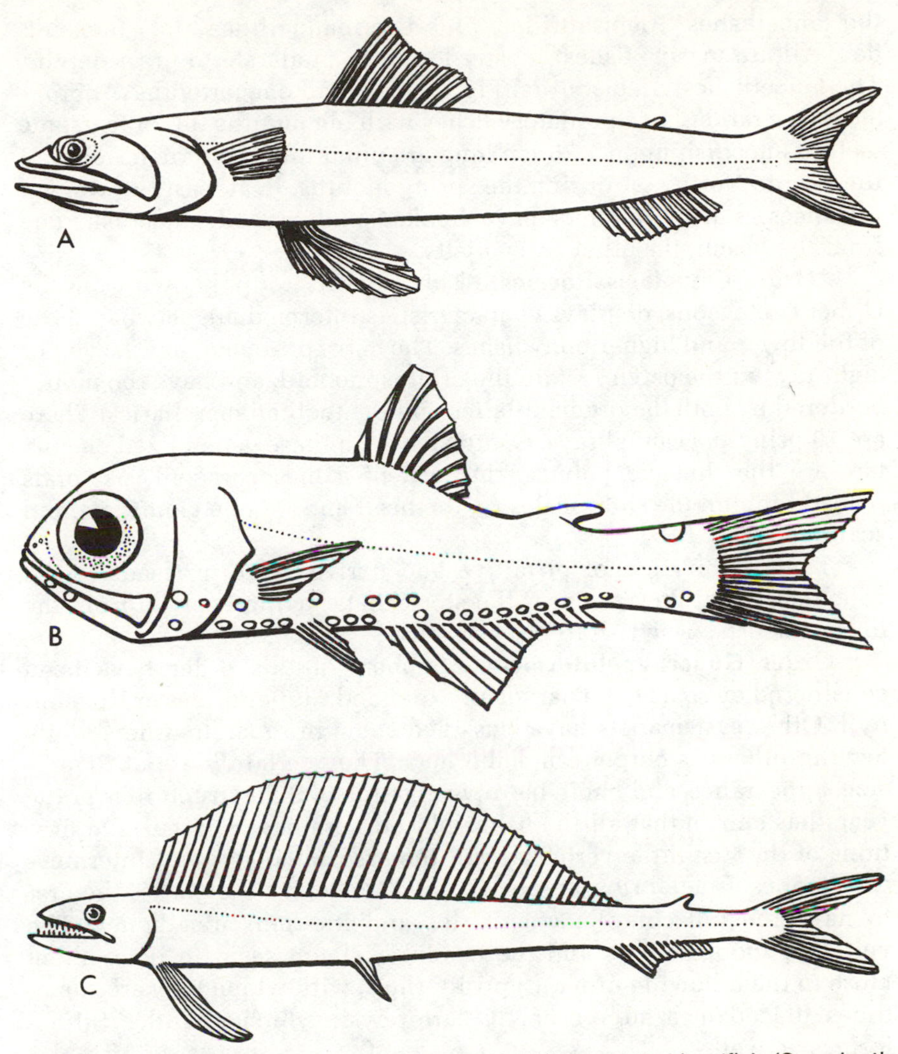

FIGURE 7–12. Representatives of order Myctophiformes. *A,* Lizardfish (Synodontidae); *B,* lanternfish (Myctophidae); *C,* lancetfish (*Alepisaurius*).

The myctophids, or lanternfishes (Fig. 7–12*B*), are numerous in many parts of the oceans and are noted for their diel vertical migrations. They move toward the surface at night and back into the lightless zone by day. There are numerous species, all small, and all set with a particular pattern of photophores. Collectively, the lanternfishes make up a considerable biomass and are fed upon by many predators. Their ability to convert plankton to food for organisms higher on the food chain gives them considerable importance in the trophic ecology of the ocean.

The Scopelosauridae, although placed here, are somewhat intermediate between this and the following suborder.

The suborder Aleposauroidei contains: the pearleyes, Scopelarchidae; the barracudinas, Paralepididae; the daggertooths, Anotopteridae;

the lancetfishes, Alepisauridae; plus Evermannellidae and Omosudidae. All are marine fishes, mostly found in moderate to great depths. The lancetfishes, *Alepisaurus* (Fig. 7–12C), and daggertooths, *Anotopterus*, are rather large predatory fishes with big mouths full of fearsome teeth. Lancetfish up to 1.25 cm long are often found on the beaches of the Pacific Northwest during the spring months. In at least two known instances, startled anglers have hooked and landed lancetfish, one from the beach, the other from a jetty.

†Order Ctenothrissiformes. This extinct group, known from the Upper Cretaceous, displays characteristics intermediate between those of the lower and higher bony fishes. They are soft-rayed, have seven or eight rays in the pelvic, retain the orbitosphenoid, and have the mouth bordered by both the premaxillaries and the toothed maxillaries. There are 19 principal caudal rays. Contrasted with these generalized characters are the thoracic pelvics, the lack of a mesocoracoid, pectorals placed high on the side, scales on the head and, in one genus, ctenoid scales.

The combination of primitive and derived features leads to the conclusion that they are near the stock that, having evolved from the lower teleosts, gave rise to the perciforms.

Order Gonorhynchiformes. Members of this order have been considered by some to be part of an expanded Clupeiformes or Isospondyli. Other systematists have classified some in separate orders while leaving others as clupeiform suborders. They certainly are of diverse size, appearance, and habit, but investigation of their structure over the years has shown that all the fishes placed here have similar modifications of the first three vertebrae and the associated ribs and intermuscular bones. Cranial ribs are present, and the fourth vertebra is the first to have the full complement of ribs and intermuscular bones. The vertebral modifications and the caudal skeleton seem to place them close to the following order, Cypriniformes, with which they are sometimes placed in a superorder, Ostariophysi, which includes Siluriformes as well.

All have small mouths, and teeth are lacking or very small. The maxillaries may be excluded from the border of the mouth, which is protrusible in some genera. The orbitosphenoid and basisphenoid are lacking. A suprabranchial organ is usually present. The pelvics have 9 to 12 rays.

The suborder Gonorhynchoidei contains one family, Gonorhynchidae, of the Indo-Pacific (Fig. 7–13*A*). *Gonorhynchus* is called sandfish by many authors because it is reported to seek refuge in sandy bottoms. It is a slender fish with a pointed snout bearing a single barbel. The dorsal, anal, and pelvic fins set far back. The scales are ctenoid.

The suborder Chanoidei includes the Kneriidae and Phractolaemidae, very small fishes of African fresh waters, and Chanidae of Indo-Pacific marine and brackish waters. *Kneria auriculata* is rarely more than 75 mm long. It is noted as an air breather that can remain out of water for some hours and can wriggle up damp vertical surfaces in order to ascend streams. There are reports of aestivation in this species.

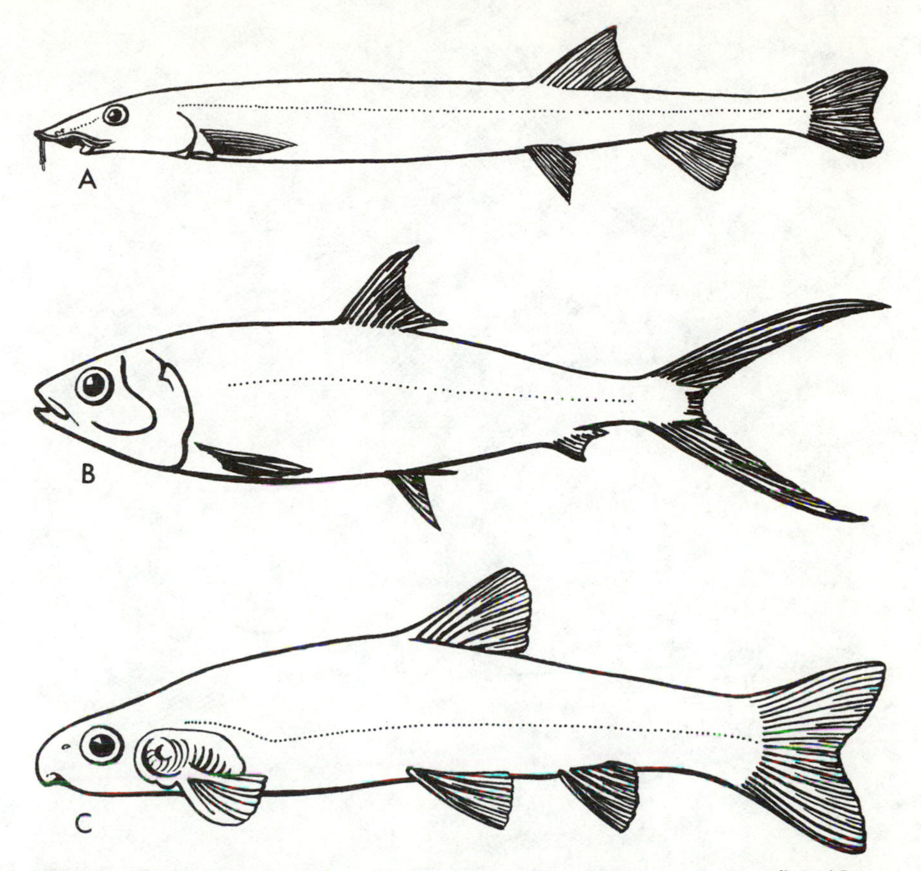

FIGURE 7–13. Representatives of order Gonorhynchiformes. *A,* Sandfish (*Gonorhynchus*); *B,* milkfish (*Chanos*); *C,* shellear (*Kneria*).

The males have a peculiar cup-shaped structure on the operculum (Fig. 7–13*B*). *Phractolaemus* has an extremely small head, and a peculiar small protrusible mouth that opens upward.

Chanidae contains the well known milkfish, *Chanos chanos* (Fig. 7–13*C*), which is the object of pond culture in many areas of Southeast Asia. Reaching a length of about 1.5 meters, it is well streamlined, with a large forked caudal fin, and is capable of great leaps. The diet is mainly planktonic algae and growth is rapid. Young are collected by a variety of means, often by providing shade along the beach in shallow water, then using a small seine or dip net to capture the fry that gather there.

Methods of culture differ from country to country, but usually involve a complex of ponds, sometimes interconnected, providing proper conditions for fry, fingerlings, and larger fishes being grown for the market. In Taiwan special wintering ponds must be provided for protection, but in spite of the generally cooler climate there than in some other places where milkfishes are cultured, crops of up to 2000 kg per hectare are possible.

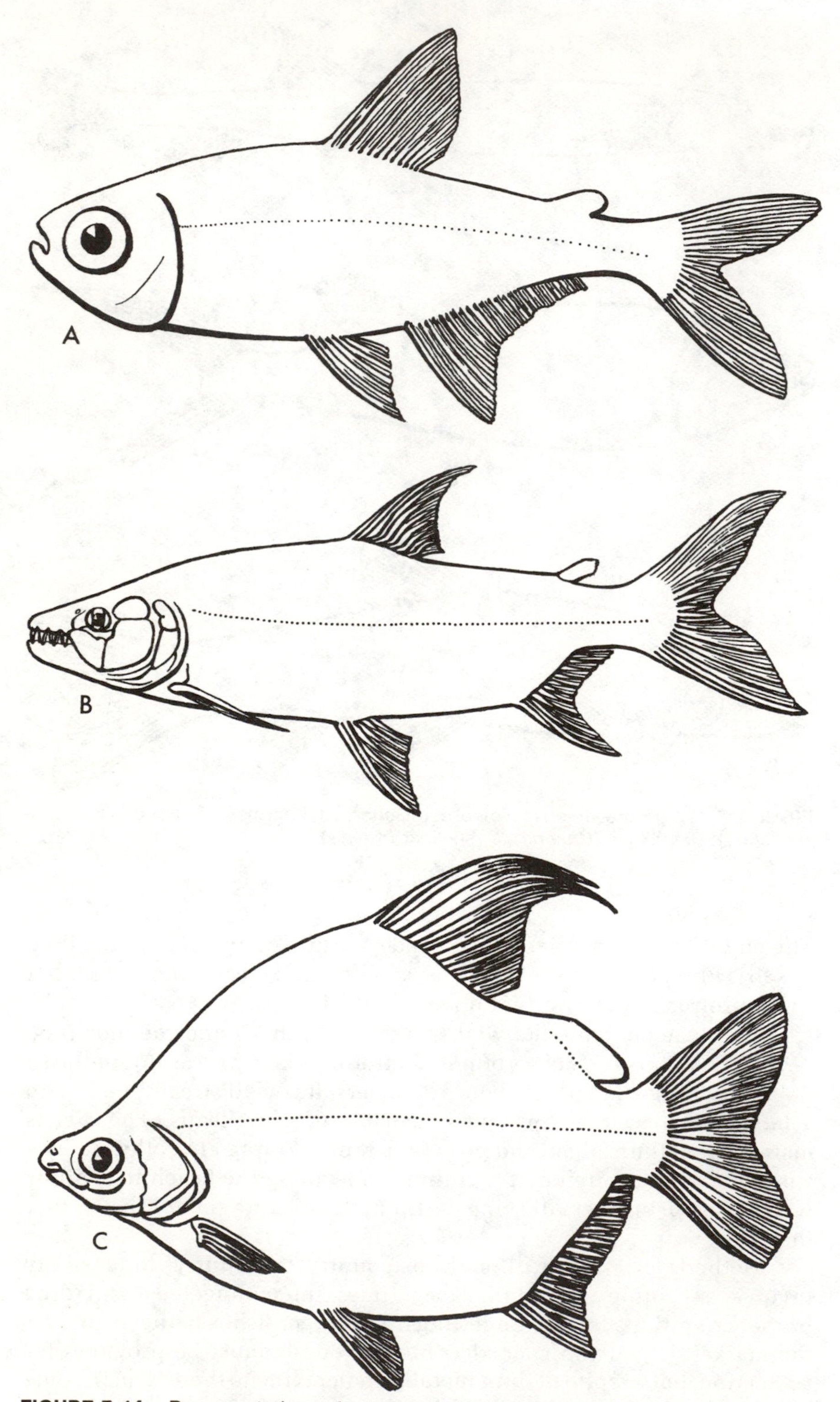

FIGURE 7–14. Representatives of suborder Characoidei. *A,* Tetra (Characinidae); *B,* tigerfish (*Hydrocynus*); *C,* moonfish (Citharinidae).

Order Cypriniformes. This is one of the most important orders of fishes. In numbers, there are about 3000 species; in importance to man, there are numerous species subject to culture or capture for food; in ecological importance, this is the dominant freshwater group in many parts of the world. Included here are the characins and allies of the suborder Characoidei, the electric eel, knifefishes, and relatives of the suborder Gymnotoidei, and the carps, minnows, and loaches of Cyprinoidei. The related catfishes, often placed with the above, are considered as forming a separate order.

Cypriniforms and catfishes are now generally placed in a superorder called Ostariophysi or Cyprinimorpha because they share the unique characteristic of the Weberian apparatus, a structure made up of the first four or five vertebrae. The modified vertebrae form the attachment points for a series of small ossicles, the most posterior pair of which can remain in contact with the gas bladder while the anterior pair is in contact with the membranous labyrinth.

The Cypriniformes usually have scales, although some are naked. Some have teeth in the jaws but others lack these. Most have teeth on the lower pharyngeal bones, modified according to the food habits of the particular species. The third and fourth vertebrae are not fused to each other; parietals, symplectics, and subopercular bones are present. These are physostomous and usually lack acanthopterygian-type spines, although many have spinous, enlarged and hardened rays at the front of the dorsal and anal fins. The orbitosphenoid is present. These fishes are widespread, but do not occur natively in Australia.

The suborder Characoidei contains the characins and relatives of South America and Africa (Fig. 7–14); most of these are considered by some authorities to be in a single family, Characidae, although there have been over 20 families designated in the group. Currently, the trend is to recognize 16 families. These have normal, opposing pharyngeal teeth. Most have teeth on the jaws and have an adipose fin. There are about 25 genera and 150 species in Africa, and over 200 genera in South, Central and North America. The range of the characoids, Africa and the Americas as far north as the Rio Grande, indicates dispersal at a time when South America and Africa were closer or joined.

Even in the restricted sense Characidae contains numerous genera of diverse habits in both Africa and the Americas. Many are small and colorful and are kept in home aquaria, forming the basis for a remarkable import-export trade. The tetras, such as *Hyphessobrycon* and *Hemigrammus*, are examples. The piranhas are well known characins of the subfamily Serrasalminae, famed for their sharp teeth, strong jaws, and voracious habits, although some close relatives are harmless fruit-eaters. The tigerfish *Hydrocynus* of Africa is another fierce predator (Fig. 7–14*B*).

Other groups that are variously accorded family or subfamily status, depending on the authority, follow: Erythrinidae, of South America, noted for the lunglike gas bladder; Hepsetidae, of Africa, a primitive form, predaceous; Ctenoluciidae, South America, pikelike, predaceous; Cynodontidae, South America, predaceous, some sought as game fish; Lebiasinidae, South America, includes some important aquarium fishes; Parodontidae, of South America; Gastropelecidae, of South America, the freshwater flying fishes or hatchet fishes, noted for their ability to flap the pectoral fins in flight; Prochilodontidae, South America, non-predaceous, cyprinid-like; Curimatidae; Anostomidae; Hemiodontidae; Chilodontidae (all of South America, non-predaceous, some with reduced or no dentition in jaws); Distichodontidae and Ichthyboridae (fin-biters), both of Africa and related to the following family.

The Citharinidae (moonfishes) of Africa are deep-bodied, resembling the lake whitefishes of North America in appearance. They are food fishes that reach a weight of about 2.5 kg (Fig. 7–14C).

The suborder Gymnotoidei is characterized by the lack of the dorsal fin, although some may have a long, slender adipose fin. The anus is placed far forward, generally in advance of the pectorals, and the anal fin usually begins close behind it and stretches to the end of the tail. The caudal is typically lacking, as are the pelvics. These are elongate fishes of South America (Fig. 7–15).

Although some ichthyologists recognize only one family, Gymnotidae, four will be considered here. The Gymnotidae, or gymnotid eels, Electrophoridae, electric eel, Apteronotidae, and Ramphichthyidae, knifefishes.

The electric eel, *Electrophorus electricus*, is rather heavy-bodied but elongate, having about 250 vertebrae. Only about 20% of the body length contains the internal organs; the remainder is the caudal region in which the large electric organ is located. This is one of the strongest organs of its kind among the fishes, capable of producing 650 volts, although the average maximum is about 350 volts.

Gymnotids have electric organs that function in electrolocation of objects, much as in the mormyroids. Correspondingly, the knifefishes are strange in structure and parallel the mormyrids in appearance. The snout, or head and snout together, may be elongated and curved downward. The anus may be as far forward as the chin. The posterior part of the body can be tapered and slender. Median fins are elongate.

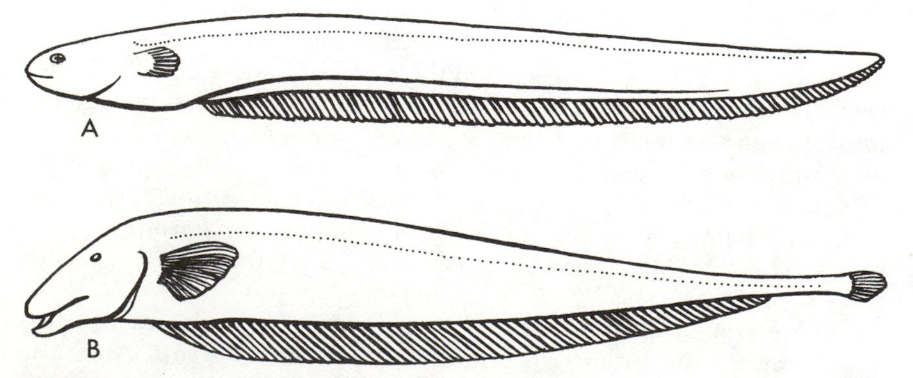

FIGURE 7–15. Representatives of suborder Gymnotoidei. *A*, Electric eel (*Electrophorus*); *B*, knifefish (*Apteronotus*).

The suborder Cyprinoidei differs from the previous two groups in having enlarged, sickle-shaped lower pharyngeal bones, which in most species bear teeth. These usually bite upward against a pad borne on a posterior extension of the basioccipital bone. There are no teeth on the jaws and no adipose fin. The dorsal fin is present and the anus is in a normal position.

Best known of the suborder are the members of the Cyprinidae, which seems to have had its origin in southeast Asia but now has over 1500 species in the fresh waters of Europe, Asia, Africa, and North America. Many important ornamental and aquarium fishes belong to this family, ranging from the small "barbs" to the larger goldfish, *Carassius auratus*, and the carp, *Cyprinus carpio*. The latter has great importance in fish culture in many parts of the world to which it has been transplanted from its native Asia. It has been the subject of selective breeding that has produced high-backed, deep-bodied individuals with few scales, rapid growth, and great efficiency in food utilization. The carp is well accepted as food in Europe and Asia and is used to some extent elsewhere. In parts of North America it is considered a pest that destroys the habitat of game fish by stirring up the bottom and aquatic vegetation during foraging.

Some of the cyprinids reach a large size, weighing up to 130 kg or more. The mahseer of India, *Barbus tor*, is one of the largest. Others are *Catlocarpio siamensis* of Thailand and *Catla catla* of India, which is a favored food fish (Fig. 7–16*A*). Other food fishes cultured in India are *Cirrhinus mrigala*, the mrigal, *Labeo rohita*, the rohu, and *L. calbasu*. Carps cultured in China include *Aristichthys nobilis*, the bighead carp, *Hypophthalmichthys molotrix*, the silver carp, and *Ctenopharyngodon idella*, the grass carp, which has been introduced to Europe and North America for use in control of aquatic vegetation.

There are about 235 species of minnows in North America, mostly east of the continental divide. The majority are small fishes that serve as forage for larger predators. Two genera with large numbers of species are *Notropis* (Fig. 7–16*B*) and *Hybopsis*. These are found throughout the eastern part of the continent except for *Hybopsis crameri*, the Oregon chub, which exists as a relict in the Willamette and Umpqua rivers.

Giants among the North American minnows are members of the genus *Ptychocheilus*, or squawfishes. Found in the Colorado, Sacramento, Columbia, and contiguous drainages, they are noted as voracious predators. The Colorado squawfish is reported by early ichthyologists to reach 1.5 meters long. The northern squawfish now reaches about 75 cm though Pliocene ancestors in Oregon and Idaho reached 1.5 m.

Catostomidae, the suckers, are closely related to the minnows and are thought to have had their origin in Asia, even though only one representative of the ancient suckers, *Myxocyprinus*, is found there now. Except for this and *Catostomus catostomus* (Fig. 7–17*A*), which has invaded Siberia from Alaska, the suckers are North American.

FIGURE 7–16. Representatives of family Cyprinidae. *A*, Catla (*Catla*); *B*, shiner (*Notropis*).

Many of the members of the family have ventral mouths with thick papillose lips, exemplified by the genus *Catostomus*, which has species in most major drainages. The largest of the genus is *C. luxatus*, the Lost River sucker of the Klamath drainage in Oregon and California, which reaches about 1 meter in length.

Buffalofishes of the genus *Ictiobus* (Fig. 7–17*B*) and the quillbacks and carpsuckers of the genus *Carpiodes*, are large, carp-like fishes of the Mississippi and contiguous river systems; they are used as food and occasionally are cultured in the southern United States.

Cobitidae, the loaches, are small, slender fishes of Eurasia and Africa (Fig. 7–17*C*). They have three or four pairs of barbels around the mouth and the gas bladder is encapsulated in bone. Some (*Misgurnus*) are so sensitive to barometric pressure that they are known as "weatherfishes" in Europe because of their reactions to changing pressure. Others are kept in aquaria because of their striking color patterns and vigorous activity. Many species swallow air to supplement their oxygen supply.

The Gastromyzontidae are flattened fishes of torrential mountain streams of Southeast Asia. Their entire ventral surface and the pectoral and pelvic fins all combine to form an adhesive structure by which they can cling to rocks. Gyrinocheilidae and Homalopteridae of the same area are similar, but less modified.

Order Siluriformes (Nematognathi). This order contains the catfishes, which are placed by some ichthyologists with the preceding order because, like the Cypriniformes, the Weberian apparatus is present. However, the catfishes have no true scales, the skin being bare or covered by bony plates, which may bear dermal denticles. The subopercles, symplectics, and parietals are absent. The Weberian apparatus may involve up to five vertebrae, with the second, third, fourth, and in some, fifth vertebrae fused to one another. The premaxillary bears

FIGURE 7–17. Representatives of families Catostomidae and Cobitidae. *A,* Sucker (Catostomidae, *Catostomus*); *B,* buffalo (Catostomidae, *Ictiobus*); *C,* loach (Cobitidae, *Cobitis*).

teeth, but the maxillary, except in the most primitive family and one higher genus, is modified or is the basal skeletal unit of the maxillary barbel.

The pectorals and the dorsal fins have large spines at the leading edges. These are not homologous with spines of actinopterygians. The spines are provided with a locking mechanism that holds them erect. All have barbels around the mouth, and an adipose fin is usually present. Fossil catfishes are known from the Paleocene and were diverse by the Eocene.

Various authorities recognize from 25 to over 30 families of catfishes and estimate that the total number of species is around 2000, of which over half are found in South America. They are known from all continents. Those found in Australian fresh waters belong to the marine families Ariidae and Plotosidae, the former widespread in warm seas, the latter in the Indo-Pacific. Most of the remainder of the catfish families are strictly freshwater fishes, rarely being found in water of more than a few parts per thousand salinity.

The most primitive living catfishes are in the family Diplomystidae of South America. In these the maxillae are normally developed and bear teeth. The fifth vertebra is not fused or firmly connected to the fourth, as in the remaining families.

One of the best known families of old world catfishes is the Siluridae, which contains the giant wels of the Danube, *Siluris glanis;* this may reach a weight of 130 kg or more (Fig. 7–18*A*). Many interesting silurids occur in Asia. One, *Wallagonia attu,* may reach 2 meters in length and, although slender, is a fierce predator; others are small and innocuous. In this family the dorsal fin may be small or absent, and there is no adipose fin.

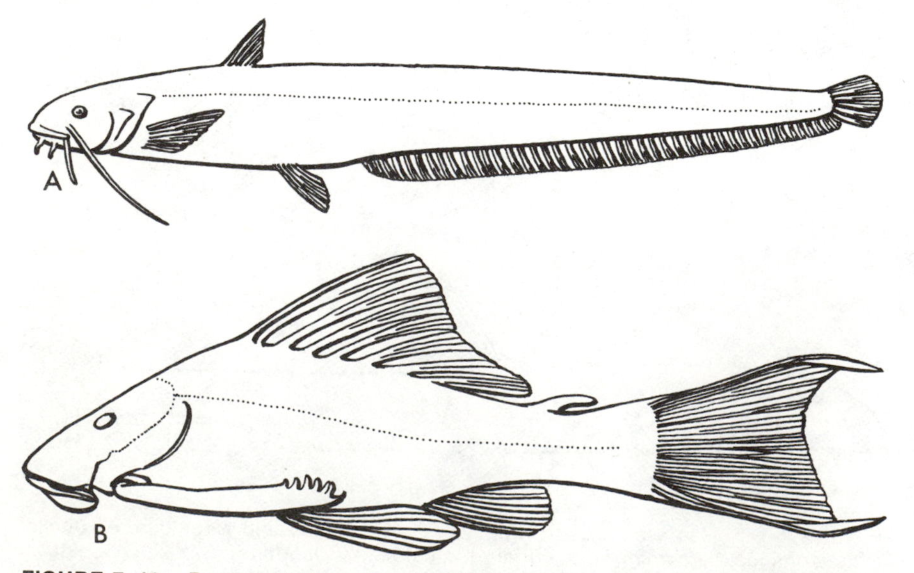

FIGURE 7–18. Representatives of order Siluriformes. *A,* Wels (*Silurus*); *B,* armored catfish (Callichthyidae).

Another catfish family of interest is the Pangasiidae of Asia, which contains *Pangasianodon gigas*. The closely related Schilbeidae occurs in both Asia and Africa. This is one of the largest of freshwater fishes, reaching a length of 2.5 meters. Clariidae of Asia and Africa and Heteropneustidae (Saccobranchidae) of Asia both contain air-breathing fishes. The former is equipped with an arborescent accessory breathing organ in the gill chamber, and the gill cavity of the latter communicates with large paired air sacs in the musculature of the body. These families are placed together in Clariidae by some ichthyologists because the two groups have similar structure except for the air breathing apparatus.

The Malapteruridae of Africa contains the electric catfish. *Malapterurus electricus,* which can deliver a severe shock. Another African family, Mochokidae, contains a fish that habitually swims upside down. The Bagridae occurs in both Asia and Africa and there is a closely related family, Pimelodidae, in South America. Some pimelodids may reach nearly 2 meters in length.

The Trichomycteridae of South America contains small fishes that can parasitically attack the gill cavities of larger fishes. One, the candiru, is attracted to nitrogenous wastes such as are normally produced by fish gills, but may mistakenly enter the urinary opening of mammals, including man. The armored catfishes, Loricariidae and Callichthyidae (Fig. 7–18*B*), are both South American and contain some rather bizarre forms, some of which are used as aquarium fishes. The South American Doradidae contains some armored forms.

Other South American catfish families are Bunocephalidae (= Aspredinidae), Cetopsidae, Helogeneidae and Hypophthalmidae. Additional Asian families are Amblycipitidae, Chacidae, Cranoglanididae and Akysidae. Amphiliidae is an African family.

The North American catfishes, Ictaluridae, which were native to western North America from Eocene to Pliocene, are now native only east of the Rocky Mountains but are established in most parts of the continent with suitably warm climate. The blue catfish, *Ictalurus furcatus,* is the giant of the group, weighing up to 70 kg. The flathead catfish, *Pylodictis olivaris,* is nearly as large, and the channel catfish, *I. punctatus,* may reach over 20 kg. The channel catfish is of great economic importance. It has long been a favored food fish and the object of a commercial fishery, but in the past decade it has been cultured extensively throughout the warmer parts of the United States. Modern methods of fish culture and the availability of dry pelleted fish food has made commercial rearing of channel catfish a profitable business.

(The following three orders represent, along with three orders of higher fishes, the superorder Paracanthopterygii of Greenwood et al. (1966). The superorder is viewed as representing a natural line showing similarities in the structure of the caudal fin supports.)

Order Polymixiiformes. These are tropical marine fishes with subabdominal pelvic fins. There are a few spines in both the dorsal and anal fins, but the pelvics have no true spines. The single family, Polymixiidae, has three species, living in mid-depths in the western Pacific and the Atlantic. They are called beardfishes or barbudos because of their long barbels. An alternate placement of the group is as a suborder of Beryciformes.

Order Percopsiformes (Salmopercae). This is a small order of North American freshwater fishes characterized by subabdominal or subthoracic pelvic fins, a caudal with 16 or 17 branched rays, and the maxillary excluded from the border of the mouth. Some members are unique in that they possess an adipose fin and have spines in the dorsal and anal fins. The orbitosphenoid and basisphenoid are absent. There is no duct connecting the gas bladder and the esophagus. Scales are ctenoid or cycloid. These fishes are considered to be middle or intermediate teleosts because of the blend of characteristics. They are known from the Eocene.

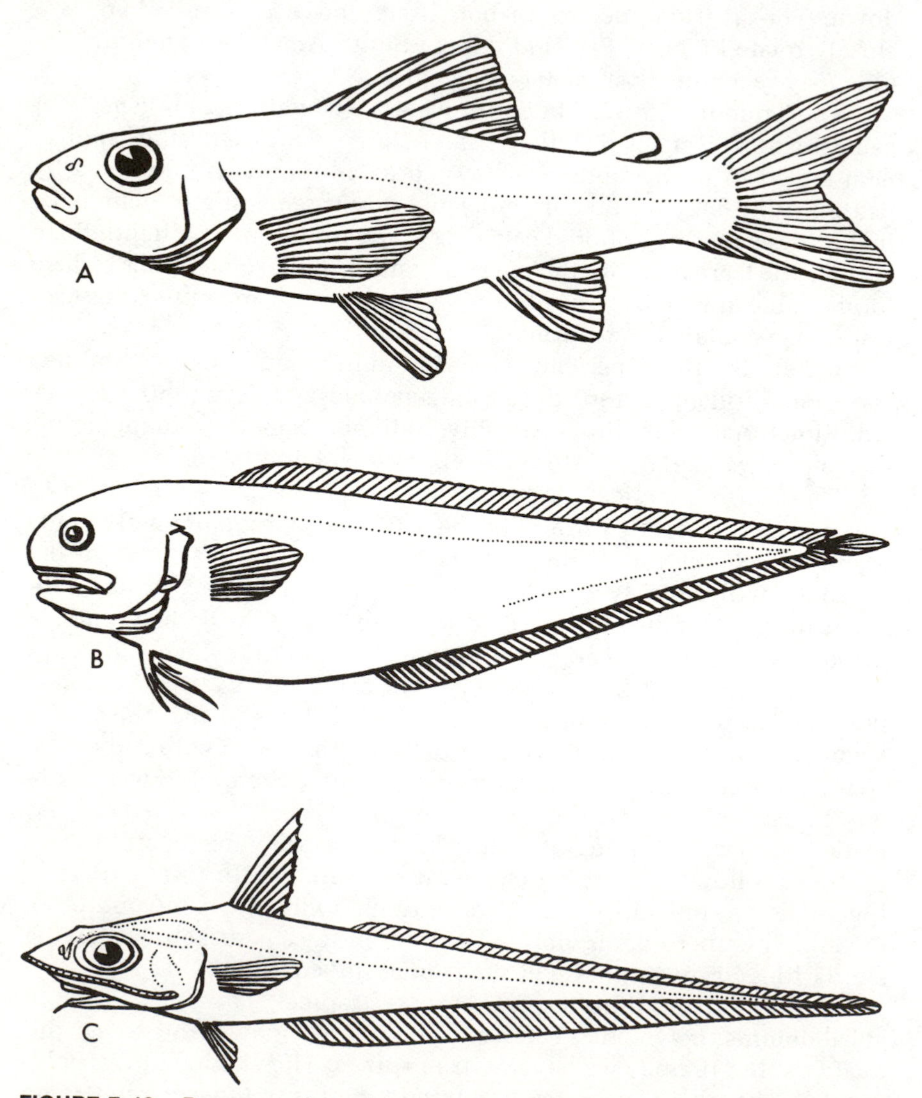

FIGURE 7–19. Representatives of *A*, order Percopsiformes, trout-perch (*Percopsis*); *B*, suborder Ophidioidei, cusk-eel (*Pteridium*); *C*, suborder Macrouroidei, grenadier (*Macrourus*).

The suborder Percopsoidei has true spines in the dorsal and anal, an adipose fin, ctenoid scales, and has the anus in a normal position. The single family Percopsidae contains two species, *Percopsis omiscomaycus*, the trout-perch of eastern North American drainages (Fig. 7–19*A*), and *P. transmontana*, the sand roller of the Columbia River system. These are small fish of still water or slow streams. The sand roller seems to be nocturnal in habit.

The suborder Aphredoderoidei lacks the adipose fin and has the anus in a jugular position. There is a single species, *Aphredoderus sayanus*, the pirate perch.

The suborder Amblyopsoidei, sometimes placed with the cyprinidontiform fishes, contains the single family Amblyopsidae which lacks strong spines and has cycloid scales and a jugular vent. These are the cavefishes and allied forms of the southeastern United States. The eyes may be reduced, and the acousticolateralis sensory system is well developed.

Order Gadiformes (Anacanthini). These are soft-rayed physoclistic fishes, usually with cycloid scales and thoracic or jugular pelvics, so that they show both primitive and derived characteristics and are representative of middle teleosts. Additional advanced features are the lack of the orbitosphenoid and mesocoracoid. The premaxillaries exclude the maxillaries from the gape. The caudal usually is isocercal and has a distinctive internal skeleton. This order contains the cods and allies, including the grenadiers, cusk-eels, and brotulas, all marine with a few exceptions. They were well-developed by the Eocene.

The suborder Muraenolepoidei contains one family, Muraenolepidae, characterized by 10 to 13 actinosts, a narrow gill opening entirely below the base of the pectoral, and the confluence of the second dorsal, caudal, and anal fins. The first dorsal is made up of a single ray. They are found in cold marine waters of the Southern Hemisphere.

The suborder Gadoidei includes fishes with four or five actinosts and normally placed gill openings. There are two or three dorsal fins. The caudal fin is of the "pseudocaudal" type in that it consists mostly of dorsal and anal elements, with a very small true caudal making up the central part. Many cods have diverticula of the gas bladder in connection with the inner ear. Members of the Gadoidei are found in temperate and cold waters of both the Northern and Southern Hemispheres.

By far, the most important family in the entire order is Gadidae (Fig. 2–13*B*), containing the cods, haddocks, pollocks, lings, and whitings. These all have commercial value as food or as a source of high quality fish meal. The famous cod fisheries of the western North Atlantic, off New England and the Maritime Provinces, have been operated by many nations for more than two centuries. The object of most of the effort has been *Gadus morhua*, a large fish that may reach 120 cm in length and a weight of 45 kg. Longlining has been the primary method employed over the years but trawling, especially since the advent of steam-powered vessels, forms the basis of the fisheries.

Another important commercial fish in the North Atlantic is the haddock, *Melanogrammus aeglefinus*, which may be suffering from overfishing by trawlers. The species reaches about 1.1 meters in length and 16 kg. The pollock, *Gadus pollachius*, and the whiting, *G. merlangus*, are additional species that enter the commercial fisheries. The Pacific Ocean counterpart of the Atlantic cod is *G. macrocephalus*, which was once fished in Canadian and Alaska waters by sailing vessels from San Francisco. The burbot, *Lota lota*, is a freshwater fish of holarctic distribution. Closely related to the cods is the hake family Merlucciidae, which also inhabits seas of both hemispheres and has commercial value.

The Moridae are deepsea fishes with very large diverticula from the gas bladder passing through wide foramina in the exoccipitals. The Bregmacerotidae are pelagic fishes of subtropical and tropical waters. They have exceptionally long pelvic fins and a first dorsal consisting of a single ray that rises from the occipital region.

The suborder Ophidioidei (Fig. 7–19*B*) was once placed among the perch-like fishes, but study of several aspects of their morphology has led recent investigators to propose a close relationship to the gadoids. The jugular pelvics of Ophidiidae are slender and barbel-like. They are widely distributed shore fishes, mostly small, but *Genypterus blacodes* of Australia and New Zealand may reach 120 cm in length.

The Brotulidae generally live at moderate to great depths, except for two genera that live in caves in Cuba and Yucatan. Possibly they are descended from ancestors that remained in situ while the land was elevated, the process taking long enough to accommodate the adaptations required for survival in that peculiar habitat. Otherwise, they could be descendants of forms such as *Stygnobrotula*, shallow water coral crevice inhabitants that could have invaded fresh water.

The Carapidae (Fierasferidae) are small, usually delicate fishes that live commensally (some parasitically) in sea cucumbers or molluscs. The anus is jugular, and the body tapers posteriorly to a point. The anal and dorsal fins are long. The Pyramodontidae, fishes of the warm Indo-Pacific, have a similar shape but are distinguished by peculiar large teeth in the anterior part of the jaws.

The suborder Zoarcoidei contains the eelpouts of the family Zoarcidae, long placed as close relatives of the blennies in the Perciformes. These are soft-rayed elongate fishes with confluent anal and dorsal fins and rather restricted gill openings. Some of the features of cranial osteology and the patterns of certain nerves appear to place them close to the gadoids. The eelpouts are viviparous fishes usually found in deep water.

The suborder Macrouroidei is sometimes considered to form a separate order because of the presence in some fishes of a spinous first dorsal ray, ctenoid scales, and some differences in placement of the olfactory lobes and nerves. These are deepsea fishes usually with a large head, often with large eyes, and with a long tapering afterbody fringed by long dorsal and anal fins (Fig. 7–19*C*).

Order Beloniformes (Synentognathi or Exocoetoidei). This order, which includes the flyingfishes, halfbeaks, and needlefishes, is sometimes combined with the two following orders because of similarities that may be due to independent adaptations for a life near the water's surface. Beloniforms are soft-rayed, physoclistic, and have abdominal pelvic fins with six rays. The lower pharyngeals are fused to each other. The dorsal is set far back over the anal, the caudal has only 13 rays, and the pectorals are set high on the sides. There is no orbitosphenoid or mesocoracoid and the maxillae are excluded from the gape. The lateral line is situated very low on the body. They have been referred to as intermediate teleosts by some ichthyologists, but in some recent works have been allied to the acanthopterygians. These are active fishes, mostly of warm seas, although some enter fresh water and some are found in temperate marine waters. Fossil beloniforms are known from the Eocene.

The family Belonidae contains the needlefishes, slender, elongate fishes with the jaws produced into a long beak-like mouth set with sharp, fine teeth (Fig. 7–20*A*). The largest members of the family may reach over 1.5 meters long. Needlefishes are noted for their jumping ability. Schools may scatter with great leaps at the swift approach of a boat, or they may jump over objects as if in play.

The family Scomberesocidae contains the sauries, which have a very short beak and a series of mackerel-like finlets behind the dorsal and anal fins. The leaping habit is somewhat developed. *Cololabis saira* of the North Pacific, a plankton feeder, occurs in large schools and is the object of a commercial fishery.

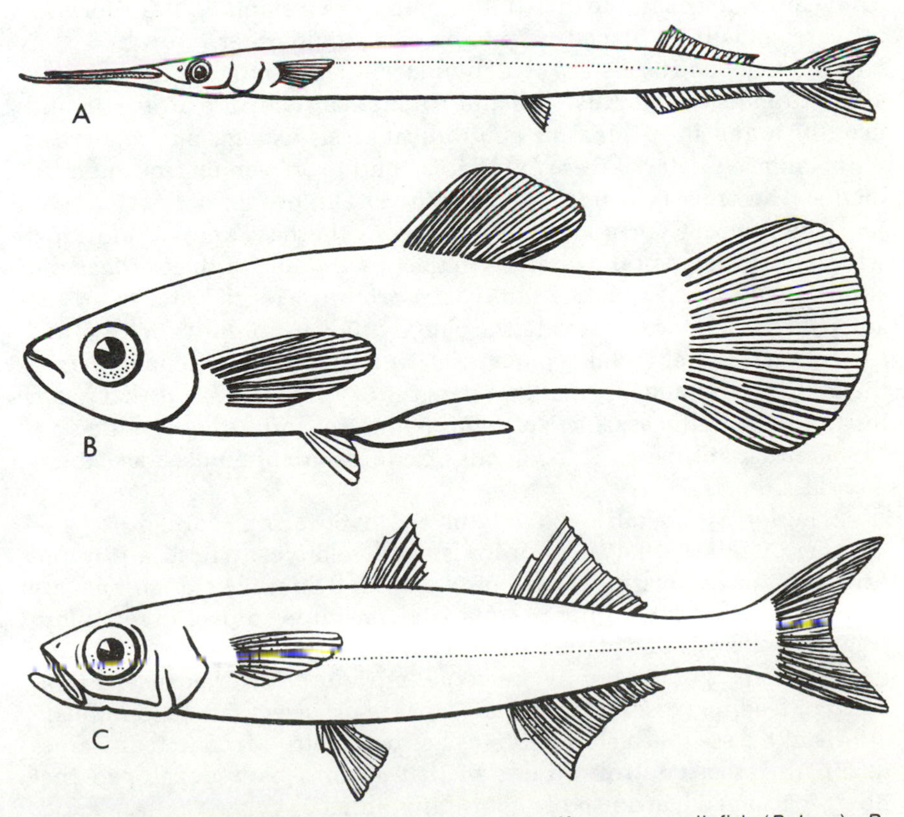

FIGURE 7–20. Representatives of *A*, order Beloniformes, needlefish (*Belone*); *B*, order Cyprinodontiformes, guppy (*Poecilia*); *C*, order Atheriniformes, silverside (*Atherinus*).

The Hemirhamphidae, the halfbeaks, also are noted as being remarkable jumpers. In this family the lower jaw is greatly extended but the upper jaw is fairly short. The lower lobe of the caudal is larger than the upper. Their food is plankton.

The Hemiramphidae, the halfbeaks, also are noted as being rethe flyingfishes that are seen gliding over the water's surface in many warm seas (Fig. 2–12C). The pectoral and pelvic fins are enlarged, the pectoral sometimes greatly so, and these act as gliding planes. The lower lobe of the caudal is usually quite elongate. Careful observation of a flying fish on takeoff or in the process of gaining additional gliding time when near the crest of a wave will reveal the lower part of the caudal being employed in a sculling motion, it being the only part of the fish in contact with the water. The California flyingfish, *Cypselurus californicus,* is one of the largest of the family, reaching a length of about 45 cm.

Order Cyprinidontiformes (Microcyprini). This is the order of killifishes, topminnows, and relatives. They are soft-rayed physoclists with abdominal pelvic fins. The single dorsal is set behind the middle of the body. The caudal has a varying number of rays and is never forked. The orbitosphenoid, mesocoracoid and basisphenoid are lacking, and the maxillary is excluded from the border of the mouth by the premaxillary. The lateral line system is incomplete both on the head and body. They show some similarities to the Beloniformes and Atheriniformes, and are sometimes placed with them in a single order.

Cyprinodontiforms are widely distributed, occurring naturally in all continents except Australia and Antarctica. They are all small, and usually found in tropical or subtropical fresh waters, but many can withstand salt water and are found in estuarine or even marine environments. The order is known from the lower Oligocene.

The family Cyprinodontidae is one of the best known and most widely distributed of the families in the order. Many of the tropical and subtropical members of the family are brightly colored and enter the aquarium fish trade. These are the egg-laying cyprinidonts or "toothed carps" such as the lyretails, panchaxes, and medakas. Some members of the family are annual species that pass the dry season as eggs that hatch the following rainy season (*Nothobranchius* sp. and *Aphyosemion* sp.). The familiar killifishes of the genus *Fundulus* are often used as experimental animals.

The family Poeciliidae contains the livebearing toothed carps — guppies, mollies, platys, swordtails, and relatives. These are found primarily in the warmer parts of North and South America and are among the greatest favorites of aquarists. The guppy, *Poecilia reticulata* (Figure 7–20*B*), is one of the easiest livebearers to keep and breed. Many of the members of the family are good "mosquito fish" because of their habit of feeding close to the water's surface on insect larvae. *Gambusia affinis* of the southern United States has been introduced to many areas to aid in insect control. Strains of that species have developed that now live somewhat north of the original range.

The foureye fishes of the family Anablepidae actually have only two eyes, but each is divided into an upper section for aerial vision and a lower one for aquatic vision. The eyes protrude from the top of the head so that the upper half can be above the surface of the water as the fish swims in the usual position just below the surface. Another peculiarity of the family is the orientation of the intromittent organ of the male and genital opening of the female. About 60% of the males have the intromittent organ oriented to the right; the remainder are sinistral. Fortunately the reverse is true for the orientation of the genital orifice of the females. Foureye fishes are found in tropical America.

Other families are: Adrianichthyidae, fishes of Celebes, in which there is internal fertilization with the eggs hatching upon extrusion; Jenynsiidae (viviparous), of South America; Horaichthyidae, of India, in which the gonopodium of the males is deflected to the right and the genital opening of the female to the left; and Goodeidae, viviparous topminnows endemic to the Mexican plateau.

Order Atheriniformes. The silversides and priapium fishes have been shifted considerably in their placement by modern ichthyologists. Silversides hold several important features in common with the mullets, and formerly were placed with them in a suborder of Perciformes or in the separate order Mugiliformes. On the other hand, some consider Atheriniformes to include the two previous orders, Beloniformes and Cyprinodontiformes. Priapium fishes have been placed with Cyprinidontiformes, Mugiliformes, or in their own order, Phallostethiformes.

This order is characterized by two dorsal fins, the first weak but made up of true spines. Pelvic fins are small, subthoracic in silversides, but in an anterior position, if present, in the priapium fishes. In silversides the pelvics consist of a spine and five soft rays. Both ctenoid and cycloid scales are present in the order. The lateral line is poorly developed or absent.

The suborder Atherinoidei contains the silversides, Atherinidae (Fig. 7–20C), and the related Melanotaeniidae and Isonidae. These are small fishes, mostly marine, with some freshwater representatives. The silversides are found along most of the warm coasts and often invade fresh water in the absence of varied freshwater faunas. *Leuresthes tenuis*, the grunion of California, is famous for its nocturnal spawning on the beaches during spring and summer high tides. The eggs are deposited in sand, between waves, by the half-buried female, which is accompanied by males. Eggs remain in the sand above the water line until the next series of high tides 2 weeks later, and then hatch when washed out by the waves. *Atherinopsis californiensis* is one of the largest silversides, reaching a length of more than 30 cm. Its spawning is more typical of the family as a whole. The eggs have filaments which attach to vegetation or to other objects.

The suborder Phallostethoidei contains two families, Phallostethidae and Neostethidae, both of Thailand, the Philippines, and the Indo-Malayan Archipelago. These are called priapiumfishes because of the strange intromittent organ of the male (Fig. 7–21*A*). This device is situated in the region of the isthmus and is supported by a skeletal

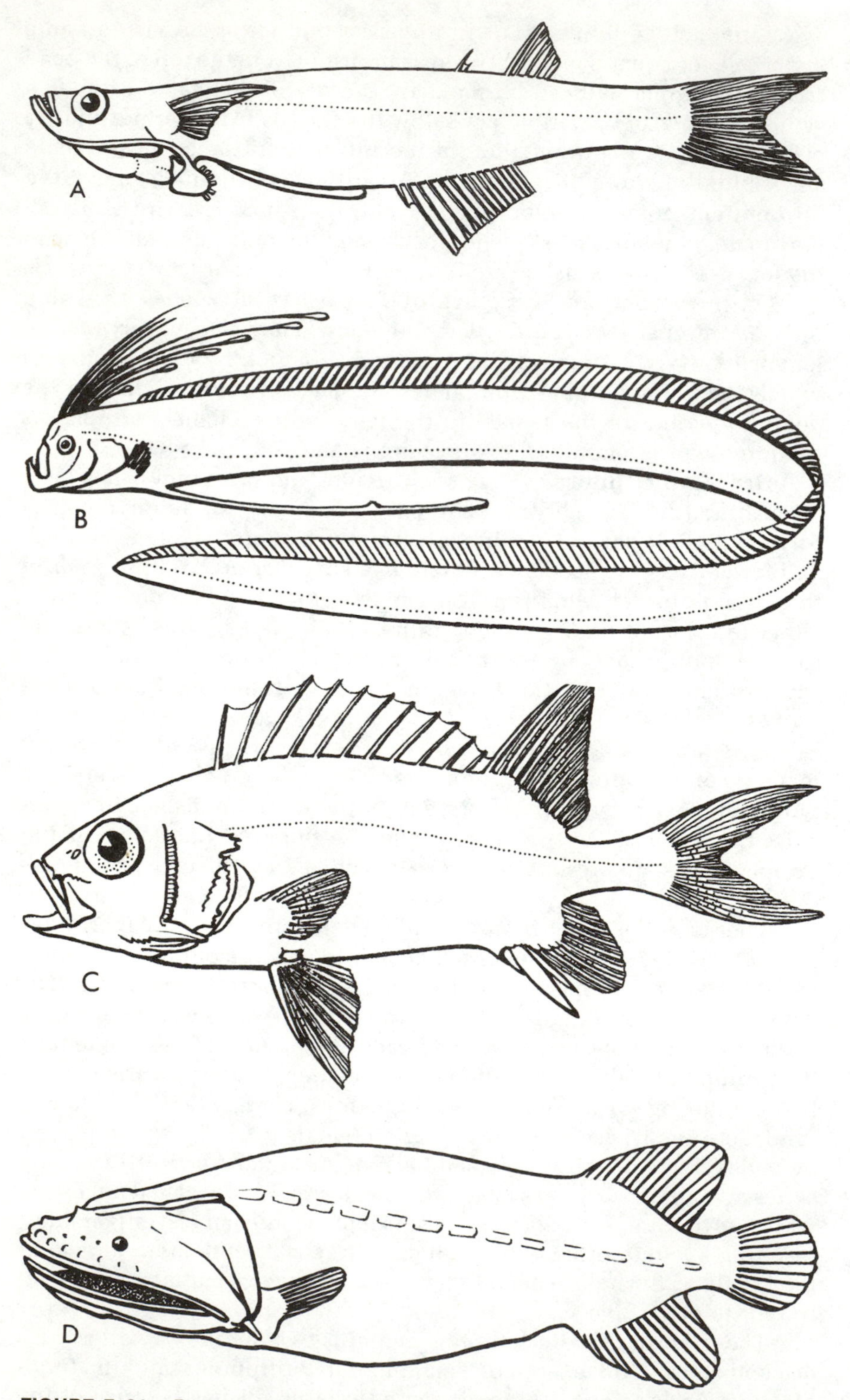

FIGURE 7–21. Representatives of *A,* suborder Phallostethoidei, priapiumfish (*Neostethus*); *B,* order Lampridiformes, ribbonfish (*Regalecus*); *C,* Beryciformes, squirrelfish (*Holocentrus*); *D,* whalefish (Cetomimidae).

structure made up of elements from the pelvic and pectoral girdles, plus the first pair of ribs. The anus and genital pore are jugular in position.

Order Lampridiformes (Allotriognathi). This order includes the opah and the greatly elongate and ribbon-like oarfishes and ribbonfishes plus some smaller fishes with uncertain relationships. Generally regarded as intermediate teleosts, they have been placed among the perch-like actinopterygians by recent investigators. They are mostly softrayed, although up to two dorsal fin spines may be present. Pelvic fins are thoracic and may have from 1 to 17 rays. The orbitosphenoid is present, but the mesocoracoid and opisthotic are absent. The protractile maxillaries have an outer blade and an inner process that meshes with a similar structure on the premaxillaries. Scales are cycloid or lacking. They are physoclistic. Lampridiforms are marine; some live in deep water. Fossils date from the Eocene.

The suborder Lampridoidei includes the widely distributed opah, *Lampris regius*, the only member of the family Lampridae. Because of its greatly compressed, ovate body the opah is also known as the moonfish. It is noted for its color pattern of blue or blue-gray on the back, silver on the sides, reddish silver on the belly, and red jaws and fins, all with an overlay of silver or whitish spots. Opahs may reach about 2 meters in length and may weigh up to 270 kg.

The suborder Veliferoidei includes the family Veliferidae, called sailbearers because of the very large dorsal and anal fins. The body is compressed as much as in the opah.

The suborder Trachipteroidei includes fishes that are greatly compressed but also elongate (taeniform; Fig. 7–21*B*). Many have bizarre coloration or fin shapes so that they are thought to be the basis of sea serpent stories. The Lophotidae, the crestfishes, have a crest which extends forward on the head and bears the anterior part of the long dorsal fin. The Radiicephalidae contains one species similar to the crestfishes. The ribbonfishes, Trachipteridae, also have an elongate dorsal, as do the oarfishes, Regalecidae. In the latter family the first several dorsal rays may be very high and the pectorals are elongate and oarlike. The suborder Stylephoroidei contains the single family Stylephoridae, the tube-eyes, which live at greater depths than do the other members of the order.

Suborder Ateleopoidei has only one family, Ateleopidae, some members of which may reach 50 cm long. These have jugular pelvics and a protrusible mouth. Some ichthyologists place them in an order of their own.

Suborder Mirapinnatoidei contains two families of strange oceanic fishes that may be more closely related to the Beryciformes, but have been placed in Lampridiformes by some recent authors. The Mirapinnidae of the Atlantic are scaleless and have skin covered with small villous projections that give the appearance of fur. No mature specimens have been captured; all have been less than 50 mm long. Family Eutaeniophoridae contains three species of elongate fishes that are called tapetails because of a long ribbonlike extension of the caudal fin in juveniles. These are found in the deepwaters of both the Atlantic and Pacific.

Suborder Megalomycteroidei contains one family, Megalomycteridae, with four genera and five species. These show a great development of the olfactory apparatus, and are represented in the Atlantic and Pacific.

Order Beryciformes (Berycomorphi). This order embraces the pricklefishes, soldierfishes, and allies. They are characterized by the presence of spines in the fins, ctenoid or cycloid scales, thoracic or subthoracic pelvics with one spine and 6 to 13 soft rays, and the orbitosphenoid, but lack the mesocoracoid. The gas bladder may be absent or physostomous but most are physoclistic. The caudal has 18 or 19 principal rays. These are marine fishes slightly more primitive than the perciforms, but showing trends characteristic of that order; thus they are usually considered among the higher teleosts. Their fossil history reaches back to the Cretaceous at which time they were diverse.

The suborder Stephanoberycoidei contains deepsea fishes, including Stephanoberycidae, the pricklefishes, Melamphaeidae, called melamphids or scalefishes, and Gibberichthyidae.

The suborder Berycoidei includes about eight families, mostly of shallow warm waters. Perhaps the best known are the Holocentridae, the brilliantly colored soldierfishes and squirrelfishes of tropical reefs (Fig. 7–21*C*). Another well known group is the Anomalopidae, or lanterneyes, some of the few shallow-water fishes with photophores. The Monocentridae, or pinecone fishes, have their large scales firmly united to one another. Other families are Diretmidae, Trachichthyidae, Korsogasteridae, Anoplogasteridae, and Berycidae.

Suborder Cetomimoidei (Cetunculi) contains the Cetomimidae or whalefishes, largemouthed little fishes, shaped like whales (Figure 7–21*D*). Other whalefishes are Barbourisiidae, Rondeletiidae and Kasidoridae.

Suborder Giganturoidei contains some small deepsea fishes with large mouths and distensible stomachs. Some have teloscopic eyes.

Order Zeiformes (Zeomorphi). This is the order of the dories and boarfishes. Characteristics include thoracic pelvics with a spine and 5 to 9 soft-rays, ctenoid scales, and lack of an orbitosphenoid. The caudal has only 12 or 13 principal rays, and the spinous anal sometimes forms a nearly distinct fin anterior to the soft-rayed anal. These are marine, mostly from deep water. Fossils are known from the Eocene.

The family Zeidae, the dories, are fishes of moderate depths and are often a part of commercial landings. The john dory, *Zeus faber*, has a round dark spot on each side, and fishermen of some European countries refer to it as "St. Peter's fish," perpetuating a legend that the marks are prints of the saint's thumb and forefinger (Fig. 7–22*A*). The species may reach a maximum length of close to 1 meter. Other families are Caproidae, the boarfishes, noted for their greatly protractile mouths, and Grammicolepidae and Oreosomatidae.

Order Gasterosteiformes (Thoracostei plus Aulostomiformes and Solenichthyes or Syngnathiformes). Fishes of this order are characterized by more or less elongated snouts and spines in the dorsal fin. Pelvic

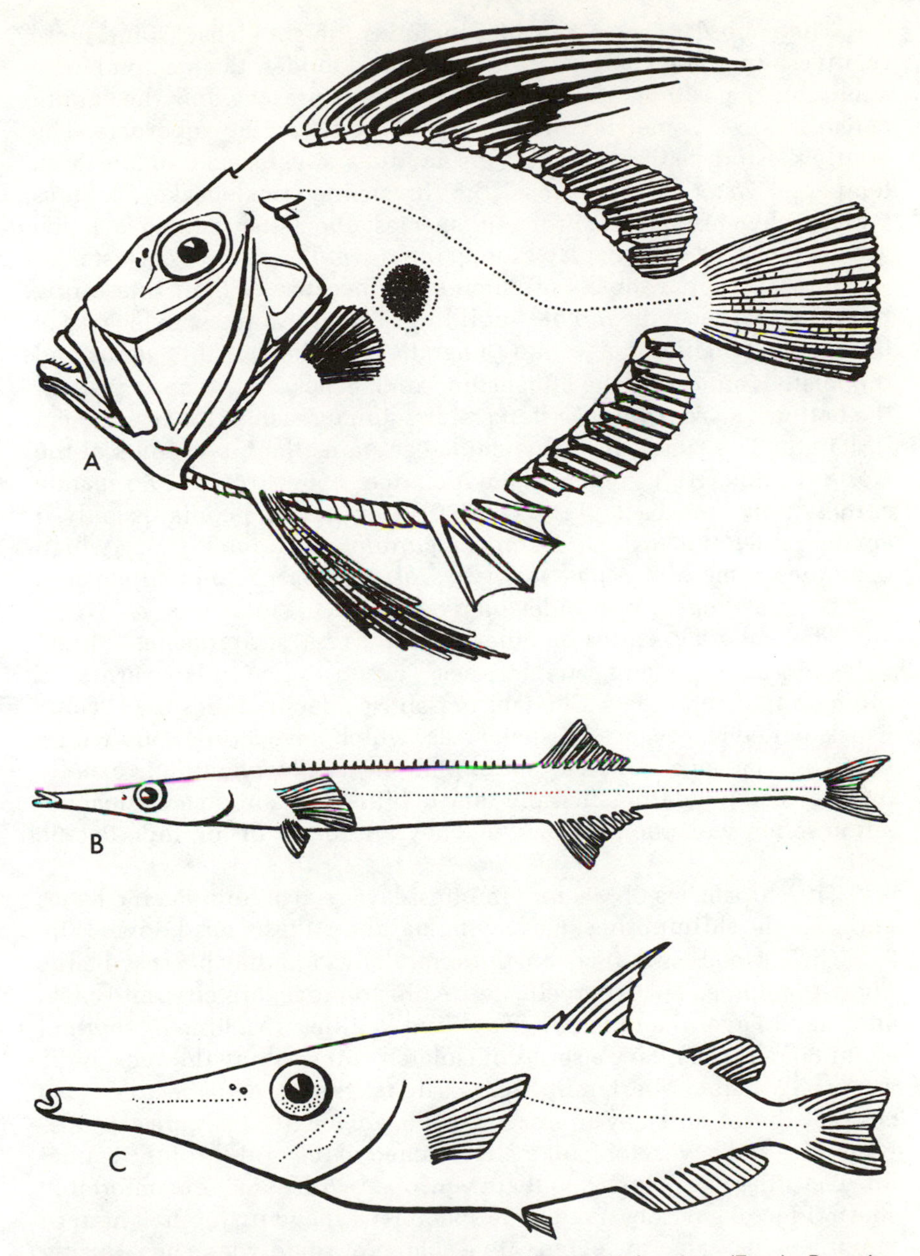

FIGURE 7–22. Representatives of *A*, order Zeiformes, john dory (*Zeus*); *B*, order Gasterosteiformes, tube-snout (*Aulorhynchus*); *C*, order Gasterosteiformes, snipefish (Macrorhamphosidae).

fins are abdominal to subthoracic and are spinous in some species. The gas bladder is physoclistic. There are sufficient differences among the three suborders that some ichthyologists prefer to consider each as a separate order. Fishes included are sticklebacks, tube-snouts, pipefishes, seahorses, trumpetfishes, and others possibly including the Pegasidae. They are mostly marine and prefer warm waters, but a few species reside in fresh waters. Fossils are known from the Eocene.

The suborder Gasterosteoidei includes the sticklebacks and close relatives. In these the pelvic fin is placed under the pectoral in a subthoracic position. Ribs and parietals are present and the mouth construction is somewhat different from the following suborders. The family Gasterosteidae contains the familiar sticklebacks of northern temperate waters. *Gasterosteus*, the threespine sticklebacks, includes both freshwater and anadromous species, the latter having a much greater covering of bony plates along the sides. The species of *Gasterosteus* have been the subject of much study because of their interesting reproductive behavior and nest-building. The ninespine sticklebacks of the genus *Pungitius* range into the arctic. Members of this genus and the related *Spinachia*, the fifteenspine sticklebacks, build nests well off the bottom in vegetation. Gasterosteids seldom reach as much as 10 cm in length. The family Aulorhynchidae contains the tube-snouts of the North Pacific. *Aulorhynchus flavidus* has about 18 to 20 separate spines along the back (Fig. 7–22*B*). Indostomidae contains only a small, rather rare fish, *Indostomus paradoxus*, from Burma, which combines some of the characteristics of sticklebacks and pipefishes. The true relationships of *Indostomus* are difficult to assess.

The suborder Aulostomoidei contains the shrimpfishes, snipefishes, cornetfishes, and trumpetfishes. These are all found in warm seas, often around coral reefs. The family Centriscidae includes the shrimpfishes, called razorfishes in some areas, which have a thin body with a sharp ventral edge and an armor of plates. Their locomotion is usually by means of undulating fins, and they habitually maintain a vertical position, often with the head down. They are found in the Indo-Pacific region.

The snipefishes of Macrorhamphosidae are also Indo-Pacific fishes and, like the shrimpfishes, may swim mainly with the head down (Fig. 7–22*C*). The body is quite deep, the snout long, and they possess a large dorsal fin spine. The trumpetfishes, Aulostomidae, are circumtropical and much larger than the two preceding families, reaching a length of about 60 cm. They have a series of isolated spines along the back and a very long snout. The family Fistulariidae, the cornetfishes, are also circumtropical, and may approach 2 meters in length. They are extremely slender and have a long filament attached to the middle of the caudal fin. The sides of the long snout are equipped with some uncomfortably sharp ridges. I have observed large specimens spend up to a half hour in moving 60 or 70 cm to get close to a school of small goatfishes. During this time the fish seemed almost completely motionless.

The suborder Syngnathoidei includes the pipefishes and seahorses of Syngnathidae, peculiar little fishes with no pelvic fins and the body enclosed in bony rings. The pipefishes (*Syngnathus*) are elongate fishes, but the seahorses (*Hippocampus*) have the head flexed ventrally so that it is at a right angle to the body (Fig. 2–9*C*). The caudal portion is flexible and prehensile. Both pipefish and seahorse males are equipped with brood pouches in which the eggs are incubated. These fishes are often well camouflaged by both color and shape. Perhaps the best example is the seadragon of Australia, *Phyllopteryx foliatus*, which bears many leaf-like lobes. The family Solenostomidae, the ghost pipefishes, are less well known than the syngnathids. They have a long tube-like snout, large pectoral and caudal fins, and a high, flexible spinous dorsal. The female incubates the eggs between the pelvic fins. The single genus, *Solenostomus,* is found in the Indo-Pacific.

Order Perciformes (Percomorphi). This is the largest order of fishes, containing 16 or more suborders, depending on the breakdown by various ichthyologists. Some investigators have set up some of the suborders as separate orders. These typify the spiny-rayed or advanced bony fishes. In addition to the fin spines, they have thoracic or jugular pelvic fins, with the pelvic girdle usually connecting to the cleithra. The caudal fin has 17 principal rays and the pectoral fins are placed high on the side, with almost vertical bases. Branchiostegal rays number 5 to 8, usually 6 or 7, with 4 being placed on the outer surface of the upper portion of the ceratohyal and epihyal. The others attach to the edge of the lower section of the ceratohyal. Perciforms are physoclistic and usually have ctenoid scales. The orbitosphenoid and mesocoracoid are absent, and the maxillary is excluded from the gape by the premaxillaries. They are mostly marine, but have numerous important freshwater species in many parts of the world. Perciforms are known from the Upper Cretaceous.

The suborder Mugiloidei (Mugiliformes) contains the mullets, barracudas, and threadfins, and formerly included the silversides, which are now separated into the order Atheriniformes. The suborder is characterized by abdominal or subabdominal pelvics, with the pelvic girdle attached to the cleithra by a ligament; this condition is judged to be a reversal of the general evolutionary trend in the character, justifying placement in Perciformes. Ichthyologists who consider the placement of the pelvics in mullets to be primitive and unreversed usually place these fishes in an order preceding the Perciformes. Mugiloids are generally elongate and have two well separated dorsal fins.

The family Sphyraenidae, the barracudas, have very strong, sharp teeth set in sockets. The pectoral fins are placed low and the lateral line is normally developed. These are fierce, medium to large predators of warm seas. One species, the great barracuda, *Sphyraena barracuda,* reaches more than 2 meters in length. Barracudas, although implicated at times in "fish poisoning," are considered good food.

Mullets are in the family Mugilidae, and are found in warm oceans, bays and estuaries, and even fresh waters. The striped mullet, *Mugil cephalus,* is nearly circumtropical in distribution. Unlike the barracudas, mullet have feeble teeth and the gill rakers and pharyngeals are modified for straining microscopic plants and other organic materials from the water. They commonly feed over muddy bottoms.

Mullets are considered to be excellent food in some areas and are cultured in ponds. In most places where mullet culture is practiced the young fish or "seed" are gathered from natural spawning or simply allowed to enter enclosures that are then closed off. Italy, Hong Kong, and Taiwan are examples of mullet-rearing areas. Taiwanese mullet culturists have recently learned how to produce mullet "seed" through artificial propagation.

The family Polynemidae, the threadfins, are also tropical shore fishes. The snout is prominent, reaching well beyond the large mouth, and the pectoral fins are in two sections — an upper, rather normal part, and a lower section composed of four to eight long filaments that apparently serve as tactile organs (Fig. 2–12*A*). I have watched individuals swim a spiral course up one piling and down the next with the pectoral filaments fanned out to sense prospective food items over about a 40-cm wide area. Examples of genera are *Polynemus, Polydactylus,* and *Eleutheronema.* Species of the latter genus may reach 2 meters in length.

The suborder Anabantoidei includes the climbing perches, snakeheads, and gouramies. These are distinguished by a labyrinth organ which is developed from the upper part of the first gill arch and occupies much of the gill chamber. The bone is expanded and folded so as to present a great surface in a small space. Oxygen can be extracted from air trapped in this structure, so these fishes are at home in warm waters that may be very low in oxygen. They are freshwater fishes of Asia and Africa. Many species are known for their remarkable territorial courtship, and nesting behavior.

The family Anabantidae contains the climbing perches, *Anabas* sp. of Asia and *Ctenopoma* sp. of Africa. *Anabas* is equipped with stout spines on the operculum that aid in pulling the fish along over the ground when it migrates to a suitable habitat during the dry season (Fig. 7–23*A*).

The kissing gourami is in the family Helostomidae. The giant gourami, *Osphronemus goramy,* is placed in Osphronemidae. It is the largest of the group, reaching 60 cm, and is a favorite food fish in many sections of southeast Asia, where it is the subject of fish culture. Combfishes are in Belontiidae, along with the Siamese fighting fish, *Betta splendens,* the paradise fish, *Macropodus,* and other favorite aquarium fishes such as *Trichopsis* and *Trichogaster.* All these families are placed by some ichthyologists into Anabantidae.

Snakeheads, Channidae (Ophiocephalidae), are often assigned to a separate suborder or order because of the structure of the air-breathing organ, subabdominal pelvics, and lack of spines. Snakeheads are much more elongate than other Anabantoids; some species approach 1 meter in length. They are voracious predators, and although they are prized as food they cannot be cultured with food fishes susceptible to predation. A great advantage of snakeheads is that they can be held alive in the markets for days with proper care. The eggs of most anabantoids incubate at the surface, some floating of their own accord, with others placed in nests of bubbles. Usually the male attends the eggs.

Luciocephalidae, the pikeheads, are sometimes placed in their own suborder. These are small fishes of the Indo-Malayan region, distinguished by the lack of a gas bladder, the very protractile mouth, and the simple suprabranchial organ.

The suborder Percoidei, the typical perch-like fishes, is an immense

FIGURE 7–23. Representatives of order Perciformes. *A*, Climbing perch (*Anabas*); *B*, Rio Grande perch (*Cichlasoma*); *C*, butterfish (*Peprilus*).

group, containing over 60 families and about 4000 species. They are mostly marine, with a few very successful freshwater families. Members of some of the families are important as food fishes, others are sport fishes, and others are used in aquaria. Many of the families are ecologically important in their respective ecosystems. Most percoids have not diverged greatly from the basic, successful morphological groundplan, but there are diverse and important ecological, behavioral and reproductive adaptations. Relationships within the group have not been well worked out, but workers recognize several superfamilies.

Among the freshwater representatives of Percoidei are the black basses and sunfishes of Centrarchidae. These are small- to medium-sized fishes of North America, being especially abundant in the southern half of the United States. The largemouth bass, *Micropterus salmoides*, is a popular game fish that has been introduced to many areas outside its native range (Fig. 1–1G). It may reach a weight of 5 kg or more. The other thirty-odd members of the family include the pygmy sunfishes, *Elassoma* spp., the crappies, *Pomoxis* spp., and several members of *Lepomis*, of which *L. macrochirus*, the bluegill, is probably best known. The latter is a good pan fish and is often cultured with the largemouth in ponds. The only sunfish found native to the Pacific drainage of the U.S.A. is *Archoplites interruptus*, the Sacramento perch, a representative of a genus that was found in Oregon, Idaho, and Washington in the Pliocene.

Although the family Percidae is present in Europe and Asia, it reaches its greatest development in North America, where there are about 115 species, mostly darters of the genus *Etheostoma*. The familiar yellow perch of North America, *Perca flavescens*, has a close relative in the Eurasian *P. fluviatilis*. The walleye and sauger, *Stizostedion* spp. have their counterpart in the Old World *S. lucioperca*.

Another important freshwater family is Cichlidae, which ranges through warm freshwaters from India, Africa, South America, and Central America north to the Rio Grande. Their presence on the southern continents and islands such as Madagascar and Ceylon suggests that either ancestors could withstand brackish or even marine waters, or the distribution pattern dates from a time (Cretaceous) when those landmasses were closer together.

Only one genus, *Etroplus*, occurs in Asia, where *E. suratensis* of India, the pearl spot, is taken as a food fish. In the Americas, the Rio Grande perch, *Cichlasoma cyanoguttatum*, is the most northerly representative (Fig. 7–23*B*). Some members of *Cichlasoma* enter the aquarium trade, as do many others in the family. *C. biocellatum* is known as the Jack Dempsey, *Pterophyllum scalare* is the freshwater angelfish, and *Symphysodon discus*, the discus, is one of the favorites of aquarists.

A well known food fish is *Tilapia mossambica*, which has been brought to many parts of the world from its native Africa. Although capable of producing great populations in a short time with a high yield of protein per unit of area, the species often stunts and, due to the small size, those harvested are not readily accepted by some peoples. There are parts of Asia where *T. mossambica* is considered a pest. Other cultured species are *T. nilotica* and *T. zillii*. The elevation of subgenus *Sarotherodon* to generic rank to include the mouthbrooders of *Tilapia* has been suggested, but will not be accepted here.

In some of the great rift lakes of Africa, there are "species flocks" of cichlids, usually of the genus *Haplochromis* and close relatives. There is some evidence that numerous species and some new genera have evolved within certain lakes from a few founding species. Some flocks include well over 100 species, showing a remarkable range of trophic adaptations. Included are sand-plowers, detritivores, herbivores, insectivores, carnivores, eye-biters, fin-biters, scale-eaters, thieves of eggs and larvae and other specialists. Even among species with rather similar feeding habits, subtle differences in reproductive behavior have apparently allowed development of closely related forms that may occupy only slightly different niches in the same lake. Courtship and nesting activities in each species of cichlid apparently follow set patterns of behavior that have been of great interest to ethologists. Many cichlids are oral incubators; others build nests in the substrate.

Percoid families that contain important food fishes include Lutjanidae, the snappers, Sciaenidae, the drums and croakers, Carangidae, the jacks and cavallas (Fig. 2–19*A*), and Bramidae, the pomfrets. The numerous sea basses, hamlets, and groupers are in the family Serranidae, from which the temperate basses, including the striped bass, are sometimes separated into the separate family Percichthyidae.

Other families present in North America usually considered to be in the suborder include: Centropomidae, the snooks; Grammistidae, the soapfishes; Priacanthidae, the bigeyes; Apogonidae, the cardinalfishes; Branchiostegidae, the tilefishes; Pomatomidae, the bluefishes; Rachycentridae, the cobias; Coryphaenidae, the dolphins; Emmelichthyidae, the bonnetmouths; Lobotidae, the tripletails; Gerreidae, the mojarras; Pomadasyidae, the grunts; Sparidae, the porgies; Mullidae, the goatfishes; Pempheridae, the sweepers; Kyphosidae, the sea chubs; Ephippidae, the spadefishes; Pentacerotidae, the armorheads; Embiotocidae, the surfperches; Pomacentridae, the damselfishes, Cirrhitidae, the hawkfishes; Labridae, the wrasses; Scaridae, the parrotfishes; and Opisthognathidae, the jawfishes.

Some other families are Badidae, Nandidae, the leaf fishes and allies (including Pristolepidae); Arripidae, the Australian salmon; Scorpididae, sweeps; Nemipteridae, threadfin breams; Toxotidae, archerfish; Monodactylidae, moonfish, kitefish; Lethrinidae, emperors; Enoplosidae, oldwife; Oplegnathidae, beak perches; Plesiopidae, longfins; Acanthoclinidae, the scotties; Kuhliidae, flag tails and aholeholes; Cepolidae, bandfishes; Lactariidae, false trevallies; Labracoglossidae, false butterfishes; and Sillaginidae, "whitings."

Some rather specialized members of the suborder are the Trichodontidae, the sandfishes (Fig. 2 – 2*D*), which have fringed lips that allow them to lie concealed in sand and bring in a respiratory current without bringing in sand, and the Chiasmodontidae, the swallowers, which live in the deep ocean and can swallow fishes much larger than themselves. Both these families are placed with the Trachinoidei by some authors.

The Bathyclupeidae, which retains a pneumatic duct, and has the maxillaries included in the gape, is often placed in the Percoidei, but is considered by some scientists to constitute a separate order.

The suborder Stromateoidei is composed of marine fishes that have papillose lateral sacs extending from the pharynx or esophagus. Pelvic fins are thoracic or subthoracic. The Stromateidae, the butterfishes, are known as good food fishes. They have teeth in the expanded esophagus and some lack pelvics as adults. Examples of genera are *Poronotus* and *Peprilus* (Fig. 7 – 23*C*).

The man-of-war fishes, Nomeidae, have teeth in the esophageal sacs, but retain the pelvics as adults. Some members of *Nomeus* associate with large jellyfishes, swimming among the tentacles. The squaretail family, Tetragonuridae, is widely distributed in temperate and tropical seas.

The ragfishes, suborder Icosteoidei, are found in deep water in the North Pacific. The pelvics disappear in the adults as in some Stromateidae, but they have no modification of the esophagus. These are flabby, limp fishes with reduced scales. *Icosteus aenigmaticus* reaches a length of more than 2 meters (Fig. 7–24*A*). Icosteidae is considered to constitute a separate order by some ichthyologists and are placed with the previous suborder by others.

The suborder Blennioidei is considered to contain a large variety of marine fishes with jugular pelvics, and with each radial of the dorsal and anal fins corresponding with a neural or haemal spine. Most of the blennioids are rather elongate, with many being either taeniform or eel-shaped. The group may not be a natural one, and there seems to be no agreement as to which fishes should be included.

The Blenniidae (combtooth blennies, scaleless blennies) includes numerous small shore fishes of tropical and subtropical seas (Fig. 7–24*B*). They are common in the intertidal and one genus, *Salarius*, contains species that will leap from one tide pool to another when disturbed. Clinidae contains the familiar kelpfishes of the Pacific coast of North America, as well as numerous blennies from other shores. These are quite often perch-like in appearance. The Tripterygiidae, including the cockabullies of Australia, are similar in appearance to the foregoing two families, and like them are noted for the male's care of the eggs during incubation.

Anarhichadidae, the wolffishes, contains at least one species that is taken commercially, *Anarhichas lupus*. The wolf-eel, *Anarrhichthys ocellatus*, reaches 240 cm in length and is noted for its strong dentition of canine teeth in the front of the jaws, and wide molariform teeth in back. The diet is principally shellfish, and the species is often found in the crab pots of Pacific coast fishermen.

Pholidae, the gunnels, and Stichaeidae, the pricklebacks, are shallow-water and intertidal fishes of the North Atlantic and the North Pacific, with several species on the Pacific coast of North America.

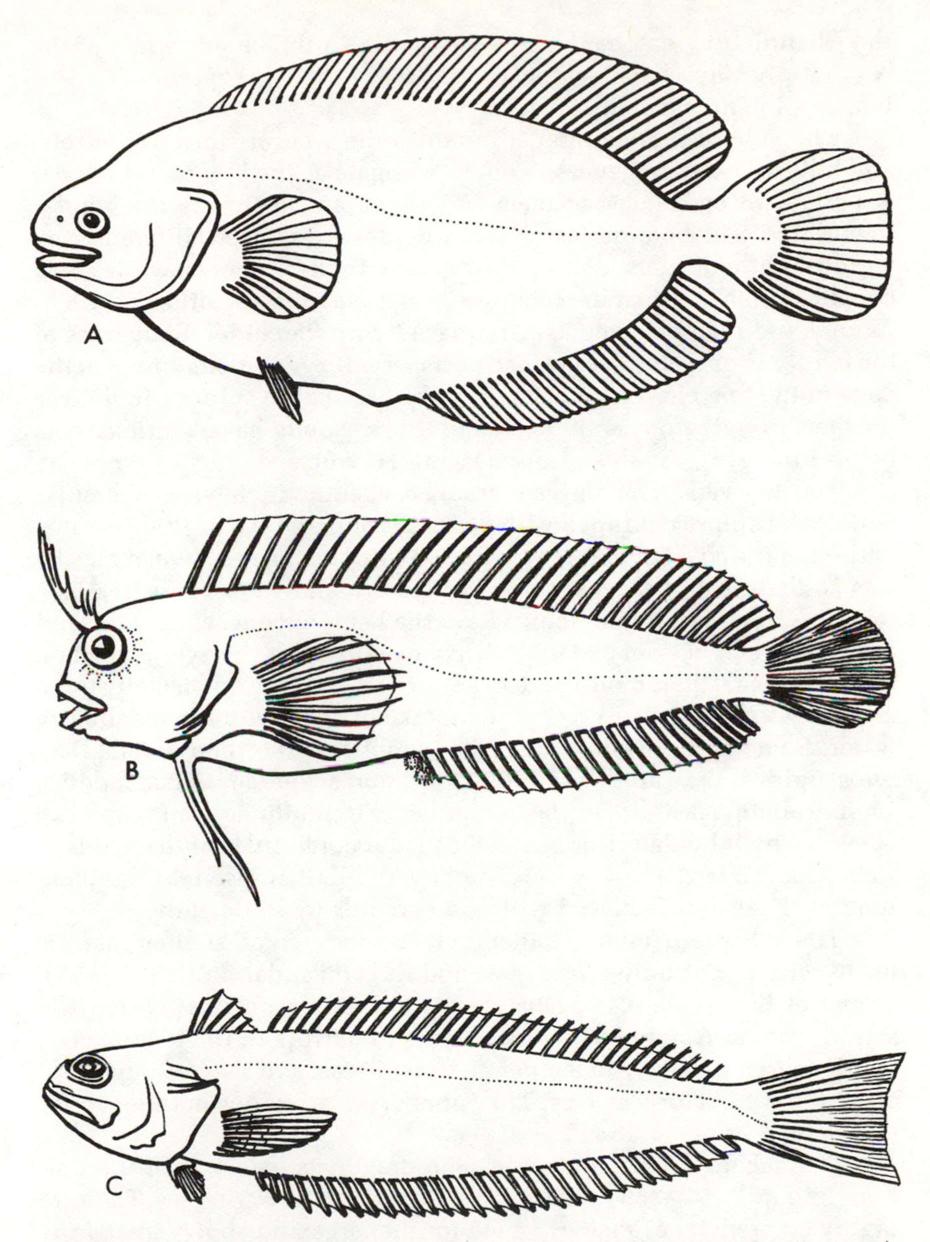

FIGURE 7–24. Representatives of order Perciformes. *A*, Juvenile ragfish (*Icosteus*); *B*, combtooth blenny (*Blennius*); *C*, weever (*Trachinus*).

Other families usually placed in Blennioidei include Xenocephalidae, Congrogadidae (mudblennies), Notograptidae, Peronedysidae, Ophioclinidae (snakeblennies), Chaenopsidae, Ptilichthyidae (quillfish), Scytalinidae (graveldivers) and Zaproridae (prowfish).

Bovichthyidae, Nototheniidae, Bathydraconidae and Channichthyidae (Chaenichthyidae) of the seas surrounding Antarctica are of uncertain relationships. They are variously placed in their own suborder (Notothenoidei), with the percoids, or with the blennioids. Ice-fishes of the Channichthyidae have no haemoglobin in the blood. Some of the Antarctic groups have a kind of antifreeze as an adaptation against freezing in supercooled waters.

The suborder Trachinoidei includes the weeverfishes, stargazers, and allies. These are generally not as elongate as the blennies, but bear structural resemblances to them. An important feature is the jugular pelvic fins. The boundaries of Trachinoidei are viewed differently by various ichthyologists. Some place them with the blennies, while some consider them a separate suborder but leave some families in Blennioidei and include some usually placed with Percoidei. Stargazers of the family Uranoscopidae are marine fishes with venomous spines at the edge of the opercle. Venom glands at the base of the spines can deliver poison through grooves in the spine. *Uranoscopus* has electric organs behind the eyes, capable of discharging 50 volts.

The weevers, Trachinidae, are also venomous, having opercular and dorsal spines equipped with venom-producing tissue (Fig. 7–24C). Although the effect of the weever's poison upon man is severe, it may be less so than that of the stargazers, which are known to cause death.

Most families of trachinoids have the habit of concealment in sand or in other soft bottom materials. They usually have the eyes placed on top of the head, their mouths are in a superior position, and usually there are fringes or flaps that prevent the intake of sand with the respiratory water. Some of these families are Trichonotidae, the sand divers, Dactyloscopidae, the sand stargazers, and Leptoscopidae. Other families include Bathymasteridae (the ronquils), Mugiloididae (sandperches), Cheimarrhichthyidae (torrent fishes), Percophididae ("flatheads"), Creediidae, Limnichthyidae, Oxudercidae, and Champsodontidae (bent tooths). The latter is placed with the percoids by some authors.

The suborder Ammodytoidei includes some small slender, marine fishes with a protruding lower jaw and forked caudal fin (Fig. 7–25*A*). They lack the gas bladder. Scales are cycloid and the pelvics jugular. The sand lances of Ammodytidae are known to burrow in the bottom. The family is found mainly in temperate to cool seas, and may be important as food for predatory species. They are harvested in northern Japan for use as food in fish culture.

The suborder Callionymoidei, the dragonets, consists of the Callionymidae (Fig. 7–25*B*) and Draconettidae, both of warm seas. They are highly colored fishes that are noted for the large and showy fins of the males.

The suborder Gobioidei consists of only six families but about 1000 species, mostly of tropical shores. The pelvic fins of some families are united into a sucking disc that adheres to the substrate. Gobiidae is the largest family in the suborder and includes species from all warm seas, some tropical fresh waters, and a few temperate marine and estuarine localities (Fig. 7–25C). These are mainly small fishes, mostly less than 10 cm long. The smallest (shortest) vertebrate is a freshwater goby of the Philippines, *Pandaka pygmaea*, which is mature at about 6 mm. Some gobies of southeast Asia and the Philippines occur in great enough

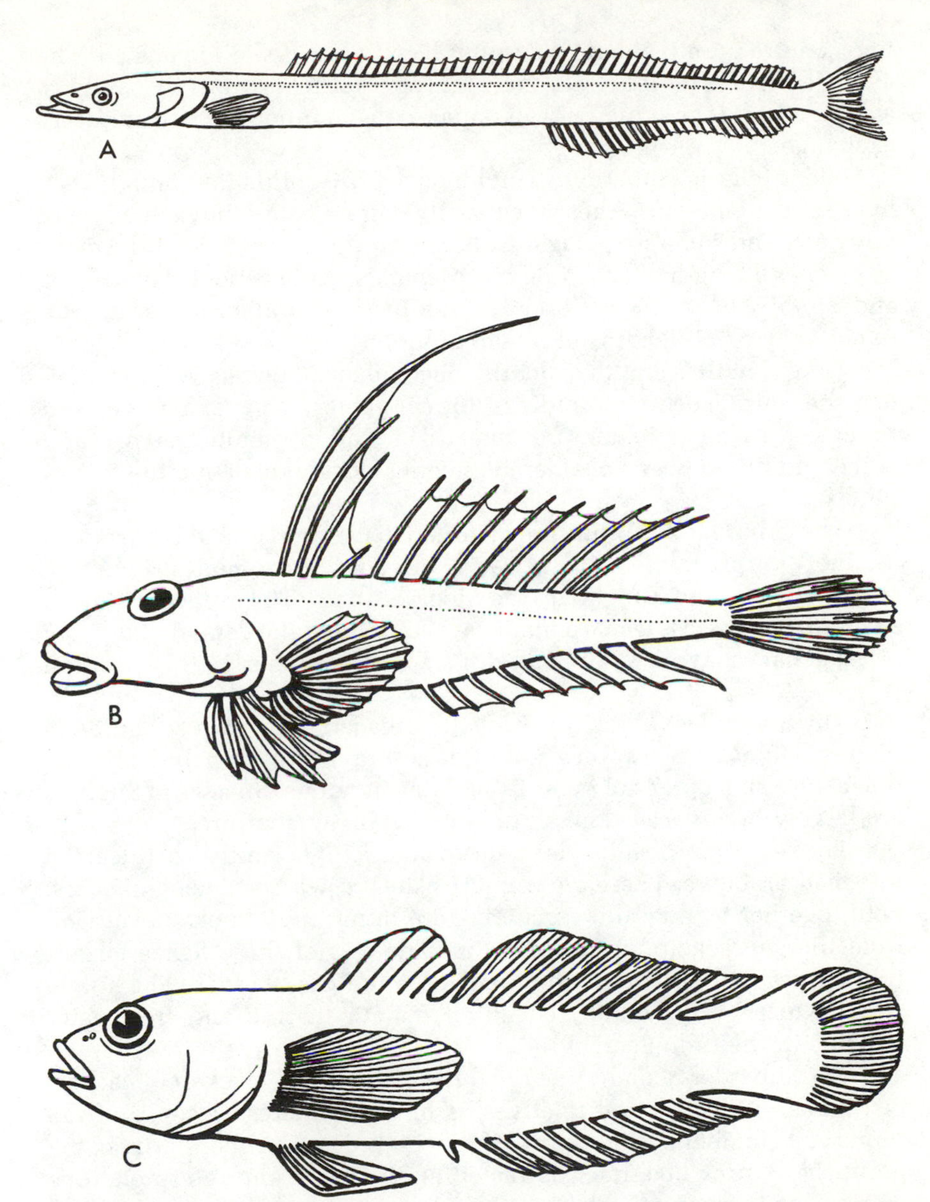

FIGURE 7–25. Representatives of order Perciformes. *A*, sand lance (*Ammodytes*); *B*, dragonet (*Callionymus*); *C*, goby (*Coryphopterus*).

numbers that they can be harvested, mixed with salt, and fermented to make "bagoong," or a sauce that is eaten with vegetables or rice.

The skipping gobies or mudskippers are often considered as constituting a separate family, Periophthalmidae. These inhabit tropical shore areas with soft bottoms and are usually seen at the water's edge or on the mud, rocks, or mangrove roots along the shore. They can pull themselves along with their arm-like pectoral fins, or can flip around with great rapidity by flexing the body. I once dropped one indoors and failed to recapture it until about 10 minutes later, when it was exhausted. The eyes of *Periophthalmus* are set high on the head, the gill cavity is expanded, and the skin is very vascular so that it functions in respiratory exchange.

Eleotridae is sometimes combined with Gobiidae, although they have separate pelvics. These are usually bottom fishes, but some live in midwater and some are pelagic. *Gobiomorus dormitor* of Central America reaches a length of nearly 60 cm. Members of the genera *Oxyeleotris* and *Bunaka* of southeast Asia and the Indo-Australian areas, respectively, may reach lengths of 50 cm or more.

Other families are Rhyacichthyidae, the loach gobies, Microdesmidae, the wormfishes, Taenioididae, the eel-gobies, Trypauchenidae, and Kraemeriidae. The latter are sometimes called sandfishes or sand lances, and have been considered as being closely related to the trachinoid fishes.

The suborder Acanthuroidei (surgeonfishes and allies) is a tropical marine group typical of coral reefs. Included are Acanthuridae, the surgeonfishes, and Zanclidae, the Moorish idols. The surgeonfishes of *Acanthurus* possess a sharp blade on each side of the caudal peduncle; other genera may be equipped with a series of spines in this position (Fig. 7–26*A*).

The Siganidae, or rabbitfishes, often placed in its own suborder, is distinguished by two spines and three soft rays in each pelvic fin. The dorsal and anal spines of rabbitfishes are venomous. Species of *Siganus* are largely herbivorous and show promise in aquaculture.

The suborder Kurtoidei is made up of the single family Kurtidae, the forehead brooders. These are tropical fishes in which the egg clusters, held together by a cordlike material, are hung upon hooks developed from the supraoccipital of the males. The ribs of these fishes form a tubular ossified structure which encloses the gas bladder. The single genus *Kurtus* (Fig. 7–26*B*) is found in both marine and freshwater habitats in the Indo-Pacific.

The suborder Scombroidei, including the snake mackerel, mackerels, tunas, billfishes, and relatives, is of great interest because of the important commercial and sport fishes it includes. The boundaries of the group have stretched and contracted according to the interpretations made of the structural relationships of the fishes included, which all have in common nonprotractile premaxillaries with maxillaries firmly attached to them.

The Gempylidae, the escolars or snake mackerels, are usually found in deep water, but some are found near the surface. They are usually equipped with strong teeth and have a slender, streamlined form. One species, *Ruvettus pretiosus*, the purgative fish or oilfish, reaches about 2 meters in length.

Trichiuridae, the hairtails and cutlassfishes, have exceptionally strong teeth and a compressed body with a long tapering tail, with that of *Trichiurus* ending in a fine point. Hairtails are harvested as food in many tropical countries.

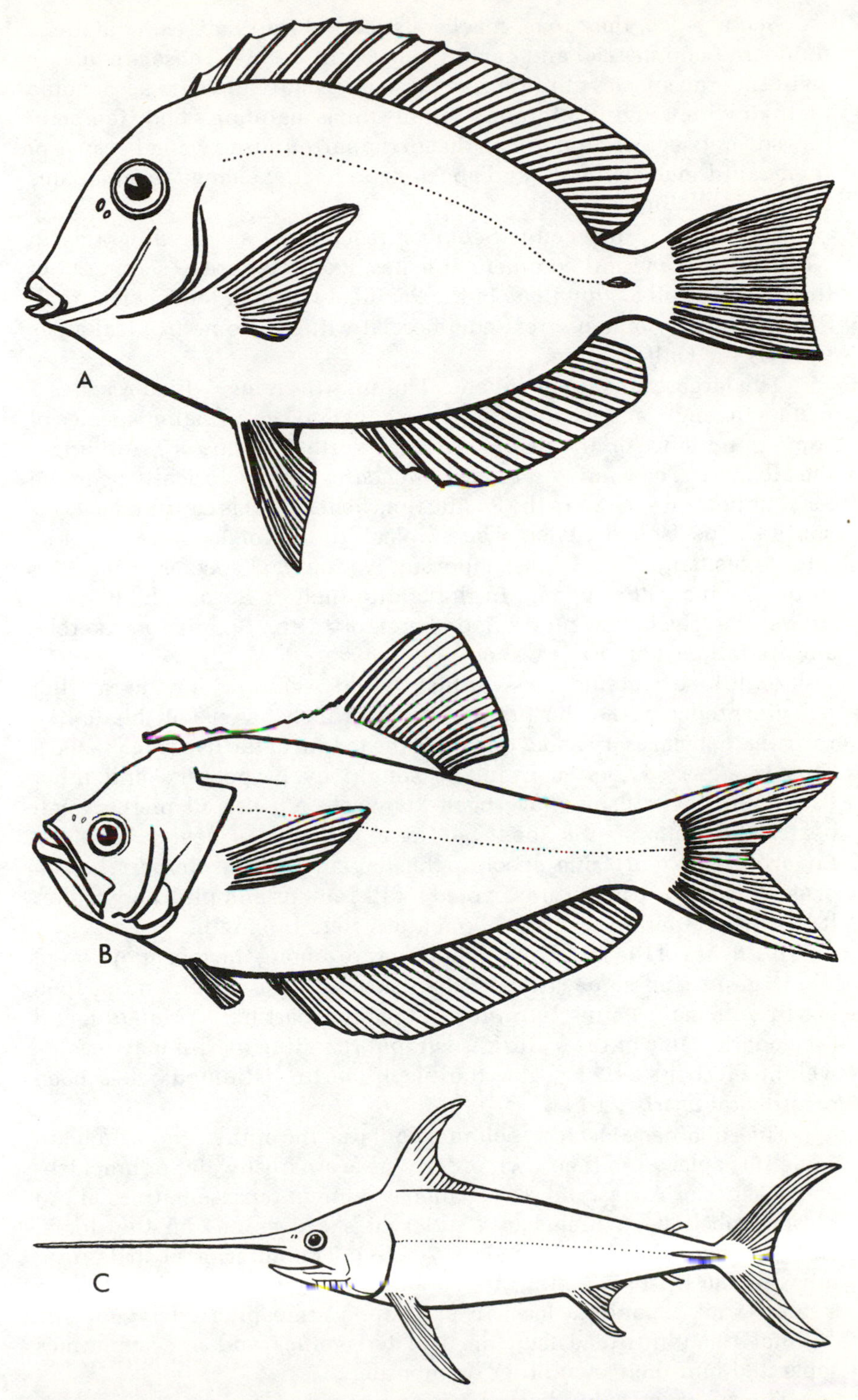

FIGURE 7–26. Representatives of order Perciformes. *A*, Surgeonfish (*Acanthurus*); *B*, forehead brooder (*Kurtus*); *C*, swordfish (*Xiphias*).

Scombridae, the tunas, mackerels, and close relatives, contains a number of commercial and game species (Fig. 2–8*A*). These are mostly swift-moving species of the surface waters, usually in warm seas, noted for their wide-ranging migrations. Many tunas maintain a body temperature several degrees higher than that of the surrounding water because of their rapid metabolic rate and specialized heat exchange mechanisms in the circulatory system.

Mackerels of the genus *Scomber* reach only a few kilograms in weight, but are useful commercial fishes. Examples are *S. scombrus* of the Atlantic and *S. japonicus* of the Pacific. Even smaller species are in *Rastrelliger*, a genus of great commercial value in some tropical areas such as the Gulf of Siam.

The largest tuna is the bluefin, *Thunnus thynnus*, which reaches 4 meters in length and may weigh up to about 800 kg. A smaller species of some importance on the Pacific Coast of North America is *T. alalunga*, the albacore. The yellowfin tuna, *T. albacares*, has historically been the most important species in the Pacific tuna fishery, but is captured mainly south of the United States. The skipjack tuna, *Katsuwonus pelamis*, although small, is one of the important commercial species, especially in the Pacific. Other genera in the family include *Sarda*, the bonitos, *Auxis*, the frigate mackerels, *Scomberomorus*, the Spanish mackerels, and *Acanthocybium*, the wahoo.

Xiphiidae contains the swordfish, *Xiphias gladius*, a wide-ranging pelagic predator like the tunas. The swordfish has a flat, blade-like rostrum that makes up about one third the length of the fish (Fig. 7–26*C*). This is a prized food fish and is sought by harpooners and other fishermen. Swordfishes have been known to attack and pierce small boats. Maximum size is about 540 kg at a length of nearly 5 meters. Luvaridae has only one species, the louvar, *Luvarus imperialis*, an oceanic fish of warm latitudes noted for its red fins and pink body color. It has been reported to reach about 1.8 meters in length.

Billfishes of the family Istiophoridae are among the most popular of the large marine game fishes. Spearfishes and the white and striped marlins are in the genus *Tetrapturus*, the blue marlin is in *Maikara*, and the generic name of the sailfish is *Istiophorus*. Blue marlin may reach a weight of about 640 kg, and the black marlin, *Istiompax*, has been recorded at nearly 710 kg.

The suborder Mastacembeloidei contains the spiny eels, which are sometimes placed in their own order. These are freshwater or brackish-water fishes of Africa and Asia. The best known representatives are in Mastacembelidae, which has a series of sharp spines on the dorsal surface before the dorsal fin (Fig. 7–27*A*). The anterior nostrils form tubes at the tip of the elongate snout. Many mastacembelids live in swamps and depend, at least in part, upon atmospheric oxygen. The oriental Chaudhuriidae lack the isolated spines and are sometimes separated into their own order or suborder.

The suborder Schindleroidei is composed of one family, Schindleriidae, with two species. These are small larvoid marine fishes of the

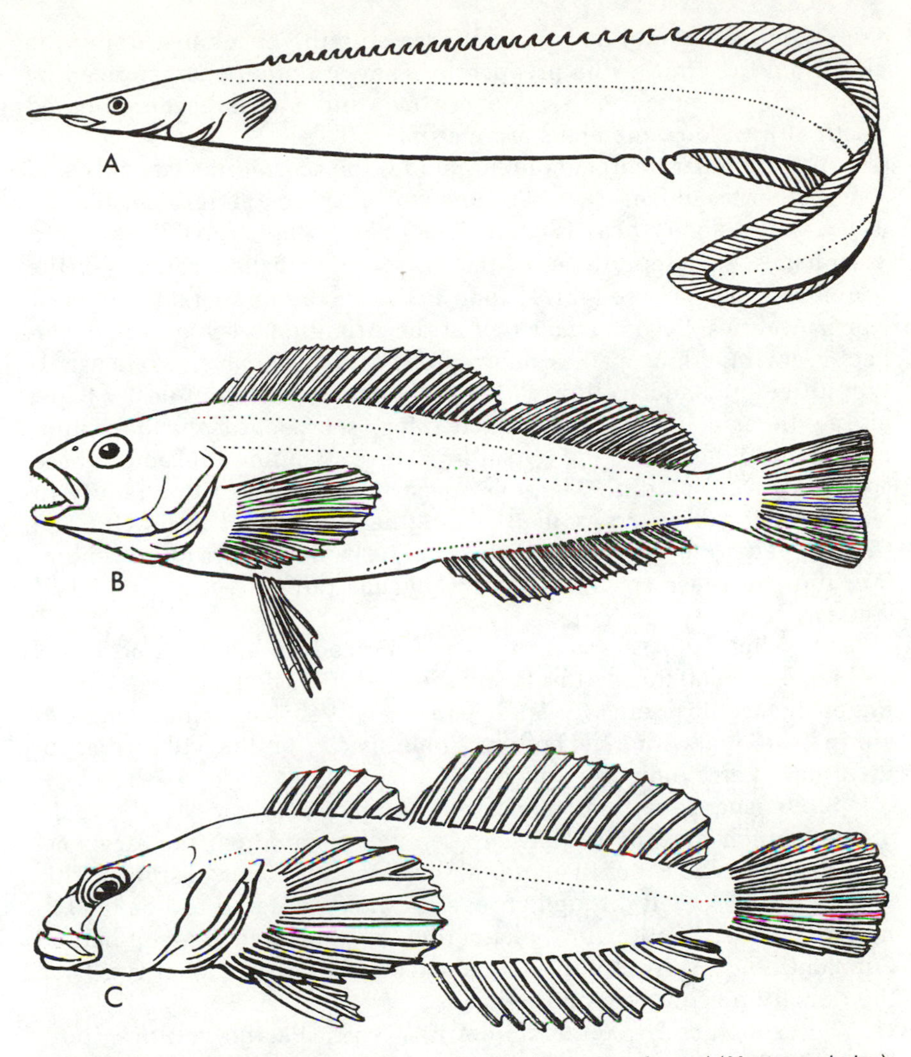

FIGURE 7–27. Representatives of *A*, order Perciformes, spiny eel (*Mastacembelus*); *B*, order Scorpaeniformes, greenling (*Hexagrammos*); *C*, order Scorpaeniformes, sculpin (*Cottus*).

tropical Pacific that resemble the Ammodytoidei. They are usually less than 3 cm long.

The suborder Echeneoidei, containing the remoras, is often considered a separate order called Echineiformes or Discocephali because of the sucking disc formed by the modified dorsal fin (Fig. 2–15*D*). These interesting fishes, all in Echineidae, are found in all warm seas and are noted for their habit of clinging to large fishes or turtles and riding along.

Order Scorpaeniformes (Scleroparei). This rather large and important group is sometimes considered a suborder of Perciformes. Its members share the common character of a "suborbital stay," formed by the second suborbital and typically crossing the cheek just under the skin from the orbit to the preopercle. They are often characterized by bony plates or spines on the head and body and have large, broad-based pectoral fins. Most members are marine.

The suborder Scorpaenoidei includes the scorpionfishes and rockfishes (Scorpaenidae), the searobins and gurnards (Triglidae, Fig. 2–12*B*), the stonefishes (Synanceidae) and the velvetfishes and waspfishes, Congiopodidae, of the Pacific. Rockfishes, especially the genus *Sebastes*, are sought as food fishes in the temperate waters of North America. *Sebastes marinus* of the Atlantic and *S. alutus* of the Pacific are important species. Scorpionfishes, *Scorpaena* spp., are usually multicolored and well provided with spines, cirri, and fleshy flaps, giving them a bizarre appearance. Most scorpaenids have venom-producing tissue along the dorsal fin spines. Wounds caused by these spines are very painful. There are over 80 members of this family in North American waters, some reaching nearly 1 meter in length. The tropical genus *Pterois*, the lionfishes and turkeyfishes, are quite venomous and can cause severe illness in persons punctured by the dorsal spines.

The triglids or searobins are characterized by bony-plated heads and large pectoral fins, with a few of the lower rays detached as separate finger-like tactile organs. The pelvics are relatively large and strong and aid in "walking" along the bottom. Some species of this widely distributed family are commercial fishes.

Stonefishes take their name from their resemblance to the rocks among which they live. They are reputed to produce the strongest venom among the fishes in glands at the bases of the dorsal spines. This venom is discharged through grooves in the spines and can be fatal to humans. Other families in the suborder are the tropical, deep-bodied Caracanthidae, Aploactinidae, and Pataecidae. Both the latter two have the dorsal fin beginning on the head.

The suborder Hexagrammoidei is a North Pacific marine group. Hexagrammidae includes the carnivorous lingcod, *Ophiodon elongatus*, and the greenlings, *Hexagrammus* spp., the latter characterized by several lateral lines on each side (Fig. 7–27*B*). Anoplopomatidae contains the skilfish, *Erilepis*, and the sablefish, *Anoplopoma*. Zaniolepidae contains the combfishes, *Zaniolepis*.

The suborder Platycephaloidei is composed of two families, the Platycephalidae, the flatheads, and the Hoplichthyidae, both from the Indo-Pacific region. They are noted for the extremely protractile mouth.

The suborder Cottoidei includes several groups of related fishes that are constantly being shuffled and reshuffled by ichthyologists. The listing given here will lump 11 of these groups,* which occasionally are accorded family status, with the Cottidae. Cottids, or sculpins, are

---

*Icelidae, Ereunidae, Blepsiidae, Scorpaenichthyidae, Cottocomephoridae, Rhamphocottidae, Synchiridae, Psychrolutidae, Cottunculidae, Hemitripteridae, Ascelichthyidae.

typically marine fishes of the temperate and cold waters of the Northern Hemisphere. They are usually large-headed with large, fan-like pectoral fins, and many are characterized by strong spines on the gill cover. They often have bony plates on the skin, and seldom have more than a few rows of scales. Freshwater members are mainly in the genus *Cottus*, whose species are common inhabitants of cold streams (Fig. 7–27C).

Comephoridae are viviparous pelagic fishes of Lake Baikal in Siberia. Normanichthyidae are fully-scaled marine fishes of the Pacific coast of South America. Agonidae contains the poachers and alligatorfishes, curious armored fishes of cold marine waters. They are present along the coasts of southern South America and are especially numerous in the North Pacific.

Cyclopteridae is considered to contain the lumpfishes and snailfishes, although the latter are often placed in their own family, Liparidae. Most of these fishes have the pelvic fins modified into a round suctorial disc with which they cling to rocks and vegetation (Fig. 2–14C). They are marine and are found in all cold seas.

Order Pleuronectiformes (Heterosomata). The flatfishes are offshoots of the perciform fishes that have acquired the habit of swimming with the laterally compressed body oriented horizontally instead of vertically. Early in their development, after a period of "normalcy," they begin the side-swimming and an eye migrates from what becomes the bottom side to the upper side. The side toward the bottom is blind and colorless. They are marine fishes, common on most coasts. A few enter fresh water and some are found at great depths. They are important food fishes.

The suborder Psettodoidei is composed of one family, Psettodidae, which has spinous rays in the dorsal and anal fins, and bears strong resemblance to some of the percoids, the group from which this order is probably derived. The suborder Pleuronectoidei has no spines in the fins and there are usually six pelvic rays. Flounders, halibuts, and sanddabs are found in this group (Figure 7 – 28*A*). The family Bothidae includes the lefteye flounders — species with both eyes on the left side. Representative genera are *Citharichthys*, the whiffs and sanddabs, and *Paralichthys*, which includes the summer flounder and California halibut.

The family Pleuronectidae, or righteye flounders, includes some well known commercial fishes such as the halibuts of the genus *Hippoglossus*. These are the largest of the flatfishes, the Atlantic halibut weighing over 300 kg and the Pacific halibut attaining about 210 kg. The history of the fishery for the latter is one of early over exploitation followed by careful biological study and finally, rational management. Some of the members of this family are called "soles" even though they do not show the characteristics of true soles. Examples of pleuronectids called "soles" are the petrale sole, *Eopsetta jordani*, the rex sole, *Glyptocephalus zachirus*, and the rock sole, *Lepidopsetta bilineata*. In the starry flounder, *Platichthys stellatus*, half or more of the specimens have the eyes on the left side.

The suborder Soleioidei includes the true soles and tonguefishes.

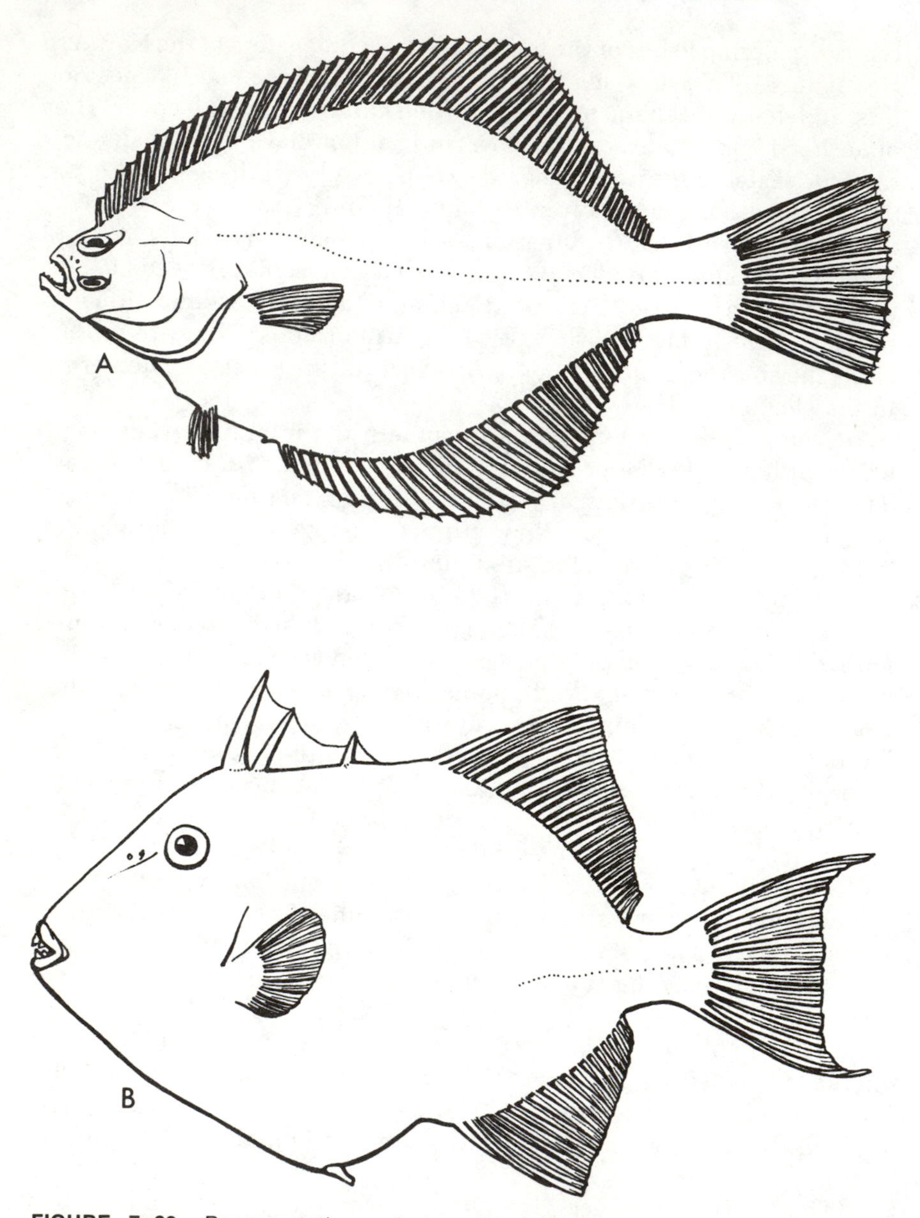

FIGURE 7–28. Representatives of *A,* order Pleuronectiformes, starry flounder (*Platichthys*); *B,* order Tetraodontiformes, triggerfish (Balistidae);

(*Illustration continued on opposite page.*)

These are generally smaller fishes than the flounders and halibuts, but those species large enough to sustain a fishery are highly prized as food. Members of Soleidae are righteyed and tend to prefer warm waters, but a few are found in temperate seas. The European *Solea solea* is the species that made filet of sole a popular dish. Tonguefishes of Cynoglossidae also tend to be warmwater fishes. These are lefteyed, rather slender fishes used as food wherever large enough species exist. Most North

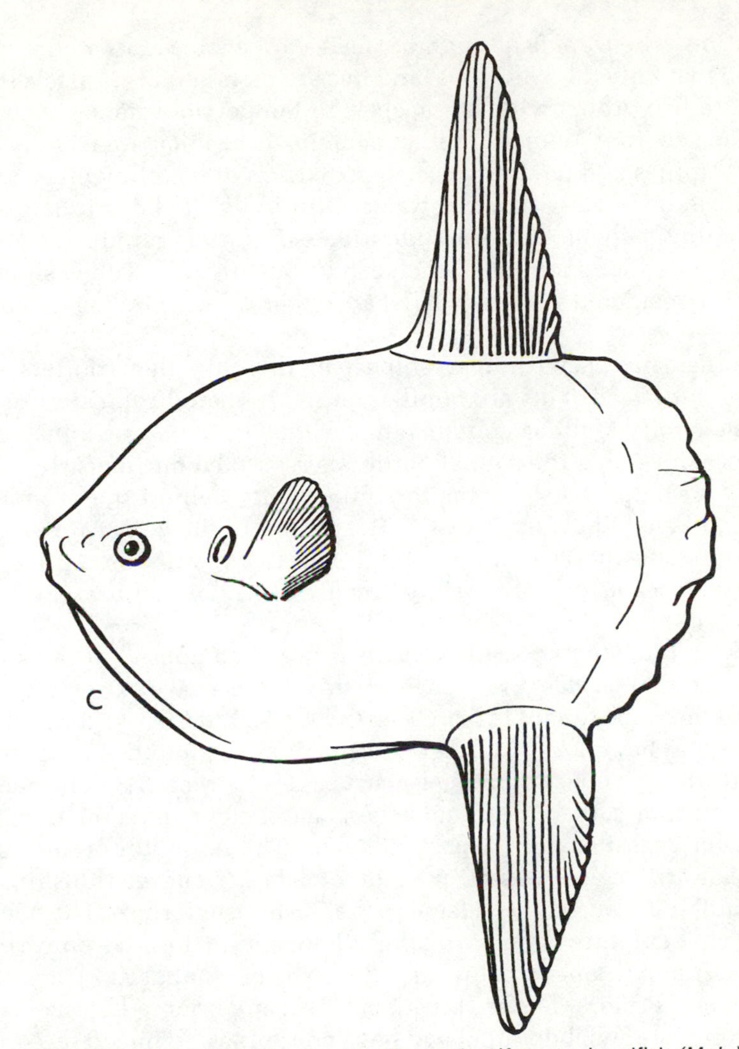

FIGURE 7–28 *Continued.* *C,* order Tetraodontiformes, headfish *(Mola).*

American species represent the genus *Symphurus*, and are mostly too small to be taken commercially.

Order Tetraodontiformes (Plectognathi). The headfishes, boxfishes, puffers, and relatives are distinguished by strong jaws and a small mouth with strong incisors or a sharp beak composed of modified teeth. Scales are modified into plates or spines.

The suborder Balistoidoi, all members of which have separate well-developed teeth, contains the spikefishes and relatives, Triacanthodidae and Triacanthidae, the triggerfishes and filefishes, Balistidae (including Monacanthidae; Fig. 7 – 28*B*) and the boxfishes and trunkfishes, Ostraciontidae. With the exception of the latter, these are all characterized by large dorsal spines which, in the balistids, can be locked erect. These are mainly shore fishes of the tropics; some have brilliant coloration. The trunkfishes and boxfishes are enclosed in a bony carapace so that only the eyes, jaws, and fins are mobile. This is also a tropical shore fish group, with few species in temperate waters.

The suborder Tetraodontoidei contains fishes that are usually covered by spines and possess a sac, an extension of the alimentary canal, which can be filled with air or water so an individual can inflate itself like a spiny balloon. The Tetraodontidae, or puffers, produce a strong poison, called tetrodotoxin, which can be fatal to man. The flesh of the puffer is considered a delicacy in Japan, but great care is taken to remove

porcupinefishes, have longer spines on the skin than puffers. The inflated and dried skins are familiar curios in many tropical areas.

The family Molidae contains the headfishes or ocean sunfishes. In these oceanic fishes the caudal portion is restricted during early development so that the fish seems to end abruptly behind the prominent dorsal and anal fins (Fig. 7 – 28C). Headfishes of the genus *Mola* range into temperate waters and may weigh nearly a metric ton. *M. mola* is called the ocean sunfish, possibly because of its habit of basking at the surface.

Order Dactylopteriformes. This order is composed of the single family Dactylopteridae, the flying gurnards. These have extremely large pectoral fins and are said to leap from the water, but their gliding ability is in doubt. They are tropical marine fishes, usually found at the bottom. This peculiar group is sometimes placed as a suborder in Perciformes or Scorpaeniformes, there being no general agreement on its relationship.

Order Pegasiformes (Hypostomides). The sea moths of the family Pegasidae are among the most peculiar of fishes. Their relationships are uncertain, but they may be related to the Gasterosteiformes. The body is covered by sculptured bone, with the tail portion enclosed in bony rings. The small mouth opens below an odd rostrum composed of modified nasal bones. Pectoral fins are broad, fan-like, and oriented horizontally. The pelvics are subabdominal and have one spine and one to three softrays. There are no other fin spines. The gas bladder is lacking. These are small shore fishes of the Indo-Pacific (Fig. 2 – 9A).

Order Batrachoidiformes (Haplodoci). This and the following two orders show certain skeletal relationships to the percopsiforms and gadiforms and are regarded by some ichthyologists to constitute, with these groups and others, a lineage (Paracanthopterygii) separate from the perciforms and near relatives. This is the order of the toadfishes and midshipmen, usually broad-headed big-mouthed bottom fishes of tropical and temperate seas. The pelvic fins are jugular and have one spine and two or three soft rays. Both the premaxillaries and maxillaries form the border of the mouth, but only the former bears teeth. There are three pairs of gills. A short spinous dorsal, of two to four spines, precedes a long soft dorsal. Venom is produced in glands at the bases of the dorsal spines, and some (*Thalassophryne* and *Daector*) have hollow spines through which the painful venom can be injected.

There is only one family, Batrachoididae. The genus *Porichthys* takes the name midshipman from the rows of photophores that course

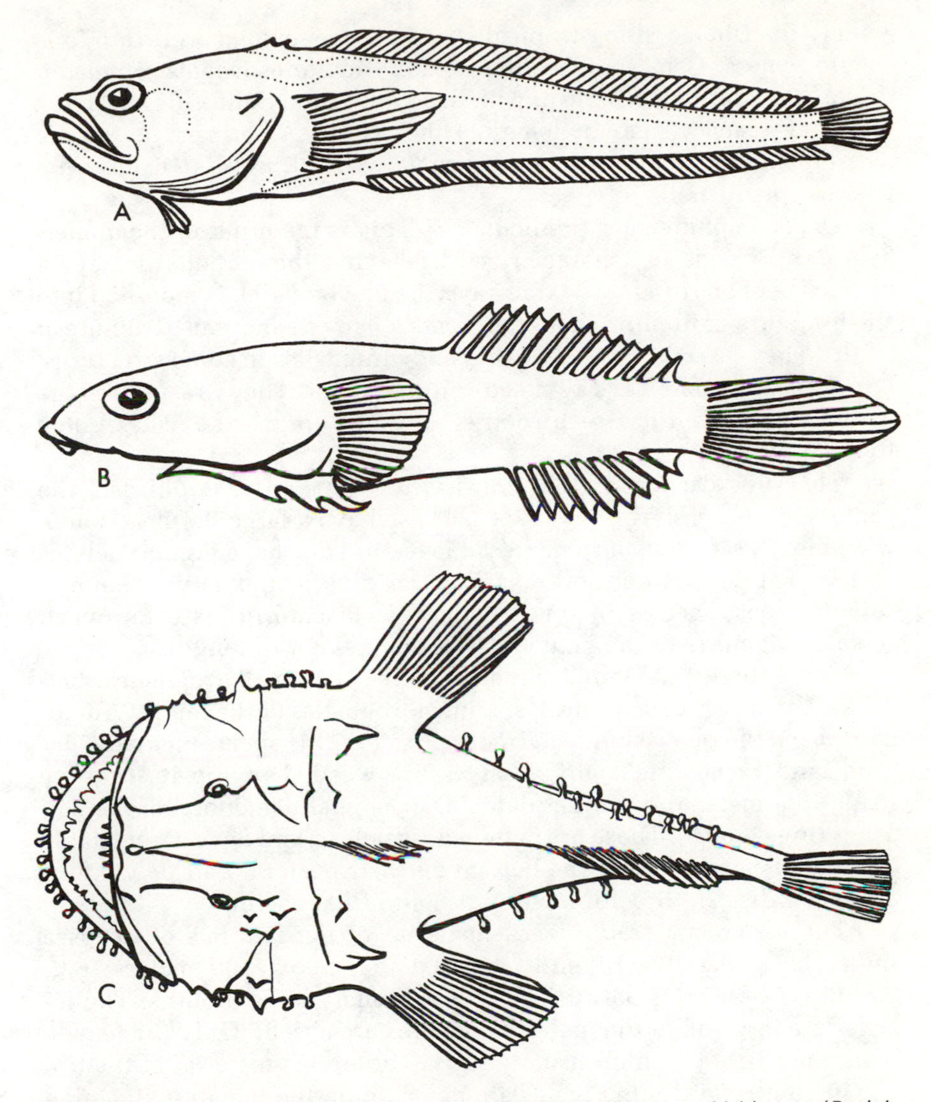

FIGURE 7–29. Representatives of *A,* order Batrachoidiformes, midshipman (*Porichthys*); *B,* order Gobiesociformes, clingfish (Gobiesocidae); *C,* order Lophiiformes, goosefish (*Lophius*), dorsal view.

along the body like buttons (Fig. 7–29*A*). These are among the few shallow-water fishes which produce light.

Order Gobiesociformes (Xenopterygii). This order is regarded by some to be part of the paracanthopterygian line. It is structurally related to the preceding and following orders, but is distinguished by a suctorial disc on the ventral surface below the pectorals (Fig. 2 – 14*A*). This consists of the modified pelvic fins and skin folds, and can exert a powerful suction so that specimens are sometimes very difficult to remove from a smooth surface. The disc is large in Gobiesocidae but reduced in Alabetidae. These are small, soft-rayed, scaleless fishes widely distributed along tropical shores. A few extend well into temperate waters. Gobiesocidae (Fig. 7–29*B*) includes *Gobiesox maeandricus*, the northern clingfish, which is found from California to Alaska and may reach 10 cm in length. Other North American genera are *Acyrtops* and *Rimicola*. Alabetidae contains small, eel-like fishes from the seas of Australia.

Order Lophiiformes (Pediculati). This is the order of the anglerfishes and frogfishes, characterized by having the spinous dorsal fin composed of one to a few flexible rays, the first of which is modified into a fishing lure or illicium that often bears a flap or a bulbous structure at its tip. They resemble the toadfishes and clingfishes in some structures, but are ribless and have reduced gill openings. They are part of the Paracanthopterygii of Greenwood et al. These are marine fishes, some living at great depths.

The suborder Lophioidei consists of one family, Lophiidae, the anglers or goosefishes (Fig. 7–29*C*). These have a large flat head and a wide mouth set with sharp depressible teeth. They have jugular pelvics with one spine and five soft rays. *Lophius americanus* is the common Atlantic coast species in the Americas; *L. piscatorius* is a European species. Members of the genus may reach 130 cm in length.

The suborder Antennarioidei includes the small frogfishes, batfishes, and allies. One of the best known frogfishes of the family Antennariidae is the sargassum fish, *Histrio histrio*. This species may live far from land in floating mats of sargassum weed, hanging its eggs by string-like material onto the plants. Ogcocephalidae contains the batfishes (Fig. 7–30*A*). These are flattened, scaleless, and have a retractable illicium. They walk over the bottom on their pectoral and pelvic fins. Other families are Brachionichthyidae and Chaunacidae.

The suborder Ceratioidei is made up of ten families of deepsea anglerfishes that live in midwater, usually below 300 meters (Fig. 7 – 30*B*). These are small fishes with extremely large mouths. The females are larger than the males. Pelvic fins are absent. Only the female bears the illicium, which usually carries luminous material at the tip.

In some families the males are parasitic on the females, attaching firmly with their jaws and becoming, in most cases, completely dependent upon the female for their blood supply. Parasitic males are known in Ceratiidae, Caulophrynidae, and Linophrynidae (including Aceratiidae and Photocarynidae). In Melanocetidae, Himantolophidae, and Oneirodidae the males are known to be free-swimming, but have a remarkably large olfactory apparatus and jaws suited to clamping onto the female. Other families are Diceratiidae, Gigantactinidae, Neoceratiidae and Centrophrynidae.

Order Synbranchiformes (Synbranchii). The swamp eels are elongate fishes without pectorals. The pelvics, if present, are jugular and have no more than two rays. They are soft-rayed, have no gas bladder, and have the gill openings confluent ventrally. The gills are greatly reduced and respiration is pharyngeal or intestinal. Many of the species move overland in or near the swamps where they live. The order contains one family, Synbranchidae, which has several species in tropical areas. The members have no fin rays except for the caudal, and lack

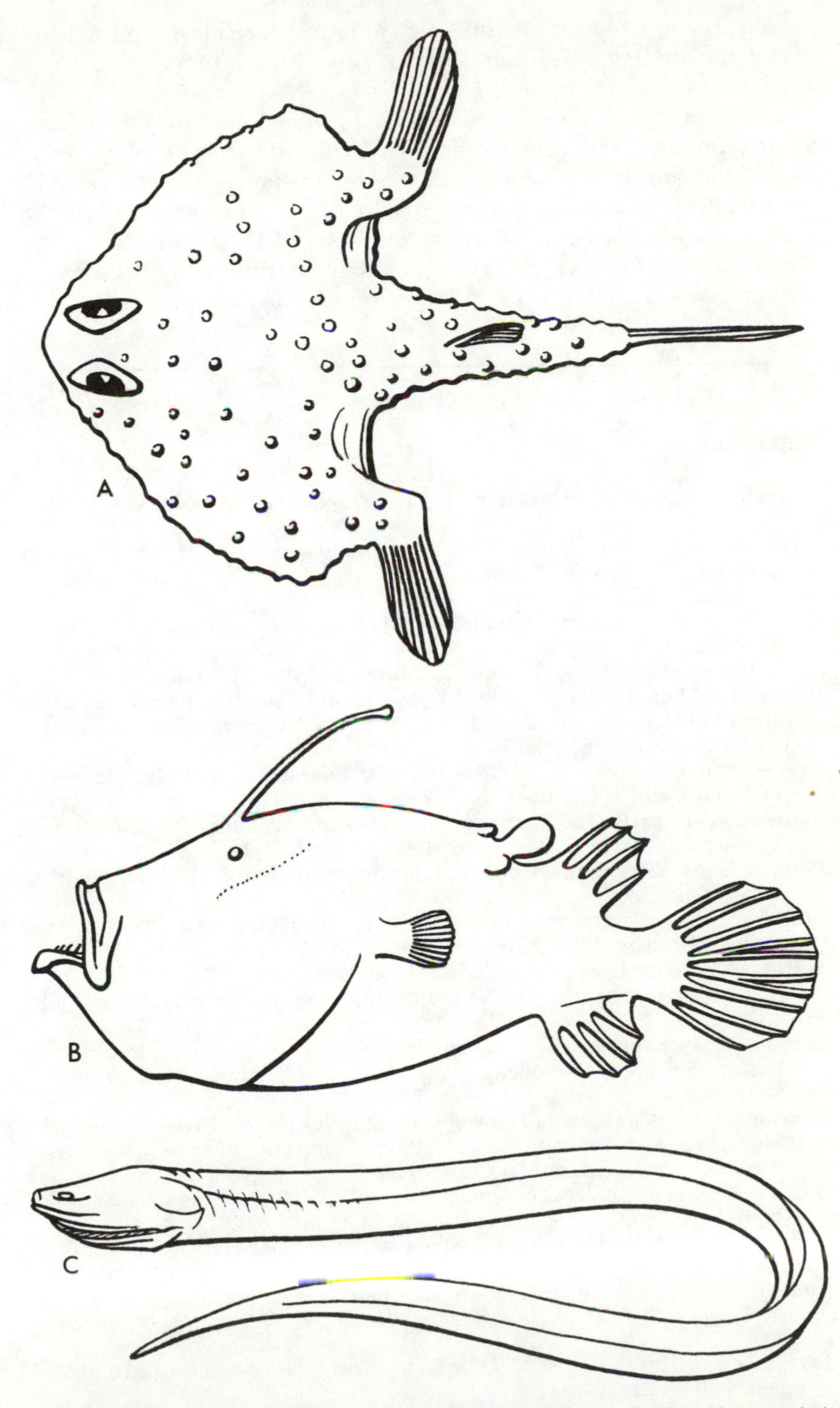

FIGURE 7–30. Representatives of *A,* order Lophiiformes, batfish (*Ogcocephalus*), dorsal view; *B,* order Lophiiformes, deepsea angler (Ceratioidei); *C,* order Synbranchiformes, rice eel (*Monopterus*).

pelvics. Some live in caves in Africa and Yucatan; most are found in freshwater or brackish-water swamps; one, *Macrotrema caligans* of Malaya, is marine.

Synbranchus is the most widespread genus, found in Asia, Africa, South America, and Mexico. The rice eel, *Monopterus albus* (Fig. 7 – 30C) of southeast Asia and Indonesia, can spend the dry season in holes in the bottom when the swamps dry up. *Amphipnous cuchia*, the cuchia of India, differs from other members of the family in having special air sacs for aerial respiration communicating with the pharynx. *Amphipnous* is often placed in its own family, Amphipnoidae.

References

The following is a complete listing for all references in Section Two.

Adam, H., and Strahan, R. 1963. Systematics and geographical distribution of myxinoids. *In:* Brodal, A., and Fänge, R. (eds.), The Biology of *Myxine*, Oslo, Universitetsforlaget, pp. 1–8.

Andrews, S. M. 1973. Interrelationships of crossopterygians. *In:* Greenwood, P. H., Miles, R. S., and Patterson, C. (eds.), Interrelationships of Fishes, New York, Academic Press, J. Linn. Soc. (Zool.) 53 (Suppl. 1): pp. 137–177.

Bailey, R. M., et al. 1970. A List of Common and Scientific names of Fishes from the United States and Canada, 3rd ed. Washington, D.C., American Fisheries Society, Special Publications No. 6.

———, and Cavender, T. M. 1971. Fishes. *In:* McGraw-Hill Encyclopedia of Science and Technology. New York, McGraw-Hill.

Banister, K. E. 1970. The anatomy and taxonomy of *Indostomus paradoxus* Prashad and Mukerji. Bull. Brit. Mus. Nat. Hist. (Zool.), 19(5):179–209.

Barlow, G. M. 1961. Causes and significance of morphological variation in fishes. Syst. Zool., 10:105–117.

Berg, L. S. 1947. Classification of Fishes, Both Recent and Fossil (transl.). Ann Arbor, Michigan, J. W. Edwards, pp. 87 – 517.

Bertelsen, E. 1951. The ceratioid fishes. Dana Rep., 39:1 – 276.

Bertin, L., and Arambourg, C. 1958. Super-ordre des Teleosteens. Trait. Zool., 13(3):2204 – 2500.

Bertmar, G. 1968. Lungfish phylogeny. *In:* T. Ørvig (ed.), Current Problems of Lower Vertebrate Phylogeny. New York, Wiley-Interscience, pp. 259–283.

Bjerring, H. C. 1973. Relationships of coelacanthiforms. *In:* Greenwood, P. H., Miles, R. S., and Patterson, C. (eds.), Interrelationships of Fishes. New York, Academic Press, J. Linn. Soc. (Zool.) 53 (Suppl. 1):179–205.

Blot, J. 1966. Holocephales et elasmobranches systématique. *In:* Piveteau, J. (ed.), Traité de Paleontologie, Tome 4. Paris, Masson et Cie, 2:702–776.

Bohlke, J. E. 1966. Lyomeri, Eurypharyngidae, Saccopharyngidae. *In:* Fishes of the Western North Atlantic. Mem. Sears Found. Mar. Res., 1(5):603–628.

Boulenger, G. A. 1904. Teleostei (systematic part). *In:* Harmer, S. F., and Shipley, A. E. (eds.), Fishes. London, Cambridge Natural History, Vol. 7, pp. 539–727.

Briggs, J. C. 1955. A monograph of the clingfishes (order Xenopterygii). Stanford Ichthyol. Bull., 6:1–224.

Cavender, T. M. 1970. A comparison of coregonines and other salmonids with the earliest known teleostean fishes. *In:* Lindsey, C. C., and Woods, C. S.

(eds.), Biology of Coregonid fishes. Winnipeg, University of Manitoba Press, pp. 1–32.

Chardon, M. 1968. Anatomie comparée de l'appareil de Weber et des structures connexes chez les Siluriformes. Mus. R. Afr. Cent. Ann. (Ser. 8, Zool.), 169:1–277.

Compagno, L. J. V. 1973. Interrelationships of living elasmobranchs. *In:* Greenwood, P. H., Miles, R. S. and Patterson, C. (eds.), Interrelationships of Fishes. New York, Academic Press, J. Linn. Soc. (Zool.) 53 (Suppl. 1): 15–61.

Dean, B. 1895. Fishes, Living and Fossil. An Outline of Their Forms and Probable Relationships. New York, Macmillan.

Forey, P. L. 1973*a*. Relationships of Elopomorpha. *In:* Greenwood, P. H., Miles, R. S., and Patterson, C. (eds.), Interrelationships of Fishes. New York, Academic Press, J. Linn. Soc. (Zool.) 53 (Suppl. 1): 351–368.

———. 1973*b*. A revision of the elopiform fishes, fossil and Recent. Bull. Brit. Mus. Nat. Hist. (Geol.; Suppl. 10):1–222.

Fowler, H. W. 1964–1973. A catalog of world fishes. Q. J. Taiwan Mus., Vols. 1–19.

Fraser, T. H. 1972. Some thoughts about the teleostean fish concept–the Paracanthopterygii, Jap. J. Ichthyol., 19:232–242.

Freihofer, W. C. 1970. Some nerve patterns and their systematic significance in paracanthopterygian, salmoniform, gobioid, and apogonid fishes. Proc. Cal. Acad. Sci., 38:215–264.

———. 1973. Trunk lateral line nerves, hyoid arch, gill rakers, and olfactory bulb location in atheriniform, mugiloid and percoid fishes. Occas. Pap. Cal. Acad. Sci., 95:1–31.

Gardiner, B. G. 1973. Interrelationships of teleostomes. *In:* Greenwood, P. H., Miles, R. S., and Patterson, C. (eds.), Interrelationships of Fishes. New York, Academic Press, J. Linn. Soc. (Zool.) 53 (Suppl. 1):105–135.

Gill, T. N. 1891. The characteristics of the Dactylopteroidea. Proc. U.S. Nat. Mus., 13:243–248.

Goody, P. C. 1969. The relationships of certain Upper Cretaceous teleosts with special reference to the myctophoids. Bull. Brit. Mus. Nat. Hist. (Geol.; Suppl. 7):1–255.

Gosline, William A. 1960*a*. Hawaiian lava-flow fishes. Part IV. *Snyderidia canina* Gilbert, with notes on the osteology of ophidioid families. Pacific Sci., 14:373–381.

———. 1960*b*. Contributions toward a classification of modern isospondylous fishes. Bull. Brit. Mus. Nat. Hist. (Zool.), 6(6):325–365.

———. 1961. Some osteological features of modern lower teleostean fishes. Smithsonian Misc. Coll., 142(3):1–42.

———. 1963. Considerations regarding the relationships of the percopsiform, cyprinodontiform, and gadiform fishes. Occas. Pap. Mus. Zool. Univ. Mich., (629):1–38.

———. 1965. Teleostean phylogeny. Copeia, 1965 (2):186–194.

———. 1968. The suborders of perciform fishes. Proc. U.S. Nat. Mus., 124:1–78.

———. 1969. The morphology and systematic position of the alepocephaloid fishes. Bull. Brit. Mus. Nat. Hist. (Zool.), 18(6):183–218.

———. 1970. A reinterpretation of the teleostean fish order Gobiesociformes. Proc. Calif. Acad. Sci., Ser. 4, 37(19):363–382.

———. 1971. Functional Morphology and Classification of Teleostean Fishes. Honolulu, University Press of Hawaii.

———. 1973. Considerations regarding the phylogeny of cypriniform fishes, with special reference to structures associated with feeding. Copeia, 1973:761–776.

———, Marshall, N. B., and Mead, G. W. 1966. Order Iniomi. Characters and synopsis of families. *In:* Fishes of the Western North Atlantic. Mem. Sears Found. Mar. Res., 1(5):1–18.

Greenwood, P. H. 1968. The osteology and relationships of the Denticipitidae, a family of clupeomorph fishes. Bull. Brit. Mus. Nat. Hist. (Zool.), 16:213–273.

———, Miles, R. S., and Patterson, C. (eds.). 1973. Interrelationships of Fishes. New York, Academic Press, J. Linn. Soc. (Zool.) 53 (Suppl. 1).

———, Myers, G. S., Rosen, D. E., and Weitzman, S. H. 1967. Named main divisions of teleostean fishes. Proc. Biol. Soc. Wash., 80:227–228.

———, and Rosen, D. E. 1971. Notes on the structure and relationships of the alepocephaloid fishes. Am. Mus. Nov. 2473:1–41.

———, Rosen, D. E., Weitzman, S. H.,and Myers, G. S. 1966. Phyletic studies of teleostean fishes, with a provisional classification of living forms. Bull. Am. Mus. Nat. Hist., 131:339–456.

Haedrich, R. 1967. The stromateoid fishes: systematics and a classification. Bull. Mus. Comp. Zool., Harv., 135:31–319.

Harrison, C. M. H., and Palmer, G. 1968. On the neotype of *Radiicephalus elongatus* Osorio with remarks on its biology. Bull. Brit. Mus. Nat. Hist. (Zool.), 16(5):187–211.

Herald, E. S. 1961. Living Fishes of the World. London, Hamish Hamilton (Garden City, N.Y. Doubleday).

Hubbs, C. L., and Potter, I. C. 1971. Distribution, phylogeny and taxonomy. *In:* Hardisty, M. W. and Potter, I. C. (eds.). The Biology of Lampreys. London, Academic Press, pp. 1–65.

Hulley, P. A. 1972. The family Gurgesiellidae (Chondrichthyes, Batoidei) with reference to *Pseudoraja atlantica* Bigelow and Schroeder. Copeia, 1972(2):356–359.

Jarvik, E. 1968a. The systematic position of the Dipnoi. *In:* Ørvig, T. (ed.) Current Problems of Lower Vertebrate Phylogeny. New York, Wiley-Interscience, pp. 223–245.

———. 1968*b*. Aspects of vertebrate phylogeny. *In*: Ørvig, T. (ed.), Current Problems of Lower Vertebrate Phylogeny. New York, Wiley-Interscience, pp. 497–527.

Jordan, D. S. 1923. A classification of fishes including families and genera as far as known. Stanford Univ. Publ. Biol. Sci., 3:77–243.

———, Evermann, B. W., Clark, H. W. 1930. Checklist of the fishes and fishlike vertebrates of North and Middle America north of the northern boundary of Venezuela and Colombia. Washington, D.C., Report of the U.S. Commission of Fisheries for 1928, Appendix X (1962 reprint).

Lagler, K. F., Bardach, J. W., Miller, R. R., and Passino, D. R. M. 1977. Ichthyology, 2nd ed. New York, John Wiley and Sons.

Liem, K. F. 1963. The comparative osteology and phylogeny of the Anabantoidei (Teleostei, Pisces). Ill. Biol. Monogr., No. 30:1–149.

Lindberg, G. U. 1971. Families of The Fishes of the World: a Checklist and a Key. Leningrad, Zoological Institute, Akademii Nauk SSSR (in Russian). [1974. Fishes of the World: A Key to Families and a Checklist. New York, John Wiley and Sons, 1974 (English translation).]

Marshall, N. B. 1962. Observations on the Heteromi, an order of teleost fishes. Bull. Brit. Mus. Nat. Hist. (Zool.)., 9(6):249–270.

———. 1965. The Life of Fishes. London, Weidenfeld and Nicolson.

McAllister, D. E. 1968. Evolution of branchiostegals and classification of teleostome fishes. Bull. Nat. Mus. Can., 221:1–239.

McDowell, R. M. 1969. Relationships of galaxioid fishes with a further discussion of salmoniform classification. Copeia, 1969(4):769–824.

McDowell, S. B. 1973. Order Heteromi (Notacanthiformes). Family Halosauridae. Family Notacanthidae. Family Lipogenyidae. *In:* Fishes of the Western North Atlantic. Mem. Sears Found. Mar. Res., 1(6):1–228.

Moy-Thomas, J. A., and Miles, R. S. 1971. Palaeozoic Fishes. Philadelphia, W. B. Saunders.

Myers, G. S. 1958. Trends in the evolution of teleostean fishes. Stanford Ichthyol. Bull., 7(3):27–30.
Nelson, G. J. 1969*a*. Gill arches and the phylogeny of fishes, with notes on the classification of vertebrates. Bull. Am. Mus. Nat. Hist., 141(4):475–552.
———. 1969*b*. Infraorbital bones and their bearing on the phylogeny and geography of osteoglossomorph fishes. Am. Mus. Nov., 2394:1–37.
———. 1969*c*. Origin and diversification of teleostean fishes. Ann. N.Y. Acad. Sci., 167:18–30.
———. 1972. Cephalic sensory canals, pitlines, and the classification of esocoid fishes, with notes on galaxiids and other teleosts. Am. Mus. Nov., 2492:1–490.
———. 1973. Relationships of clupeomorphs, with remarks on the structure of the lower jaw in fishes. *In*: Greenwood, P. H., Miles, R. S., and Patterson, C. (eds.), Interrelationships of Fishes. New York, Academic Press, J. Linn. Soc. (Zool.) 53 (Suppl. 1):333–349.
Nikolskii, G. V. 1961. Special Ichthyology. Washington, D.C., Department of Commerce, Israel Program for Scientific Translation.
Norman, J. R. 1934. A systematic monograph of the flatfishes (Heterosomata). Brit. Mus. Nat. Hist., 1:1–459.
———. 1957. A Draft Synopsis of the Orders, Families and Genera of Recent Fishes and Fish-like Vertebrates. London, British Museum of Natural History, unpublished photo offset copies.
———, and Greenwood, P. H. 1975. A History of Fishes, 3rd ed. New York, Halsted Press.
Obruchev, D. V. (ed.). 1967. 'Class Placodermi.' *In:* Fundamentals of Paleontology, Vol. XI. Agnatha, Pisces. Jerusalem, Israel Program for Scientific Translations.
Orlov, Yu. A., and Obruchev, D. V. (eds.). 1967. Fundamentals of Paleontology, Vol. XI. Agnatha, Pisces. Jerusalem. Israel Program for Scientific Translations.
Ørvig, T. (ed.). 1968. Current Problems of Lower Vertebrate Phylogeny. New York, Wiley-Interscience.
Patterson, Colin. 1965. The phylogeny of the chimaeroids. Philos. Trans. R. Soc. Lond. [Biol. Sci.], 249:101–219.
———. 1967*a*. Are the teleosts a polyphyletic group? Coloq. Int. Cent. Nat. Res. Scient., 163:93–109.
———. 1967*b*. Classes Selachii and Holocephali. *In:* Harland, W. B., et al. (eds.), The Fossil Record. London, London Geological Society, pp. 666–675.
———. 1967*c*. *Menaspis* and the bradyodonts. *In:* Ørvig, T. (ed.), Current Problems of Lower Vertebrate Phylogeny. New York, Wiley-Interscience, pp. 171–205.
Penrith, M. J. 1972. Earliest description and name for the whale shark. Copeia, 1972(2):362.
Rass, T. S., and Lindberg, G. U. 1972. Modern concepts of the classification of living fishes. J. Ichthyol., 11:302–319.
Regan, C. T. 1929. Fishes. Encyclopaedia Britannica. 14th ed., 9:305–329.
Roberts, T. R. 1969. Osteology and relationships of characoid fishes, particularly the genera *Hepsetus, Salminus, Hoplias, Ctenolucius,* and *Acestrorhynchus*. Proc. Cal. Acad. Sci., Ser. 4, 36(15):391–500.
———. 1971. The fishes of the Malaysian family Phallostethidae (Atheriniformes). Breviora, 374:1–27.
———. 1973. Interrelationships of ostariophysans. *In*: Greenwood, P. H., Miles, R. S., and Patterson. C. (eds.), Interrelationships of Fishes. New York, Academic Press, J. Linn. Soc. (Zool.) 53 (Suppl. 1):373–395.
Rosen, D. E. 1962. Comments on the relationships of the North American cave fishes of the family Amblyopsidae. Am. Mus. Nov., 2109:1–35.
———. 1964. The relationships and taxonomic position of the halfbeaks, killi-

fishes, silversides, and their relatives. Bull. Am. Mus. Nat. Hist., 127(5):217–268.

———. 1973. Interrelationships of higher euteleostean fishes. *In*: Greenwood, P. H., Miles, R. S., and Patterson, C. (eds.), Interrelationships of Fishes. New York, Academic Press, J. Linn. Soc. (Zool.) 53 (Suppl.1): 397–513.

———, and Bailey, R. M. 1970. Origin of the Weberian apparatus and the relationships of the ostariophysan and gonorynchiform fishes. Am. Mus. Nov. 2428:1–25.

———, and Patterson, C. 1969. The structure and relationships of the paracanthopterygian fishes. Bull. Am. Mus. Nat. Hist., 141(3):357–474.

Schaeffer, Bobb. 1967. Comments on elasmobranch evolution. *In*: Gilbert, P. W., Mathewson, R. F. and Rall, D. P. (eds.), Sharks, Skates and Rays. Baltimore, Johns Hopkins Press, pp. 3–35.

———. 1968. The origin and basic radiation of the Osteichthyes. *In*: Ørvig, T. (ed.), Current Problems of Lower Vertebrate Phylogeny. New York, Wiley-Interscience, pp. 207–222.

———. 1969. Adaptive radiation of the fishes and the fish-amphibian transition. Ann. N.Y. Acad. Sci., 167:5–17.

———. 1973. Interrelationships of chondrosteans. *In:* Greenwood, P. H., Miles, R. S., and Patterson, C. (eds.), Interrelationships of Fishes. New York, Academic Press, J. Linn. Soc. (Zool.) 53 (Suppl. 1):207–226.

———, and Rosen, D. E. 1961. Major adaptive levels in the evolution of the actinopterygian feeding mechanism. Am. Zool., 1(2):187–204.

Smith, J. L. B. 1939. A living coelacanthid fish from South Africa. Trans. R. Soc. S. Afr. 28:1 – 106.

Stahl, B. 1967. Morphology and relationships of the Holocephali with special reference to the venous system. Bull. Mus. Comp. Zool. Harv., 135(3):141–213.

Stensio, E. A. 1968. The cyclostomes, with special reference to the diphyletic origin of the Petromyzontida and Myxinoidea. *In:* Ørvig, T. (ed.). Current Problems of Lower Vertebrate Phylogeny. New York, Wiley-Interscience, pp. 13–71.

Svetovidov, A. N. 1948. Gadiformes. *In:* Pavlovskii, E. N. and Shtakel'berg, A. A. (eds.), Fauna of the U.S.S.R., Fishes, 9(4):1–304. Zoological Institute, Akademii Nauk SSSR. (Translated for the National Science Foundation, and Smithsonian Institution, Washington, D.C., 1962.)

Watson, D. M. S. 1937. The acanthodian fishes. Philos. Trans. R. Soc. [Biol. Sci.], 228:49–146.

Weitzman, S. H. 1967. The origin of the stomiatoid fishes with comments on the classification of salmoniform fishes. Copeia, 1967(3):507–540.

Zangerl, Rainer. 1973. Interrelationships of early chondrichthyans. *In*: Greenwood, P. H., Miles, R. S. and Patterson, C. (eds.), Interrelationships of Fishes. New York, Academic Press, pp. 1–14.

———, and Case, G. R. 1973. Iniopterygia, a new order of chondrichthyan fishes from the Pennsylvanian of North America. Fieldiana (Geol. Mem.), 6:1–67.

Section Three

BIOLOGY AND SPECIAL TOPICS

8
ENVIRONMENTS, HABITATS, AND LIFE STYLES

Any attempt to cover the ecological relationships of a group as diverse as the fishes in a single chapter is bound to fall short. The members of the group and their habitats are of seemingly endless variety, and their adaptations and reactions to the environment are often difficult to study and understand. Usually the fishes that are best known from an ecological standpoint are those with economic value as recreational or commercial resources, but those with economic impact as pests also receive considerable attention. These types of fishes are at the forefront because knowledge of their life history and ecology is necessary to the management of the populations. Many interesting nongame species of no food value have been ignored or have been the subjects of sporadic studies, even though they might have unsuspected importance in their habitats. In recent years, interest in the protection of rare fishes and those threatened with extinction has brought about increased interest in the ecology of all fishes.

The following sections will include characterizations of the habitats in which fishes live, some dynamic features of these habitats, and how fishes make a living in them. For the reasons previously given, many examples will be of food or sport fishes.

The Aquatic Medium

Discussion of the habitats of fishes must inevitably begin with a discourse on some properties of water, the medium to which they are bound. There are such obvious relationships between the structure and life of fishes and certain chemical and physical aspects of water that they do not need to be discussed at length, but a brief coverage is necessary.

The density of water, about 800 times that of air, demands that streamlining be developed in bodies that must move rapidly in relation to the medium. Water is sufficiently viscous so that objects with a high surface-to-volume ratio will be retarded in rate of sinking even though their specific gravities are somewhat more than that of water, which is equal to 1 at 4° C. The great support given by water, along with the possibility of achieving nearly neutral bouyancy by incorporating a bit of lighter material such as gas or oil into the body, relieves fishes of the necessity of expending excessive amounts of energy in maintaining position.

Pressure, which increases about 1 atmosphere per 10 meters of depth, has profound effects upon the life and structure of deepwater fishes, involving the firmness of flesh, the amount of calcium in the skeleton, and the ability to maintain a gas bladder. The specific gravity of water changes with temperature, becoming greater as 4° C is approached. The fact that the greatest density is at 4° C warmer than the freezing point of fresh water is of great importance, since ice forms on the surface instead of on the bottom. Water's high specific heat allows the extension of warm or cold currents far beyond the latitudes in which they form, thus affecting both dispersal and distribution of fishes.

Although clear natural water is quite transparent, light absorption is rapid and differential, so that at depths of more than 100 meters only a fraction of the blue end of the spectrum is present. Most red is absorbed in the upper 5 meters, and orange is mostly gone at 15 meters. Green and yellow penetrate to about 20 meters. Green plants generally thrive and produce organic matter in depths of less than about 30 to 50 meters. In stained or turbid water, light penetration can be reduced to the point that photosynthesis can occur only close to the surface, thus reducing plant productivity. Another important aspect of differential light penetration is the effect upon the apparent color of fishes. Many species living at depths of 20 meters or more have red or orange pigment. As these wavelengths do not penetrate to these depths, the fishes must appear to be colorless in their accustomed habitats. Fishes that live at great depths where no sunlight can penetrate are usually dark in color, and many produce light by means of luminous photophores.

Water has remarkable powers as a solvent. In falling through the air as rain or in running over or percolating through the ground, it takes on numerous substances, including carbon dioxide, oxygen, and chlorides, sulfates and carbonates of calcium, sodium, magnesium, and potassium. Silicon and phosphorus compounds, as well as many organic substances, are among the biologically important materials carried by water. There are many more or less complex relationships involving hydrogen ion concentration and the solubility of certain kinds of compounds, such as carbonates.

Dissolved substances have both direct and indirect effects upon fishes. The amounts and balance of plant nutrients can govern the production of the vegetation that forms the food base of the body of water. Toxic substances can act upon food organisms or directly upon the fishes. Dissolved oxygen, which is affected by pressure, temperature, and other materials in solution, can determine what kinds, if any, of fishes can occupy a body of water. Whereas air contains nearly 21% oxygen by volume, fresh water can dissolve very little, holding 10.23 cc/liter at 0° C and less as the temperature rises. Sea water of 30 ppt salinity holds about 8.8 cc/liter at 0° C. Usually a varied freshwater fish fauna can exist at a dissolved oxygen concentration of 3.5 cc/liter, which is about 55% of saturation at 20° C. Salmonids and many other cold-water fishes thrive best at dissolved oxygen concentrations near saturation, which at 20° C is 6.4 cc/liter. Although nitrogen is inert, supersaturations of it under certain conditions can cause emboli within the bodies of fish, bringing about injury or death. Where spillways of great dams allow air to be carried deep into plunge pools, many fish become subject to "gas bubble disease." Carbon dioxide, dissolved in the free state — not bound as carbonate or bicarbonate — can affect the ability of fish blood to take up oxygen.

Dissolved salts such as those encountered in the ocean or in various lakes and mineral springs place an osmotic burden on fishes. As will be pointed out in detail later, some fishes have developed physiological methods of coping with that burden, and can inhabit waters with a much greater salt concentration than that of their blood. Freshwater fishes have adapted differently and maintain a higher salt concentration than that of the surrounding medium.

Habitats and Adaptations

Fresh Water

The term "habitat" will be used to indicate where a population of a given fish species makes a living in a particular environment. "Role" will indicate how the living is made, and "niche" will be used to describe the combined habitat and role.

Populations are considered to be groups of individuals of the same species. Communities are comprised of populations of different species living in the same general environment. Ecosystems are combinations of communities and habitats, and of course are dynamic in nature.

Fresh waters compose only a small proportion of the available surface water of the earth (0.01%), but harbor about 41% of the known fish species. The reasons for this are probably related to the numerous niches available over the great ranges of latitude and altitude in both still and running water, and to the great opportunity for geographic isolation. Fresh waters differ greatly in such factors as temperature, current, depth, suspended materials, dissolved materials including oxygen and nutrients, and substrate and temporal constancy. All of these, along with many other factors, may have profound effects upon the abilities of fishes to colonize and to find food, shelter, and other necessities.

An outstanding feature of freshwater environments is that they are subject to change by geological and biological processes.

Changes in small streams and ponds may be readily observable over the course of a few years, but larger bodies may change imperceptibly. Because the processes and consequences in running and still water differ considerably, the two series of environments will be discussed separately.

Running Water (lotic or fluviatile environments). Due to the erosional force of water, the natural evolution of streams is from brook to creek to river, the tendency being to cut down to the level of the sea or lake into which the system empties. Given sufficient time, the system will consist of a large, slow, meandering river with a bed near base level and, above this, tributaries with increasing gradients as one proceeds headward. These tributaries receive creeks as tributaries, with brooks and rivulets feeding the creeks. Considered from the standpoint of a fish species adapted to a brook habitat, the species is inexorably destined to move inland and higher in elevation, while the increasing creek and river environments will be colonized by immigrants. Of course, given enough time, there is the possibility of evolution of creek and river species from the brook species.

Upstream environments are characterized by swift current, rocky substrate, and high levels of dissolved oxygen. In addition, temperatures are usually lower than in the downstream sections of the same system. Exposure to sunlight may be great in small streams above timberline, but brooks in forested areas can be covered by a canopy of vegetation that will effectively limit the light reaching the water, and that will hold the temperature down. Small streams draining granitic areas may be extremely low in dissolved nutrients, whereas those in volcanic mountains may have a better complement of minerals in solution. Depending on altitude, latitude, and exposure, brooks may be subject to ice and snow cover for part of the year. Flow in some upstream situations may be interrupted by the summer disappearance of the snow pack that feeds the stream, or by freezing in severe winters.

Development of a fish fauna in headwater streams is dependent upon the ability of fish species to cope with the many related factors mentioned, but probably of foremost importance is the capacity of the environment to produce food. In water, as in terrestrial habitats, the bulk of food production depends on sunlight, green plants, and plant nutrients. In small exposed rocky streams with steep gradient, there is limited opportunity for rooted plants to develop, so various algae, particularly diatoms, are the important primary food producers. The simplest trophic relationships among fishes in these swift watercourses involve species adapted to clinging to rocks and scraping off the diatoms and other attached material (*Aufwuchs*). Fishes of this nature have developed in Asia, especially in the Himalayan region, and in the Andes, as well as in other tropical mountain regions.

Some of the best adapted fishes for torrential life are seen in the family of Asian hill-stream loaches, Homalopteridae (Fig. 8–1). These have flattened ventral surfaces with the pectoral and pelvic fins ex-

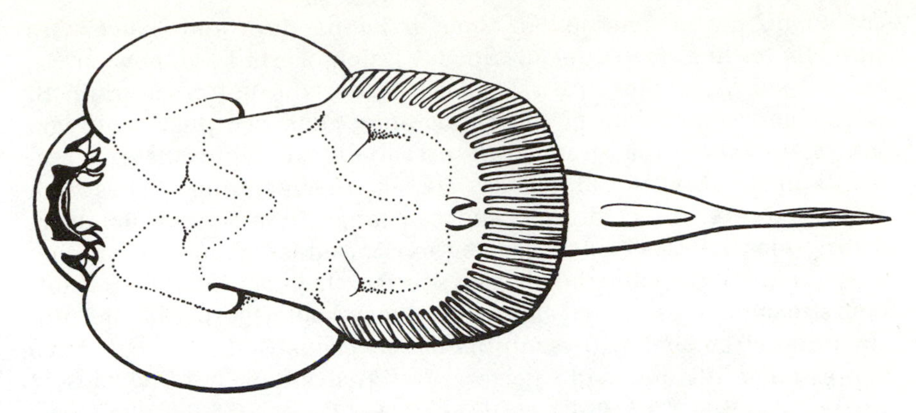

FIGURE 8–1. Sucking disc of Homalopteridae (after Nikolsky).

panded along the body margins to form, with the flat body, an effective sucker that enables the fish to hold its position on rocks where it feeds upon attached vegetation. Some species have evolved special mechanisms for introducing respiratory water to the gill chamber, a few apparently using the mouth for both incurrent and excurrent flow. In the related family Gyrinocheilidae, another group of hillstream fishes of Asia, the body is of more typical fish shape and the mouth is developed into a sucker that can hold the fish against the current. In these, the gill opening on each side has been divided into upper and lower orifices. The upper is the point of entry for respiratory water, which is expelled from the lower after having passed over the gills. Other torrent fishes are in the South American catfish family Loricariidae, which includes *Plecostomus*, a genus well-known to aquarists. *Cheimarrichthyes* (Mugiloididae) of New Zealand and the Rhyacichthyidae of Asia are also fishes of swift waters.

Actually, comparatively few fishes are strictly vegetarian in habit, so the simple feeding relationship in which the fish is the primary consumer is seldom realized except under special circumstances. More usual are the multistep food chains in which, for instance, the fish feeds upon insects or crustaceans that graze diatoms off rocks. In small streams significant amounts of food may fall into the water from streamside vegetation. As the gradient of a mountain stream diminishes, there is greater opportunity for diversification in all elements of the biota. Eddies and interstices among the rocks can collect fine materials, including organic matter produced higher in the stream or from terrestrial sources. Mosses, rooted plants, and filamentous algae join the diatoms as producers, forming substrata upon which an increasing variety of invertebrates can forage. Included now are scavengers that utilize the organic detritus. Very small streams with heavy cover may depend on terrestrial plant detritus as the main base of the food web.

In northern regions, salmonids are typical of upland streams, but are by no means confined to them if temperature and oxygen are suitable farther downstream. Species such as the rainbow trout, *Salmo gairdneri*, the brown trout, *S. trutta*, the cutthroat trout, *S. clarki*, and some chars of the genus *Salvelinus* can push upstream into areas where the streams are essentially all rapids having gradients up to 75 meters/km. These are active, streamlined fishes with strong powers of locomotion, and can range into rapid water from resting places to feed upon drifting food or attached aquatic invertebrates. Like other active headwater species, they can invade intermittent streams on a seasonal basis. They may share the permanent upstream environment with sculpins of the genus *Cottus*, which have depressed heads and wide pectoral fins that aid in holding closely to the substrate. The Piute sculpin, *C. beldingi*, and the shorthead sculpin, *C. confusus*, are examples of headwater species in the western part of North America. The sculpins, like the trout, feed upon various invertebrates but are adapted to forage among the stones of the bottom.

In upstream sections where there is a gradient low enough to permit the alternation of pools with riffles, fishes can be more numerous. Cover is increased, softer bottom materials can be deposited, and the available niches for both vertebrates and invertebrates are increased. For instance, good rainbow trout streams have a gradient of about 20 meters/km. In various parts of the Northern Hemisphere the following fishes can be found in such habitats: the trouts and chars; the graylings, *Thymallus* spp., and the sculpins. In North America, typical headwater species are the longnose dace, *Rhinichthys cataractae*, the blacknose dace, *R. atratulus*, the creek chub, *Semotilus atromaculatus*, and various sculpins including the mottled sculpin, *Cottus bairdi*, and the torrent sculpin, *C. rhotheus*. The fantail darter, *Etheostoma flabellare*, has been noted in headwaters. In tropical Asia, swift-water fishes include less highly modified homalopterids than those found farthest upstream, loaches of the genus *Nemacheilus*, catfishes of the family Sisoridae, and minnows such as *Lobocheilus sp.*

Swift-water fishes of large creeks and small rivers of North America can include, in addition to American fishes mentioned above, a variety of streamlined species capable of moving over riffles or holding position in swift runs. Examples are the mountain whitefish, *Prosopium williamsoni*, the stoneroller, *Campostoma* sp., the creek chubsucker, *Erimyzon oblongus*, the torrent sucker, *Moxostoma rhothoecum*, and the mountain sucker, *Catostomus platyrhynchus*. Additional species of *Rhinichthys* and *Cottus*, and darters of the genus *Etheostoma* join the above, but are tied closer to the bottom.

Upstream fishes are usually solitary in habit and are characterized by a high metabolic rate and a consequent high demand for oxygen. Those that live on or close to the bottom may lack the gas bladder, have it reduced, or have it encapsulated in bone. Also, their reproductive habits must ensure that the eggs will not be swept away from suitable incubation sites. Many trouts, chars, and salmons seek out gravelly bottom at the heads of riffles or in springs where a nest is dug by the female for reception of the eggs during spawning. The eggs are demersal and temporarily adhesive so that they can remain in place until covered. Other stream spawners, such as stonerollers, bury the eggs in gravel as do the salmon and trout. Others, such as some whitefish, grayling, and suckers, spawn with vigorous activity that disturbs the gravel sufficiently so that most of the eggs are covered or lodge in spaces among the rocks. Sculpins and darters attach adhesive clumps of eggs on the underside of stones. In some of these species the eggs are protected by the male parent.

Many fishes typical of the higher reaches of streams extend down into the middle areas if rapid water of suitable temperature and substratum persists. They may thus be associated with an increased number of species exploiting the greater variety of niches in the middle course. Here the stream may be characterized by more pools and runs than riffles, the current will be variable but slower than that upstream, and bottom materials will vary from coarse particles in the swifter water to considerable deposits of sand and silt in pools. Vegetation can range from diatoms and associated material on stones to beds of rooted aquatic plants in depositing areas. Cover along the banks can be relatively insignificant in wide stretches, concealment being afforded by features of the stream itself, which may have considerable depth in places.

This environment provides opportunities for fish species less dependent upon high concentrations of dissolved oxygen and low temperatures than those confined to the upper reaches, and permits the existence of those types having deeper, less well streamlined bodies and weaker powers to sustain locomotion. Over much of the world members of the family Cyprinidae abound in creeks and rivers of moderate gradient, some species being more typical of small waters than others, but with considerable overlap. In the Mississippi drainage, genera such as *Notropis*, *Hybopsis*, *Semotilus*, and *Pimephales* are representative. In various drainages of western North America, the squawfish, *Ptychocheilus* (a predator), the roach, *Hesperoleucus*, chubs, *Gila* spp., the peamouth, *Mylocheilus*, and the redsides, *Richardsonius* spp., can be encountered in running water of moderate to slow speed. In Europe typical running water cyprinids are the barbel, *Barbus barbus*, the chub, *Squalius cephalus*, and the dace, *Leuciscus leuciscus*. In North America, several sucker species can be found in the middle sections of streams. *Catostomus commersoni* and *C. macrocheilus* are examples.

In lower sections of streams, currents are sluggish, the bottom materials tend to be finer, and the depth greater. Large species equipped to feed on the bottom or in weed beds are able to do well here. Suckers of several species representing the genera *Catostomus*, *Hypentelium*, and *Moxostoma* are found in the slow water, along with catfishes and bullheads, *Ictalurus* spp. Numerous small cyprinids can live here, and preying on them can be such predators as the pikes, *Esox* spp., and various members of the sunfish family Centrarchidae. Of the latter group, the smallmouth bass, *Micropterus dolomieui*, tends to be found where gradients are about 2 to 4 meters/km, so may be encountered farther upstream than the largemouth bass, *M. salmoides*, the rock bass, *Ambloplites rupestris*, and many species of sunfishes, *Lepomis* spp.

In Europe, typical downstream forms are said to be the bream, *Abramis brama*, the carp, *Cyprinus carpio*, and the tench, *Tinca tinca*. The carp and tench have been introduced to North America, where the carp especially has become a familiar component of the slow-water fauna.

Obviously, some scheme for naming fish habitats in streams would be useful, but the difficulty in developing classifications pertaining to wide areas is also obvious. Differences in altitude, latitude, and other factors bring about differing temperature regimes in streams. Fish faunas vary greatly in composition among geographic and climatic regions, and stream courses and water quality are profoundly influenced by geomorphology and geology. Furthermore, the diversity of fishes in some areas and the ecological versatility of many species add to the difficulty of classifying streams by fish habitat, especially in view of the fact that microhabitats occurring all along the course of a stream might allow certain species with relatively exacting requirements to be present, at least in low numbers, from small to large sections.

Various researchers have classified streams by volume of flow, width, gradient, size of substrate particles, vegetational types, and geological features, including formation and age. A scheme that has been used in fish distribution studies, and seems to fit certain local situations, refers to the smallest headwater streams as order I streams. The result of confluence of these primary tributaries is an order II stream, with two order II streams merging to form an order III.

Most studies of stream fishes and their ecology have been reported in terms of associations of fishes, or stream zones have been named for the dominant fish species or association found there. Often these zones or associations have mainly local applications but in Europe, where climatic and physiographic features allow the spread of an essentially similar fish fauna over a wide area, a zonation of streams proposed by Dr. Marcel Huet of Belgium seems to have rather broad application. He has recognized: the trout zone, where brown trout and grayling predominate; the grayling zone, where a mixed fauna develops, but where trout and grayling are more abundant than the minnows and the few pike, perch, or eels that can occur here; the barbel zone, in which trout and grayling are not abundant and the minnows, especially those adapted to running water, dominate; and the lowest zone, the bream zone, where the bream, carp, and tench live with many smaller minnows and the predatory species.

A similar system suggested for North America by Dr. K. F. Lagler et al. lists the following succession of fishes from upstream to down: (1) grayling; (2) stream char; (3) flowing-water minnows and pike; (4) basses; and (5) catfishes, suckers, and quiet-water minnows. This scheme is descriptive of a series of habitats that exist in North America, but cannot be applied generally, mainly because of the diversity encountered on this continent. There is greater utility in naming associations according to the species present in a local faunal area, as has been done by Dr. John Hopkirk for the Sacramento River system. He has recognized a trout zone, roach *(Hesperoleucus)* zone, sucker zone, and hitch *(Lavinia)* zone. Similar zones have been noted in the San Joaquin drainage by Dr. Peter Moyle, except that the fish association in the lowest zone included mainly introduced fishes, especially centrarchids.

Still Water (lentic environments). Unlike streams, where evolution is from small to large, bodies of still water change from large to small, from deep to shallow. Once a lake has been created by one of the many geological processes that can cause basins to form in the surface of the earth, it begins to fill up with materials washed in by streams, blown in by winds, or produced in the lake itself. The natural progression, then, is from lake to pond to swamp, and finally to dry ground if the processes can continue for a sufficient time. Small shallow lakes such as might form in irregularities in glacial moraines, or be caused by a landslide damming a small valley, have relatively short lives and may become extinct in a few centuries. Lakes formed by the gouging action of major portions of continental glaciers, or formed in gigantic rifts in the earth's surface, have very long lives and change relatively little in the course of a few centuries, but the processes of change are inevitably at work. Lakes such as Baikal in Siberia and the deep rift lakes of Africa are more permanent features, lasting for millions of years.

Typical youthful lakes in a temperate region are low in nutrients and organic materials. These oligotrophic lakes, as they are called, are characterized by an abundant supply of dissolved oxygen at all depths so that fishes have access to the entire body of water. Thermal stratification, which, in summer, layers the lake into a warm surface epilimnion, a middle mesolimnion in which the temperature drops rapidly with depth, and a lower, cool hypolimnion, allows the year-round existence of cold-water fishes (Fig. 8–2). Whitefishes, chars, and trout are the usual inhabitants of oligotrophic lakes in the northern hemisphere. Depending upon the summer surface temperature and the tolerances of the individual species, the salmonids can seek suitable temperature layers in the mesolimnion (thermocline), or descend to the hypolimnion and still find sufficient dissolved oxygen and food to sustain themselves.

Fishes of open water in lakes are typically strong swimmers that can seek out and capture the crustacean or insect prey that forms most of the food base. Some of the whitefishes and the sockeye salmon, *Oncorhynchus nerka*, are equipped with fine, long gill rakers for capturing small plankton organisms. The young of the latter species spends its early life in lakes before migrating to the ocean, and has a freshwater phase called kokanee that completes its life cycle entirely

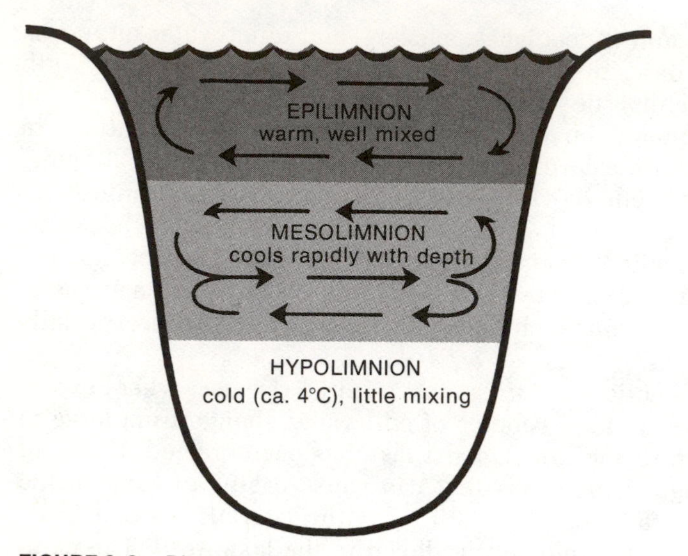

FIGURE 8–2. Diagram of temperate oligotrophic lake, showing thermal stratification.

in fresh water. Large predators of the open deep water include the rainbow trout, *Salmo gairdneri,* and the lake trout, *Salvelinus namaycush.* The northern pike, *Esox lucius,* and the yellow perch, *Perca flavescens,* are typical of shallow water in oligotrophic lakes, but can be found in warmer shallower lakes than the salmonids. The pikes are lurkers, capturing prey by means of short lunges.

Oligotrophic lakes support a relatively low biomass per unit of area or volume. Steep rocky banks support a sparse representation of rooted aquatic plants and the nutrient-poor water supports a sparse supply of phytoplankton, so that the primary productivity is generally low. Because the recycling of nutrients proceeds slowly, growth of fish can be slow in this situation, and replacement of fish biomass removed may take considerable time.

Nutrients accumulate in a basin with age, by minerals entering in solution, by organic matter washing or falling in, and through various other processes. Nutrients can be trapped in the lake by conversion to organic matter in the flora and fauna, and sinking to the bottom on the death of the organisms. From here they can be recycled through consumption by scavengers or decomposition by microorganisms. Any nutrients reaching a soluble state in the hypolimnion have a small chance of reaching the upper, well-lighted regions where photosynthesis occurs until there is a general circulation of the lake. Circulation occurs upon cooling in the autumn and warming in the spring when the lake is close to the same temperature from top to bottom. Of course, there are many exceptions to this twice-a-year circulation pattern depending upon such factors as depth, altitude, latitude, and exposure to wind.

As bottom deposits increase and the nutrient content (and consequently the productivity) of the lake grows, it is generally said to be passing from the oligotrophic to the eutrophic condition, and its capacity to support a fish fauna changes accordingly. Accumulation of soft bottom deposits, including much organic material, can increase the amounts of rooted vegetation in the shallow waters and can provide niches for many burrowing organisms that at some stage in their life histories are available to foraging fish. The rich water can support denser populations of phytoplankton, and the zooplankton community is thus provided with a greater food resource. In a situation in which nutrients are recycled rapidly, production of fish biomass can be high. Eutrophic lakes have the capacity to sustain a large biomass per unit of area, but more important is their capacity for rapid replacement of biomass that is removed.

There are some trade-offs involved in the eutrophication process. Decomposition of organic material requires much oxygen, and the hypolimnetic waters of a rich lake can be low or lacking in dissolved oxygen. Permanent occupancy of the deep waters of such a lake by fish species with a high oxygen requirement is not possible, so lake trout and certain other salmonids disappear unless the lake is in a geographical location that allows for cold surface temperatures. In some rather rich lakes salmonid populations are restricted during the summer to a narrow stratum somewhere in the mesolimnion, because the water above is warmer than their tolerance and the water below is too low in dissolved oxygen to sustain them.

Other changes that occur during eutrophication affect color and transparency, due to organic compounds in solution and suspended materials, including plankton. The amount of carbon dioxide can increase significantly, as can that of ammonia and ammonium compounds. All these changes can be intensified and accelerated artificially by fertilization or organic pollution.

Fishes found in lakes that have evolved away from the deep, rocky, salmonid-supporting situation vary with climate and faunal area. In northern areas pike, perch, and pike-perches, *Stizostedion* spp., can coexist with or succeed the salmonids. In warmer waters of North America, many centrarchids are encountered. These include black basses, *Micropterus,* crappies, *Pomoxis,* sunfishes, *Lepomis,* and rock basses, *Ambloplites.* Most centrarchids are compressed fishes adapted to life around cover such as vegetation or logs. Some tend to be solitary while other species move in schools, but they are not wide-ranging in habit. Typical bottom feeders in warm lakes and ponds are catfishes and bullheads of the Ictaluridae, which, like some of the suckers and buffalofishes of the Catostomidae, frequent large rivers as well as lakes. Numerous species of cyprinids are adapted to life in still waters, and form an important part of the food base for predatory species. The introduced carp can be very numerous in North American lakes, often interfering with the natural ecological relationships of native species.

As mentioned earlier, the natural succession of still-water habitats leads to extinction of open water, and some of the last stages are bog ponds and swamps. These are characterized by shallow acid water, emergent vegetation, and soft organic bottoms. These waters are usually well past the peak of productivity as far as fish are concerned, and in North America do not usually support a varied fish fauna. In the far north, Alaska blackfish, *Dallia pectoralis*, are found in shallow tundra ponds that freeze to the bottom, entrapping the fish, which can survive if their temperature does not drop much below 0° C. Sticklebacks, Gasterosteidae, are other examples of northern swamp fishes.

Other fishes that can be found in swampy habitats are mudminnows, *Umbra* spp., bullheads, some of the sunfishes, and, especially in the warmer regions, gars, *Lepisosteus*, and the bowfin, *Amia calva*.

The examples given above are composites of conditions and faunas of temperate and cold North American lakes, and actually represent a great simplification. There are many kinds of lakes classified by limnologists according to such features as temperature, numbers of circulation periods per year, productivity, and salinity. Each kind is capable of supporting a distinctive fauna. Some large and ancient lakes and lake systems such as the rift lakes of Africa and Lake Baikal in Siberia have developed, over their long existence, remarkable and extensive fish faunas, based upon the Cichlidae in Africa and upon cottoids in Baikal.

Amazing adaptive radiation has occurred in lakes such as Malawi (Nyasa), Victoria, and Tanganyika, so that there are cichlids able to exploit most of the niches one could expect to exist in a still-water environment. Examples are phytoplankton eaters, various *Tilapia*, zooplankton collectors, *Haplochromis intermedius* and *Limnochromis permaxillaris*, mollusc eaters, *Macropleurodes bicolor*, and piscivores, *Rhamphochromis* spp. There are many other types, some feeding upon algae scraped from rocks or from higher plants, some eating the higher plants, and others feeding on organic material in bottom deposits. All these and still others have body form and mouth and tooth structure suited to their specific way of life.

The sculpin subfamily Cottocomephorinae of Lake Baikal contains both benthic and open-water members. The former are adapted to various bottom types, some at considerable depths, and divide a food resource consisting largely of midge and caddis larvae, and numerous species of copepods. Pelagic members of the family are said to specialize in feeding upon copepods, with each fish species having a preferred prey species. The related Comephoridae are pelagic fishes that can be found as deep as 1000 meters. They are colorless and are distinguished by large mouths and broad pectoral fins. Because they are viviparous and feed upon pelagic crustaceans, they have adapted completely to the open-water existence.

In tropical regions where wet and dry seasons alternate, numerous kinds of swamp-dwelling fishes have evolved habits and structures that permit them to occupy environments that are often lacking in dissolved oxygen or even water. Several methods of air breathing can be seen in swamp fishes; some are capable of moving overland to escape desiccation, and the African lungfish aestivates in a burrow during the dry season. A few species that have an annual life cycle pass the dry season in the egg stage.

Caves form very special habitats for freshwater fishes, presenting conditions that range from nearly representative of the exterior, near the openings, through a twilight area of decreasing light and slight temperature fluctuations, to the lightless interior where temperatures are usually uniform. Some fish species that normally live outside caves occasionally enter them for food or shelter. These are called trogloxenes and may include members of several families, including Ictaluridae, Cyprinidae, and Cyprinodontidae. Fishes that normally spend part of their life cycle in, and part out, of caves are termed troglophiles and include the springfish, *Chologaster agassizi*, of North America and representatives of *Chondrostoma* in Europe. These, of course, have well-developed eyes.

Those species that are confined to caves and live in constant darkness are known as troglobites. The Amblyopsidae of North America are well-known cavefishes. Included are the genera *Amblyopsis* and *Typhlichthys*. Other representative families with troglobite members are Ictaluridae *(Satan, Trogloglanis)* of North America; Characidae *(Anoptichthys)* of Mexico; Clariidae of Africa; Pimelodidae *(Typhlobagrus)* of Brazil; Cyprinidae *(Aulopyge, Coecobarbus)* of Europe and Africa; Synbranchidae *(Pluto)* of Yucatan; and Brotulidae *(Stygicola)* of Cuba. *Poecilia sphenops*, a livebearer of Mexico, has been noted as having normally sighted forms outside a cave near Tabasco, but with some sightless populations and intergrades inside. Some of the blind individuals are born without sight, while others are secondarily blinded by the growth of circumorbital tissue over the eyes. Cephalic lateralis canals are modified in the cave dwellers, being enlarged and partially open. In addition, those living in the dark depend on tactile stimulation in the courtship behavior rather than the typical visual stimulation generally seen in the species.

Virtually all organic matter in the inner cave habitat must originate outside. Streams flowing through may bring debris and plants that can provide food for invertebrates directly, or can nurture fungi upon which invertebrates can feed. In some caves bat droppings provide considerable organic material. Invertebrate troglobites include collembolans, crickets, ants, millipedes, spiders, isopods, and crayfishes.

True cavefishes are usually sightless and lack pigment in the skin. The lack of vision is compensated by development of other senses, particularly the acousticolateralis system, and olfaction. The cave-dwelling catfishes utilize the sense of touch (and taste?) on their barbels. The lateral line organs of amblyopsids are set out on the surface.

Cavefishes have no predators and many show no strong escape

reactions. Some show an annual rhythm of reproduction even though there are no seasonal cues in the deep caves. Amblyopsids have developed the habit of incubating their eggs in the branchial chamber.

Marine Habitats and Adaptations

The oceans are characterized by great size and depth, continuity in time and space, and diversity of bottom type, water motion, temperature, and salt content. Marine fishes can live on or near the bottom, in what is called the benthic realm, or in the open sea, the pelagic realm. Obviously all this has led to the formation of ecological associations and communities ranging from simple to intricate, and much effort has been directed toward identifying, describing, and analyzing these associations and communities. The usual framework within which marine ecologists work is one of zonation of the two realms mentioned. Zone boundaries are crossed by numerous animals, but because the zones are set up according to such factors as light penetration, temperature, and the extent of the continental shelf and slope, they have considerable biological significance.

The benthic realm is divided into the shelf zone, which reaches down to about 200 meters; the upper slope zone, which extends to about 1000 meters; the lower slope, reaching to about 3000 meters; the abyssal, which extends to about 6000 meters; and the hadal zone, which includes the deep trenches. The pelagic realm is divided into the epipelagic zone, to about 200 meters (this roughly corresponds to the depth of effective light penetration and the edge of the continental shelf); the mesopelagic zone, to about 1000 meters, which is the limit of all surface light; the bathypelagic zone, which is aphotic and reaches to the 6000-meter depth; and the hadopelagic zone, in the deep trenches below 6000 meters. Some ecologists find it convenient to divide the ocean into the neritic system (on and above the continental shelf) and the oceanic system (beyond the shelf) (Fig. 8–3).

Benthic Realm. The part of the shore that is periodically covered and uncovered by the tides, the intertidal or littoral environment, constitutes a very special section of the shelf zone. Depending on exposure to waves and currents, the shore can be the site of erosion or deposition. Eroding shores are typically rocky and rugged, and the irregular substrate usually has depressions that retain water as the tide recedes so that a relatively permanent intertidal habitat is provided. These tide pools are often strewn with boulders and harbor a variety of marine algae and invertebrates. Thus, they furnish shelter and food to fishes that are adapted to this sometimes severe environment. Fishes found in the intertidal must be able to withstand the strong wave motion by swimming, hiding, or clinging. Many take refuge in the pools at low tide but some move with the receding water and take refuge below the low tide level. Others can maintain positions under boulders or in wet vegetation for the few hours between tides.

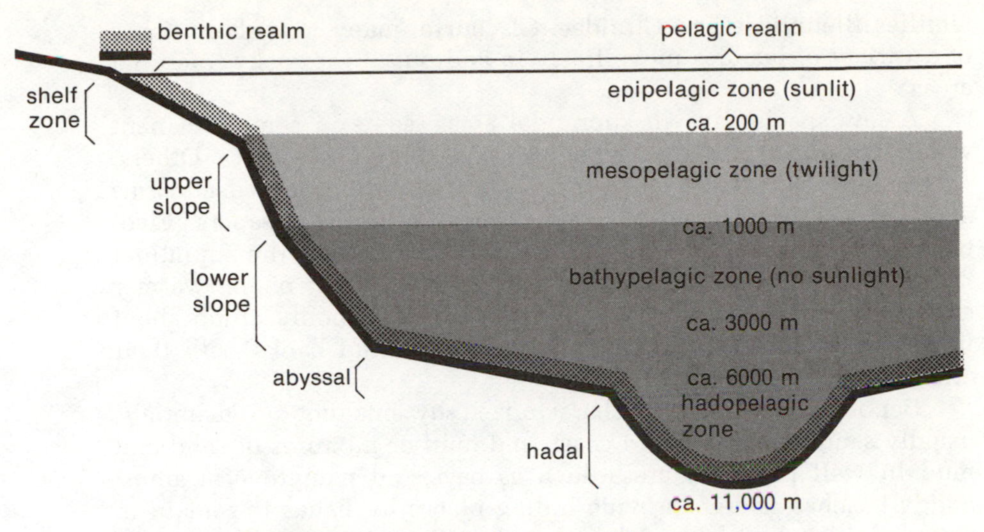

FIGURE 8–3. Ecological zonation of the ocean.

In temperate regions intertidal fishes must be tolerant of fluctuating temperature and salinity, especially if they frequent shallow pools close to the high tide level. Examples of both eurythermic (having wide temperature tolerance) and euryhaline (wide salinity tolerance) fishes can be found in the Cottidae and in blennioids, including Stichaeidae and Pholidae. Among the sculpins, members of the genera *Clinocottus* and *Oligocottus* are well represented in the intertidal and are well adapted to conditions there. *O. maculosus*, the tidepool sculpin, is able to withstand temperatures up to about 30° C and is quite resistant to rapid change. Experimentally it has been shown to endure salinities from nearly fresh to about twice that of sea water to which it is usually exposed.

Sculpins are cryptically colored, appearing in mottled and barred patterns and in colors that make them inconspicuous on rocks, green vegetation, coralline algae, and a variety of other materials. *Clinocottus globiceps*, among others, can rapidly change colors to fit the substrate, and this species resembles others in having cirri on the skin. This aids further in blending into the background. The wide pectoral fins allow them to move slowly across the bottom in search of food and can be used to wedge the animals into rock crevices for protection.

Gunnels (Pholidae), and pricklebacks (Stichaeidae), are examples of elongate fishes that can wriggle through interstices in the rocks or similar small spaces to exploit resources and hiding places there. Other exposed-coast intertidal fishes throughout the world include the clingfishes (Gobiesocidae), which have a strong ventral sucking disc and can hold position in turbulent waters, gobies (Gobiidae), and snailfishes (Liparidae), both of which have ventral sucking devices somewhat weaker than those of the preceding family, and the blennies of the families Blenniidae and Clinidae. Of course, many near-shore fishes of a variety of families move freely in and out of the rocky intertidal areas.

A few species live in supratidal areas, some in semipermanent waters that are replenished by splashes or spray from waves. Others, which can spend some of their time out of water, make excursions among rocks or vegetation. In rocky areas sculpins, sleepers (Eleotridae), blennies, gobies, and clingfishes often inhabit the supratidal "spray zone" or supralittoral. In soft-bottomed areas of the warmer shores, the mudskippers (Periophthalmidae) commonly climb about in mangrove roots or haul themselves along mud banks with their arm-like pectoral fins.

Depositing shores are characterized by fine bottom materials, usually sand along the open coast, and mud or mixtures of sand and mud in well protected areas such as bays and mangrove swamps. Sandy beaches seldom provide hiding places for fishes to remain in between tides, but a few fishes are adapted for burrowing or hiding in the sand. The sandfish, *Trichodon,* has fringed lips that allow respiratory water to pass but which strain out the sand. Flounders (Pleuronectidae), sanddabs (Bothidae), and various skates and rays can cover themselves with sand. *Ammodytes,* the sandlance, is said to dive into the sand for protection. Weeverfishes (Trachinidae) and sand stargazers (Dactyloscopidae) habitually bury themselves in sand.

In general, sandy shores are not as rich in food resources as are the muddy areas where burrows can be more permanent, a greater variety of vegetation can attach and grow, and where accumulated organic material in the bottom can foster the development of a diverse invertebrate fauna. These soft-bottomed areas may not host the variety of intertidal fishes seen in the rocky areas, but may provide daily food to a much greater biomass of fish per unit area. Most fishes that feed in these areas move in from near shore as the tide covers the sand or mudflats and seek the great variety of crustaceans, worms, molluscs, and other invertebrates found there. A few species, especially gobies, remain in the flats, hiding in clam holes or shrimp burrows.

Sculpins, flounders, tonguefishes (Cynoglossidae), soles (Soleidae), sanddabs, surfperches (Embiotocidae), mullets (Mugilidae), and silversides (Atherinidae) are some examples of the many kinds of fishes that might be expected to move inshore at high tide. Some piscivorous species take advantage of the food resources in the intertidal. The striped bass, *Morone saxatilis,* is a familiar North American example.

Below the limit of low tide the environments are more stable than in the intertidal area. But, at shallow depths, there is water motion due to wave action and tidal currents, and seasonal fluctuations of light, temperature, and salinity are possible. Near shore there are often great beds of marine algae such as the great kelp forests of California, and organic productivity is high. Exceptionally high productivity is encountered where upwelling currents impinge upon the shore. Fish faunas are at their richest in the shelf zone, especially in the upper 50 meters, the "inner sublittoral" part of the shelf. Scores of families and hundreds of genera have species adapted to niches in this series of environments. Included are the families encountered in the intertidal plus — for example, skates (Rajidae) and stingrays (Dasyatidae), sharks (Carchariidae and others), guitarfishes (Rhinobatidae), many eels (Anguillidae, Ophichthyidae, Muraenidae), cods (Gadidae), pipefishes (Syngnathidae), numerous perch-like families (such as goatfishes, Mullidae; drums, Sciaenidae; nibblers, Girellidae; wrasses, Labridae), scorpionfishes (Scorpaenidae), and triggerfishes (Balistidae).

One of the most interesting sections of the shallow shelf is the coral reef environment, best developed in the Indo-Pacific and Caribbean regions at temperatures of 23° to 25° C. Here there are nooks, crannies, and hiding places of many shapes and sizes providing shelter, and great numbers of invertebrate species that provide food for a diverse fish fauna. Many fishes of the reefs are highly colored, with bright patterns of stripes, bars, ocelli, and other outline-breaking markings. Thin, compressed shapes and fins that allow rapid turns and access to narrow spaces are common in the angelfishes (Pomacanthidae), the damselfishes (Pomacentridae), the surgeonfishes (Acanthuridae), the butterflyfishes (Chaetodontidae) and others. Various families of eels, such as morays (Muraenidae), snake eels (Ophichthidae), and worm-eels (Echelidae), with their elongate bodies, can utilize holes and grottoes denied to heavier bodied forms. Fishes with armor of plates or spines include the curious shrimpfish (Centriscidae), seahorses and pipefishes (Syngnathidae), boxfishes (Ostraciidae), triggerfishes (Balistidae), and porcupinefishes (Diodontidae).

Modifications for food gathering range from the small mouths and long snouts of the long-snouted coralfish, *Forcipiger,* and the Moorish idol, *Zanclus,* to the coral-crushing beak of the parrotfishes (Scaridae). The wrasses (Labridae) usually have small- to moderate-sized mouths with anterior canine-like teeth adapted for grasping and posterior teeth modified for crushing. The numerous sea basses (Serranidae) have wide mouths suited to predatory habits. Food resources of the reefs are tremendous, including not only the coral polyps but numerous forms of algae, hosts of invertebrate species and, of course, the plankton upon which the coral polyps depend. Reef fishes take advantage of all levels of production — some specialize in biting off coral polyps, some are herbivorous, and others seek mainly invertebrates.

To catalog the myriads of adaptations encountered in the dozens of families inhabiting the reefs is beyond the scope of this discussion, but the fishes here range from beautiful to bizarre, from clown-like to deadly dangerous. Some are adapted to life in the surf and in the channels that dissect the reefs. Others seek quieter water just off the reef or in the lagoon areas.

The continental shelf, here restricted by definition as that part of the sea bottom shallower than 200 meters, is variable in width, ranging from 0 to over 1200 km and averaging about 70 km wide. The shallow parts of the shelf are characterized by high productivity, large vegetation, and a great variety of fish species. Depending upon the clarity of the water, the remainder of the shelf, the "outer sublittoral," retains some capacity to support plant life and therefore to produce organic matter. However, much of the organic material available in the water or sediments has its origin on the land, in the inner sublittoral, or in the neritic waters above the shelf. Daily and seasonal fluctuations of temperature and light are modulated by the depth, and the bottom tends to be more uniform over larger areas, especially in sedimented regions.

This environment, although offering a considerable variety of habitats, can provide only a fraction of the niches encountered in shallower water. Thus, instead of great numbers of species, each with a small or moderate biomass, there are relatively fewer species, some with greater biomasses. Whether these are strictly benthic, spending all their time on the bottom, or benthopelagic, swimming off the bottom but feeding on it, they are subject to mass capture by trawls and many of the world's great fisheries are centered on the shelf. Numerous soles and flounders, rockfishes (*Sebastes*), and gadids (*Gadus, Pollachius, Melanogrammus, Urophycis*) are involved in this harvest.

Many shelf fishes reach lengths of 30 cm to 1 meter. They are adapted to feed on the worms and molluscs of the bottom, and on the crustaceans that are found both on and near the bottom. Some are relatively broad in their selection of food, taking other fishes or acting as scavengers. Eyes are well developed here, with some enlargement in the dimmer light. Benthic species are usually without gas bladders and some, such as the poachers (Agonidae) and a few sculpins (such as *Radulinus*) are armored. Species that maintain position above the bottom are equipped with gas bladders. Barbels are present in the cods, poachers, and some of the sculpins.

Moving from the shelf onto the upper slope, a nearly lightless environment is found, where the last weak rays of light from the surface eventually disappear at about 750 to 1000 meters. The light would seem insignificant, but the sensitive eyes of fishes can make use of incredibly small amounts of light energy. Consequently, many fishes of this environment have enlarged eyes. Lateral line systems tend to become highly developed. Temperature decreases steadily over the 200- to 1000-meter span and the water movement is generally slight.

This zone depends, as do the deeper zones, upon the shallower parts of the sea for virtually all the available food. Remains of plankton and other pelagic organisms accumulate on the bottom where scavengers attack and recycle them. Many of the fish families from the upper slope are also encountered on the shelf. Flatfishes, including the halibut, *Hippoglossus,* are common, as are the hagfishes, cods, morids, scorpaenids, skates, squaloid sharks, eelpouts (Zoarcidae), seasnails (Liparidae), and brotulas (Ophidiidae). Fishes with elongate shapes appear to be successful in these deep waters. The latter three families named are typical of a very common shape that includes a rather large head, large pectoral fins, long dorsal and anal fins, and a tapering afterbody. The deeper parts of this zone and the zones below this contain several families of this shape. Chimaeras, rattails (Macrouridae), Halosauridae, and Notacanthidae are examples. In many of these chimaeriform or rat-tailed fishes a prominent snout is developed and the anal fin is longer than the dorsal, thought to aid in positioning the body with tail up and snout to the mud so that foraging can be more efficient. Many of these fishes reach lengths of 30 to 45 cm and more.

Although bioluminescence is important in the pelagic realm at depths corresponding to the upper slope, only a few benthic fishes produce light. A few macrourids display luminescence at these depths.

Reproductive adaptations in some appear to prevent young from being removed too far from suitable habitat. Hagfishes, skates, and chimaeras deposit large eggs from which hatch juveniles essentially like the adults. Some brotulids and electric rays are ovoviviparous and a number of eelpouts are viviparous. Agonids and some eelpouts deposit demersal eggs, and the latter guard the spawn. Cods, macrourids, and eels have pelagic eggs that can drift freely with the deep currents. Because these eggs are small and may hatch pelagic larvae that must drift for a while, the numbers must be great. A sufficient number must be spawned to balance losses to predators and to the drift of young away from optimum living areas. In many species the eggs and larvae are hydrostatically balanced to float not at the surface but at some intermediate depth. In some flatfish, eggs and larvae may drift inshore where good nursery areas exist. Juveniles subsequently return to the deeper water. A few liparids place their eggs in the gill cavities of crabs.

In the lower slope and abyssal zones physical conditions become quite uniform except for increasing pressure with depth. These zones are dark and cold, with slow currents. The bottom contains less and less material from the continents, being made up of calcareous or siliceous deposits from the pelagic realm or, in the very deep areas, of red clay. Food is a product of the upper layers of water. Particulate matter that reaches the bottom, if not eaten by scavengers, can be broken down by bacteria, which might then be eaten by mud-ingesting invertebrates, which in turn can be eaten by fishes.

Although the fish fauna of the benthic regions below 1000 meters contains many elements of the upper slope, modifications for the deeper life are evident. Eyes are generally small, even vestigial in a few groups, and are so highly modified as to be considered absent in at least one genus (*Ipnops*). Pigmentation, though usually black or otherwise dark, may be lacking in some fishes. Lateral line systems are greatly developed, with the neuromasts often set out on stalks. Many gadiform species here have the ability to produce sounds that may be useful in locating mates. Bioluminescence is common among the macrourids.

Gas bladders are present in about half the species below 2000 meters and are often of large size. Skeletons and scales are reduced greatly in weight and many fishes are flabby and watery. The tripod-fishes, Bathypteroidae, have elongate lower rays on the pelvic and caudal fins. These are used to prop the animal above the soft ooze bottoms upon which it lives.

Not many fishes occur in the hadal zone. A brotulid, *Bassogigas profundissimus,* and a liparid, *Careproctus amblystomopsis,* have been collected deeper than 7000 meters, and the deep-sea explorers Piccard and Walsh observed a "flatfish" at 10,800 meters in the Marianas Trench. The identity of the fish is not known.

Pelagic Realm. The pelagic realm is constituted of the great volume of the interconnected oceans consisting of almost all the earth's water. Fishes that live a pelagic life are more or less independent of the sea bottom. Those that are completely independent, carrying out their entire life cycle in open waters, are termed holopelagic. Others that swim free of the bottom most of the time but which may depend on it for reproduction or food are generally called meropelagic or hemipelagic.* Some pelagic species are found mostly above the continental shelf in the neritic division of the pelagic, while others live a far-ranging existence away from land in the oceanic division.

Independence of the bottom is achieved in several ways by pelagic fishes. Usually some means of gaining a favorable hydrostatic balance is evident. The gas bladder is an almost universal organ among epipelagic fishes, and is common in the mesopelagic. Inclusion of fats and oils in muscles, body cavity, gas bladder, or liver is an effective flotation device, but requires a greater volume of material than does gas. Pelagic sharks may have livers constituting a quarter of their total weight. Specific gravity of a fish can be brought close to that of water simply by including a large proportion of water somewhere within the confines of the skin. Many deep pelagic fishes have flabby, watery flesh; the epipelagic ocean sunfish (*Mola*) is extremely watery, and many snailfishes are equipped with a thick layer of dilute gelatinous fluid between muscles and skin. Reduction in the weight of the skeleton obtains some hydrostatic advantage. Many deep-living species that lack gas bladders have very thin papery bones, and lack scales.

Body conformation giving a high surface-to-volume ratio can reduce the rate of sinking. This is extremely important in pelagic invertebrates and in larval fish, and is no doubt helpful to the much larger manta rays. Independence of the bottom, or maintenance of position at a given midwater level is, in part, gained by powers of locomotion. Sharks

*The Russian scientist N. V. Parin has recognized a number of categories of epipelagic components, based upon such factors as reproduction, nutrition, and migration.

and mantas are constant swimmers, and the swimming feats of tunas and other mackerel-like fishes are well known.

Pelagic life requires special adaptations for food gathering. The basis of the food web is phytoplankton, upon which numerous invertebrates feed and, although some fishes take the small plants, these invertebrates are sought by fishes large and small. Fishes may be equipped to pick off zooplankters one by one, or may have fine strainers and consolidators for use in mass capture of the tiny prey. Of course, the system includes predatory invertebrates of various sizes and predatory fishes that can take advantage of a wide range of primary and secondary consumers.

Great swimming powers are common among epipelagic fishes, being necessary for capture of prey, avoidance of predators, and migrations. Deeper regions may not require or allow the development of such powers, and the deep predators are usually softer and weaker.

The Epipelagic Zone. This thin surface layer of the oceans is generally considered to extend to about 200 meters, although conditions typical of the epipelagic may disappear at shallower depths or extend deeper, depending upon location and numerous interacting physical factors. It is subject to daily and seasonal changes in many physical features; it is in general the warmest zone and is the best-lighted layer. Wherever the light and sufficient nutrient materials coincide, plants can live and produce the organic material that supports the life not only of the upper layers, but indirectly that of the lower layers as well. Dissolved oxygen is high in the epipelagic, near saturation in the upper layers. Salinity ranges from about 33 to 37 ppt.

Close to the surface, usually in the upper 10 to 25 meters in high latitudes, but sometimes reaching to 200 meters or more in the subtropics, there is a stratum of uniform temperature. This may reach or exceed 20° C in the tropics and fluctuates only a few degrees during the year. Proceeding polewards, through temperate to polar latitudes, the upper temperatures become lower, with greater seasonal fluctuations in temperate areas than in colder regions. Below the surface layer there is a zone of rapidly falling temperature, the thermocline, in which the temperature can drop nearly 1° C per 10 meters of depth. This is a constant feature of warm oceans, but develops only during the summer in high latitudes. Below the thermocline (and well below the epipelagic) the sea is cold, in some places reaching below 0° C. Stenothermic fishes are confined to certain depths, water masses, or latitudes by their narrow tolerances, whereas eurythermic species can range from one temperature to others, depending on overall tolerance and requirements.

About 70 families of fishes are represented in the epipelagic. These range from dwarf sauries (Scomberesocidae) of 15 cm up to the gigantic whale shark of 18 meters, but most representatives are from 30 cm to 1 meter in length. The swift swimmers, such as tunas and other mackerel-like species, pelagic sharks, and some carangids are finely streamlined, with narrow, keeled, caudal peduncles and crescent-shaped caudal fins. The powerful billfishes (marlins, swordfish) probably derive a considerable streamlining advantage at high speeds from their elongate bills;

these animals are reported to reach 130 km per hour. Many families have somewhat less of the classic fusiform shape, some retaining a lunate or forked caudal but having a more compressed body and others being more elongate, even arrow-shaped, as the barracudas.

There are weak swimmers and drifters among epipelagic fish, some of which have such weak powers of locomotion that some authorities consider them as part of plankton, rather than as representative of nekton, the swimming animals. Among these are the ocean sunfishes (Molidae) and the ribbonfishes (Regalecidae), which include rather large species. Some of the smaller sauries and flyingfishes might also be classed as plankton because of their inability to sustain strong motion. The 2 cm *Schindleria* is a plankton component.

The remarkable color pattern typical of epipelagic species is well known. The dark back, often green or blue, blends into the ocean's color from above, and the silvery sides reflect whatever colors the sea has to offer. Variations of this, sometimes with stripes or bars, are seen in pelagic fish of the neritic region.

The richest parts of the pelagic are usually the neritic division, close to the land-derived mineral nutrients, and subpolar regions, where vertical interchange of water occurs each winter. Also, upwelling currents impinging upon the shore can enrich the surface waters with nutrients liberated by the breakdown of organic matter in the depths. This rich water is brought up also in regions of current divergence, as along the equator. Great abundances of fishes can be supported in these areas. Much of the wide open ocean is a "wet desert" with minimal primary productivity.

A special community of the pelagic takes essentially littoral demersal fishes into areas far from land in floating masses of Sargasso weed. This alga carries a remarkable complement of invertebrates, and thus supplies both food and shelter to the sargassumfish (*Histrio histrio*), the sargassum pipefish (*Syngnathus pelagicus*), some species of filefishes (*Alutera*) and the young of several true pelagic species. Sargasso weed is found not only in the Sargasso Sea but also in the western Pacific. Flotsam such as *Sargassum*, other algae, and debris from the land attracts not only small animals that seek shelter in it but also larger fishes that lurk under it, feeding on the smaller organisms attracted there. Large floating jellyfishes attract certain small fishes that can live unharmed among the tentacles. The best example appears to be the man-of-war fish, *Nomeus gronovii*, which lives with the siphonophore *Physalia*.

Reproductive adaptation of epipelagic bony fishes involves high potential fecundity, with most larger species releasing hundreds of thousands to millions of eggs. Tunas, depending on size, release from about 1 to 10 million eggs. The ocean sunfish, *Mola mola*, is reported to spawn up to 300 million. In holopelagic species the eggs are released in the open ocean, mostly to drift freely with the winds and currents, but some sauries and flyingfishes attach eggs to flotsam by means of thread-like structures on the shell. Duration of incubation is short in most of the drifting eggs, especially in the tropics. Usually they hatch in from one to three days, and liberate drifting larvae. Some species, especially sauries, halfbeaks and flyingfishes, require an incubation time of one to two weeks. Some holopelagic species are intermittent spawners, extending their reproductive activity over several months.

Meropelagic fishes that deposit demersal eggs must do so where the larvae can drift to proper nursery areas before transforming into juveniles. They have therefore evolved spawning migration patterns, as have the holopelagic species that require them, that place a large proportion of the drifting eggs and larvae in favorable currents. The larvae usually are modified for flotation, with a great surface-to-volume ratio, or with inclusions of oil. Epipelagic sharks, for the most part, are ovoviviparous or viviparous, giving birth to young well able to swim and find food. Consequently, broods are small. The giant whale shark is oviparous, depositing eggs surprisingly small for such a great animal.

Although many epipelagic fishes have been mentioned, a listing of typical families and some of their representatives appears in Table 8–1. Outline drawings of some of these are presented in Figure 8–4.

The Mesopelagic Zone. Although sunlight attenuates and disappears almost completely somewhere within the mesopelagic, live phytoplankton can be found in small numbers, along with debris from epipelagic plankton. This forms the food base for a sometimes abundant and varied zooplankton community that includes some species with the habit of moving upward to shallower depths during the night. These vertical migrations are shared by numerous mesopelagic fishes, notably the Myctophidae or lanternfishes. Like many mesopelagic fishes, these are generally small (less than 20 cm long), dark in color, and supplied with photophores. These are arranged, especially in lanternfishes, in specific patterns for each species, with slight differences between the sexes.

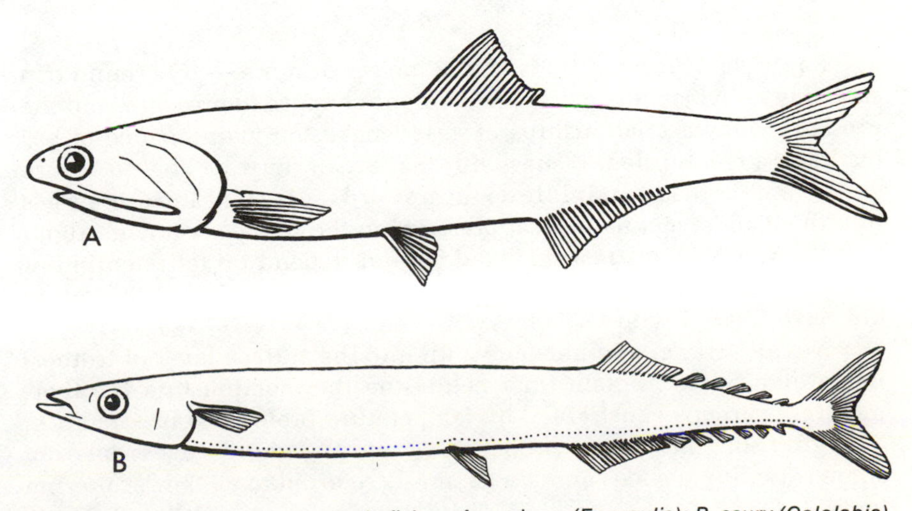

FIGURE 8–4. Examples of epipelagic fishes. *A*, anchovy *(Engraulis)*; *B*, saury *(Cololabis)*.

Table 8–1. FAMILIES AND REPRESENTATIVE SPECIES PRESENT IN THE EPIPELAGIC ZONE

Family	Species	Remarks
Lamnidae	White shark, *Carcharodon carcharias*	Holopelagic, neritic
	Salmon shark, *Lamna ditropis*	Holopelagic
Cetorhinidae	Basking shark, *Cetorhinus maximus*	Holopelagic, neritic
Alopiidae	Thresher shark, *Alopias vulpinus*	Holopelagic, neritic
Rhiniodontidae	Whale shark, *Rhiniodon (=Rhincodon) typus*	Meropelagic
Carcharinidae	Tiger shark, *Galeocerdo cuvieri*	Holopelagic, neritic
Clupeidae	Herring, *Clupea harengus*	Meropelagic
	Sardine, *Sardinops sagax*	Holopelagic, neritic
Engraulidae	Anchovy, *Engraulis encrasicolus*	Holoepipelagic, neritic
	Northern anchovy, *E. mordax*	Holoepipelagic, neritic
Salmonidae	Atlantic salmon, *Salmo salar*	Meropelagic
	Chinook salmon, *Oncorhynchus tshawytscha*	Meropelagic
Myctophidae	(Numerous species of lanternfishes – migrate into epipelagic at night from below)	
Exocoetidae	Oceanic flyingfish, *Exocoetus obtusirostris*	Holopelagic
	California flyingfish, *Cypselurus californicus*	Holopelagic, neritic
Scomberesocidae	Pacific saury, *Cololabis saira*	Holopelagic
	Atlantic saury, *Scomberesox saurus*	Holopelagic
Lampridae	Opah, *Lampris regius*	Holopelagic
Echeneidae	Remora, *Remora remora*	Holopelagic
Carangidae	Pilotfish, *Naucrates ductor*	Holopelagic
	Scad, *Trachurus trachurus*	Holopelagic, neritic
Coryphaenidae	Dolphin, *Coryphaenus hippurus*	Holopelagic, neritic
Bramidae	Bigscale pomfret, *Taractes longipinnis*	Holopelagic
Gempylidae	Escolar, *Lepidocybium flavobrunneum*	Holopelagic
Scombridae	Frigate mackerel, *Auxis thazard*	Holopelagic, neritic
	Albacore, *Thunnus alalunga*	Holopelagic, neritic
Xiphiidae	Swordfish, *Xiphias gladius*	Holopelagic
Luvaridae	Louvar, *Luvarus imperialis*	Holopelagic
Istiophoridae	White marlin, *Tetrapturus albidus*	Holopelagic
Stromateidae		
Centrolophinae	Medusafish, *Icichthys lockingtoni*	Holopelagic
Nomeinae	Silver driftfish, *Psenes maculatus*	Holopelagic
Tetragonurinae	Bigeye squaretail, *Tetragonurus atlanticus*	Holopelagic
Molidae	Ocean sunfish, *Mola mola*	Holopelagic
	Slender mola, *Ranzania laevis*	Holopelagic

Rapid vertical migrations of 200 meters or more — one round trip each day — subject the fish to tremendous changes of pressure, and the pressure-volume relationships of gases make the maintenance of gas bladders a real burden. Fishes with that organ must be able to absorb gases from the bladder rapidly on the upward movement and secrete gas into the bladder against a great pressure on the way back down. About half the species have lost this bladder and depend on fat retention or other means for hydrostatic balance, or else have retained the bladder but have filled it with fat.

Not all vertical migrants move up into the surface layer of isothermic water. Some migrate from below the thermocline into the thermocline, perhaps constrained by temperature preference or tolerance. This group of fishes has the property of scattering sound waves sent from some sounding devices and are generally referred to as the "scattering layer."

Although sunlight is dim in the mesopelagic, the amounts are no doubt significant to the remarkably sensitive eyes of fishes there. Eyes tend to be large, or equipped with a large area for gathering light. Visual pigments are those suited to the blue section of the spectrum, and the visual cells consist largely of rods — entirely of rods in most fishes that never leave the deeper areas. In addition to the dim sunlight, there are many sources of light among the luminous fishes and invertebrates in the mesopelagic. Black and red are common colors of mesopelagic animals.

Predatory fishes in the mesopelagic range from small creatures such as stomiatoids of less than 20 cm to the slender javelinfish (*Anotopterus*) and lancetfishes (*Alepisaurus*) of 1 meter or more, with most species being in the small range. These usually have large eyes, large mouths, and impressive teeth, which in many species can be dagger-like or even barbed. Some of the species have greatly distensible stomachs and can swallow organisms larger than themselves. The swallower, *Chiasmodon*, is an example.

Modifications of the lateralis system, apparently for greater acuity, appear in some mesopelagic fishes. These involve placement of the neuromasts on the surface of the body or actually on papillae or pedicels, as in snipe eels and some of the ceratioid anglers. Other groups such as whalefishes have enlarged pores opening from large canals. The auditory organs of deep-sea species differ greatly in the relative size of the sacculus and lagena, those parts concerned with hearing. Some investigators believe that the different sizes may be better attuned to certain frequencies. There is probably abundant sound in the deep waters for the ears to respond to, as many fishes and crustaceans have what are obviously sound-producing structures.

Reproduction of mesopelagic fishes is not well known. Most apparently release pelagic eggs, some of which float at the surface and some at density layers below or in the thermocline. The larvae are also pelagic and usually drift at the same level as the eggs.

There is a great diversity of life in the mesopelagic. Zooplankton species here usually outnumber those of the epipelagic, and there are about 1000 fish species represented, some of which are interzonal. Resources are available for the establishment of many niches, and most fish species are specialized in some way. Soft-rayed fishes dominate; there are around 200 species of Myctophidae and numerous salmoniforms. Table 8–2 shows examples of species encountered in the mesopelagic zone. Some are shown in Figure 8–5.

The Bathypelagic and Hadopelagic Zones. Below the limit of light the environment is uniformly dark and cold. Generally food is scarcer than in the upper zones, and fish populations are sparser and consist of fewer species. The cold temperatures and great pressures affect the ability of fish to carry out the physiological reactions necessary to build strong skeletons, so their bodies are characteristically soft. Gas bladders are usually absent in species below 1000 meters. Luminous organs are present in many deep pelagic fishes.

Modifications for feeding are remarkable in these zones. The

Table 8–2. FAMILIES AND REPRESENTATIVE SPECIES PRESENT IN THE MESOPELAGIC ZONE

Family	Species	Remarks
Squalidae	*Etmopterus hillianus*	
	Isistius brasiliensis	Interzonal (shallow)
Alepocephalidae	Slickhead, *Alepocephalus bairdi*	
Argentinidae	Pacific argentine, *Argentina sialis*	
Bathylagidae	California smoothtongue, *Bathylagus stilbius*	
Opisthoproctidae	Barreleye, *Macropinna microstoma*	
Gonostomatidae	Lightfish, *Gonostoma*	
	Anglemouth, *Cyclothone microdon*	Interzonal (deep)
Sternoptychidae	Hatchetfish, *Argyropelecus olfersi*	Interzonal
Melanostomiatidae	Longfin dragonfish, *Tactostoma macropus*	Interzonal
Chauliodontidae	Pacific viperfish, *Chauliodus macouni*	
Stomiatidae	Boafish, *Ichthyococcus ovatus*	
Idiacanthidae	Stalkeyed fish, *Idiacanthus fasciola*	
Chlorophthalmidae	Shortnose greeneye, *Chlorophthalmus agassizi*	
Paralepididae	*Paralepis atlanticus*	Interzonal
Alepisauridae	Longnose lancetfish, *Alepisaurus ferox*	Interzonal
Anotopteridae	Daggertooth, *Anotopterus pharao*	Interzonal
Myctophidae	Lanternfish, *Myctophum punctatum*	Interzonal
	Northern lampfish, *Stenobrachius leucopsaurus*	Interzonal
	(Numerous additional genera and species)	
Scopelarchidae	Northern pearleye, *Benthalbella dentata*	
Synaphobranchidae	Atlantic deep-sea eel, *Synaphobranchus infernalis*	
Nemichthyidae	Slender snipe eel, *Nemichthys scolopaceus*	
Bregmacerotidae	Antenna codlet, *Bregmaceros atlanticus*	
Trachipteridae	Dealfish, *Trachipterus arcticus*	
Gempylidae	Oilfish, *Ruvettus pretiosus*	

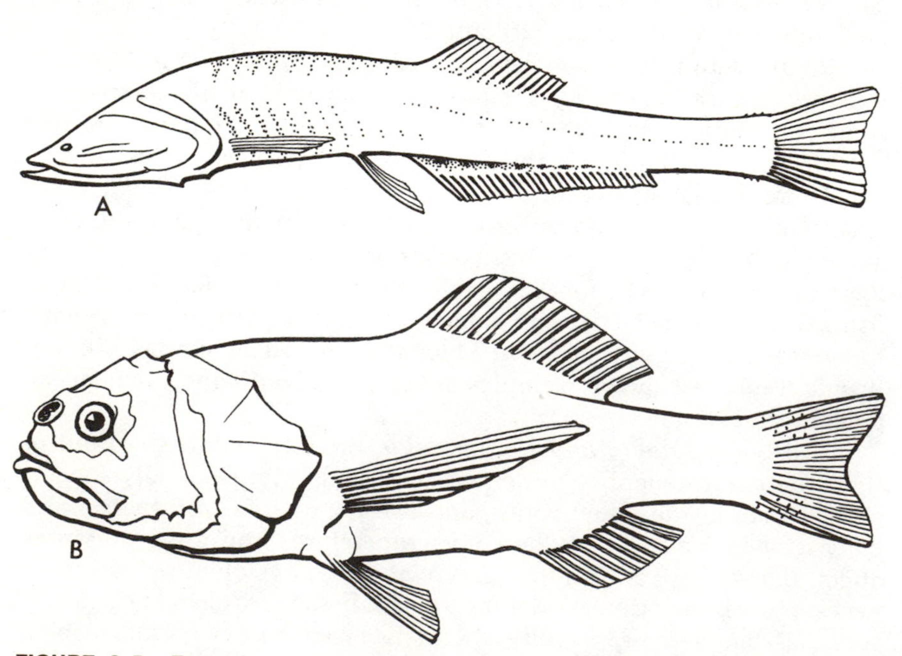

FIGURE 8–5. Examples of mesopelagic fishes. *A*, bristlemouth *(Cyclothone)*; *B*, bigscale *(Poromitra)*.

mouths of such animals as the gulpers and anglerfishes are unbelievably capacious. Many species display long pointed teeth. The swallowers can distend their stomachs to accept prey of greater bulk than themselves. Anglers apparently attract prey by displaying a luminous lure called the esca, borne on a "fishing rod" or illicium on the head. Many have luminous barbels below the mouth as well. The viperfishes, *Chauliodus*, have an elongate first ray in the dorsal fin that appears to function as a lure.

The ceratioid anglerfishes are well known for their obviously efficient method of ensuring that the sexes will be together at spawning time. The tiny males of many species are parasitic on the females. They grasp the female with their jaws and subsequently become something like testes-bearing lobes on her body, depending on her circulatory system. Little is known of reproduction in the deep sea except that many species have pelagic eggs and larvae, and that males of many species are smaller than the females.

Table 8–3 lists some families and representative species found in the deep pelagic zones below 1000 meters. Some outline drawings of deep pelagic fishes are shown in Figure 8–6.

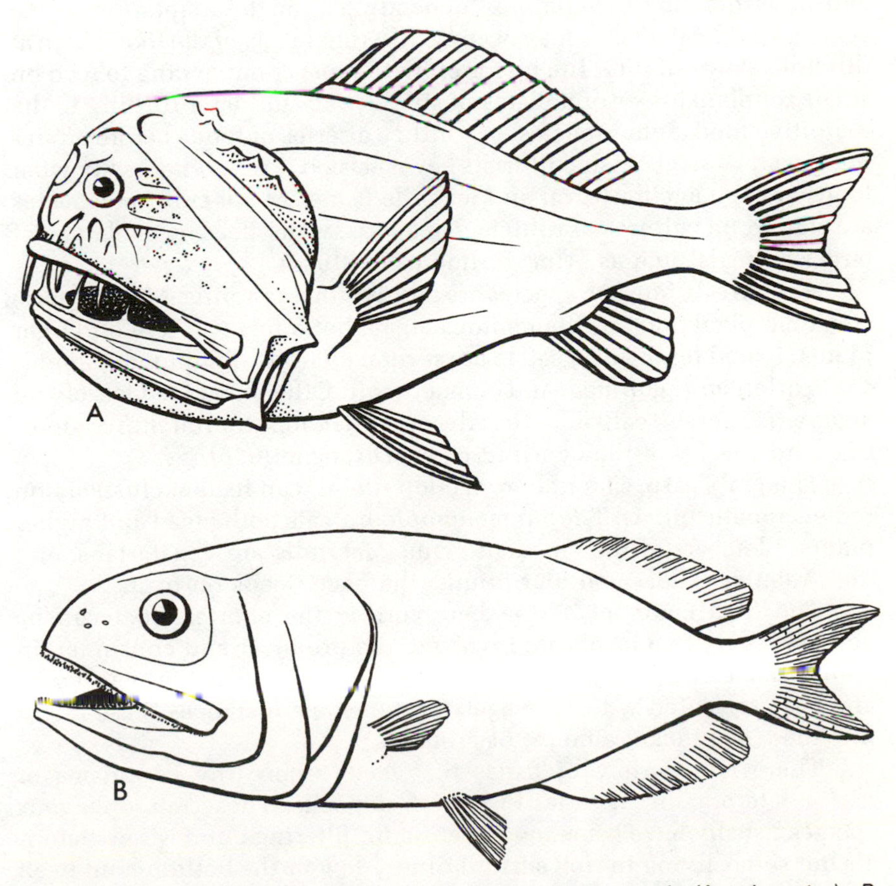

FIGURE 8–6. Examples of deep pelagic fishes. *A*, fangtooth *(Anoplogaster)*; *B*, whalefish *(Rondeletia)*. (*A* after Fitch and Lavenberg; *B* after Jordan and Evermann.)

Table 8–3. REPRESENTATIVE FAMILIES AND SPECIES IN PELAGIC ZONES BELOW 1000 METERS

Family	Species
Bathylagidae	Blacksmelt, *Bathylagus euryops*
Gonostomatidae	Veiled anglemouth, *Cyclothone microdon*
Malacosteidae	Loosejaw, *Malacosteus niger*
Idiaçanthidae	Stalkeyed fish, *Idiacanthus fasciola*
Evermannellidae	*Evermannella atrata*
Paralepididae	Slender barracudina, *Lestidium ringens*
Giganturidae	Gianttail, *Gigantura vorax*
Saccopharyngidae	Gulper, *Saccopharynx ampullaceus*
Eurypharyngidae	Gulper, *Eurypharynx pelecanoides*
Nemichthyidae	Snipe eel, *Nemichthys scolopaceus*
Chiasmodontidae	Great swallower, *Chiasmodon niger*
Caulophrynidae	Seadevil, *Caulophryne polynema*
Ceratiidae	Warted seadevil, *Cryptopsaras couesi*

Trophic and Other Relationships of Fishes

Who Consumes Whom?

As previously mentioned, trophic relationships of fishes are diverse and vary from simple to complex, depending upon the adaptations of the species involved. Fishes may occupy different levels of the food chain at different stages of their life histories, with most commencing to feed on small zooplankton or organisms of similar size and later turning to the definitive food. Some species may utilize a series of foods before attaining the adult stage. Fish culturists have learned through necessity what foods are suitable for larval and juvenile fishes. Many cultured species are started on cultures of rotifers, copepods, water fleas or the larvae of larger animals such as brine shrimp or molluscs.

A relatively few fish species are herbivorous as adults. Most of these consume planktonic or filamentous algae, but a few can live on larger plants. Examples of algae eaters are several cichlids, *Tilapia mossambica, T. nilotica, T. galilaea,* and *T. macrochir.* Others that feed largely on algae are loricariid catfishes, the silver carp, *Hypophthalmichthys molotrix,* and species of the cyprinid genus *Discognathichthys.*

The grass carp, *Ctenopharyngodon idella,* can feed exclusively on rooted aquatic plants. *Tilapia melanopleura* eats both algae and higher plants. Members of the characid genus *Metynnis* are vegetarians, and their relative *Colossoma nigripinnis,* the pacu, feeds on fruit.

The food chain of a true herbivore is the shortest that can be obtained; only two levels are involved, the producer and consumer. In reality, fishes that eat plants must take in and digest the very small animals associated with the vegetation. In some instances these might provide a significant amount of protein.

There is a variety of fishes that feeds upon tiny organisms or particulate organic material in bottom deposits. These microphagous species usually have some modification for filtering out desired materials but some, living in rich surroundings, devour the bottom mud itself and digest the living and dead organic stuff contained in it. The larvae of lampreys are well known for their microphagous habits, living in organically rich bottoms and straining out their food at the mud-water interface by means of an intricate oral sieve. Other microphages are the grey mullet, *Mugil auratus,* the South American hemiodontines and anodontines, and the mud carp, *Cirrhinus molitorella. Tilapia leucosticta* and *Trichogaster trichopterus* eat plankton and decomposing plant materials.

Fishes that feed willy-nilly on bottom deposits may be fulfilling the combined roles of herbivore, carnivore, and scavenger. Algae and small invertebrates may be prominent in the organic component of the bottom.

Most fishes are equipped to feed upon small animals of some sort, from a variety of small invertebrates to small fish. Some species may be specialized and feed only on certain kinds of prey but most are opportunists, taking a wide selection of organisms, usually of similar size and habit. Considerable overlap of food habits may exist in places where many species live in a rich environment, but this overlap can be changed into competition if food resources are scarce. In such cases species will resort to the foods and living spaces to which they are best suited, and the population of each species will be limited to the numbers that can fit into its prime habitat.

A few fishes may be truly omnivorous. The common carp, *Cyprinus carpio,* although well suited for feeding on bottom insects and other small invertebrates, can feed predominantly on vegetation and will take small fish on occasion. This species has dominated many bodies of water to which it has been introduced, sometimes to the exclusion of a variety of native species which had lived in balance with the environment.

Whereas an insectivorous fish is usually a first-level carnivore, feeding on animals that eat the primary producers, many fishes can be second-level carnivores and a few are on the third level. Many of these are piscivorous to a large extent, but other carnivores, such as squids, frogs, snakes, crustaceans, and even mammals and birds are eaten. The top carnivore among fishes is the white shark, *Carcharodon carcharias;* its great size and capacious jaws allow it to take prey the size of sea lions or men. Other high level carnivores are the billfishes and the large tunas.

Naturally, the waters cannot support great numbers or great weight of large carnivores according to the familiar "pyramid of numbers" or "pyramid of biomass" concept. In any ecosystem the greatest biomass, the greatest amount of energy, and the largest numbers of organisms will be represented by the producers — largely green plants. The herbivores convert a small percentage of the biomass of plants eaten into their biomass, and so on up through the various levels of carnivores. The energy required to produce a large carnivore — a bluefin tuna for example — is tremendous, as represented in Figure 8–7. Figure 8–8 shows a simplified hypothetical food web.

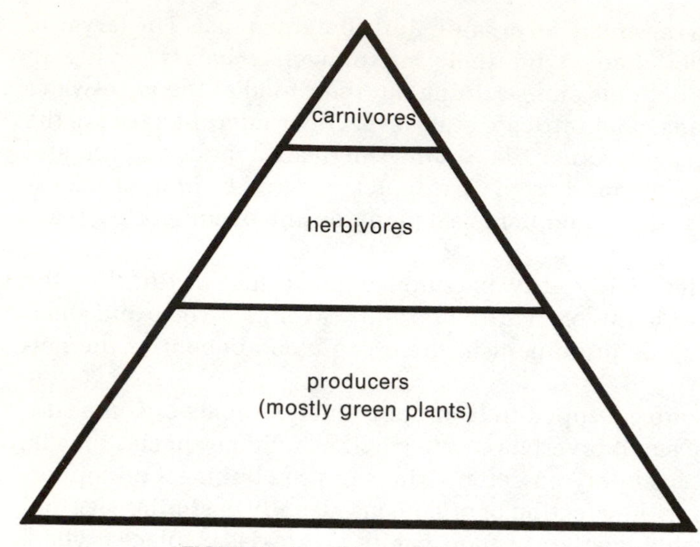

FIGURE 8–7. Pyramid of biomass.

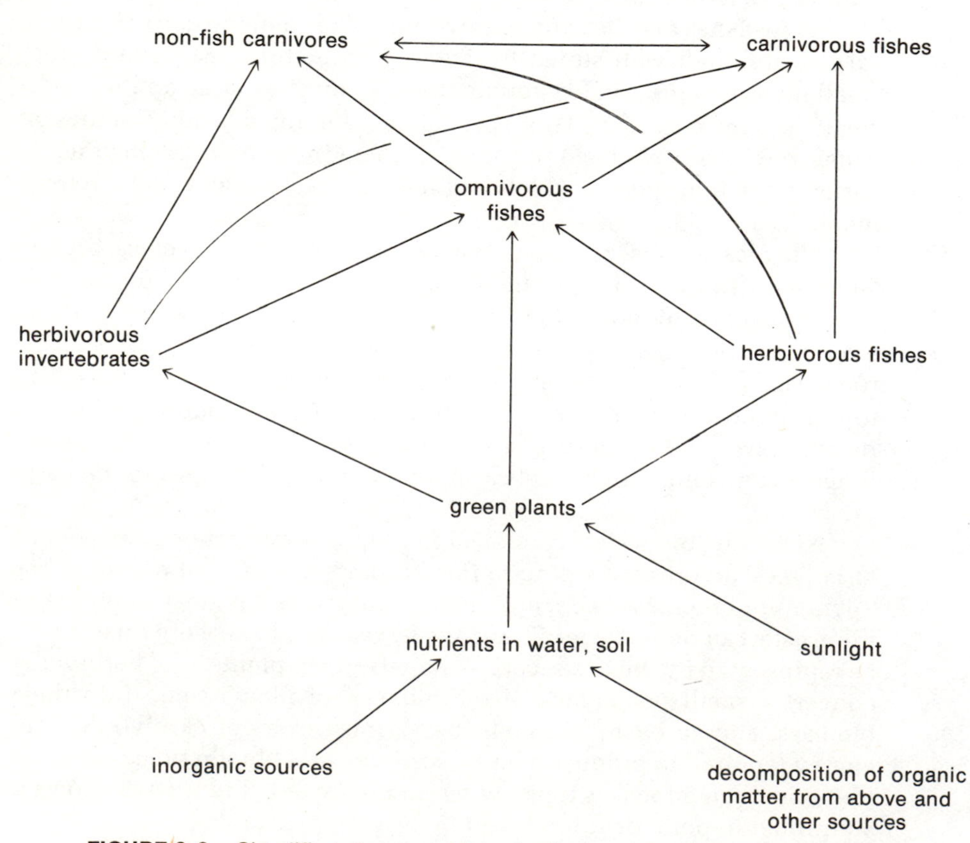

FIGURE 8–8. Simplified diagram of hypothetical trophic relationships of fishes.

Fish provide a source of food for many other animal groups. Fish larvae and very small juveniles are part of the food of aquatic insects such as naiads of dragonflies, larval and adult water beetles, and the backswimmers. A few spiders have become fishermen as well. In the oceans many kinds of invertebrates prey upon fish. Coelenterates — sea anemones, medusas, siphonophores, and jellyfishes — are all capable of catching and consuming fish. Arrowworms (Chaetognathidae) are known to eat fish larvae, and some of the larger carnivorous prawns are fish eaters. Squid are in part piscivorous. Vertebrates that eat fish include frogs, salamanders, snakes, lizards, and turtles, but perhaps the most successful fishermen are found among the birds and mammals. Birds have evolved many types of fish eaters, some that wait along shore, others that attack from the sky, and a few that join the fish in the water and outswim their prey. Piscivorous mammals range from tiny water shrews through minks and otters to seals and sea lions and small and great whales. Man, however, with modern technology, is efficient to the point of being destructive to some fisheries.

Ecological Relationships Among Fishes

Fishes often set up relationships with others in which two or more species are in regular contact. Often the juveniles will school together during the time of life during which they all are seeking the same kind of foods in nursery areas. For instance, one scoop of a dip net into a school of small fry in a backwater of the Columbia River might capture the young of the largescale sucker, *Catostomus macrocheilus*, the northern squawfish, *Ptychocheilus oregonensis*, the speckled dace, *Rhinichthys osculus*, and the redside shiner, *Richardsonius balteatus*. In the adult stage the sucker is a bottom feeder, the squawfish a predator, the dace a riffle dweller that feeds among stones, and the shiner a feeder on drift food. Although they do not associate as adults, they seem to derive some benefits from schooling together at a very young stage.

The habit of the pilotfish, *Naucrates ductor*, of swimming with large sharks and manta rays is well known. By swimming very close to the shark, the pilotfish's own swimming efforts can be reduced and, in addition, it has opportunity to feed on bits and pieces of the shark's meals. A closer association between large and small fishes is encountered in the habits of the sharksuckers or remoras of the family Echeneidae, which have an oval sucking device on top of the head by which they cling to sharks or other large swimming creatures. They can thus be carried along with little effort of their own, and can detach simply by swimming forward when an opportune time for feeding arises.

One of the most remarkable and perhaps most important form of symbiosis is "cleaning," in which certain marine species remove parasites, bacteria, and infected tissue from other, usually larger, species. Most cleaners are tropical and represent "higher" groups; among them are gobies, angelfishes, butterflyfishes, and wrasses. A surfperch, *Brachyistius frenatus*, is a temperate ocean cleaner-fish. The cleaners may not engage in this activity on a full time basis, and probably do not depend on it for a complete livelihood; some clean other fishes only as juveniles and forsake the habit as adults. Some clean only on occasion. The habit is of great importance to some of the cleaners in that they have evolved special colors and patterns, and seem to solicit business at certain stations by various displays that appear to be recognized by customers. Fishes coming to the stations to be cleaned cooperate fully with the cleaners, remaining still, opening the mouth and gill covers when necessary and, most important, refraining from eating the cleaner. Large barrucudas and groupers allow the smaller fishes to swim around their mouths and gills with impunity. While the cleaner performs a service for the larger fishes, there are a few "con men" among the blennies that mimic closely the color and actions of the cleaners but actually feed on bits of skin from the unsuspecting customer.

Another type of relationship is parasitism, in which one species gains its food from the body of another while closely associated with it, ideally without killing the host. The relationship of a small lamprey to a

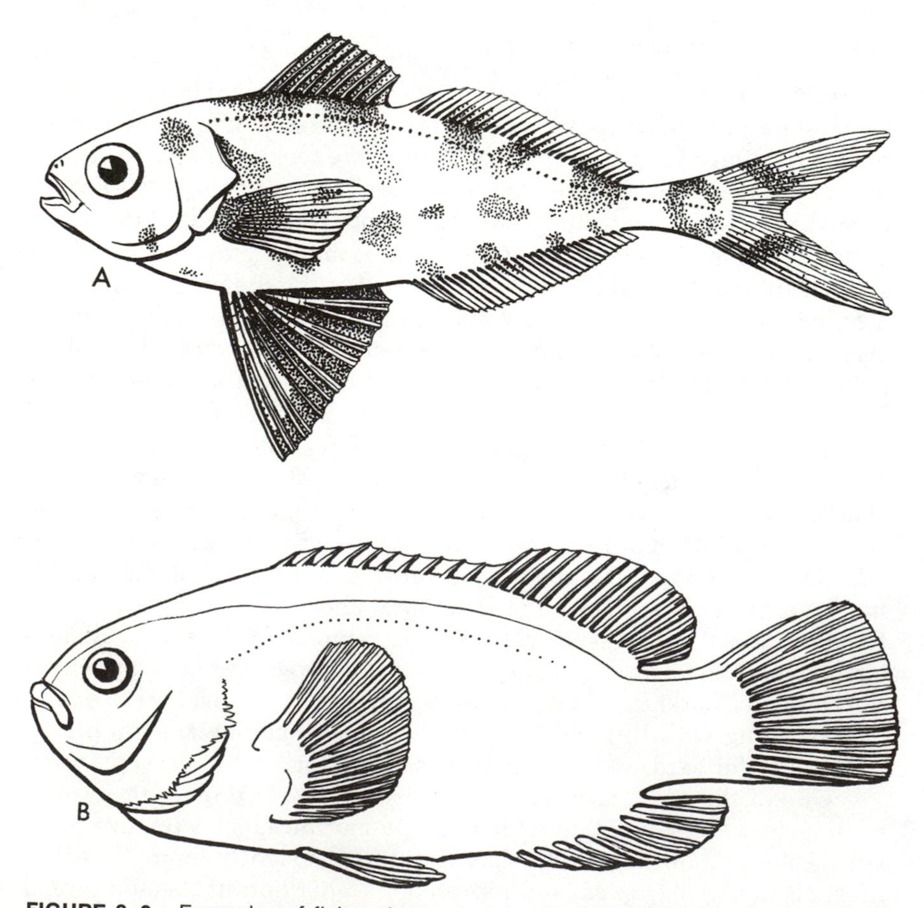

FIGURE 8–9. Examples of fishes that associate with other animals. *A,* man-of-war fish *(Nomeus)* swims among tentacles of jellyfish; *B,* anemonefish *(Amphiprion akallopisos);*

(Illustration continued on opposite page)

large fish could be viewed as parasitism, but the converse would be more like predation. Hagfishes can bore into the bodies of large fish and feed on them, possibly without causing immediate death. This ability is shared with the deep-living snubnose eel, *Simenchelys parasitica*, which has been found to be "parasitic" on large halibut. Some of the trichomycterid catfishes of South America are parasitic upon larger species, usually living in the branchial chamber of the host. The truly parasitic males of the ceratioids have already been mentioned.

A kind of "nest parasitism" occurs among some North American minnows that deposit their eggs in nests made by other species of minnows or by centrarchids. Survival of the brood appears to be enhanced by having another, perhaps larger, individual guard the eggs.

There are numerous examples of close relationships between fish and other animals, one of the commonest being the hosting of parasites. Fish may harbor protozoans, coelenterates, spiny-headed worms (Acanthocephala), flukes (Trematoda), tapeworms (Cestoda), roundworms (Nematoda), isopods, amphipods, and the larvae of freshwater mussels. Some individual fish can qualify as swimming communities

FIGURE 8–9 *Continued.* *C*, bitterling *(Rhodeus)* deposits eggs in siphon of mussels.

with many of the above groups represented in or on their bodies. The life cycles of many parasitic worms involve several hosts, and the fish may serve as an intermediate, or as the definitive host, so in these cases the life of the fish may be linked with the lives of copepods, snails, birds, or other vertebrates.

Pilotfish and remoras do not always associate with other fishes; they may accompany, or hitch free rides on, whales or turtles. There are some remarkable symbiotic relationships between coelenterates and fish. Members of the stromateid genus *Nomeus* are known as man-of-war fish because they live with the large jellyfish *Physalia,* swimming among the tentacles without injury (Fig. 8–9*A*). *Icichthys lockingtoni,* the medusafish (another member of the Stromateidae), has similar habits. Larvae and small juveniles of other marine families such as Carangidae and Gadidae are known to shelter under jellyfish.

Species of the genus *Amphiprion* (Pomacentridae) are able to associate closely with sea anemones without suffering harm, apparently protected by their mucus (Fig. 8–9*B*). Similar immunity has also been noted in juveniles of the sculpin genus *Artedius.* The pearlfishes, Carapidae, are often associated with echinoderms or molluscs. Most commonly, certain species hide inside sea cucumbers, entering through the anus tail first. Some species hide in the mantle cavities of pearl oysters; others enter starfish. There is an Indian scorpionfish of the genus *Minous* that carries at least a partial covering of hydroids on its skin; apparently this helps to conceal the fish. The blind goby, *Typhlogobius californiensis,* usually lives with the ghost shrimp, *Callianassa,* in the shrimp's burrows. A closer relationship is seen between crustacea and the sea snail genus *Careproctus,* members of which deposit eggs on, or in the branchial region of, certain large crabs. The habit of the bitterling, *Rhodeus,* of ovipositing in the incurrent siphon of freshwater clams is well known (Fig. 8–9*C*).

References

Allen, K. R. 1969. Distinctive aspects of the ecology of stream fishes: a review. J. Fish. Res. Bd. Can., 26(6):1429–1438.

Boddeke, R. 1963. Size and feeding of different types of fishes. Nature, 197:714–715.

Brett, J. R. 1969. Resume: temperature and fish. Chesapeake Sci., 10:275.

Bruun, A. F. 1955. Life and conditions in the deep sea. Proc. 8th Pacific Sci. Congr., 1:399–408.

Burton, G. W., and Odum, E. P. 1945. The distribution of stream fish in the vicinity of Mountain Lake, Virginia. Ecology, 26(2):182–193.

Coker, Robert E. 1947. This Great and Wide Sea. Chapel Hill, University of North Carolina Press.

Cushing, D. H., and Walsh, J. J. 1976. The Ecology of the Seas. Oxford, Blackwell.

Fryer, G., and Iles, T. D. 1972. The Cichlid Fishes of the Great Lakes of Africa: Their Biology and Evolution. Neptune, N.J., T.F.H. Publications.

Gerking, S. D. (ed.). 1967. Biological Basis for Freshwater Fish Production. Oxford, Blackwell.

Gorbunova, N. N. 1971. Vertical distribution of eggs and larvae of fish in the western tropical Pacific. *In:* Vinogrado, M. E. (ed.), Life Activity of Pelagic Communities in the Ocean Tropics. Academy of Sciences U.S.S.R., Israel Program for Scientific Translations, pp. 256–269, English translation, 1973.

Gunther, K., and Deckert, K. 1956. Creatures of the Deep Sea. New York, Scribners.

Hardin, Garrett. 1960. The competitive exclusion principle. Science, 131:1292–1297.

Hardy, A. C. 1959. The Open Sea: Its Natural History. Part II, Fish and Fisheries. St. James Place, London, Collins, pp. 35–190.

Hedgpeth, Joel W. (ed.). 1957. Treatise on Marine Ecology and Palaeoecology Vol. 1. Ecology, New York, The Geological Society of America.

Hopkirk, J. D. 1973. Endemism in fishes of the Clear Lake Region of central California. University of California Publ. in Zoology, 96:1–135.

Huet, Marcel. 1959. Profiles and biology of western European streams as related to fish management. Trans. Am. Fish. Soc., 88:155–163.

Hynes, H. B. N. 1970. The Ecology of Running Waters. Toronto, University of Toronto Press.

Isaacs, John D. 1969. The nature of oceanic life. Sci. Am., 221(3):146–162.

Jenkins, R. E., and Freeman, C. A. 1972. Longitudinal distribution and habitat of the fishes of Mason Creek, an upper Roanoke River drainage tributary, Virginia. Va. J. Sci., 23(4):194–202.

Kuehne, R. A. 1962. A classification of streams: illustrated by fish distribution in an eastern Kentucky creek. Ecology, 43(4):608–614.

Lagler, K., Bardach, J. E., Miller, R. R., and Passino, D. R. M. 1977. Ichthyology, 2nd ed. New York, John Wiley and Sons.

Larkins, H. A. 1964. Some epipelagic fishes of the North Pacific Ocean, Bering Sea and Gulf of Alaska. Trans. Am. Fish Soc., 93(3):286–290.

Lowe-McConnell, R. H. 1964. The fishes of the Rupununi savanna district of British Guiana, South America. J. Linn. Soc. Lond. (Zool.), 45:103–144.

———. 1975. Fish Communities in Tropical Freshwaters. London and N.Y., Longmans.

Ludwig, G. M., and Norden, C. R. 1969. Age, growth and reproduction of the northern mottled sculpin (*Cottus bairdi bairdi*) in Mt. Vernon Creek, Wisconsin. Milwaukee Public Mus. Occas. Pap., (2):1–67.

MacArthur, R. 1968. The theory of the niche. *In:* Lewontin, R. C. (ed.), Population Biology and Evolution. Syracuse, Syracuse University Press, pp. 159–176.

Marshall, N. B. 1958. Aspects of Deep Sea Biology. London, Hutchinson & Co. (Publishers).

Moyle, Peter B., and Nichols, R. D. 1973. Ecology of some native and introduced fishes of the Sierra Nevada foothills in central California. Copeia, 1973 (3):478–490.

Nikolsky, G. V. 1963. The Ecology of Fishes. New York, Academic Press.

Odum, Eugene P. 1971. Fundamentals of Ecology, 3rd ed. Philadelphia, W. B. Saunders.

———. 1975. Ecology, The Link Between the Natural and the Social Sciences, 2nd ed. New York, Holt, Rinehart and Winston.

Oglesby, R. T., Carlson, C. A., and McCann, J. A. 1975. River Ecology and Man. New York, Academic Press.

Parin, N. V. 1968. Ichthyofauna of the Epipelagic Zone. Academy of Sciences of the U.S.S.R., Institute of Oceanography. Israel Program for Scientific Translations, translated 1970.

Pennak, Robert W. 1971. Toward a classification of lotic habitats. Hydrobiologia, 38(2):321–334.

Ricker, W. E. 1934. An ecological classification of certain Ontario streams. Publ. Ont. Fish. Res. Lab., 49:1–114.

Roberts, T. R. 1972. Ecology of fishes in the Amazon and Congo basins. Bull. Mus. Comp. Zool., 143(2):117–147.

Shelford, V. E. 1911*a*. Ecological succession. I. Stream fishes and the method of physiographic analysis. Biol. Bull., 21(1):9–34.

———. 1911*b*. Ecological succession: II. Pond fishes. Biol. Bull., 21(3):127–151.

Stauffer, J. R., Dickson, K. L., Cairns, J., and Cherry, D. S. 1976. The potential and realized influences of temperature on the distribution of fishes in the New River, Glen Lyn, Virginia. Wildl. Monogr., 50:1–40.

Strahler, A. N. 1957. Quantitative analysis of watershed geomorphology. Trans. Am. Geophys. Union, 38:913–920.

Sverdrup, H. U., Johnson, M. W., and Fleming, R. H. 1942. The Oceans, their Physics, Chemistry and General Biology. Englewood Cliffs, N.J., Prentice-Hall.

Trautman, Milton B. 1942. Fish distribution and abundance correlated with stream gradients as a consideration in stocking programs. Trans. 7th N. A. Wildl. Conf., 7:221–223.

Weatherley, A. H. 1963. Notions of niche and competition among animals, with special reference to freshwater fish. Nature, 197:14–17.

———. 1972. Growth and Ecology of Fish Populations. New York, Academic Press.

Zaret, T. M., and Rand, A. S. 1971. Competition in tropical stream fishes: support for the competitive exclusion principle. Ecology, 52(2):336–342.

Zenkevitch, L. 1963. Biology of the Seas of the U.S.S.R. London, George Allen and Unwin, pp. 1–955.

9
DISTRIBUTION AND MIGRATIONS

There are two main aspects to the study of fish distribution: the descriptive aspect, simply finding out what species are where, and the often difficult aspect of learning why the species are where they are, and how they arrived there. The descriptive aspect involves much work with collecting gear, ranging from hand nets to deepwater trawls, all over the sea and land masses. At the time of collection many ecological data are gathered, for, as pointed out earlier, the lives of fishes are governed by temperature, salinity, current speed, dissolved oxygen, pressure, light, available food, and numerous other factors, a knowledge of which is vital to the analytical phase of fish zoogeography. Physical and ecological barriers play an important part in the dispersal of fishes, so geography, geology, and the historical aspects of these sciences enter the analyses. The drift of continents, their submergence or emergence, glaciation, volcanism, climatic cycles, mountain building and erosion, and many other factors must be considered. Following will be descriptions of the fish faunas of the continents and oceans, with comments and speculation on the historical aspects.

As far as distribution is concerned, there are two kinds of fishes — those that seem to depend on land masses for their distribution, and those that seem to have dispersed mainly through the seas. Thus, the discussion will be divided into freshwater and marine sections.

Distribution of Freshwater Fishes

The term "freshwater fishes" is an elastic one, with different uses for various purposes. It can mean "fishes present in freshwater" and can therefore include, for instance, migratory species that are found in fresh water at some stage of their life history. Or, it can be restricted in meaning to something like "fishes that cannot enter marine waters," thus bringing up discussion of how much salt makes marine water, and consideration of whether the fossil history of the group under discussion includes any finds in marine sediments.

Ecologists and zoogeographers have subdivided freshwater fishes according to their abilities to tolerate salt water, with the most detailed scheme in general use being that of Dr. George S. Myers. Table 9–1 lists his divisions, with brief explanatory notes.

Table 9–1. DIVISIONS OF FRESHWATER FISH GROUPS*

Division	Remarks
I. Primary	Groups with little or no tolerance for sea water, currently or historically; examples are lungfishes, paddlefishes, bichirs, pikes, minnows, catfishes (with exceptions), characins, sunfishes, black basses, percids (yellow perch, darters, etc.).
II. Secondary	Groups usually restricted to fresh water but with enough salt tolerance so that members can enter the ocean and sometimes cross narrow saltwater barriers; examples are garpikes, killifishes, livebearers, cichlids.
III. Diadromous	Fishes with regular migration between fresh and salt water; examples are several lampreys, many salmons and trouts, shad, American and European eel, mountain mullet, striped bass.
IV. Vicarious	Nondiadromous members of marine groups living in fresh water; examples are the burbot (*Lota*), brook silverside, sculpins (*Cottus*), certain cardinalfishes, freshwater drum, tule perch (*Hysterocarpus*).
V. Complementary	Migratory members of marine groups that dominate freshwater habitats in the absence of a varied fauna of primary and secondary fishes; certain silversides, mullets, gobies, clingfishes.
VI. Sporadic	Fishes that enter fresh water sporadically (not on regular migration) or can live and breed in either medium; certain anchovies, mullets, snappers, croakers, sea basses.

*According to Myers; 1938, 1949, 1951.

Naturally there are some difficulties involved in placing fishes in Divisions IV, V, and VI, as many families and genera might qualify for all three, depending on circumstances. Some investigators simplify the divisions, as Dr. Philip J. Darlington has done, by considering freshwater fishes to be primary, secondary, or peripheral. Others consider that there are those fishes that have dispersed by means of continental connections or proximity, and those that have dispersed through the seas. This essentially sets the primary fishes apart from all others, except that some secondary species seem to have been restricted in their travels through salt water.

Continental Distribution of Fishes

Each continent has a distinctive fish fauna in that each has a unique combination of primary families found there exclusively, plus other primary, secondary, or peripheral families shared with other continents. Strong resemblances between certain faunas may be the result of past land connections or proximity, the abandonment of marine habits by formerly widespread groups, or simply the presence of many freshwater representatives of salt-tolerant groups. Furthermore, physiological tolerances and ecological adaptations may produce real or superficial resemblances over similar climatic zones. Climatic zonation is important in the distribution of fishes, and some earlier ichthyologists began their treatment of the subject with the recognition of a Northern Zone, an Equatorial Zone, and a Southern Zone.

Günther's Northern Zone includes the area roughly north of the Tropic of Cancer, except that its boundary crosses Asia at 30° to 35° north. The Southern Zone takes in only part of Victoria in Australia, Tasmania, New Zealand, and a strip of Chile, plus Patagonia. In between the two is the Equatorial Zone, divided into cyprinoid (Africa and Asia) and acyprinoid (South America and Australia) regions. The utility of this old system in describing general and ecological resemblance between fish faunas is evident in that the fresh waters of the north are characterized by the salt-tolerant salmonids, sturgeons, smelts, northern lampreys, and several primary families, including pikes, leuciscine cyprinids, and perches. The warm-water zone contains cichlids, characins, lungfishes, osteoglossoids, etc., and the southern bits of land have a sparse fauna of salt-tolerant galaxiids, haplochitonids, retropinnatids, and southern lampreys.

Refinement of zoogeographical divisions based upon fish distribution leads to a better foundation from which to consider both the descriptive and historical aspects of fish distributions and dispersal. Fish fit reasonably well into the world divisions usually accepted by zoogeographers. There are, of course, certain exceptions, which will be pointed out. The world is divided into three faunal realms; the first, encompassing most of the world, has historically been called Arctogea, but Darlington's term Megagea will be used here. Subdivisions of Megagea are as follows: (1) Ethiopian Region, which consists of Africa except for the Atlas mountain area of northwest Africa (part of southern Arabia is generally included), (2) Oriental Region, which includes tropical Asia plus some of the continental islands to the southeast; (3) Palearctic Region, including temperate and cold Eurasia plus northwest Africa; (4) Nearctic Region, consisting of North America south to the tropical areas of Mexico. The second realm is Neogea, consisting of the Neotropical Region (South and Central America and the tropics of Mexico). The third realm is Notogea, containing the Australian Region (Fig. 9–1).

Formerly, the Nearctic and Palearctic were considered subregions of a larger Holarctic Region. Even if this usage is abandoned, the adjective "holarctic" is very useful in describing those fishes that span the northern parts of both the old and new worlds, and will be used for that purpose here.

Patterns of fish distribution in and among the world's regions are due in part to physical barriers and in part to temperature adaptations. For example, the Australian Region is now isolated by marine waters, and the Neotropical Region nearly so except for the narrow connection with the Nearctic, which allows for a transitional fauna in tropical

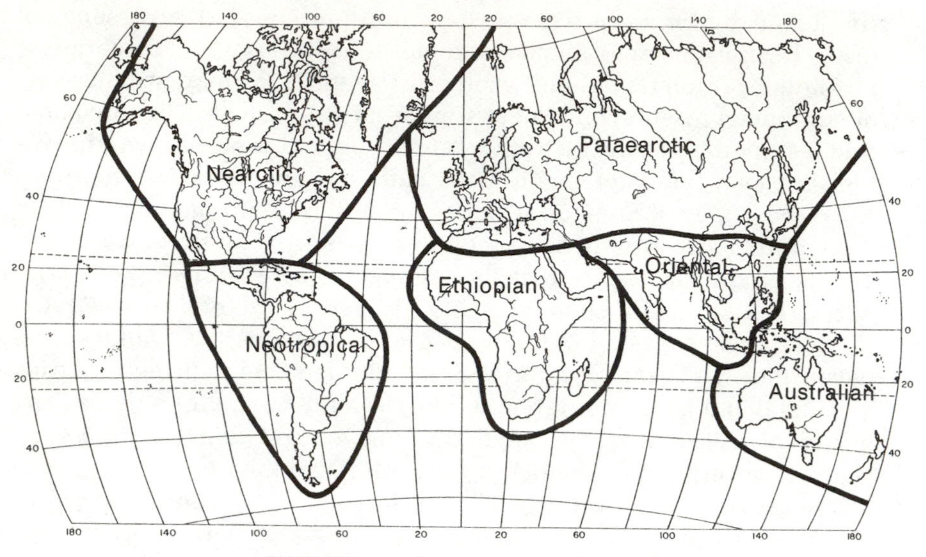

FIGURE 9–1. Zoogeographical regions.

Mexico. The Oriental Region is separated in the northeast from the Palearctic mainly by climate, and a transitional zone exists along the Yangtze. The Nearctic and Palaearctic are now separated by salt water but have been connected in the past, notably in the Bering Sea area. The Nearctic fauna is limited in southern dispersal both by the narrowing of the Central American region and by the temperature of the tropics.

The "Holarctic" Regions. The fish faunas of the Nearctic and Palearctic Regions show close relationships among the different parts of the region in that a few primary families and a number of other families are shared. As might be expected, the similarity is greatest in the northern parts and becomes less as the transitional areas of the south are approached. In both the two large land masses involved, North America and Eurasia, the fish faunas are richer in the eastern parts than in the western, the Rocky Mountains being a boundary of importance in North America, and the Urals in Eurasia.

Some primary families are found only in the holarctic regions. These are Polyodontidae, Esocidae, Umbridae, Dalliidae, Catostomidae, and Percidae. Polyodontidae has only two species, *Psephurus gladius* of the Yangtze and *Polyodon spathula* of the Mississippi, and appears to represent an early connection between the two continents. Esocidae has a wide distribution, especially in the northern part of the holarctic area, although members range rather far south in the Mississippi, and into the Caspian area. *Esox lucius* has truly a circumpolar distribution, and is the only pike found in many areas of North America and Europe. The related mudminnows, Umbridae, have disjunct distributions. *Umbra krameri* lives in the Danube, *U. pygmaea* on the Atlantic coast of North America, *U. limi* in the Great Lakes and Mississippi, and *Novumbra hubbsi* in a few drainages of the Olympic Peninsula in western Washington.

Dalliidae contains one species, *Dallia pectoralis,* which is found in eastern Siberia, northwestern Alaska, and some intervening islands. Its former range was probably diminished by glaciation. Catostomidae shows evidence of two periods during which fishes could have interchanged between the two land masses. *Myxocyprinus* of China represents an ancient type of sucker with a close relative, *Cycleptus elongatus,* in eastern North America, suggesting Tertiary exchange of fish faunas. *Catostomus catostomus* ranges from the St. Lawrence to Alaska in North America and has crossed to Siberia, apparently during the later stages of the Wisconsin glaciation. The Percidae are represented in the two subregions by two shared genera, *Stizostedion* and *Perca,* plus three other genera in Eurasia and three rather speciose genera of darters in North America. Except for one darter in western Mexico, the perches are absent from the Pacific coast of North America.

The remaining primary family found in the holarctic is Cyprinidae, one of the largest and most widespread of the freshwater groups. It forms an important component of all continental faunas except those of Australia, South America and, of course, Antarctica, which harbors no freshwater fishes (Fig. 9–2). The subfamily Leuciscinae is well represented in Asia, Europe, and North America, and one genus, *Phoxinus,* is shared by the latter two.

The killifish family, Cyprinodontidae, is found in both parts of the holarctic regions, being distributed in eastern North America from the St. Lawrence system south, and present on the Mediterranean coastline, but the salt tolerance of the group makes it only partially depend-

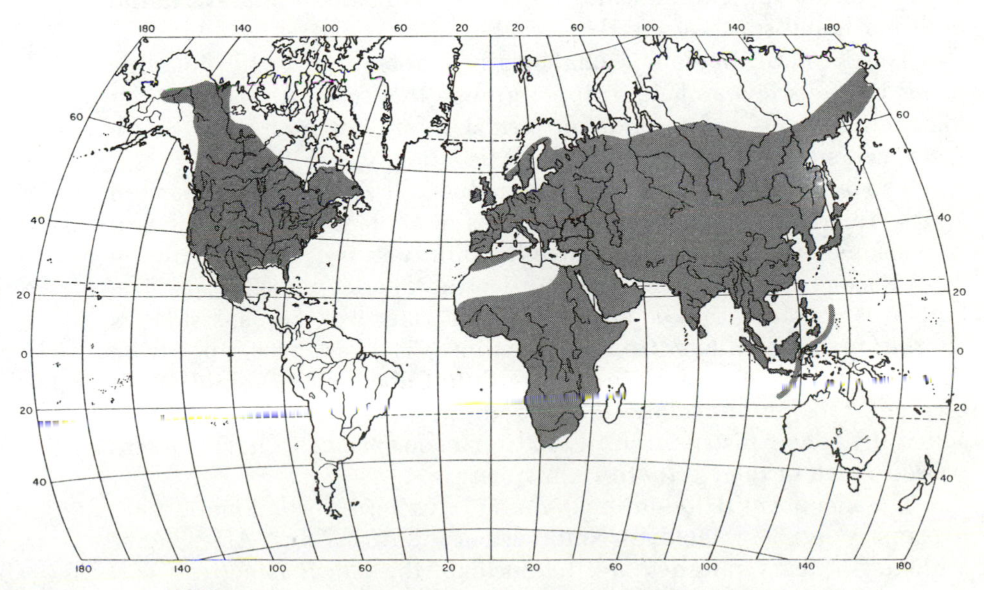

FIGURE 9–2. Distribution of Cyprinidae, showing Wallace's Line.

ent upon fresh waters for dispersal. It is found in South America and Africa as well. The lampreys, Petromyzontidae, and the eels, Anguillidae, are widespread salt-tolerant families found in the holarctic. Other salt-tolerant families prominent in the fresh waters of the holarctic are Acipenseridae, Osmeridae, Salmonidae, Gasterosteidae, and Cottidae. These are not found in fresh water elsewhere except where introduced. The cod family, Gadidae, has a freshwater genus, *Lota,* which is widely distributed in northern waters. There are other peripheral groups that will not be covered here.

The Palearctic Region claims no primary families exclusively, sharing all strictly freshwater groups with other land masses. There are three families peculiar to the Palearctic, all with rather restricted distribution. The Plecoglossidae has but one species, the anadromous ayu *(Plecoglossus altivelis)* of Japan and Taiwan. Salangidae is another family found along the east coast of Asia; the apparently neotenic icefish, *Salangichthys microdon,* enters the rivers of China. Another endemic family, Comephoridae, is found in the Lake Baikal region, and although the members live only in fresh water they are considered peripheral (vicarious) fishes because of their close relationship to the salt-tolerant Cottidae.

Other than the Cyprinidae, the Palearctic shares a number of primary families with tropical Africa and the Oriental Region, which encompasses most of the Asian tropics. Loaches (Cobitidae) occur in all three areas, including Europe. Climbing perches (Anabantidae), snakeheads (Channidae [=Ophiocephalidae]), the spring swamp eels (Mastacembelidae), and the bagrid catfishes range from tropical Asia throughout the transitional area into temperate Asia. The hill-stream catfishes (Sisoridae), absent from Africa, are found in both tropical and temperate Asia. The "Old World catfishes," or sheatfishes (Siluridae) are shared by the Oriental and Palearctic. Homalopterids are found both north and south of the Himalayas.

The Nearctic Region is characterized not only by the holarctic fishes but by a few archaic or relict groups that formerly had a wider distribution, plus a few endemic primary families and some South American species that have invaded Mexico.

The bowfin family Amiidae, now reduced to *Amia calva* of eastern North America, once had representatives in ancient seas (Jurassic and Cretaceous) but later seems to have lived only in fresh water, judging from Eocene and Miocene fossils in North America and Europe. The bowfin is considered to be a primary freshwater fish. Lepisosteidae is another family held over from former times. The gars now range from the Great Lakes–St. Lawrence area south to Cuba and Nicaragua. Various members, but especially the alligator gar, have some tolerance for salt water. Their distribution in the Cretaceous seems to have encompassed much of the Northern Hemisphere.

The mooneyes (Hiodontidae), found in eastern North America, are a primary family related to Notopteridae of Africa and Asia. Other endemic primary families are: Ictaluridae, the North American catfishes, of eastern North America south into Central America; Percop-

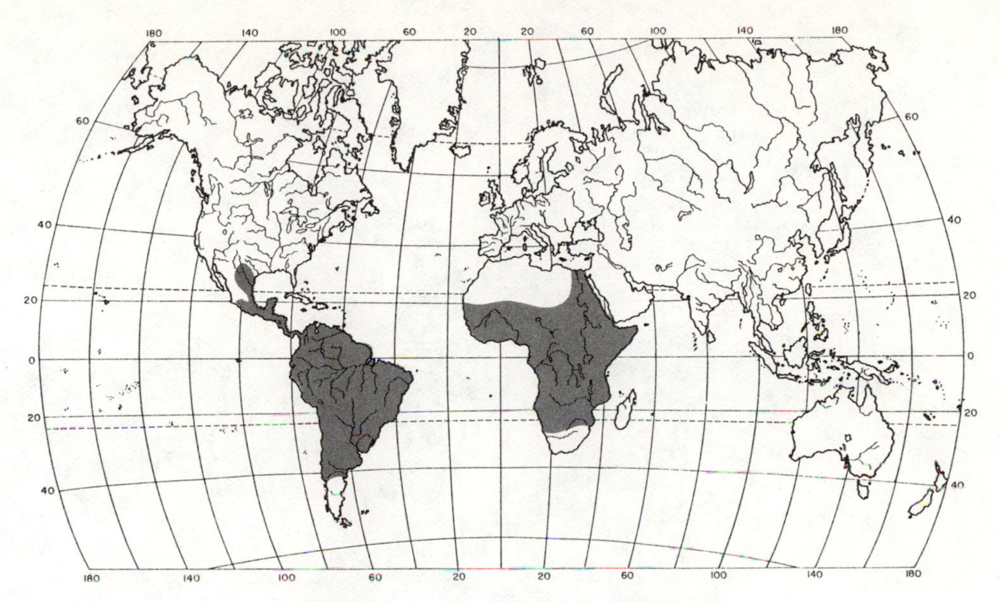

FIGURE 9–3. Distribution of Characidae.

sidae, the trout-perches, with a representative in both the eastern and western sections of the continent; the related cavefishes (Amblyopsidae) and pirate perch (Aphredoderidae) found mainly in the eastern United States (fossils known from west of the Rocky Mountains); and Centrarchidae, the sunfishes, distributed from Canada to Mexico east of the Rockies, but with only one species in the western United States.

Although the perches (Percidae) are shared with the Palearctic, the North American endemic tribe Etheostomatini is worth separate mention. There are over 100 species formerly placed in several genera but now considered to represent only three, *Etheostoma, Percina,* and *Ammocrypta.* They inhabit streams from the Hudson Bay drainage to northern Mexico, with one species reaching the Pacific drainage in Mexico.

Primary families shared with South America are Characidae (also in Africa), with representatives in Mexico and as far north as the Rio Grande in Texas and New Mexico (Fig. 9–3), and Pimelodidae, which reaches central Mexico. A few other catfish families extend from South into Central America, and gymnotids reach to Guatemala.

Salt-tolerant families of interest other than holarctic groups include: Cichlidae, which ranges from South America to the Rio Grande (Fig. 9–4); Goodeidae, of Mexico; Poeciliidae, ranging from the eastern United States to Argentina; Cyprinodontidae, with the North American representatives distinct from those of South America; Percichthyidae, the temperate basses; and Sciaenidae, with a freshwater species, *Aplodinotus grunniens,* ranging from the Hudson Bay drainage to Guatemala.

Pleistocene glaciation, in which ice drove south and receded north

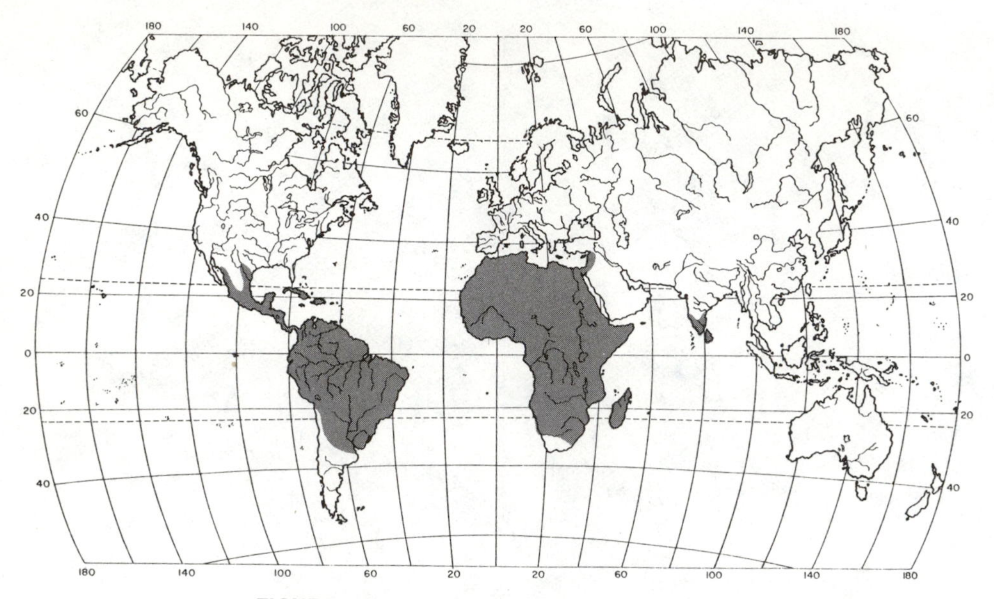

FIGURE 9–4. General distribution of Cichlidae.

several times, exerted great and sometimes lasting effects upon the distribution of freshwater fishes in both North America and Eurasia. Advancing continental ice sheets covered the landscape and forced northern fishes to retreat before them or disappear. Unglaciated areas in the north provided refuge for some species, and heavy rains to the south of the sheets filled watercourses and basins, allowing the spread of many forms southward where some remain as relicts in suitably cold areas. Occurrences of trout in Mexico, Turkey, and Afghanistan are examples, along with the presence of *Stenodus* in the Caspian and several salmonids in the Adriatic drainage.

The amounts of water piled up as glacial ice lowered sea level to the extent that a land bridge across the Bering Sea allowed for some passage of freshwater fishes, and some of the great islands adjacent to continents became part of the larger land masses and shared drainages and fishes.

As the ice retreated, melt-water flowed or ponded along glacial margins and, connecting with lakes gouged into the face of the continent and lakes left as hollows in moraines, formed passage for many species to follow the disappearing ice north, so that many regained something of their former range, or possibly even exceeded it.

The Oriental Region. Boundaries of the Oriental Region are not easy to define from the standpoint of fish fauna because of the broad land connection with the Palearctic and the overlaps in ranges of fishes, but the Himalayas form a backstop to the north, and dry areas of Afghanistan and Iran form a partial barrier in the west. Elements of the oriental fish fauna range in general from eastern Iran through tropical Asia, with a transitional zone roughly south of the Yangtze in China.

Taiwan, the Philippines, and Indonesia are included through the Greater Sunda Islands to the famous "Wallace's line" that separates Bali from Lombok and Borneo from Celebes. Actually there is something of a transitional zone at the line, with a few primary fishes extending beyond Bali to Lombok and Sumbawa.

The Oriental Region has a rich and varied fish fauna characterized by numerous minnows, loaches, and catfishes. The Cyprinidae are especially well represented, to the point that many investigators believe they must have originated here. There are numerous endemic genera representing several subfamilies. A few of the cyprinid genera are shared with Africa; *Barbus, Barilius,* and *Labeo* are examples. Catfish families endemic to the Oriental Region are Amblycepidae, Chacidae, Cranoglanididae, and Pangasiidae. Catfishes found in both the Oriental and Palearctic are Bagridae, Siluridae, and Sisoridae; those shared with Africa are Bagridae, Clariidae, and Schilbeidae. The hillstream loach families Homalopteridae and Gyrinocheilidae are essentailly confined to the Oriental Region but their relatives, Cobitidae, are present in temperate Eurasia and Africa as well.

Other primary families of the Oriental Region are Osteoglossidae (shared with Africa, Australia, and South America), Notopteridae (shared with Africa), Nandidae (shared with Africa and South America), Anabantidae (shared with temperate Asia and Africa), Channidae (shared with Africa and temperate Asia), Luciocephalidae, Mastacembelidae (shared with Africa and temperate Asia), and Chaudhuriidae.

Salt-tolerant groups of interest in the Oriental Region include the widespread Cyprinodontidae and Cichlidae, which are found mainly in South America and Africa but have a few representatives in the Middle East and in India and the island of Ceylon. Toxotidae, the archerfishes, range from India east and south through the islands to Australia. The priapiumfishes, Phallostethidae (including Neostethidae), occur from Thailand to Borneo and the northern Philippines. Synbranchidae, the swamp eels or mud eels, are widespread in the tropics (Fig. 9–5), and a related family, Amphipnoidae, is shared between tropical Asia and Australia. The peculiar Indostomidae, endemic to Burma, is thought to be related to salt-tolerant fishes.

The Ethiopian Region. The continent of Africa is bounded by the seas, except for the narrow connection with the Middle East in the Suez area. The Ethiopian Region as generally defined does not coincide with the limits of the continent, however, because the Atlas Mountains and adjacent land in the northwest corner constitute a part of the Palearctic. This section (the Maghreb subregion of Roberts) contains a mixture of European, Asian, and African forms. Madagascar is sometimes considered to be a subregion, but its fish fauna is not distinctly African.

The equator bisects the continent, so the climate is hot to warm, being tempered to the far north and south and by altitude in the extensive uplands and mountains. Large areas of desert — Sahara, Kalahari, and others — that constitute nearly 40% of the land mass have no active drainages. Tropical rain forests follow the equator, but are

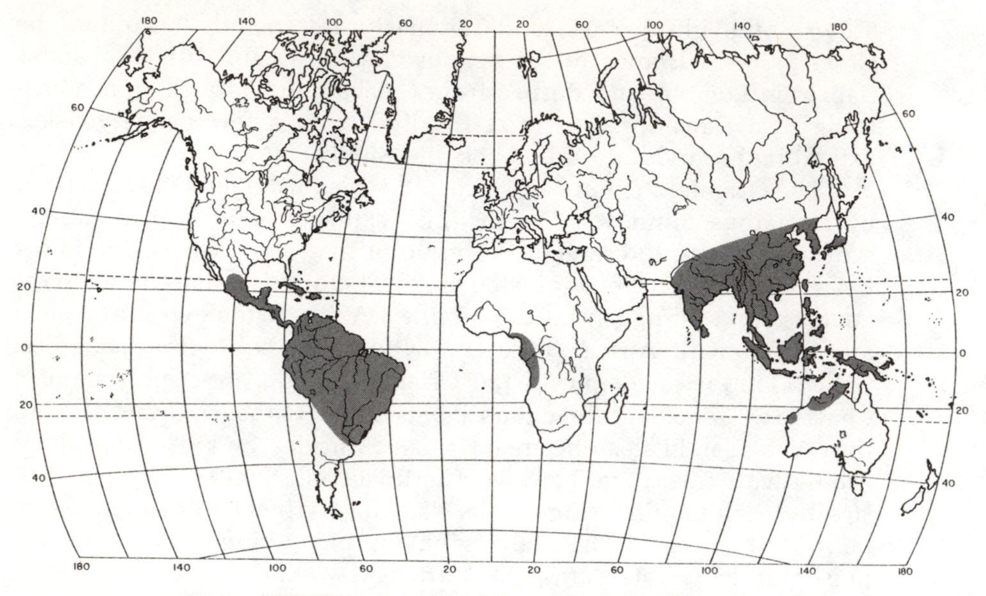

FIGURE 9–5. Distribution of Synbranchidae.

broken up in the eastern highlands. There are broad bands of savanna and steppe to the north of the equator but, in the south, scrub forest is sandwiched between the savanna and steppe.

Fish distribution has been influenced by the presence, in earlier times, of great inland drainage basins that held extensive lakes. These facilitated dispersal across the continent as they overflowed in pluvial periods. Mountains have served both as barriers to lowland forms and as highways for upland species. Extensive swamps have barred passage of species not adapted to them.

Several families of primary fishes are restricted to Africa. Some of these represent archaic groups of lower bony fishes, including the lungfishes (Protopteridae) and the bichirs (Polypteridae). Archaic teleost families endemic to Africa are Denticipitidae, Pantodontidae, Gymnarchidae, Mormyridae, Phractolaemidae, and Kneriidae. The remainder of the endemic primary families are ostariophysine. Distichodontidae, Citharinidae, Ichthyboridae, and Hepsetidae are all relatives of the characins. Amphiliidae, Malapteruridae, and Mochokidae are endemic catfish families.

There is an important relationship to Asia, for several families are shared by the two land masses. The primary Notopteridae, Channidae, Mastacembelidae, Schibeidae, and Bagridae are found only in Africa and Asia, as are the possibly secondary Anabantidae and Clariidae. A few genera of Cyprinidae are shared with Asia. An ancient relationship to South America is shown by sharing of the primary Characidae, plus the following families that are distributed in Asia, South America, and Africa: Osteoglossidae (also in Australia), Cyprinodontidae (widespread), Synbranchidae (widespread in tropics), Nandidae, and Cichlidae. (Nandidae and Osteoglossidae have usually been considered to be primary, but are thought to be secondary freshwater fishes by some investigators.) Another link with South America is Protopteridae, which is closely related to the South American Lepidosirenidae, and is sometimes placed in that family.

Disregarding the desert areas, where a few fishes may remain in suitable water holes as relicts of the most recent pluvial period, the fish fauna is poorest in the highlands of Ethiopia and in the far south. Cyprinidae, Cichlidae, and some catfish families are most typical, but Ethiopia has representatives of Cobitidae, and the Cape area has Anabantidae and Cyprinodontidae. Tropical West Africa, especially the Zaire (Congo) basin and the rain forests along the Gulf of Guinea, has the richest fish fauna, with all primary and secondary families of the continent represented, except Cobitidae and Gymnarchidae. The Nile and faunally related drainages that stretch south of the Sahara to Senegal are rich in fishes, having 24 of the 30 primary and secondary families. Other areas are somewhat more restricted in fauna.

The large lakes of Africa, mostly along the great rift line, are famed for the great proliferation of species of fish that has occurred in them. Lakes Tanganyika, Victoria, and Malawi all have species flocks of cichlids, the latter two each having over 100 species of *Haplochromis*. Tanganyika harbors over 40 genera of cichlids, mostly endemic, and 12 endemic *Mastacembelus* species. Malawi has a number of endemic clariids.

Dr. Tyson Roberts has recognized 10 ichthyofaunal provinces in Africa. These are:

1. Maghreb (northwest, along Mediterranean)
2. Abyssinian Highlands
3. Nilo-Sudan (the Nile and faunally related drainages south of the Sahara west to Senegal. Includes Juba and a few other east coast rivers. Includes lakes Albert, Edward, George, and Rudolf)
4. Upper Guinea (coastal drainages on lower part of bulge)
5. Lower Guinea (coastal drainages along Gulf of Guinea, north of Zaire drainage)
6. Zaire (includes Zaire [Congo] drainage, plus lakes Kivu and Tanganyika)
7. East Coast (Mozambique to Kenya; includes lakes Victoria, Kioga, and Eastern Rift lakes, but not Malawi or Rudolf)
8. Zambesi (includes Lake Malawi)
9. Quanza (Quanza River and adjacent drainages in Angola)
10. Cape of Good Hope (Orange River drainage and southern coastal drainages)

The Neotropical Region. South America is bounded by the oceans except for the narrow isthmus of Panama, through which a few fishes have filtered north into tropical Mexico. The continent is cut by the equator through the northern part of the Amazon basin, so the bulk of the land is in the tropics and the narrow south is temperate. The western rim of the continent is bordered by the Andes, so that streams mostly flow east to the Atlantic. The Orinoco, Amazon, and Parana river systems dominate the hydrographic features.

Despite the proximity of the Neotropical to North America, there seems to be no close relationship with the Nearctic fish fauna. There is a mixing of northern and southern forms in the transitional areas of Mexico, but truly North American fishes do not extend into South America. Rather, there is a definite relationship between the Neotropical and Ethiopian regions, evidenced by the sharing of the primary family Characidae, the possibly primary Osteoglossidae and Nandidae, and the secondary Cyprinodontidae and Cichlidae. In addition, the South American Lepidosirenidae is closely related to the African Protopteridae. Many tropical catfish families occur in each continent, although none are shared. Synbranchidae, although salt-tolerant and distributed in several tropical areas, represents another link to Africa.

South America and Africa were broadly connected as part of Gondwanaland and began to split apart in the south at the beginning of the Cretaceous. Shortly after the South Atlantic was wide enough to prevent dispersal of freshwater stocks, according to a theory proposed by M. J. Novacek and L. G. Marshall, an ancestral ostariophysine line gave rise to the early cypriniform and siluriform lines. These are thought to have spread north and to have entered west Africa, which was still attached to South America but was separated from the rest of Africa by an epicontinental sea. Following complete separation of the two plates and the disappearance of the sea across Africa, radiation of the characin and catfish stocks took place on both continents. Gymnotoids and Neotropical characoids and siluroids developed in South America and other lines, including the cyprinoids, are thought to have originated in Africa. Cyprinidae and Old World catfishes then dispersed from Africa to Eurasia at the end of the Cretaceous. This theory is of interest because the appearances and disappearances of the saltwater barriers are so timed as to have allowed passage of limited groups of the evolving cypriniforms without giving opportunity for broad connections that would have allowed all of the groups to interchange.

The radiation of characoids in South America has been remarkable; some authorities recognize 11 families other than Characidae, although others consider many of these to be subfamilies. The four families of gymnotoids, all endemic, are thought to be derivatives of the characoids. The most famous gymnotoid is the electric eel, *Electrophorus electricus* (Electrophoridae), but the other three families are of great interest because of their weak electrical capabilities, bizarre appearance, and ecology. There are 13 families of catfishes, all primary, endemic to South America. Important endemic families of nonprimary fishes are Potamotrygonidae, Jenynsiidae, and Anablepidae.

The Amazon River system is the biggest and richest drainage in South America, and has the greatest representation of both families and species. There are nearly 500 species of catfishes there, and possibly 900 others. Much of what is written about the distribution of fishes of the continent uses similarity with the Amazon as a point of departure.

Other systems have connected with the Amazon by means of stream capture in the uplands and by flooding of intervening swamps, so that from the Magdalena and Orinoco rivers of the north to the Sao Francisco and Paraguay-Parana to the east and south, there are fish faunas related to or based upon that of the great river. The Amazon, together with the Guianian coastal area, makes up the Guianian-Amazon faunal region of J. Gery.

The Orinoco-Venezuelan ichthyofaunal division, including Trinidad, has over 300 species. The north-flowing Magdalena River represents a separate division or province with about 150 species, nearly half of which are catfishes. The Trans-Andean area includes Panama, part of Colombia, and the Pacific slope of Ecuador and Peru. There are nearly 400 freshwater fishes in the area, some of which may have been isolated there by the uplift of the Andes. Others could have crossed during the uplift through some manner of headwater transfer, and some may have made an end-run around the Andes to the north. Over half the species in the Trans-Andean are found also in the Atlantic drainage.

The Andean region includes waters inhabited by fish from about 900 to 4600 meters in elevation. Most species are at the lower altitudes, but some torrent fishes live beyond the 4300 meter level. Lake Titicaca, about 3700 meters above sea level, is noted for the speciation in the orestiine Cyprinodontidae that took place there in the absence of predatory fishes. Subsequent introduction of rainbow trout and other predators has greatly altered the population structure of the fish community there. The Andean area thrusts south into the related Patagonian division, which has a relatively poor freshwater fauna that includes some southern peripheral species. The most primitive catfish, *Diplomystes*, is representative of the area.

The Paranean division includes the Parana, LaPlata, Uruguay, and Paraguay systems. The latter shares about half of its species with the Amazon, there having been considerable transfer through the swamps of the Gran Chaco and rivers to the north. Three sections of the Brazilian coast make up an area that Gery calls the East Brazilian. This is divided into three sections — Northeast Brazil, Rio Sao Francisco, and Rio Ribeiro. The fish faunas there are relatively poor in comparison to the Amazon or Paranean, and the three sections show different affinities.

The Australian Region. Most of the freshwater fishes of Australia are representatives of salt-tolerant groups, and include many that commonly move between salt and fresh water. The only primary fishes are the lungfish, *Neoceratodus*, of the Ceratodidae, and an osteoglossid, *Scleropages leichhardti*. The lungfish has widespread fossil relatives, and the genus *Scleropages* is represented in Borneo and Southeast Asia by *S. formosus*, and by fossils on Sumatra. Apparently the two groups arrived in Australia at an ancient time and remain as relicts, whatever freshwater species that were with them having long since disappeared. Both species are now restricted to Queensland, in eastern Australia, where they seem to have survived in swamps and permanent rivers.

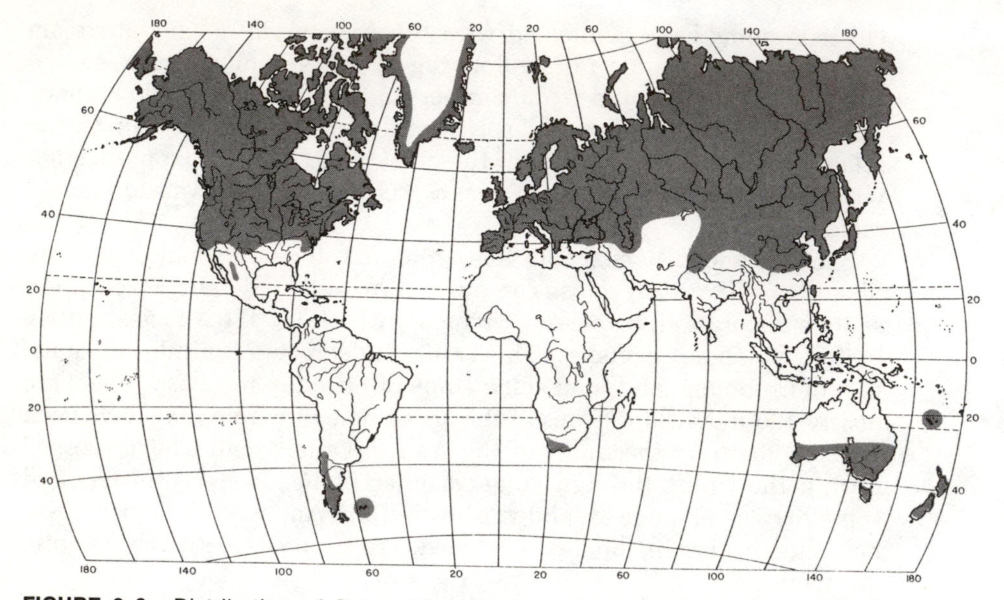

FIGURE 9–6. Distribution of Salmonidae (Northern Hemisphere) and Galaxiidae (Southern Hemisphere).

Salt-tolerant fishes in Australia include the secondary Synbranchidae, the diadromous Galaxiidae. (Fig. 9–6), Aplochitonidae, Retropinnatidae, Anguillidae, and Petromyzontidae, and numerous vicarious or complementary representatives or derivatives of marine families. Included in the latter are Melanotaeniidae (a derivative of the Atherinidae) and Gadopsidae, both of which have become essentially freshwater fishes. Others include Clupeidae, Ariidae, Plotosidae, Atherinidae, Mugilidae, Gobiidae, Eleotridae, Kuhliidae, and Serranidae. Many of these families have characteristic freshwater forms that extend from Australia into New Guinea but not beyond.

New Zealand, some 1600 km from Australia, has no primary fishes. The aplochitonids, galaxiids, retropinnatids, and eels are represented there, along with many vicarious or complementary species.

Land Bridges and Moving Continents

The listing of fish families of the various continents serves to point out that certain of the present faunas show definite intercontinental relationships that argue for land connections where there are currently none. Over these connections the ancestral forms from which contemporary species arose, or perhaps even some of the existing archaic forms themselves, may have had continuous ranges or may have traveled as emigrants. The ancient fish faunas have been modified by adaptive radiation, extinctions, and further mingling with forms dispersing later, so that the contemporary faunas may furnish only clues to directions and routes of dispersal and not many real answers.

At one time the concept of continental drift was not well accepted, as geologic evidence supporting it was scanty. The idea of land bridges was popular, and these were used to explain faunal similarities between stable continents. Some of these stretched the imagination, although there have been and are very good bridges, some of which have been important in the distribution of freshwater fish. North America has been connected to Asia more than once via the Bering bridge, and is presently linked to South America by a narrow bridge that allows some fishes to filter north. Many great islands have been tied to adjacent continents by land now submerged by shallow seas. These include the British Isles, which have several of the European species of cyprinids, and the Greater Sunda Islands of the East Indies, which have many of the Asian primary families. Japan, Taiwan, and other islands were tied to adjacent continents so that primary fish families are now present, but Madagascar, thought to have been connected to Africa into the Cretaceous, has only secondary and marine fishes.

In recent years the study of plate tectonics, involving investigation of sea floor spreading, paleomagnetism of the earth's crust, etc., has provided the evidence that places the concept of continental drift on a logical base. The earth's crust, the lithosphere, is about 100 km thick and consists of a few major plates and several smaller ones, all of which ride upon part of a mantle that is thickly plastic. Convection currents in the mantle send new rock to the surface along ridges in the oceans, causing the sea floor to spread in both directions at right angles to the ridges, and continental masses such as Africa and South America are pushed apart. At the great trenches of the oceans, plates meet and one slides down at a steep angle below the other. These zones are marked by such restless activities as volcanoes and earthquakes. Another kind of movement results from plates sliding past each other at faults in the crust.

The knowledge that these movements of the earth's crust are occurring leads to better acceptance of theories of how continents once fit together and how they split apart and moved to their present locations. Some of the imponderables inherent in the study of freshwater fish distribution are at least partially answered by the drift theories, but some remain enigmas. For the sake of this brief discussion, we can pass over the multiple millions of years during which land masses combined and recombined prior to the latest recombination about 225 million years ago. At the end of the Palaeozoic, the continents were together in a single continent referred to as Pangaea. The bulge of South America fit neatly into the concavity below Africa's bulge; the continental shelf of what is now the Atlantic coast of North America fit against that of northwestern Africa and a greatly tilted Europe; Arabia, Iran, and Turkey were aligned with Africa and well separated from Eurasia; and India, Madagascar, Antarctica, and Australia were bunched off what is now the east coast of Africa (Fig. 9–7).

Some 200 million years ago North America began to split off from its connection with Africa, so that there were two tiers of continents —

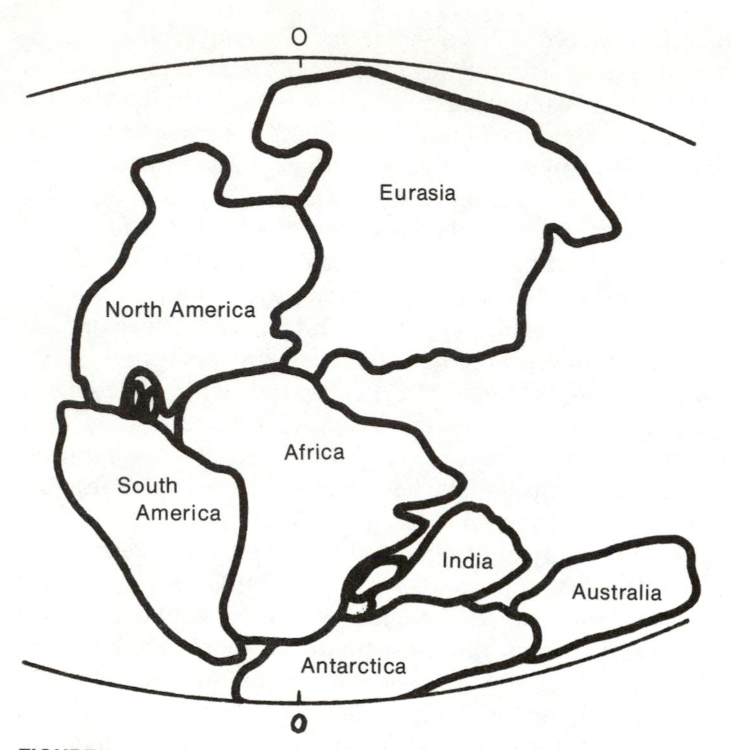

FIGURE 9–7. Pangaea. Hypothetical arrangement of the land mass 225 million years before present.

the northern group, still connected to each other, called Laurasia, and the southern ones, called Gondwanaland (Fig. 9–8). The separation of North America from Africa was complete about 180 million years ago, but North America and Europe remained in contact until a little under 50 million years before the present (mybp), and North America and Asia were in contact during a period some 30 to 50 million years ago. The connection was apparently broken in the Miocene, but reestablished about 10 mybp. North and South America separated about 70 mybp.

Estimated times of separation of other sets of land masses (in mybp) are: South America-Africa, 85 to 90; Africa-India, about 65; India-East Antarctica, about 100; East Antarctica-Australia, about 55; Australia-New Zealand-West Antarctica, more than 80. South America and West Antarctica were possibly connected until about 10 to 12 mybp, although some estimates put the separation at more than 70 mybp. The splitting of Africa and East Antarctica is estimated to have been from about 90 to 180 mybp. The rift between Australia and India is estimated to have been from 120 mybp to a time antedating the formation of Pangaea. Madagascar and Africa have been separate for more than 65 million years.

In order to place the timing of the above events in perspective with freshwater fish distribution we should refer to the antiquity of some of

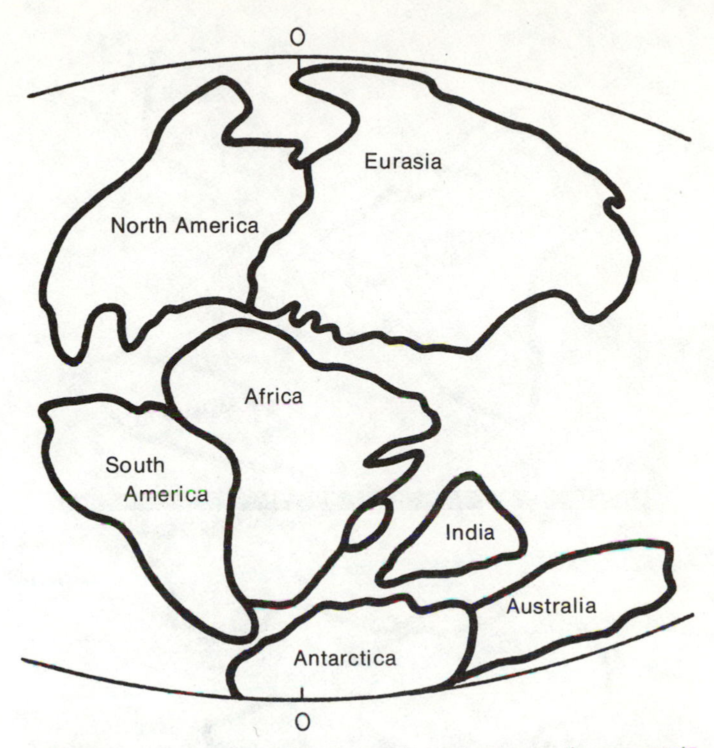

FIGURE 9–8. Breakup of Pangaea. Land to the north represents Laurasia (North America plus Eurasia); Gondwanaland, to the south, had begun to separate (about 180 mybp).

the families involved, as shown by the fossil record. Ceratodidae dates from the Triassic, about 200 mybp, Protopteridae from about 35 mybp, and Polypteridae from at least 50 mybp. Amiidae is known from about 170 mybp, and Lepisosteidae from 130 mybp. Among the teleosts, osteoglossoids are known from the Upper Cretaceous, minnows and catfishes date from the Palaeocene, which began about 60 mybp, and fossil pikes, cyprinodonts, percopsids, catostomids, and anabantids date from about 40 to 50 mybp. The histories of the various groups prior to the deposition of the earliest known fossils await the discovery of new material. Some might be extended farther back in time, while others may not.

From the present and fossil distribution of fishes, there have been many attempts at deducing centers of origins for families or higher categories. In view of the changing positions and relationships of the continents in respect to each other, pinpointing of such centers seems difficult, for fish dispersal might be overshadowed by continent dispersal. Consider India, leaving its position next to Africa, and traveling as a giant "Noah's Ark" across the ancient Tethys Sea to merge with Asia (Fig. 9–9)! Obviously, there are elements of the freshwater fish fauna that are either northern or southern in origin, and Dr. Joel Cracraft has listed these as Laurasian and Gondwanaland faunal elements,

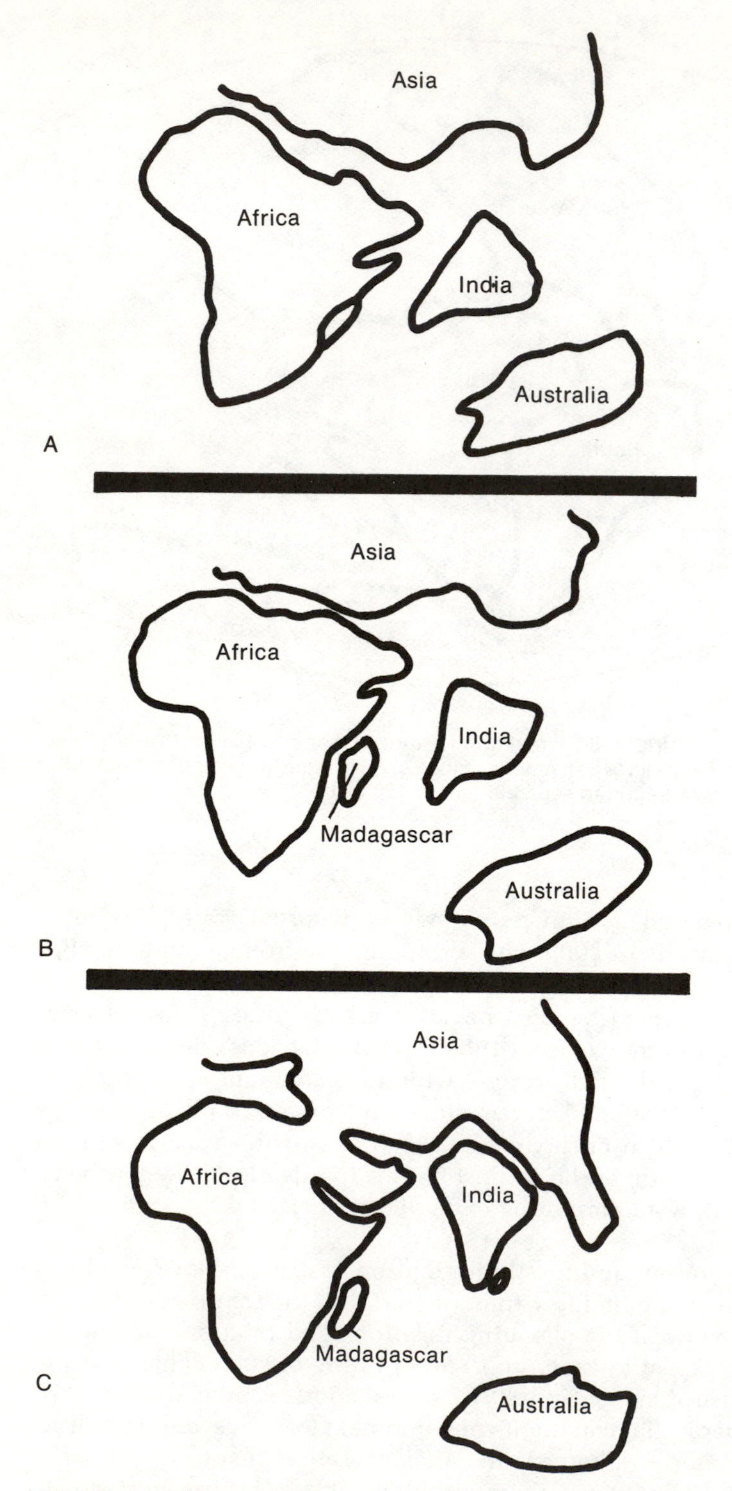

FIGURE 9–9. Movement of Indian Plate. *A*, 135 mybp; *B*, 65 mybp; *C*, 50 mybp.

respectively. Significant among the Gondwanaland elements are the lungfishes, bichirs, osteoglossiforms, mormyrids, characoids, gymnotoids, all the catfishes, cyprinodonts and allies, snakeheads and, among the perciforms, the cichlids, nandids, anabantids, and mastacembelids. The Laurasian elements include the sturgeons, paddlefishes, gars, bowfins, the cyprinoids, pikes and allies, salmonids, percopsiforms, centrarchids, and percids.

Some of the above groups have traveled out of their presumed place of origin, either by riding a moving mass of land or by their own dispersal mechanisms and abilities. Some southern forms that reached the north are ceratodid lungfishes (widespread as fossils), osteoglossids, hiodontids, several families of catfishes, cyprinodontiforms, cichlids, nandids, anabantids, mastacembelids, and channids. Characins have invaded Mexico from the south. Most of these examples involve Africa and Asia, which are joined at present by the dry Middle East. Given wetter climatic conditions, that area could have served as an avenue of dispersal. There is also the possibility that some fishes may have survived India's swift ride and strong collision with Asia.

Of the Laurasian fishes, the Cyprinidae have made the most notable incursion into a Gondwanaland continent. Minnows have spread over a wide area in Africa, and are locally numerous in some places, but they do not hold the position of dominance as firmly as in Asia.

Of special interest is the family Hiodontidae; these osteoglossomorphs of North America are the close relatives of Notopteridae of Africa and tropical Asia, with no relatives in other temperate areas. The North American catfishes, Ictaluridae, do not seem to be closely related to the Siluridae of Europe, but must be the descendants of some Palaearctic group unless they moved north through the tenuous connection with South America. These two families are representative of some of the questions left to be resolved.

In summation, the distribution of freshwater fishes is the result of numerous geological, climatological, and biological processes, some of which are obvious, some disguised, some hidden, and some possibly not yet considered. All levels of interplay among the shifting of continents, submergence and emergence, wet and dry eras, shifts from cold to hot and back, adaptive radiation, extinction, and perhaps even changes in the physiological abilities of some groups to withstand salt water make the work of the historical ichthyogeographer difficult. But, as the extant fishes become better known, and the fossil record becomes better understood, knowledge of the fishes can be considered in light of the knowledge of plate tectonics to make some of the enigmatic distributional patterns understandable. There is ample work for future ichthyologists.

Distribution of Marine Fishes

Considering that the oceans are all interconnected, and were more effectively so during the breakup of Pangaea when the Tethys Sea was in existence, marine fishes should have had a chance to swim anywhere in the salt waters, given wide physiological tolerances. Some appear to have a wide distribution; certain tunas and pelagic sharks live in the Atlantic, Pacific, and Indian oceans, ranging over great expanses of sea, but only in the warmer parts. We know that certain families — for instance Petromyzontidae, Cottidae, and Zoarcidae — have representatives in the far north and the far south, but in between they are absent from the shallow shelf, their bipolar distribution having been achieved by submergence under the warm equatorial waters. (Or by having had a wider north-south distribution in the cold Pleistocene, a distribution they were unable to maintain in warmer times.) The major barrier to dispersal of marine fishes is temperature. Although there are eurythermal fishes that can withstand relatively wide ranges of temperature, these ranges are not wide enough to span the difference between arctic and tropical conditions, so that even eurythermal fishes are essentially limited by either warmth or cold.

Another general barrier is salinity, although differences of salinity are not nearly as great and restrictive to saltwater fishes as are temperature differences. Certain areas, such as the Red Sea, may have salt content sufficiently different from surrounding seas so that distribution of fishes is affected. Land constitutes a barrier that may be permanent through millions of years, and great expanses of ocean can effectively prevent the spread of fishes adapted to shallow water. On the other hand, deepwater fishes are restricted by shallow water at ridges and rises.

Pelagic Fishes

Fishes that are free of the shores for their life requirements during all or most of their life histories can spread over wide areas of suitable temperatures and food supply. Some can exploit the resources of more than one thermal region, ranging from tropical to temperate or from temperate to cold waters but, for the most part, they are bounded by certain isotherms. Many are characteristic of certain water masses formed by interactions of currents, temperature, and salinity. These restrictions apply not only to epipelagic species but to mesopelagic ones as well, there being a remarkable correspondence between distributions of the two ecological types. Distribution of deepwater pelagic species is poorly known, but some appear to be limited to certain ocean basins. In the following discussion distribution of fishes in the epipelagic will be stressed.

Polar Waters. In the Arctic and Antarctic there are waters ranging in temperature from nearly −2° C in winter to 5° or 6° C in summer. Parts of these areas can be covered much (or all) of the year by ice, and have a poor pelagic fish fauna. In the Arctic the cold water extends south into the northern Bering Sea, reaching approximately to the 60th parallel. On the Atlantic side, the boundary of the arctic water begins at the southeast corner of Labrador at about 50° N, extends up to Iceland, and then continues northeast over Norway to a bit east of North Cape, at

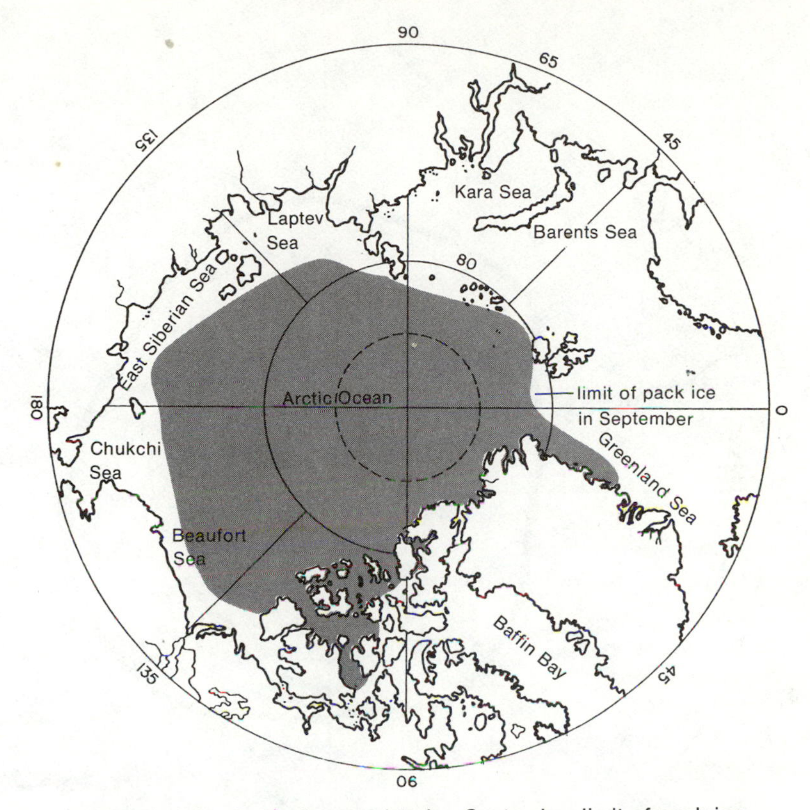

FIGURE 9–10. Arctic Ocean, showing September limit of pack ice.

about 70° N. This displacement is due to the southward flow of the Labrador Current in the west and the northeastward drift of the warm currents of the Atlantic (Figs. 9–10, 9–12).

Pelagic fishes of the Arctic waters include some anadromous salmonids such as the arctic char, *Salvelinus alpinus*, and the pink salmon, *Oncorhynchus gorbuscha;* the gadids, *Arctogadus* and *Boreogadus;* and the herring, *Clupea harengus*.

Around Antarctica, south of the antarctic convergence, located at about 60° S in the Pacific and about 50° S in the Atlantic, there is extremely cold water that remains at less than 5° C even in summer (Fig. 9–11). This frigid area supports a very small pelagic fish fauna, including representatives of many of the typical mesopelagic families and some of the nototheniiforms that leave the bottom to eat the abundant krill. The chaenichthyid *Champsocephalus gunnari* is an example.

North Pacific — cold temperate waters. In the North Pacific, cold temperate waters are found between the arctic and a line roughly extending from Tokyo to southern California, the southern limit being along the winter 13° isotherm (about 34° to 35° N) (Fig. 9–12). The

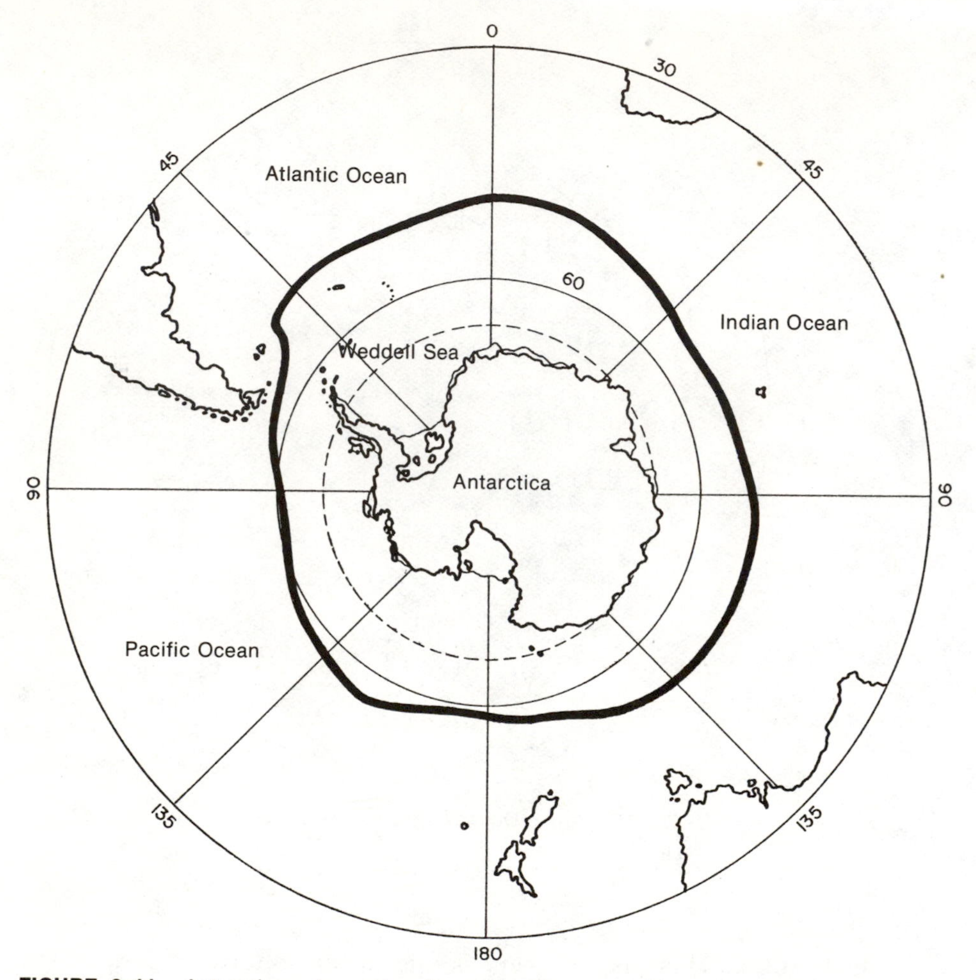

FIGURE 9–11. Antarctic waters, showing approximate position of Antarctic Convergence *(heavy black line).*

northern part (from about 42° N to the arctic) is characteristically colder than 10° C in the winter, and has a surface salinity of less than 34 ppt. This is called the Pacific Subarctic or North Boreal Region. Typical pelagic fishes of the region are anadromous salmonids of the genera *Oncorhynchus*, *Salmo*, and *Salvelinus*, the first containing the six Pacific salmons, the second the steelhead trout, the third the Dolly Varden trout. One hexagrammid, the Atka mackerel, *Pleurogrammus monopterygius*, lives a pelagic life, and the young of other greenlings, especially *Hexagrammos decagrammus*, can be found in the open waters of the Gulf of Alaska. Relatives of the greenlings in open waters are the anoplopomatids — the skilfish, *Erilepis zonifer*, and the young of sablefish, *Anoplopoma fimbria*. *Theragra chalcogramma*, the walleye pollock, is another typical fish of the region, as is the Pacific herring, *Clupea harengus pallasi*.

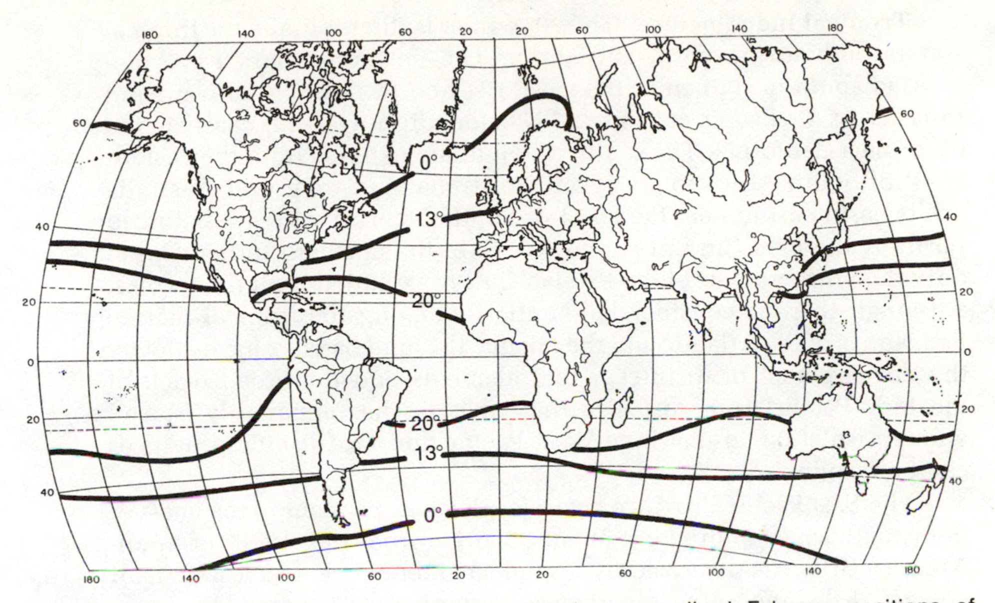

FIGURE 9–12. Pelagic zones of the oceans, showing generalized February positions of 0°, 13°, and 20° isotherms.

Other fishes of the region, generally with greater temperature tolerance than the above, include the wide-ranging Pacific saury, *Cololabis saira,* jack mackerel, *Trachurus symmetricus,* salmon shark, *Lamna ditropis,* basking shark, *Cetorhinus maximus,* and the Pacific pomfret, *Brama japonica.*

The Pacific south boreal extends from about 34° to 42° N, and is characterized by pelagic fishes that often cross regional boundaries. In fact, some scientists regard the region as a transitional area. The pomfret and the saury are much more typical of this than of the north boreal, as are the sardine, *Sardinops sagax,* the northern anchovy, *Engraulis mordax,* the medusafish, *Icichthys lockingtoni,* and the blue shark, *Prionace glauca.* Various tunas, including the albacore, *Thunnus alalunga,* and the bluefin, *T. thynnus,* are common in the area during certain times in their migrations.

North Pacific — warm temperate waters. The southern boundary of the warm temperate region in the North Pacific is the winter 20° C isotherm, or about 23°N. The northern boundary is at about 34° N, although in mid-Pacific the regional boundaries trend to the north. This region has a well developed pelagic fish fauna, with many species shared with adjoining regions. Many tunas and flying fishes live in these warm waters along with marlins, sharks, and others. Typical species are the flying fishes *Prognichthys rondeleti* and *Cypselurus pinnatibarbatus,* the albacore, the skipjack, *Katsuwonus pelamis,* the squaretail, *Tetragonurus cuvieri,* and the striped marlin, *Tetrapturus audax.* Sauries are numerous here. The rather uncommon opah, *Lampris regius,* and louvar, *Luvarus imperialis,* appear to prefer the warm temperate water.

Tropical Indo-Pacific. The 20° winter isotherm limits the tropical waters both to the north and south. The northern boundary is depressed south to Hainan in the west, rises north through Taiwan, and runs across the Pacific at about 23° N before dipping south nearly to the tip of Baja California. The southern boundary extends from the eastern coast of South Africa to about mid-Australia, and from mid-Australia on its east coast to near the coast of South America where it is forced north by the Peru Current to about 3° S. In the northern Indian Ocean only the northern part of the Persian Gulf is excluded from the tropics. Although there is considerable similarity among the tropical marine fish faunas of the Atlantic and the Pacific, the land barriers have allowed the development of distinct faunal elements and the broad reach of the Indo-Pacific from Africa to the Americas has apparently caused enough isolation so that Indo-West Pacific and East Pacific provinces can be recognized.

The East Pacific Province occupies the area adjacent to the coast of the Americas, extending out about 3500 km from the shores of South America. It is characterized by a number of endemic species of small pelagic fish, including anchovies, *Cetengraulis,* thread herrings, *Opisthonema,* flyingfishes, *Cypselurus* and *Prognichthys,* and several carangids. Many large pelagic species are shared with the rest of the Indo-Pacific.

The Indo-West Pacific includes most of the waters of the region and contains a variety of pelagic fishes, some of which are circumtropical and some of which are shared with the East Pacific. Several species of sharks are widespread in the pelagic tropical waters. These include the whale shark, *Rhiniodon typus,* blue shark, *Prionace glauca,* great white shark, *Carcharodon carcharias,* and the thresher, *Alopias vulpinus.* The silky shark, *Carcharhinus falciformis,* and the whitetip, *Carcharhinus longimanus,* are widely distributed in the Pacific. Manta rays are circumtropical. Other large fishes of the tropical Pacific are the marlins, *Makaira* spp., the sailfish, *Istiophorus platypterus,* and several of the tunas. Smaller pelagic fishes of the Indo-West Pacific include sardines, such as *Sardinella longiceps,* many flyingfishes, mackerels, *Rastrelliger kanagurta* and others, many species of Carangidae, the Pacific squaretail, *Tetragonurus pacificus,* the butterfish, *Formio,* and many more.

Southern warm temperate region. Because the warm temperate waters are continuous around southern Africa and are barely interrupted by southeastern Australia, there has been opportunity for a circumglobal fauna to develop. The major land barrier is the southern part of South America, from about 41° S on the west side and from the region of Buenos Aires on the east. Along the west coast of South America there are some pelagic or meropelagic species such as *Trachurus murphyi* and *Scomber japonicus* that do not reach the east coast, but most of the pelagic species, including the southern bluefin tuna, *Thunnus maccoyi,* the saury, *Scomberesox saurus,* the pomfret, *Palinurichthys antarcticus,* and the squaretail, *Tetragonurus cuvieri,* spread through the warm southern parts of the oceans.

Southern cold temperate region. The cold temperate area between the Antarctic and southern subtropical convergences does not seem to have a distinct pelagic fish fauna, sharing most species with the warm temperate waters to the north. Examples are the basking shark, *Cetorhinus maximus,* the saury, the albacore, the bigscale tuna, *Gasterochisma melampus,* and the gempylid, *Thyrsites atun.*

Atlantic Ocean — boreal waters. The Atlantic boreal includes the ocean between southern Labrador and northern Norway on the north and between Cape Hatteras and the English Channel on the south. The pelagic fish fauna of the region is mixed, made up of eurythermal species that can range to the north or the south or both, depending on individual tolerances. Examples are the mackerel shark, *Lamna nasus,* basking shark, *Cetorhinus maximus,* opah, *Lampris regius,* Atlantic mackerel, *Scomber scombrus,* bluefin tuna, *Thunnus thynnus,* swordfish, *Xiphias gladius,* and the mola, *Mola mola.*

Atlantic Ocean — warm temperate waters. The spread of this region is narrow on the west (Cape Hatteras to mid-Florida) but its boundaries are the English Channel and Senegal on the east. Pelagic fishes include not only forms more or less restricted to the warm temperate but many eurythermal tropical species and some, such as the saury, that are shared with the boreal. Typical epipelagic fishes of the area include such widely distributed examples as the blue shark, *Prionace glauca,* the whitetip shark, *Carcharhinus longimanus,* the pilotfish, *Naucrates ductor,* the skipjack, *Katsuwonus pelamis,* the dolphin, *Coryphaena hippurus,* and the albacore, *Thunnus alalunga.* The epipelagic fish fauna of the Mediterranean Sea is quite similar to that of the warm temperate Atlantic, but is not quite as extensive.

Atlantic tropical region. This region is broad to the west (Florida to Rio de Janeiro) and constricted by cooler currents in the east (Senegal to Angola). The pelagic fishes of the region are for the most part eurythermal and range into the adjacent wam temperate waters. Many circumtropical species are present. The familiar sharks, flyingfishes, tunas, marlins, and carangids are found here, along with widespread gempylids *(Ruvettus, Gempylus),* the Atlantic squaretail, *Tetragonurus atlanticus,* and others.

Shelf Fishes

Unlike the pelagic species that can easily travel over broad reaches of the oceans, shelf fishes are restricted by some kind of dependence upon the substrate. Some are entirely benthic, others must live close to the bottom (engybenthic), and some may live close to the surface, have strong powers of locomotion, and may even mingle with true pelagic fishes to some extent, but are bound to the waters of the shelf by trophic or reproductive needs. Dispersal along east-west oriented coastlines in a given temperature zone is possible even for many slow-moving species, but usually north-south orientation of coasts involves temperature barriers which prevent dispersal. Great expanses of open water slow

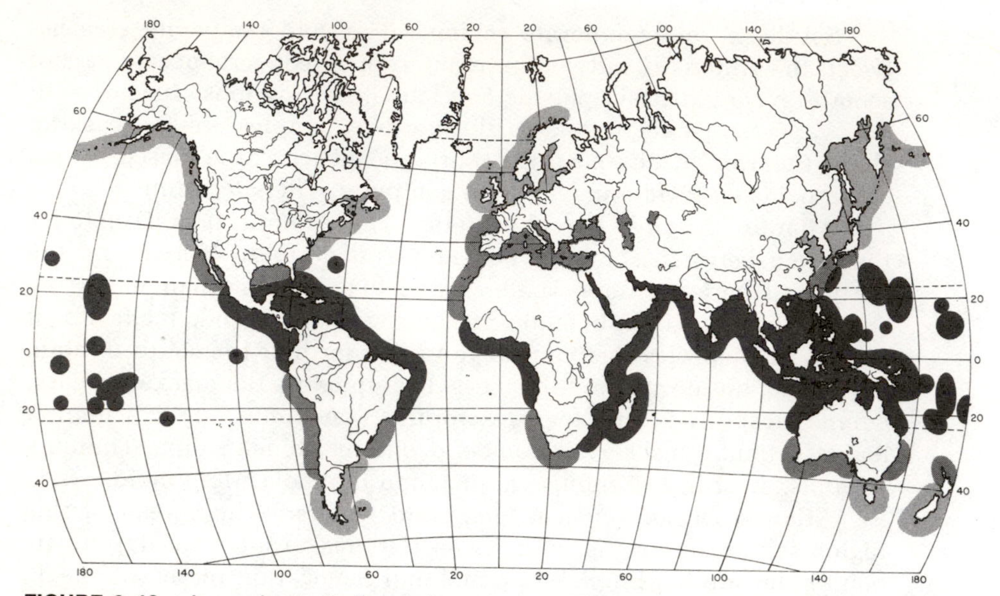

FIGURE 9–13. Approximate limits of shore faunas. *Dark gray,* tropical; *medium gray,* warm temperate; *light gray,* cold temperate; *unshaded* shore line, cold.

or prevent the spread of shore fishes, but some have achieved a wide distribution, a few because of their pelagic larvae.

Generally, there is a greater diversity in shelf fishes than in pelagic fishes over the world. The chances are better for isolation, not only because of temperature zonation but also because of the barriers of land and open water, and endemism is common. Zoogeographers recognize many subdivisions of the seas based upon the fauna of the shelf. Some of these schemes fit fish distribution better than others, but most have similar major divisions or regions. The regions and subdivisions used here (Fig. 9–13) are essentially those of Dr. J. C. Briggs.

Arctic Region. The Arctic Region includes the shores of North America from southeast Labrador to Nunivak Island near Bethel, Alaska on the Bering Sea. In Eurasia it includes the shore from the Murmansk Peninsula around to the northern part of the Bering sea at about 62° N. All of Greenland is included, but only the north coast of Iceland. The arctic fish fauna consists of a relatively few species that are really restricted to the region and some that are shared with the boreal shelf. Typical cold-water forms are *Boreogadus saida, Arctogadus borisovi,* the arctic flounder, *Liopsetta glacialis,* the fourhorn sculpin, *Myoxocephalus quadricornus,* the arctic sculpin, *M. scorpioides;* and the eelpout, *Lycodes frigidus.* Boreal species entering the Arctic from the Atlantic include the sleeper shark, *Somniosus microcephalus,* the skate, *Raja radiata,* the wolffish, *Anarhichas lupus,* the snailfish, *Liparis liparis,* and the sand lance, *Ammodytes hexapterus.* Boreal species from the Pacific include the capelin, *Mallotus villosus,* the pond smelt, *Hypomesus olidus,* the eelpout, *Gymnelis viridis,* and the sculpin, *Artediellus scaber.*

Pacific — western boreal region. Included is the coast of Asia north of the Formosa Strait to Cape Olyutorsky in the Bering Sea, except for the tip of Korea and southern Japan, which are influenced by warm currents. Three subdivisions or provinces have been recognized, the Oriental to the south, the Kurile to the north, and the Okhotsk in the northwest. In the more northerly provinces the fish fauna includes such migratory salmonoids as the Pacific salmons, *Oncorhynchus* spp., the icefish, *Salangichthys*, the ayu, *Plecoglossus*, and the "Chevitsa," *Hucho perryi*. There are a few species of smelts (Osmeridae), gunnels (Pholidae), eelpouts, (Zoarcidae), several rockfishes (Scorpaenidae) and greenlings (Hexagrammidae), numerous sculpins (Cottidae), snailfishes, lumpsuckers (Cyclopteridae), and flounders (Pleuronectidae).

Pacific — eastern boreal region. This region includes the Aleutian Islands and the coast of North America from the southern part of the Bering Sea (about 60° N) to Point Conception in southern California. A northern Aleutian province and a southern Oregon province are recognized, divided in southeastern Alaska. The northern fauna is broadly comparable in composition to that of the Kurile and Okhotsk of the west Pacific Boreal, consisting of many species of sculpins, liparines, rockfishes, eelpouts, and flounders, as well as cods, gunnels, and poachers (Agonidae). The Oregon province is rich in species, especially in the surfperches (Embiotocidae) and the genus *Sebastes*. Part of the diversity is due to the overlapping of Aleutian elements in the north and warm-water species in the south. A few fishes are widely distributed on both sides of the Pacific.

Examples of these fishes and their distribution are: sixgill shark, *Hexanchus griseus* (Baja California to northern British Columbia); chum salmon, *Oncorhynchus keta* (Japan to California); eulachon, *Thaleichthys pacificus* (Northern California to Bering Sea); longnose skate, *Raja rhina* (California to southeast Alaska); plainfin midshipman, *Porichthys notatus* (Baja California to Sitka); blackbelly eelpout, *Lycodopsis pacifica* (Ensenada to Afognak I.); jacksmelt, *Atherinopsis californiensis* (Baja California to Yaquina Bay, Oregon); yellowmouth rockfish, *Sebastes reedi* (Crescent City, California to Sitka); speckled rockfish, *S. ovalis* (Baja California to San Francisco); whitespotted greenling, *Hexagrammos stelleri* (Oregon to Japan); buffalo sculpin, *Enophrys bison* (Monterey to Kodiak).

North Pacific warm temperate — Japan region. This region includes the southern tip of Korea, the coast of Honshu south of about 35° N, Shikoku, Kyushu, the west coast of Taiwan, and the coast of China from about Wenchou to Hong Kong. Examples of fishes common to the region are lizardfishes (Synodontidae), flyingfishes, mullets (Mugilidae), mackerels (Scombridae), jacks (Carangidae), sea basses (Serranidae), porgies (Sparidae), and grunts (Pomadasyidae).

North Pacific warm temperate — California region. Included here is the California coast from Point Conception to Magdalena Bay of Baja California, and nearly all of the Gulf of California, with the tip of Baja California in tropical waters forming a barrier between the two parts of the temperate fauna. This has led to the recognition of two subdivisions, called San Diego and Cortez provinces by Briggs. The fauna of the Gulf of California has been long isolated from the outer coast and consists mainly of tropical elements adapted to temperate waters. Cortez Province fishes include some northern species shared with the outer coast, such as the leopard shark, *Triakis semifasciata,* the California skate, *Raja inornata,* the white seabass, *Cynoscion nobilis,* and the giant sea bass, *Stereolepis gigas.* The warm temperate fauna includes rockfishes, sculpins, pricklebacks, clinids (Clinidae), combtooth blennies (Blenniidae), gobies, lefteye flounders (Bothidae), pipefishes (Syngnathidae), flyingfishes, damselfishes (Pomacentridae), several drums or croakers (Sciaenidae), and others.

Tropical Pacific — Indo-West Pacific region. Superlatives are required in describing this large and complex region that includes the Pacific and Indian oceans from Africa through Polynesia, and is inhabited by more than 3000 species of shore fishes, representing numerous families. It includes shores that lie between the winter 20° C isotherms, and is the region of reef-building corals. Several subdivisions or provinces have been proposed, including the western Indian Ocean, Red Sea, Indo-Polynesian, Northwest Australia, Hawaiian, Easter Island, and others.

The Indo-Polynesian, especially the Malay Peninsula and the Indo-Malayan Archipelago, appears to be richest in fish species, but the coasts of Asia and Africa are also extremely plentiful. Throughout the islands of the Pacific, the fauna becomes less diverse from west to east, with nearly 2000 fish species in the Indo-Malayan area and fewer than 400 in Hawaiian waters.

Typical Indo-Pacific fish families include the carpet sharks (Orectolobidae), angel sharks (Squatinidae), eagle rays (Myliobatidae), herring and round herring (Clupeidae), milkfish (Chanidae), marine catfishes (Ariidae and Plotosidae), moray eels (Muraenidae), needlefishes (Belonidae), seahorses and pipefishes (Syngnathidae), shrimpfishes (Centriscidae), and others among the soft-rayed groups. Among the spiny-rayed fishes, especially the perciforms, the variety is staggering, not only in numbers of families but in the great diversity within families. Examples of families are pinecone fishes (Monocentridae), squirrelfishes (Holocentridae), sea basses and relatives (Serranidae), cardinalfishes (Apogonidae), goatfishes (Mullidae), butterflyfishes (Chaetodontidae), damselfishes (Pomacentridae), wrasses (Labridae), triggerfishes (Balistidae), puffers (Tetraodontidae), and many others.

Tropical eastern Pacific region. This includes the shelf from the southern tip of Baja California south to about 3° to 5° S. It includes only a few offshore islands, the most notable being the Galapagos, which are considered to be in a distinct faunal province. The other provinces recognized are the Mexican, north of the Gulf of Tehuantepec, and the Panamanian. Because of its isolation from the western Pacific by a great expanse of open water, this region lacks the diversity of the Indo-West Pacific, and only a small number of species from that area are encountered along the tropical American coast. Many fishes of the region have close relatives in the Atlantic, the relationships apparently dating from periods when portions of Central America were submerged. The resemblance to the tropical West Atlantic is notable at the generic level. Over half the genera shared by the East Pacific and West Atlantic are circumtropical, but nearly half are endemic to the American tropics. About a dozen noncircumtropical species are said to be found on both sides of Central America.

There appears to be only one family, the Dactyloscopidae, that is an amphi-American endemic. Most families in the Eastern Pacific are those typical of other tropical areas, but many of the well known Indo-West Pacific families are absent.*

Indo-Pacific warm temperate — Southern Australian region. Except for the cooler waters of southern Australia adjacent to Tasmania, that continent south of about 25° S constitutes a warm temperate region. The two parts of the region separated by the cool waters of Victoria, Southwestern and Southeastern Australia, are regarded as provinces. The region shares many fish families with tropical waters, but shows evidence of having been long isolated from other warm temperate regions. Some families — Peronedysidae, Pataecidae, and Gnathanacanthidae — are endemic to the region, as are certain genera of other families restricted to southern warm temperate waters.

Northern New Zealand region. The northern part of North Island and the Kermadec Islands north of it make up the Northern New Zealand region, the fishes of which show a strong relationship to those of southern Australia. A few families, Odacidae, Latridae, Leptoscopidae, Arripidae, and Chironemidae, are found exclusively or mainly in these two areas.

Western South American region. The warm waters of the west coast of South America, because of the northward flow of the Peru Current, extend from about 3° to 5° S to about 41° S. Fishes of the region include many familiar subtropical and tropical families with wide distribution. A few of the families (Cheilodactylidae, Latridae, and Oplegnathidae) are shared with Australia or other southern warm-water areas. The upwelling of rich water along the coast of Peru has made possible the development of large populations of certain plankton-feeding fishes that support both large commercial fisheries and huge populations of birds. The anchovetta, *Engraulis ringens*, appears to be the most productive, and has been the object of the world's greatest single species fishery.

Southern cold-temperate waters. There are four faunal regions recognized in the southern cool waters: the Tasmanian, including Victoria and Tasmania; the southern New Zealand; the southern South American, consisting of the west coast south of about 41° S and the east coast south of Buenos Aires; and the Sub-Antarctic, made up of small islands such as Kerguelen and Macquarie.

---

*Examples of families that have not crossed the East Pacific barrier are Chirocentridae, Gonorhynchidae, Plotosidae, Solenostomidae, Centriscidae, Plesiopidae, Theraponidae, Sillaginidae, Monodactylidae, Siganidae, Kraemeriidae, and Platycephalidae.

The families Bovichthyidae and Nototheniidae, of Antarctic origin, are found throughout the area, and eelpouts (Lycodidae) of northern origin are present in some of the regions. Other families have originated in tropical or subtropical waters. These include gobies, clingfishes, blennies, and pipefishes.

Antarctic region. South of the Antarctic convergence, where the cold polar water sinks beneath the warmer cold temperate water, causing an abrupt north-south temperature change, there is a region with winter surface water temperatures mostly below 1° C. There has been limited exchange of shore fishes with warmer waters and a remarkable fauna of cryophilic shore fishes has developed in the Antarctic. The families Chaenichthyidae, Nototheniidae, Bathydraconidae, and Harpagiferidae constitute the bulk of the Antarctic shelf fauna. Chaenichthyidae are noted for the lack of hemoglobin in their blood.

Western Atlantic boreal region. Between Newfoundland and Cape Hatteras, the coast of North America is under the influence of a southwest-flowing coastal current that brings cold water from the Labrador Current between the coast and the Gulf Stream. This effectively prevents full establishment of most tropical or subtropical fish species along the coast and favors cold temperate and arctic species as year-round residents. Warm-water species are common south of Cape Cod, and may appear throughout the region in the summer.

Typical northern fishes of the region include the sea lamprey, the basking shark, the sleeper shark, sturgeons, herrings, salmonids, smelts, sticklebacks, sand lance, sculpins, poachers, lumpsuckers and sea-snails, gunnels, wolffishes, cods, flounders and others.

Eastern Atlantic boreal region. Included is the southern coast of Ireland, all of Great Britain, and the coast of northern Europe from the entrance of the English Channel through the Baltic Sea and around Scandinavia to a little beyond North Cape. The fish fauna is similar in composition to that of the Western Atlantic in that the same Arctic and Boreal families are present and some species are shared. Warm-water species commonly enter the waters of Great Britain and some groups, such as mullets, gurnards, and soles, are part of the typical fauna of the area.

Atlantic warm temperate — Carolina region. Possibly a more descriptive name for the Region would be "Carolina-North Gulf" because it includes the Carolina and Georgia Coast south of Cape Hatteras, part of the adjacent Florida Atlantic Coast, and the coast of the northern part of the Gulf of Mexico from about Tampico to southern Florida (Cape Romano), leaving the extreme southern end of the Florida Peninsula in the tropics.

The distinctness of the fish fauna is primarily at the specific level, for the families represented here are mainly typical of tropical and warm temperate areas over the world. Included are many sharks and rays, eels, herrings, anchovies, brotulas, silversides, gars, killifishes, pipefishes, and many perciforms, left-eye flounders, and gurnards.

In years that bring water temperatures in the range of 27° to 29° C to the northern Gulf of Mexico for three months or more, tropical shore species can move in temporarily. Such temperature conditions force some of the warm temperate species to seek deeper, cooler water.

Mediterranean-Atlantic region. This great region includes the Atlantic and Mediterranean coasts of Europe and Africa, from the English Channel to about 14° S at Cape Verde. The Black, Caspian, and Aral seas are included because of the derivation of their vertebrate and invertebrate faunas, even though their waters are rather cool. The Atlantic and Mediterranean coasts form one gigantic province, and the three inland seas each constitute separate provinces.

The fish fauna includes many shark, skate, and ray families typical of warm waters. Bony fish families represented are mostly widespread in warm temperate and tropical waters, but very few shore fish species are shared with the western Atlantic. The Mediterranean has the richest fauna, about 360 species, and shows considerable endemism at the species level. A few typical northern forms, such as the sea lamprey, *Petromyzon marinus*, and the sturgeon, *Acipenser sturio*, occur in the Mediterranean, and the latter has been recorded from the Black Sea.

The Black Sea has a narrow connection with the Mediterranean at the Bosporus, shares most (about 100) of its 140 fish species with the larger sea, and has derived about half of its endemic species from the Mediterranean. Others are shared with, or are derived from, the Caspian fauna. The Black Sea, which has a surface salinity of about 18 ppt, has a fauna which includes anchovies, herrings, cyprinids, needlefishes, pipefishes and seahorses, rockling (Gadidae), jacks, mackerels, gobies, mullets, gurnards, flounders, and lefteye flounders.

The Caspian Sea is a brackish body of water with a salinity of about 13 ppt. Its native fish fauna is remarkable for its complement of northern representatives. It has lampreys, several species of sturgeon, a whitefish (*Stenodus*), a trout, and several cyprinids, some of which (e.g., *Rutilus frisii*) are of considerable economic importance. Certain gobies, herrings, and pipefishes, among others, are related to the Black Sea fauna.

The Aral Sea is further diluted, both as concerns salinity (10 ppt) and marine fish fauna, with most fauna derived from fresh waters.

Tropical Atlantic — Western Atlantic region. This region includes Bermuda, southern Florida, the West Indies, and the coasts of Mexico, Central America, and South America, roughly between the Tropics of Cancer and Capricorn. It is rich in familiar tropical families and has over 900 species. Three provinces have been recognized — the West Indian, the Caribbean (south Florida and the coast south to the Orinoco), and the Brazilian (Orinoco south to Rio de Janeiro). The two more northerly provinces have a greater variety of reef fishes than the Brazilian, which has a greater abundance of families such as sea catfishes (Ariidae) and croakers (Sciaenidae).

The expanse of the Atlantic Ocean has not been a complete barrier to the spread of tropical shore fish species. About 120 species are common to both the eastern and western Atlantic.

Eastern Atlantic tropical region. This region extends along the coast of Africa from Cape Verde to Mossamedes at about 15° S, constricted by the flow of the Canary Current from the north and the Benguela Current from the south. The fishes of the area are not as well known as those of other tropical regions, but the region evidently has a poorer, less diverse fauna than the others. There are about 430 species, of which 120 or so are shared with the western Atlantic. There appears to have been little movement of Indo-Pacific species around Africa to the western Atlantic. Fewer than 20 species are known to be common to both areas.

Eastern South American warm temperate region. Warm temperate water is present along the Atlantic coast of South America from Rio de Janeiro to the mouth of the la Plata River. The fishes of the region are poorly known.

Southern African warm temperate region. The southern end of Africa from Southern Angola around to Durban is in a warm temperate zone. The Indian Ocean section of the coast is richer in species because of its proximity to the tropical Indo-Pacific and its abundance of species, some of which enter the temperate waters.

Migrations of Fishes

Although there are some definitions of the word "migration" that limit the meaning to certain annual, active movements of groups of organisms, fishes perform a variety of movements that do not fit into restricted definitions but appear to be best described as migrations. Therefore we will adopt a broad definition that will include any mass movement from one habitat to another with characteristic regularity in time or according to stages of life history. This broad view allows the inclusion of both active and passive mass movements, whether they are extensive seasonal or annual changes of habitat or short-term, short distance travels.

Most migrations are attuned to food-gathering, adjustment to temperature, or reproduction, and are sometimes referred to simply as "feeding," "breeding," and "wintering." In many fishes these take place on an annual basis but, in others, individuals are involved in certain changes of habitat only at certain developmental stages. Others perform what we will call migrations on a daily basis. Migrations have been categorized in many ways, but one of the most useful descriptive schemes is that of Dr. G. S. Myers, who has suggested the following usage.

Diadromous. Truly migratory fishes that migrate between the sea and fresh water. Included are the three following types:

Anadromous. Diadromous fishes that spend most of their lives in the sea and migrate to fresh water to breed.

Catadromous. Diadromous fishes that spend most of their lives in fresh water and migrate to the sea to breed.

Amphidromous. Diadromous fishes in which migration from fresh water to the sea or vice versa is not for the purpose of breeding, but occurs regularly at some other definite stage of the life cycle.

Potamodromous. Truly migratory fishes that migrate wholly within fresh water.

Oceanodromous. Truly migratory fishes that live and migrate wholly in the sea.

There are many other terms that will be encountered in literature on migrations, such as "adfluvial" (running to rivers), "denatant" (swimming with the current), and "contranatant" (swimming against the current).

Migrations require exposure to unusual dangers, as well as much effort on the part of the active migrants but, because migrations are part of the life histories of many fishes, the advantages seem to outweigh the disadvantages for those species. Disadvantages include not only the expenditure of great amounts of energy, but exposure to predators as the migrants mass along the migration route or on wintering grounds, losses of passively drifting eggs and larvae to wayward currents that take them to unsuitable regions and, in some instances, adjustment to changes in temperature or salinity. Evolutionary adjustments in migratory species offset most natural disadvantages. Extra energy in the form of fat is stored prior to the onset of breeding migrations so that the destination can be reached while there is sufficient energy left for spawning. Some species such as Pacific salmons and lampreys mobilize all their resources and commit them irreversibly to the acts of migration and spawning so that adults die following spawning.

Given unchanging or gradually changing current patterns species that require passive drift of eggs or larvae can adjust to the patterns by an obvious selective mechanism. The offspring of adults that spawn in the wrong place will not contribute to the gene pool. On the other hand, straying of adults or young into new suitable habitats increases the range of a species.

Advantages of migration, of course, are concerned with placing the individuals of a species in the best — or necessary — location for given biological activities at the appropriate time. Nocturnal feeding near the sea surface by species that spend the day in the twilight of the depths places them in contact with a high concentration of food organisms and also allows them to feed while avian and other diurnal predators are inactive. Feeding migrations of far-ranging species allow them to reach or to follow food resources that can support far larger populations than the resources of the spawning or nursery areas. Spawning migrations are especially important to species with restrictive requirements for reproduction and those whose eggs and larvae must drift to specific nursery grounds.

Examples of migrations range from the insignificant to the herculean. Many stream fishes, such as minnows, suckers, trout, and darters, may move only a few yards from feeding to spawning grounds. The prickly sculpin, *Cottus asper*, for instance, lives most of the year on sandy bottoms in still water but spawns where hard objects for attachment of eggs can be found. Spawners may move either upstream or down to reach proper materials, perhaps in a riffle of a stream, or perhaps in an

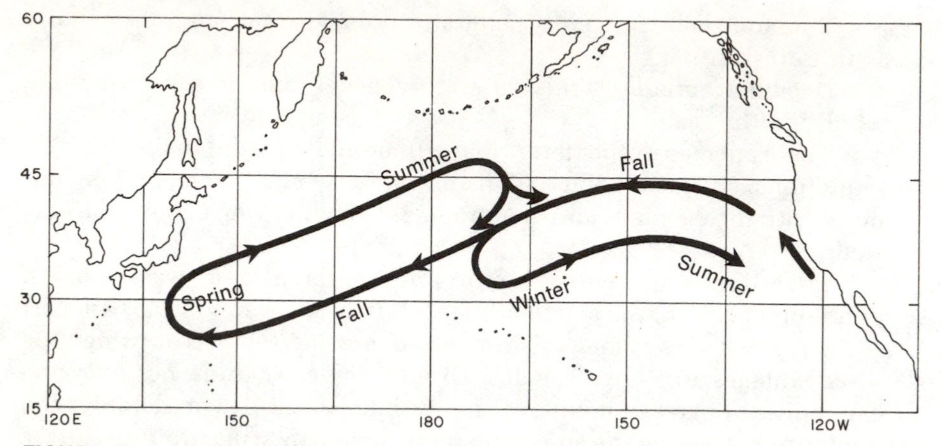

FIGURE 9–14. Probable migration routes of young albacore *(Thunnus alalunga)* in the Pacific Ocean. (After Nakamura.)

estuary where boulders and debris can be found, but the migration is never long. Other river fishes, such as the bocachica of the Amazon, are known to migrate hundreds of miles.

Short feeding migrations are made by many mesopelagic fishes, especially the lantern fishes, Myctophidae, which make nightly migrations from resting areas in the twilight area to the surface, where they feed, returning as dawn approaches. Certain lake-dwelling fishes such as the yellow perch *(Perca flavescens)* show regular movements to the shore at night, with a return to deeper water in daytime. The same species, in common with many other lacustrine fishes, usually congregates in selected shore areas for spawning.

Although some marine fishes may move only a short distance onshore or offshore during their lifetimes, astounding oceanodromous migrations are carried out by various tunas that span entire oceans in their feeding movements. In the North Pacific the albacores that have moved north along the California-Oregon coast during the summer may be in mid-ocean during the winter and back to the coast in June. Not all return, as there is another pattern sometimes followed by a portion of the population. This takes them from the mid-Pacific to Japan, where they move northeast along the coast in May and June, returning to mid-ocean by fall (Fig. 9–14). The spawning migration of mature fish occurs as a divergence from the western Pacific pattern. The bluefin tuna of the Atlantic cross temperature boundaries on their known migrations, moving between Florida and Norway and between Norway and Spain. Returns of tags from tuna indicate that some move between Baja California and Tokyo and between West Australia and Northern New Zealand. Tagged and recaptured white marlin have been logged at a minimum known speed of about 1850 km per month while traveling from off California to Hawaii. Blue sharks tagged off New York and New Jersey were recovered two years later off South America and off Africa, distances of 3200 and 4800 km, respectively.

There are many examples among herrings of adults performing contranatant spawning migrations followed by a denatant, passive movement of the larvae to nursery grounds while the spent adults swim back to their feeding area. In the fall herring migrate from feeding grounds to areas, usually quiet and protected, where they spend the winter.

Diadromous migrations are especially remarkable because a change from one medium to another is necessary, and because the distances and locations involved often require amazing feats of orientation, navigation, and precise recognition of spawning areas. Catadromy ranges from the short migrations of facultative catadromes such as *Cottus asper* to the epic travels of the eels of the genus *Anguilla.* A marginal example is seen in a salmonoid of the Southern Hemisphere, *Galaxias maculatus,* which is known to migrate downstream into estuaries for spawning, so that the larvae drift to sea. Some of the mullets, such as *Mugil dussumieri* and *Agonostomus monticola,* which live in rivers, sometimes at considerable elevations, make downstream movements to the ocean for spawning.

The longest of the catadromous migrations are seen in the eels of the genus *Anguilla* (Fig. 9–15). Several species are known, but significant life history information is available for only the American eel, *A. rostrata,* and the European eel, *A. anguilla.* Both species ascend streams from the Atlantic as small elvers, the males remaining mostly in the lower sections and the females traveling great distances upstream in large systems such as the Nile or Mississippi. Eels can surmount barriers or travel between disconnected bodies of water by wriggling overland. Freshwater life appears to last from 4 to 7 years for the females and a shorter period for the males, which do not grow as large as the females. During the freshwater feeding stage the eels are brownish-yellow in color and have small eyes that apparently suit their nocturnal habits, but as the time for migration back to the ocean approaches they take on a silvery coloration and the eyes increase remarkably in size.

After they descend the streams and enter the ocean they are virtually unheard of again. Some have been captured near the coast, but none beyond the continental shelf. The destination of the migrating adult eels is an area of the Sargasso Sea, where they appear to spawn at a depth of 400 to 700 meters but over water much deeper than that. The apparent spawning grounds were discovered by a Danish scientist, Johannes Schmidt, who backtracked the drifting larvae until he found where they seemed to be first appearing in surface waters. The larvae are delicate, leaf-shaped, transparent creatures called leptocephali, which must drift a matter of months or years before encountering the sea coast, where they enter rivers. Before (or as) they enter the streams, there is a metamorphosis from leptocephalus to glass eel to elver.

Considering the effects that the spreading of the Atlantic Ocean floor has had on the adjustment of the eels to the distances to be covered is of interest. Not only do the European eels have to swim a phenomenal distance to reach the spawning area, but the larvae must have had to adjust the onset of metamorphosis to the increased time needed to drift

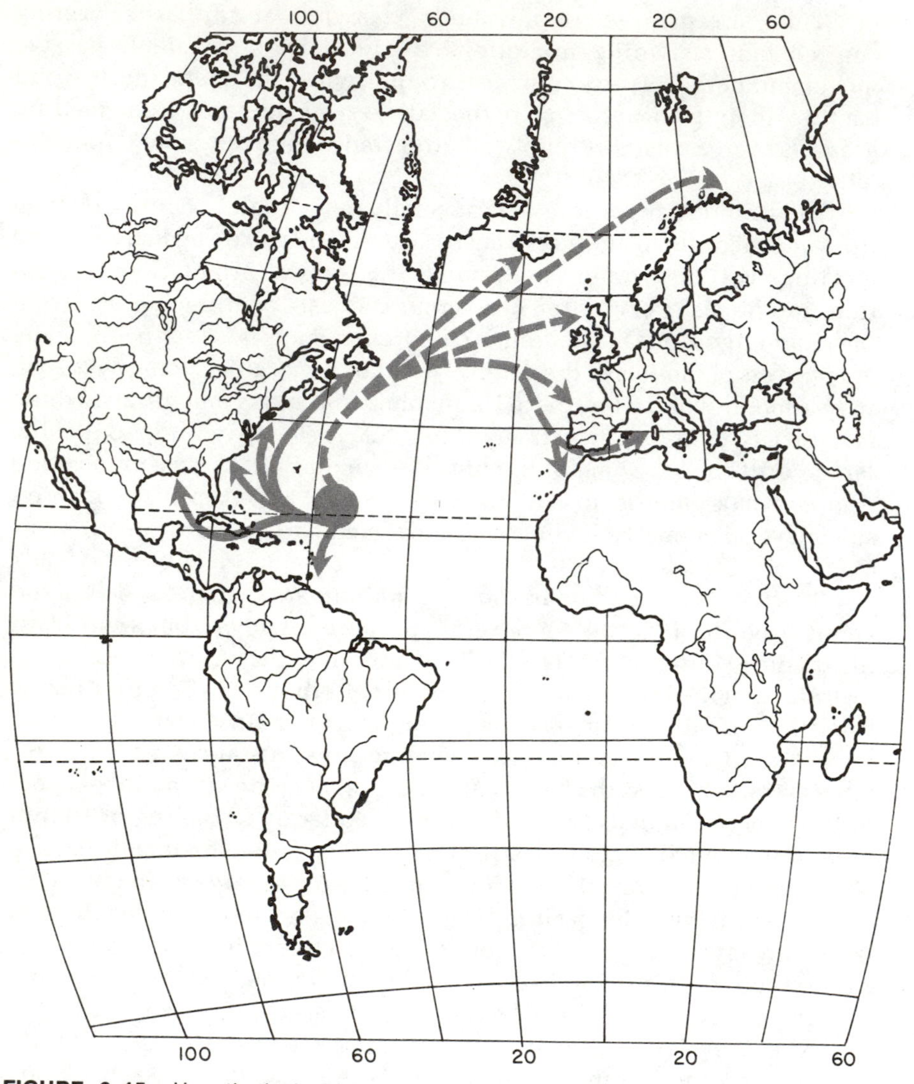

FIGURE 9-15. Hypothetical drift routes of leptocephalus larvae of American eel, *Anguilla rostrata (solid lines)* and European eel, *A. anguilla (broken lines).*

to the European shores. Some scientists maintain that the adults of the European eel do not return to the Sargasso Sea but die without spawning, and that the rivers of Europe are stocked by a larval drift which are offspring of American eels that have acquired a characteristic vertebral number during early development at different temperatures from those experienced by those destined to drift to American waters. This hypothesis will remain a possibility until adult European eels are captured in the vicinity of their spawning grounds.

Among anadromous species there are some that spend only a short period in salt water before returning to streams to mature and breed.

Some of the species of *Galaxias* in New Zealand have this life history pattern, but details of their life histories are not known. Another species with a similar migration is the ayu, *Plecoglossus altivelis*, an annual fish of some commercial importance in Japan. In this species spawning occurs in fresh water, but in the lower stretches of the rivers, so that the young drift into salt water where they remain for a few months before migrating back into the rivers. In the rivers they move to the middle and upper reaches and feed for several months before migrating downstream to spawn and die. Usually anadromy entails a life in the seas of several months to a few years, even if the freshwater stage of the life history is lengthy as well, but there are some trout that may spend only a few weeks in salt water before migrating upstream. The striped bass, *Morone saxatilis*, the shad, *Alosa sapidissima*, and the pink salmon, *Oncorhynchus gorbuscha*, are species that commonly migrate only a short distance into fresh water, depending on the availability of suitable spawning sites. Famous long-distance anadromous migrants are the chinook salmon, *O. tshawytscha*, and the sockeye salmon, *O. nerka*. Certain races of these once made heroic migrations up the Columbia River, swimming all the way to Canada. Construction of hydroelectric dams has stopped the long runs up the Columbia, but these species still make mighty journeys up Canadian and Alaskan rivers.

Ocean life of anadromous species may be spent close to shore, as in the cut-throat trout, *Salmo clarki*, or may include lengthy coastwise migration or movement to feeding grounds well offshore. Pacific salmon feed along the coasts of North America and Asia, but some species and races range far out into the Gulf of Alaska and the North Pacific, some from North America going well beyond 180° longitude (Fig. 9–16). For these, no land clues are available for orientation as they mature and migrate back to the exact stream from which they emigrated months or years earlier.

Using the broad definition of migration, as noted earlier, travel by

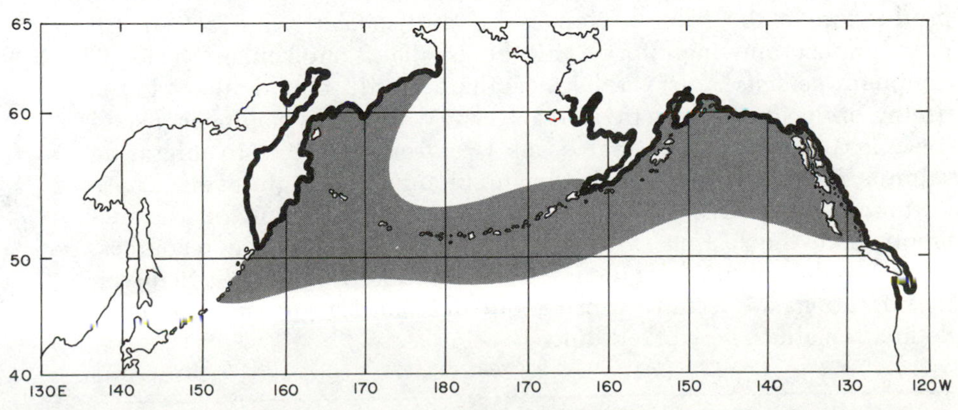

FIGURE 9–16. Main feeding and spawning areas of sockeye salmon *(Oncorhynchus nerka)*. The gray shading indicates the most important summer feeding grounds; the heavy shoreline indicates range where main spawning rivers enter the sea.

fishes during a migration might be entirely passive, as in the case of larval fishes that capture a living among the plankton while drifting along in a moving water mass that will transport them to a suitable nursery area they may enjoy as juveniles. There are even some arguments that the skipjack tuna, *Katsuwonus pelamis*, may be involved in passive migration while feeding on smaller pelagic species being carried in the Pacific North Equatorial Current. If passive drift (or active movements) takes the larvae, juveniles, or adults of the species involved far from a locality that will figure in a subsequent phase of their life history, then active migration will be necessary for the completion of the life cycle.

Active migration requires adjustment to appropriate environmental stimuli, including temperature, light, water current, salinity, alkalinity, and no doubt many others better known to the fish than to us. The adjustments to changes or gradients in these or other factors are usually mediated through the endocrine system, and involve some kind of orientation regarding the stimulus, followed by a movement to a different locality. There may be numerous complicating factors concerning the physiological readiness of the fish to react to the stimuli. Some of these are reproductive state, overall physical condition and the amount of fat stored, availability of food, and development of certain social responses to other individuals. In some salmonids there seems to be a certain size or rate of growth critical to the onset of the changes that result in the downstream migration. Endogenous physiological cycles or rhythms plus external factors may place the fish into a state of readiness as concerns sexual maturity, osmoregulatory ability, and so on; then, further external stimuli can trigger the migration.

Pacific salmon, Atlantic eels, and sticklebacks have been favorite subjects of physiologists studying fish migration. Considerable evidence points to the importance of thyroid activity at the time of diadromous migrations. Thyroxine has definite effects on the osmoregulatory ability and salinity preference of fish, as shown by numerous experiments involving the injection of the material or the blocking of its production in the fish, as well as by studies of thyroid activity in premigratory and migratory individuals. The "chicken or the egg" argument persists, however, because there is still some doubt as to the timing of the onset of the thyroid activity, which may begin *because* of the migration. The interrenal tissue is extremely active in migrating salmon, and the corticosteroids in the blood reach high levels. These hormones have strong effects upon osmoregulation and on the metabolism of the fasting salmon as it mobilizes its stored resources. Although there is ample evidence that the endocrine system is involved in migrations, especially those requiring changes of salinity, most details remain to be worked out.

References

Distribution

Andriashev, A. P. 1965. A general review of the Antarctic fish fauna. *In*: Van Mieghem, J. and Van Oye, P., (eds.), Biogeography and Ecology in Antarctic.

The Hague, Monographiae Biologicae, Vol. 15, Dr. W. Junk, pp. 491–550.

Axelrod, H. 1973. African Cichlids of Lakes Malawi and Tanganyika, 2nd ed. Neptune City, N.J., T.F.H. Publications.

Banarescu, P. 1973. Origin and affinities of the freshwater fish fauna of Europe. Ichthyologia, 5(1):1–8.

Bigelow, H. B., and Schroeder, W. C. 1953. Fishes of the Gulf of Maine. Washington, D. C., U.S. Fish and Wildlife Service, Bulletin No. 74.

Boulenger, G. A. 1905. The distribution of African freshwater fishes, Nature, 72:413–421.

Briggs, J. C. 1960. Fishes of worldwide (circumtropical) distribution. Copeia, 1960(3):171–180.

———. 1961. The East Pacific Barrier and the distribution of marine shore fishes. Evolution, 15(4):545–554.

———. 1974. Marine Zoogeography. New York, McGraw-Hill.

Cohen, D. M. 1960. Geographical distribution of fishes. The Encyclopedia Americana, 1960 edition, pp. 275–278.

———. 1973. Zoogeography of the fishes of the Indian Ocean. *In*: Zeitzschel, B. (ed.), Ecological Studies: Analysis and Synthesis, Vol. 3. New York, Springer-Verlag, pp. 451–463.

Cracraft, Joel. 1973. Vertebrate evolution and biogeography in the old world tropics: implications of continental drift and palaeoclimatology. *In*: Tarling, D. H., and Runcorn, S. K. (eds.), Implications of Continental Drift on the Earth Sciences, Vol. I. London, Academic Press, pp. 373–393.

———. 1974. Continental drift and vertebrate distribution. Ann. Rev. Ecol. Syst., 5:215–261.

Croizat, L., Nelson, G., and Rosen, D. E. 1974. Centers of origin and related concepts. Syst. Zool., 23:265–287.

Dadswell, M. J. 1974. Distribution, Ecology and Postglacial Dispersal of Certain Crustaceans and Fishes in Eastern North America. Nat. Mus. Canada, Publ. Zool., No. 11.

Darlington, Philip J., Jr. 1948. The geographical distribution of cold-blooded vertebrates. Q. Rev. Biol., 23(1):1–26.

———. 1957. Zoogeography. London, John Wiley & Sons.

deBeaufort, L. F. 1951. Zoogeography of the Land and Inland Waters. London, Sidgwick and Jackson.

Demel, K., and Rutkowicz, S. 1966. The Barents Sea. Translation from Polish of Morze Barentsa, Gdynia, Wydawnictwo Morskie, 1958. Warsaw. Published for the U.S. Department of Interior and National Science Foundation by Scientific Publications Foreign Cooperation Center of the Central Institute for Scientific, Technical and Economic Information.

Dietz, R. S., and Holden, J. C. 1970. The breakup of Pangaea. Sci. Am., 223(4):30–41.

Doak, W. 1972. Fishes of the New Zealand region. Aukland, Hodder and Stoughton.

Ebeling, A. W. 1967. Zoogeography of tropical deep-sea animals. Stud. Trop. Oceanogr. Miami, 5:593–613.

Eigenmann, C. H. 1906. The freshwater fishes of South and Middle America. Pop. Sci., LXVIII:515–530.

Ekman, S. 1953. Zoogeography of the Sea. London, Sidgwick and Jackson.

Evermann, B. W., and Radcliffe, L. 1917. The Fishes of the West Coast of Peru and the Titicaca Basin. Washington, D.C., Bulletin of the U.S. National Museum, Vol. XI, No. 95.

Fowler, H. W. 1936. The marine fishes of West Africa. Bull. Am. Mus. Nat. Hist., 70:1–1493.

Gery, J. 1969. The fresh-water fishes of South America. *In*: Fittkau, E. J., Illies, J., Klinge, H., Schwabe, G. H., and Sioli, H. (eds.), Biogeography and Ecology in South America, Vol. 2. The Hague, Monographiae Biologicae, Vol. 19. Dr. W. Junk.

Gressitt, J. L. (ed.) 1963. Pacific Basin biogeography. Honolulu, Bishop Museum Press.

Grey, M. 1956. The distribution of fishes found below a depth of 2000 meters. Fieldiana [Zool.], 36(2):75–337.

Günther, A. C. L. G. 1880. An introduction to the Study of Fishes. Edinburgh, A. and C. Black.

Hallam, A. 1972. Continental drift and the fossil record. Sci. Am., 227(5):56–66.

Heard, W. R., Wallace, R. L., and Hartman, W. L. 1969. Distributions of fishes in fresh water of Katmai National Monument, Alaska, and their zoogeographical implications. Washington, D.C., U.S. Fish and Wildlife Service Special Scientific Report, Fisheries (590):1–20.

Herre, A.W.C.T. 1924. Distribution of the true freshwater fishes in the Philippines. Phil. J. Sci., 24:249–306, 683–707.

Hesse, R., Allee, W. C., and Schmidt, K. P. 1951. Ecological Animal Geography. New York, John Wiley and Sons.

Hildebrand, S. F. 1946. A descriptive catalog of the shore fishes of Peru. Bull. U.S. Nat. Mus., (189):1–530.

Howden, H. F. 1974. Problems in interpreting dispersal of terrestrial organisms as related to continental drift. Biotropica, 6(1):1–6.

Hureau, J. C., and Monod, T. (eds.). 1973. Checklist of the Fishes of the Northeastern Atlantic and of the Mediterranean, Vols. 1 and 2. Paris, UNESCO.

Hurley, P. M. 1968. The confirmation of continental drift. Sci. Am., 218(4):52–64.

Inger, R. F., and Chin, P. K. 1962. The freshwater fishes of North Borneo. Fieldiana [Zool.], 45:1–259.

Jayaram, K. C. 1974. Ecology and distribution of freshwater fishes, amphibia and reptiles. *In*: Mani, M. S. (ed.), Ecology and Biogeography in India. The Hague, Monographiae Biologicae, Vol. 23, Dr. W. Junk.

Johnson, D. S. 1967. Distributional patterns of Malayan freshwater fish. Ecology, 48:722–730.

Jordan, D. S. 1901. The fish fauna of Japan, with observation on the geographical distribution of fishes. Science. New Series, XIV:545–567.

———. 1928. The distribution of fresh-water fishes. Annual Report of the Smithsonian Institution, 1927:355–385.

Jubb, R. A. 1967. Freshwater Fishes of Southern Africa. Capetown, Balkema.

Kerr, J. 1950. A Naturalist in the Gran Chaco. London, Cambridge University Press, pp. 169–194.

Kozhov, M. 1963. Lake Baikal and its Life. The Hague, Monographiae Biologicae, Vol. 11, Dr. W. Junk.

Kurten, B. 1969. Continental drift and evolution. Sci. Am., 220(3):54–64.

Lake, J. S. 1971. Freshwater Fishes and Rivers of Australia. Melbourne, Thomas Nelson.

Lindsey, C. C. 1956. Distribution and taxonomy of fishes in the Mackenzie drainage of British Columbia. J. Fish. Res. Bd. Can., 13(6):759–789.

Marshall, N. B. 1964. Fish. *In*: Priestly, R., Adie, R. J., and Robin, G. deQ. (eds.), Antarctic Research. London, Butterworthy.

McDowall, R. M. 1964. The affinities and derivation of the New Zealand fresh-water fish fauna. Tuatara, 12(2):59–67.

McKenna, M. C. 1973. Sweepstakes, filters, corridors, Noah's arks, and beached Viking funeral ships in palaeogeography. *In*: Tarling, H., and Runcorn, S. K. (eds.), 1, pt. 3 (3.5). New York, Academic Press. pp. 293–306.

Mead, G. W. 1970. A history of South Pacific fishes. *In*: Wooster, W. S. (ed.), Symposium on Scientific Exploration of the South Pacific. Washington, D.C., National Academy of Science, 1968.

Miller, R. R. 1959. Origin and affinities of the freshwater fish fauna of western North America. Zoogeography, Am. Assoc. Adv. Sci. Publ., 51:187–222.

———. 1966. Geographical distribution of Central American freshwater fishes. Copeia, 1966(4):773–802.

Moore, R. H. 1975. Occurrence of tropical marine fishes at Port Aransas, Texas 1967–1973, related to sea temperatures. Copeia, 1975(1):170–171.

Myers, George S. 1938. Freshwater fishes and West Indian zoogeography, Washington, D.C., Smithsonian Report for 1937, Pub. 3465, pp. 339–364.

———. 1941. The fish fauna of the Pacific Ocean, with special reference to zoogeographical regions and distribution as they affect the international aspects of the fisheries. Proc. 6th Pacific Sci. Cong., (3):201–210.

———. 1951. Freshwater fishes and East Indian zoogeography, Stanford Ichthyol. Bull., 4:11–21.

———. 1966. Derivation of the freshwater fish fauna of Central America. Copeia, 1966(4):766–773.

Novacek, M. J., and Marshall, L. G. 1976. Early biogeographic history of ostariophysan fishes. Copeia, 1976(1):1–12.

Pantin, C. F. A. (convenor). 1960. A discussion on the biology of the southern cold-temperate zone. Proc. R. Soc. (B), 152(949):499–677.

Parin, N. V. 1970. Ichthyofauna of the Epipelagic Zone. Translated from Russian. Jerusalem. Israel Program for Scientific Translation.

Pearson, Nathan E. 1947. The fishes of the Beni-Mamore and Paraguay basins, and a discussion of the origin of the Paraguayan fauna. Proc. Cal. Acad. Sci., 23(8):99–114.

Peden, A. E. 1973. Records of eelpouts of the genus *Lycenchelys* and *Embryx* from the northeastern Pacific Ocean. British Columbia Provincial Museum, Syesis, 6:115–120.

Rass, T. S. (ed.). 1964. Fishes of the Pacific and Indian Oceans. Biology and Distribution. Academy of Science U.S.S.R., Trans. Inst. Oceanology. Translated from Russian. Jerusalem, Israel Program for Scientific Translation.

Regan, T. 1920. The geographical distribution of salmon and trout. Salmon and Trout, 22:25–35.

Ricker, K. E. 1959. Mexican shore and pelagic fishes collected from Acapulco to Cape San Lucas during the 1957 cruise of the "Marijean." Mus. Contrib. Inst. Fish. Univ. Brit. Col., (3):1–18.

Roberts, Tyson. 1975. Geographical distribution of African freshwater fishes. Zool. J. Linn. Soc., 57(4):249–319.

Rosen, D. E. 1975. A vicariance model of Caribbean biogeography. Syst. Zool., 24:431–464.

Rosenblatt, R. H. 1967. The zoogeographic relationships of the marine shore fishes of tropical America. Stud. Trop. Oceanogr. Miami, (5):579–592.

Robins, C. R. 1971. Distributional patterns of fishes from coastal and shelf waters of the tropical western Atlantic. FAO Fish Rep., 71–2:249–255.

Schmidt, P. Yu. 1950. Fishes of the Sea of Okhotsk. Translated from Russian 1965. Jerusalem, Israel Program for Scientific Translation.

Schultz, L. P. 1957. A new approach to the distribution of fishes in the Indo-West Pacific area. Proc. 8th Pacific Sci. Congr., 3:413–416.

Scott, T. D., Glover, C. J. M., and Southcott, R. V. 1974. The Marine and Freshwater Fishes of South Australia, 2nd ed. Handbook of the Flora and Fauna of South Australia. Adelaide, Government Printer, South Australia.

Slastenenko, E. P. 1959. Zoogeographical review of the Black Sea fish fauna. Hydrobiologia, 14(2):177–188.

Smith, J. M. B. 1974. Southern biogeography on the basis of continental drift: a review. Aust. Mammal., 1(3):213–229.

Stokell, G. 1955. Freshwater Fishes of New Zealand. Christchurch, Simpson and Williams.

Tarling, D. and Tarling, M. 1975. Continental Drift. New York, Anchor Press/Doubleday.

Tortonese, E. 1964. The main biogeographical features and problems of the Mediterranean fish fauna. Copeia, 1964(1):98–107.

Valentine, J. W. 1971. Plate tectonics and shallow marine diversity and endemism, an actualistic model. Syst. Zool., 20(3);253–264.
———, and Moores, E. M. 1974. Plate tectonics and the history of life in the oceans. Sci. Am., 230(4):80–89.
Walker, B. W. 1960. The distribution and affinities of the marine fish fauna of the Gulf of California. Symposium: the biogeography of Baja California and adjacent seas. Syst. Zool., 9(304):123–133.
———. 1966. The origins and affinities of the Galapagos shorefishes. *In*: Bowman, R. I. (ed.), The Galapagos. Berkeley, University of California Press, pp. 172–174.
Weber, M., and deBeaufort, L. F. 1913–1962. The Fishes of the Indo-Australian Archipelago, 11 vols. Leiden, Neth., E. J. Brill.
Wheeler, Alwyne. 1969. The fishes of the British Isles and Northwest Europe. London, MacMillan.
Wilson, T. 1963. Continental drift. Sci. Am., 208(4):86–100.

Migrations

Baggerman, B. 1960. Factors in the diadromous migrations of fish. *In*: Jones, I. C. (ed.), Hormones in Fish. London, Symposium of the Zoological Society, No. 1, pp. 33–60.
Fage, L., and Fontaine, M. 1958. Migrations. *In*: Grassé, P. (ed.), Traité de Zoologie, Vol. 13(3). Paris, Masson et Cie., pp. 1850–1884.
Fontaine, M. 1975. Physiological mechanisms in the migration of marine and amphihaline fish. Adv. Mar. Biol., 13:241–255.
Foster , N. R. 1969. Factors in the origins of fish migrations. Underwater Nat., 6(1):27–31, 48.
Harden-Jones, F. R. 1968. Fish Migration. New York, St. Martin's Press.
Hasler, A. D. 1965. Underwater Guideposts. Madison, University of Wisconsin Press.
———. 1971. Orientation and fish migration. *In*: Hoar, W. S., and Randall, D. J. (eds.), Fish Physiology, Vol. VI. New York, Academic Press, pp. 429–510.
Hess, W. N. 1964. Long journey of the dogfish. Nat. His., 73(9):32–35.
Hoar, W. S. 1965. The endocrine regulation of migratory behavior in anadromous teleosts. Proc. 16th Int. Congr. Zool., Washington, D. C. 1963, Vol. 3. Garden City, N. Y., Natural History Press, pp. 14–20.
Leggett, W. C. 1973. The migrations of the shad. Sci. Am., 228(3):92–98.
Mather, F. J. III. 1969. Long distance migrations of tunas and marlins. Underwater Nat., 6(1):6–14.
McDowall, R. M. 1968. The application of the terms anadromous and catadromous to the Southern Hemisphere salmonoid fishes. Copeia, 1968(1):176–178.
Myers, George S. 1949. Usage of anadromous, catadromous and allied terms for migratory fishes. Copeia, 1949(1):89–96.
Nakamura, H. 1969. Tuna Distribution and Migration. London, Fishing News (Books).
Royce, W. F., Smith, L. S., Hartt, A. C. 1968. Models of oceanic migrations of Pacific salmon and comments on guidance mechanism. Washington, D. C., U.S. Fish and Wildlife Service, Fisheries Bull., 66:441–462.
Schmidt, Johannes. 1925. The breeding places of the eel. Rep. Smithson. Inst., 1924, pp. 279–316.
Seckel, G. R. 1972. Hawaiian-caught skipjack tuna and their physical environment. Washington, D.C., U.S. National Marine Fish Service, Fishery Bull., 72(3):763–787.

10
COLOR, LIGHT, SOUND, AND ELECTRICITY

Color

A trip to a public aquarium or a tropical fish shop — or simply a glance through a tropical fish book — is sufficient to impart the idea that fishes rank among the world's most colorful inhabitants. The multitude of patterns involving two or more colors in many tints and shades, arranged in spots, stripes, patches, and blotches may seem superficially bewildering and random when fishes are viewed away from their normal habitats, but color is quite functional in fishes. Much experimentation and observation remains to be done, but work on many phases of the biology of fishes shows that color plays important social roles, as well as roles in concealment, advertisement, and mimicry. Coloration is remarkably labile in fishes, changing with background, mood, or stage of life history, and although much has been learned about the significance and mechanisms of these changes, there are mysteries in the relationships between color and behavior yet to be unraveled.

Only a few fishes lack pigment in the skin, or are very lightly pigmented. Cave-dwelling species are notable examples, generally appearing white or slightly colored by blood or other body fluids. The larvae of fish are often unpigmented, or have only a few color cells in the head or yolk sac. Larvae of some smelts and gunnels, for example, may remain virtually transparent up to a length of 2 cm or more, and the leptocephalus larvae of eels may, depending upon the species, reach 15 cm or more while still transparent (Fig. 10–1). Pigment is seldom free in the skin or other tissues of fish, but some blennies, greenlings, and mackerel-like fishes have a greenish pigment (related to bile pigments) in the skin, flesh, or bones. A very special type of coloration is seen in some tropical fish larvae, which have a blue carotenoprotein pigment in the muscles, fin bases, or the walls of the alimentary canal. Pigment is mostly contained in special cells called chromatophores, which are named for the color they impart or for the color of the pigment they carry.

Colors imparted by cells are of two kinds, those due to pigments — biochromes — and those due to the reflection of light from a colorless mirror-like surface and refraction by the tissues — structural colors, or

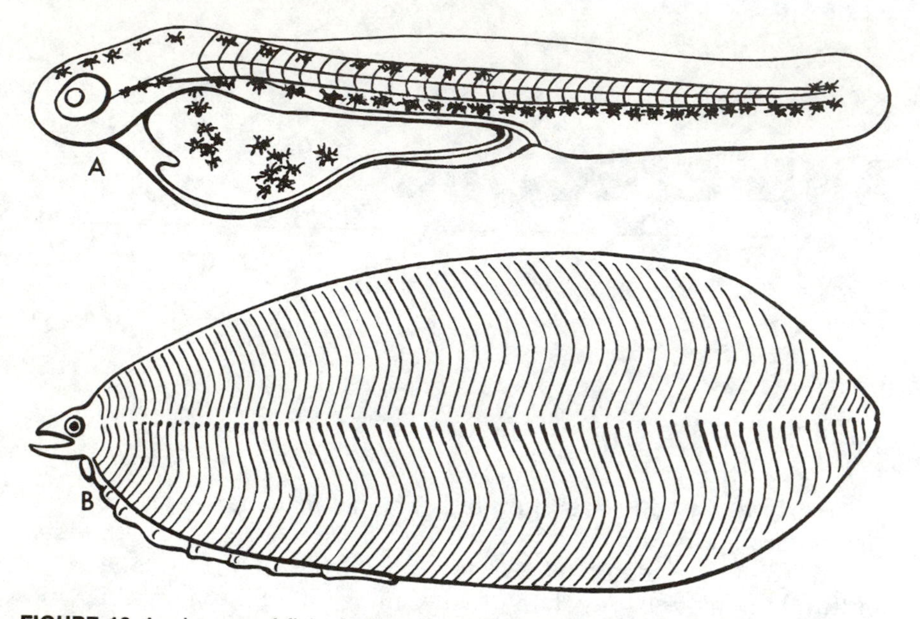

FIGURE 10–1. Larvae of fish. *A*, Showing scattered pigment cells; *B*, leptocephalus without pigment.

schematochromes. The two are often encountered in combination. Chromatophores are named as: melanophores, containing a black or brown pigment called melanin; erythrophores, containing reddish pigments (carotenoids and pteridines); xanthophores, containing yellow carotenoid pigments; leucophores, containing white, colorless purines, usually guanine, in the form of small crystals that can be moved within the cytoplasm; and iridophores, containing purines, mostly guanine, in large, nonmotile crystals. Cells carrying more than one pigment are called compound chromatophores. Chromatophores are located mainly in the skin, sometimes superficially in the epidermis, but more usually in the dermis or subdermally. They occur in the peritoneum, the eye, and sometimes around parts of the central nervous system. The study of the biochemistry of the numerous types of melanins, carotenoids, flavines, purines, and pteridines involved in the very complex matter of coloration and color changes in fishes is an important and interesting topic, but is beyond the scope of this discussion.

Color changes involving biochromes and some changes involving structural color depend upon the movement of pigments within the color cell; these short-term and often rapid changes are called "physiological." Long-term change can be due to increases or decreases in the number of chromatophores and a consequent general alteration of pigmentation, referred to as "morphological" color change. Pigment occurs in the cells in the form of granules, organelles or, as in the case of the purines, as crystals. Melanin is carried in organelles called melanosomes; pterinosomes bearing red pteridines have been identified in erythrophores. Chromatophores are usually characterized by irregular shapes involving dendritically branched processes, although

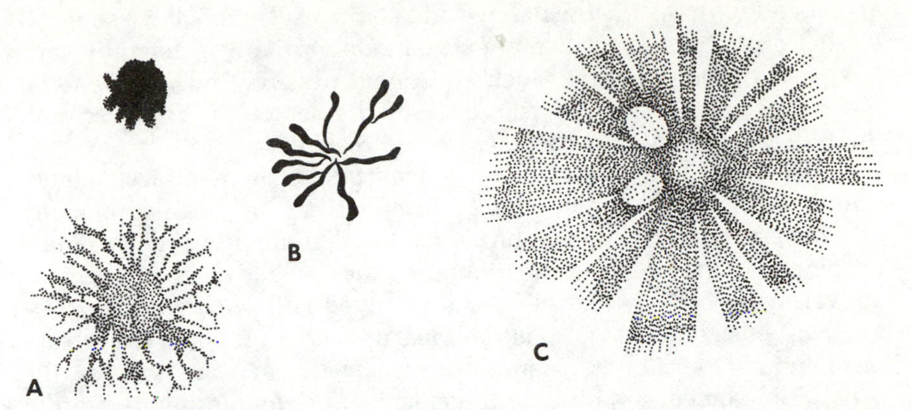

FIGURE 10–2. Chromatophores. *A,* Dendritic cell showing pigment aggregated *(top)* and dispersed *(bottom)*; *B,* pigment in finely radiate pattern; *C,* pigment in coarsely radiate pattern. (After Wunder.)

finely radiate and other shapes are known (Fig. 10–2). When the pigment granules are aggregated in the center of the cell, the skin is pale; when the pigment is dispersed throughout the processes, color is intensified. Various hues are made possible by the combinations of different chromatophores overlying each other or by the action of compound chromatophores.

Of special interest are the structural colors caused by the leucophores and iridophores. Many open-water species have a continuous subdermal sheet of iridophores on the sides and belly. This stratum argenteum, as it is called, is made up of several layers of cells, so that much of the light — up to 80% — striking the side of the fish is reflected. In addition, there are usually numerous reflecting cells in the dermis. If all these diminutive mirrors are parallel to the surface of the body, the fish generally appears uniformly silvery, but if the cells are arranged so that the reflecting crystals are differentially oriented, the apparent color of an individual plate of crystals might be dark when viewed from above and light when seen from below. Iridophores in *Fundulus* and a few others have been seen to change the color of reflected light with no apparent change in the position of the guanine crystals.

Aggregation and dispersion of pigments in chromatophores seem to be under both hormonal and neural control. In some fishes hormones are of greatest importance and in others only nervous control appears to operate, but probably most fishes combine the two types. The pituitary is of great importance in control of melanophores, secreting melanocyte-stimulating hormone (MSH) and apparently other substances that cause movement of the melanosomes. Not all species are affected in the same manner by MSH, which in most causes the pigment to disperse and the skin to darken. Some instances of the same extract causing aggregation in one species and dispersion in another are known. A melanophore-concentrating hormone (MCH) has been associated with the hypothalamus, and some researchers have suggested that other hormones affecting chromatophores are secreted by various parts of the pituitary. Much additional research must be done on the subject before an understanding of the secretions involved and their effects upon various species can be reached.

Adrenalin has the effect of concentrating pigment in melanophores in fish in general. The same effect can be caused in many species, but not all, by melatonin, a secretion of the pineal body. There is some evidence that the pseudobranch may be active in secreting or activating a hormone-like substance involved with melanophores.

Nervous control of chromatophores appears to involve the release of chemical transmitters (neurohumors) by the neurons serving the cells. Some evidence points to the existence of double innervation of color cells, so that one set of fibers releases a transmitter that causes dispersion of pigment and another set produces an aggregating neurohumor. Research has not always produced consistent results, but recent studies on the goldfish appear to confirm double innervation for that species. Iridophores do not seem to be under nervous control but leucophores, melanophores and, to some extent, erythrophores and xanthophores, are.

Physiological color changes may result from a fish's response to a changed background color in its surroundings, or may be due to any one of many responses to social, behavioral, or chemical stimuli. Numerous fishes are known to react to the color of their environment by altering their own hues to match it. Usually, two sources of light are involved in the processes that govern the changes — light penetrating the surface of the water and light reflected from the bottom or other background. The nervous system of the fish seems to take the ratio of light from the two sources (called "albedo") into consideration in mediating the response. Light from the surface impinges upon the lower part of the retina, whereas that reflected from the bottom strikes the upper part, and the eyes of fishes appear to be specialized to respond to the albedo. Responses may be rapid, taking only a few seconds, or may require hours. Darkening or paling can be effected by light striking the pineal body of most fishes, and there is evidence that other parts of the central nervous system can respond to light, as paling responses have been induced by exposure to light of individuals in which eyes and pineal have been rendered nonfunctional.

Morphological color changes can often be identified in the various life history stages of fish. For instance, young trout and salmon are usually colored to resemble the stream background and may have vertical bars on the sides, but as they physiologically prepare to enter the ocean they become silvery on the sides and blue or green on the back. Later, as they return to fresh water for breeding, the pelagic coloration is obliterated and a nuptial coloration, often involving very bright colors, is assumed. Loss of chromatophores can be induced by retaining fish under constant illumination in a light container. Similarly, keeping them on a dark background results in an increase of melanophores. Melanogenesis seems to be controlled by the pituitary, being brought about by action of the hormone ACTH.

Color has many significant functions in fish, such as protection of the central nervous system from illumination in larval stages, or as an aid to thermoregulation. Some cichlids even have yellow pigment in the cornea and lens, which apparently acts as an optical filter. However, most functions of color involve ecology or behavior and serve the purposes of concealment, advertisement, or disguise, following the terminology of Hugh B. Cott. Inasmuch as concealment and disguise sometimes serve the same purpose in similar manners, there is often a fine line between them, but the differences can be made clear. A common method of concealment among fishes is countershading (obliterative shading), generally seen in any fishes that are darker on the back than on the sides and belly. Many species of open waters show the type of countershading that is sometimes referred to as pelagic coloration, with dark backs of the general hue of the water as seen from above — usually blue or green — and silver sides and belly. With this arrangement, illumination from above lightens the back and the light color of the sides lightens the shadows on the flanks so that the entire fish can appear to have a uniform color that blends with the background. The reflectivity of the silvery sides and belly can render a fish almost invisible. Some smelts and silversides are translucent, but have a reflective layer around the body cavity and another around the red lateral muscles. In both instances the silver does not extend over the top, so the structures are essentially countershaded like the fish itself. Some brightly colored reeffishes are countershaded in that the upper sides and back are black or some other dark tone and lighter, brighter patterns cover the flanks and belly. One species of butterflyfish has been noted to turn its dark back toward approaching predators, an act that should render the conspicuous fish less so. Some of the black on the sides of the species serves to conceal an apparent social signal of yellow, which replaces the black under certain circumstances.

Dark pigment in the peritoneum or gut wall of deep-sea fishes might serve to conceal the lights of a luminescent meal. The blue pigment in the gut wall of certain tropical fish larvae could mask the presence of the orange or red crustacea upon which they feed.

Concealment can be aided by an overall resemblance to the substrate or background. For instance, some species have hues and patterns similar to those of the bottom (demersal coloration), while some patterns bear a general resemblance to vegetation (vegetal coloration). Many species can change color or pattern to match their surroundings. Trout that live in streams with a heavy overstory of trees tend to be very dark and heavily spotted, those that frequent open riffles are light, and those living in ponds opalescent with colloidal clay can be ghostly pale. Much attention has been given the flounders and other flatfishes, but various blennies, sculpins, scorpionfishes, and others are capable quick-change artists. Many demersal fishes have flaps and irregular outlines that aid in concealment. Fishes that generally resemble plants include sea snails (*Liparis*), pipefishes and seahorses (Syngnathidae), pricklebacks (Stichaeidae), and gunnels (Pholidae). Many bottom fishes show a color resemblance to substrate that consists of stones, shell, and small algae in combination. Mr. John Ratliff, a diver, has noted a species of the sculpin genus *Artedius* living on sea anemones, and accurately matching the color of the landlord.

Disruptive coloration may cause a fish out of its accustomed habitat to look conspicuous, but certain bold patterns of stripes, bars, ocelli, and other markings tend to break up the outline of the individual, or to make the eyes or other readily recognizable features less prominent (Fig. 10–3). In schooling species disruptive markings can present a

FIGURE 10–3. Examples of fish with disruptive pigment patterns. *A*, Scythe butterfly fish *(Chaetodon falcifer)*; *B*, treefish *(Sebastes serriceps)*. (*A*, after Miller and Lea.)

confusing pattern. The sight of a large aggregation of horizontally striped fish in the brightness of a coral lagoon might make you think that your eyes will not focus properly, until individual fish at the edges become evident and the illusion is destroyed. Eye stripes and "hoods" of dark color that cover the top or anterior of the head and the eye are common in fishes. Elongate fishes usually have horizontal eye stripes and those with deep bodies or blunt heads tend to show vertical stripes, or stripes that follow the contour of the head.

Advertisement by bright colors or subtle or conspicuous patterns may serve many purposes for various fishes. Differential coloration of the sexes aids recognition of potential mates, or attracts one sex to the other. Schooling species are marked with small or great signs that must aid individuals in staying with their own kind. Even countershaded species such as sardines and tunas may have rows of spots or a series of stripes arranged so as not to detract from the obliterative shading to any great degree. Color and behavior are closely related in fish and conspicuous, usually rapid, color changes are often involved. One example in a schooling species is the rudderfish, *Kyphosus elegans*, in which individuals change from stripes to spots and take on the behavior of "policemen," rounding up stragglers from the school and discouraging members of other schools from feeding in their school's area. During their short stints of policing activity, the spotted individuals are sharply different from the schooled fish. There are numerous instances wherein color functions to advertise the mood of an individual in intraspecific behavior. (Some of this will be covered in a later chapter.)

Cleaner-fishes, which remove ectoparasites from other fishes, usually have distinctive, conspicuous coloration and markings so that they are readily recognized by larger species and are allowed to approach with little danger of being eaten. Often, bright colors and bold patterns accompany unpalatability or dangerous venom. Perhaps the least conspicuous of this warning coloration is the black area on the dorsal fin of the venomous weevers (*Trachinus* spp.), which are generally cryptically colored. The black dorsal is erected, apparently as a warning, on the approach of animals larger than the weever's prey. Lionfishes (*Pterois* spp.) are very showy — and very venomous. One of the conspicuously colored saber-toothed blennies, *Meiacanthus nigrolineatus*, has venom glands in the lower jaw and is rejected by predators, probably after a venomous bite from the large lower teeth. Some surgeonfishes show a distinctive bright spot at the location of their formidable peduncular spines.

Although general resemblance and disguise may be indistinguishable in some instances, the concept of disguise is useful because it involves resemblance to specific objects and usually involves not only color but body shape, special appendages, and behavior. Adaptive advantage has apparently accrued to species that bear a close likeness to objects in the environment that elicit neutral reactions in their predators or prey. On the other hand, mimicry, a special kind of disguise, bestows some advantage on a species that looks like some

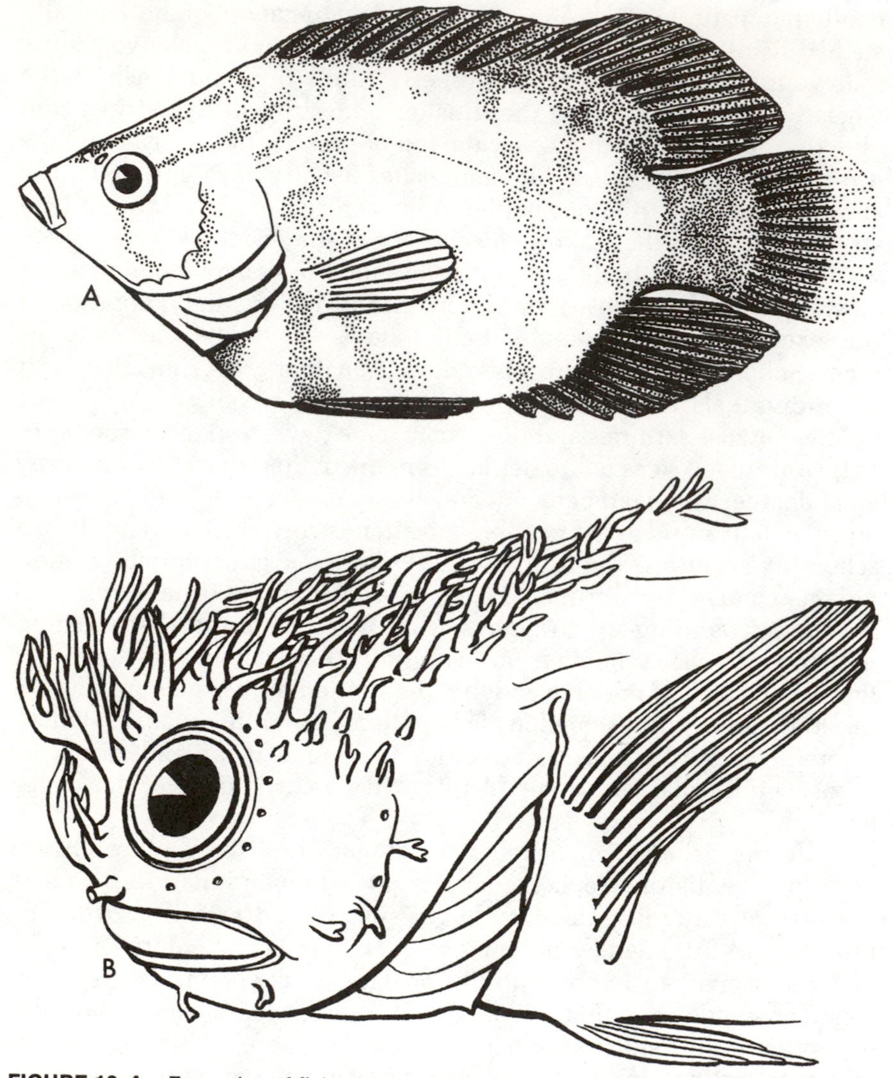

FIGURE 10–4. Examples of fish with structure and coloration resembling plant parts. *A,* Young of tripletail *(Lobotes),* with general resemblance to mangrove leaf; *B,* mosshead warbonnet *(Chirolophis nugator),* with cirri and flaps resembling marine algae;

(Illustration continued on opposite page)

recognizable entity that may be sought or avoided by prey, predators, hosts to cleaners, and so on. Most masters of disguise among fishes bear specific resemblance to plant parts (Fig. 10–4). The leaffish of South America, *Monocirrhus polycanthus,* has a barbel at the chin, simulating a leaf's stem, and has coloration, body shape, and postural behavior that complete the illusion. The young of a labrid, *Hemipteronotus pavo,* has a "stem" consisting of the first dorsal fin, and posture and coloration resembling a floating or drifting leaf. Others that use the leaf

FIGURE 10–4 *Continued.* *C,* sargassumfish *(Histrio histrio).*

disguise are the batfishes, *Platax* spp., and the young of a carangid, *Trachinotus falcatus,* both of which combine color and posture in the deceit. Other genera with leaf-like young include *Lobotes* and *Oligoplites.* The naked sole, *Gymnachirus melas,* by means of color and serrate dorsal and anal fins, looks like a dead, sunken leaf. The filefish, *Aluterus,* with its mouth to the substrate and slender tail upward, resembles eelgrass. Juveniles of many fishes are suitably colored and shaped to resemble floating or drifting plant debris, but one of the most remarkable instances is seen in the genus *Chaetodipterus,* in which the black young resemble the old, sunken seed pods of mangrove with which they share light-colored bottoms. An interesting adaptive coloration is the possession of large, realistic eye spots (ocelli), especially in the posterior part of the fish. When coupled with disruptive bars or stripes that make the fish's eyes inconspicuous, the eye spots effectively disguise the direction in which the fish is heading, and may be confusing to a predator. Ocelli may be involved in intraspecific behavior, or may be the types of camouflage or disruptive coloration. They could be considered instances of mimicry, in that they might resemble the eyes of other organisms.

Mimicry may serve many functions among various animals, and several types of mimicry have been recognized, especially among insects. Mimicry is not widespread among fishes, but there are some interesting examples. Some fishes mimic more numerous models that are distasteful or otherwise avoided, or at least not preferred by predators (Batesian mimicry). Thus a sole living among weevers lifts a black pectoral fin to mimic the warning signal given by the weever's dorsal fin or a blenny assumes the coloration of a cleaner wrasse and is able to live near predators with reduced probability of being eaten. Young *Plotosus anguillarus*, themselves venomous, may aggregate so as to resemble sea anemones. Another type of mimicry, aggressive mimicry, is seen in pairs of cleaners and skinbiters. An example of such a pair involves the wrasse, *Labroides dimidiatus*, which has access to larger fishes as an ectoparasite cleaner, and one of the saber-toothed blennies, *Aspidontus taeniatus*, which feeds on skin torn from larger fishes. Experienced fishes learn to distinguish the cleaner from the outlaw.

Display of lures to entice prey within reach is a special kind of aggressive mimicry usually involving modified body parts, such as the maxillary barbels of the angler catfish, *Chaca chaca*. The lures of anglerfishes commonly resemble worms or other edible invertebrates.

Another type of mimicry is that in which two or more dangerous or unpalatable species assume similar warning coloration, and thus increase both their chances of avoiding being tried out by inexperienced predators and benefiting from the learned response of the predator (Müllerian mimicry). This is a rare phenomenon among fishes; the closest qualifying instance involves some of the saber-toothed blennies (Fig. 10–5). The venomous *Meiacanthus nigrolineatus* is a model not

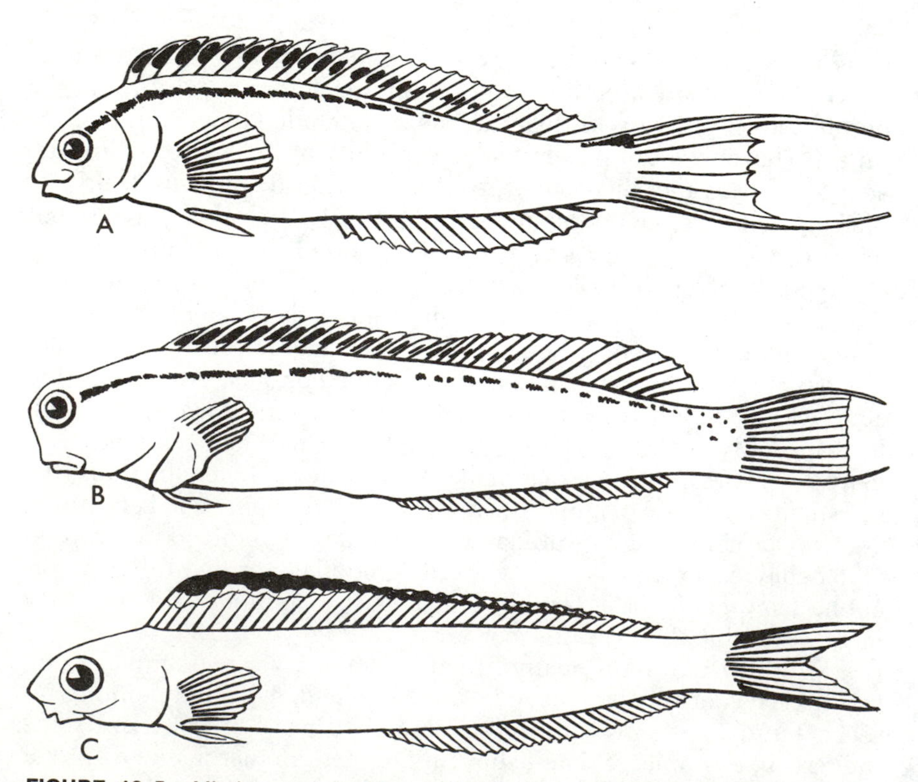

FIGURE 10–5. Mimicry among blennioid fishes. *A, Meiacanthus nigrolineatus*, a venomous species; *B, Ecsenius gravieri*, a nonaggressive species; *C, Plagiotremus townsendi*, an aggressive species. (After Springer and Smith-Vaniz.)

only for the nonaggressive *Ecsenius gravieri*, but also for *Plagiotremus townsendi*, a fierce biter that, like *M. nigrolineatus*, is usually rejected by predators.

Some mouth-brooding cichlids display an interesting mimicry that ensures fertilization of the eggs, which are taken into the female's mouth after deposition and before fertilization. The male has a series of egg-like spots ("dummy eggs") on his anal fin. He brings these to the attention of the female in the area where she has been depositing eggs, and as she attempts to pick them up he releases sperm, fertilizing the eggs carried in her mouth.

Bioluminescence in Fishes

Many terrestrial and marine organisms are capable of producing light. Included are bacteria, fungi, flagellates, comb jellies, molluscs (including snails, clams, and squids), crustaceans, polychaete worms, insects, ophiuroid starfishes, and fishes, as well as others not mentioned. Only one freshwater organism is known to be luminescent — the limpet, *Latia*, of New Zealand. At the surface of the ocean, the "phosphorescence" caused by various small organisms, especially the dinoflagellate *Noctiluca* and the ostracod *Cypridina*, is well known. There are a few luminous fishes that are permanent residents of shallow water, but most live in or over moderate to great depths, and many move into surface waters as part of a nightly feeding migration. Apparently most light-bearing fishes live in the twilight areas of the ocean, being found mostly at depths of from 300 to 1000 meters. At some localities more than 60% of the fish species and more than 90% of the individuals collected have been luminescent.

Ability to produce living light has been noted in at least 42 families of fish (Table 10–1). Most of these are teleosts, as only two families of elasmobranchs and no lampreys or hagfishes are known to be luminous. Among the teleosts, a wide range of families from soft-rayed salmoniforms up through the perciforms and lophiiforms display bioluminescence. The family of lantern-fishes, Myctophidae, has more luminescent genera and species than any other, but Ophisthoproctidae, Macrouridae, Ogcocephalidae, and the suborders Ceratioidei and Stomiatoidei have numerous light-bearing representatives, and in some areas the last-named group may outnumber lanternfishes in terms of individuals.

Light is produced chemically, without heat, in an interaction of the enzyme luciferase with the substance luciferin, (a heterocyclic phenol) usually in the presence of an oxygen source and ATP. There are apparently great differences among fishes in their luciferin-luciferase systems, and there is much yet to be learned about all the biochemistry involved in light production. Light production usually takes place in special organs called photophores (Fig. 10–6).

Although most luminous fishes produce their own light (self-luminous) many depend on symbiotic bacteria nurtured in special

Table 10–1. FAMILIES CONTAINING LUMINESCENT FISHES

Order	Family
Squaliformes	Squalidae (including Dalatiidae)
Torpediniformes	Torpedinidae
Salmoniformes	Alepocephalidae Searsidae Bathylagidae Opisthoproctidae Gonostomatidae Sternoptychidae Chauliodontidae Astronesthidae Malacosteidae Melanostomiatidae Idiacanthidae Bathylaconidae
Myctophiformes	Myctophidae Neoscopelidae Scopelarchidae Paralepidae
Anguilliformes	Saccopharyngidae
Gadiformes	Gadidae Macrouridae
Beryciformes	Trachichthyidae Anomalopidae Monocentridae Cetomimidae
Perciformes	Acropomatidae Leiognathidae Pempheridae Sciaenidae Chiasmodontidae
Batrachoidiformes	Batrachoididae
Lophiiformes	Chaunacidae Ogcocephalidae Melanocetidae Himantolophidae Diceratiidae Oneirodidae Centrophyrnidae Ceratiidae Gigantactinidae Linophyrnidae

gland-like structures. Many fishes with bacterial luminescence have the luminous structures associated with the ventral body wall, the anus, or the digestive organs. Grenadiers (Macrouridae), some cods (Gadidae), and roughies (Trachichthyidae) are examples of groups with luminescent structures at or around the anus or along the ventral surface close to the anus. In some of these gelatinous tissue under the skin allows the light to diffuse over a relatively wide area. Some of the light glands

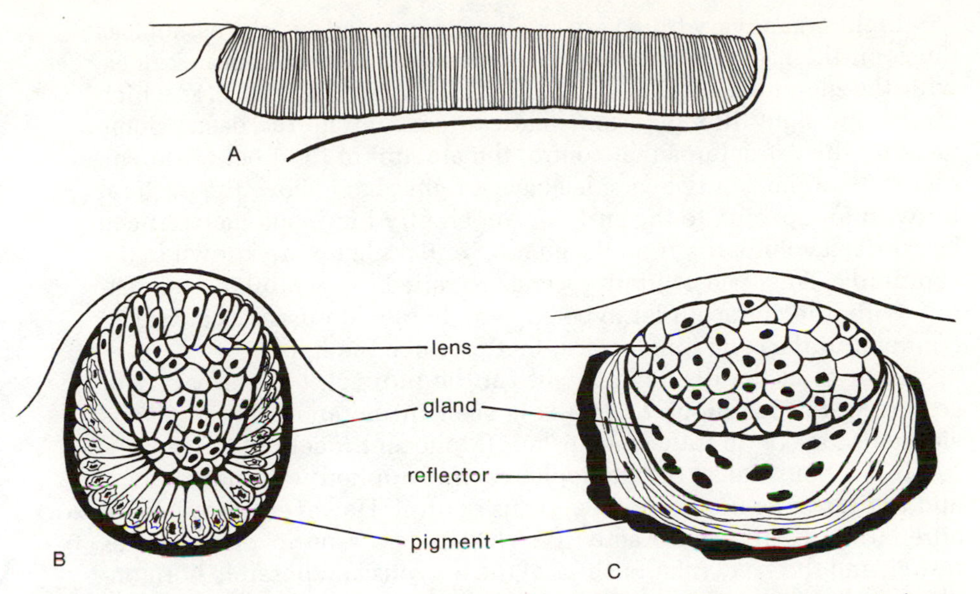

FIGURE 10–6. Diagrams of examples of photophores. *A,* Type in which luminous bacteria are nurtured in tube-like structures; *B,* self-luminous organ with lens and pigment sheath; *C,* self-luminous organ with lens, reflector, and pigment sheath.

release luminous material that can spread along the exterior of the fish. Slipmouths (Leiognathidae) have a bacterial light gland around the esophagus, a lens that concentrates the light and directs it into the gas bladder, and a pigmented shutter that can conceal the light. Others with bacterial luminescence associated with the digestive system, anus, or ventral musculature are the lanternbellies (Acropomatidae) and some of the cardinalfishes (Apogonidae), although some of the latter appear to be self-luminous. Most of these can light up the thoracic or ventral areas because of internal reflectors and translucent tissue under the skin.

The lanterneyes (Anomalopidae) bear bacterial light organs under the eyes. The two best-known species are *Photoblepharon palpebratus* and *Anomalops kaptotron,* both of which are shallow-water fishes found mainly in the Banda Sea area of Indonesia. *Anomalops* has been observed to flash light on and off in a rather regular manner by rotating its light organ inwards and down into a black-pigmented pocket under the eye. *Photoblepharon* shows a more steady light, but can extinguish it by drawing a fold of black tissue over it. Most members of the suborder of deepwater anglerfishes (Ceratiodei) are luminous, having light on the barbels, the esca, and elsewhere on the body. Recently the source of light has been proven to be luminous bacteria.

Self-luminous fishes commonly possess a series of light organs along the ventral aspect of the body that direct the light downward. However, a few have some dorsal lights, and many of the stomiatioids have an abundance of small, simple organs along the dorsal surface that enable the entire fish to be silhouetted in light. Some have light organs within the mouth as well as on the jaws.

Light organs vary from very small unpigmented structures, such as those on the back and fins of the stomiatioids, to very complex ones, with the glandular portion surrounded by an efficient reflector which directs the light through a lens that can concentrate the beam. Some have iris-like structures that control the amount of light being shown. Members of the family Searsidae have a light gland above the pectoral fin with an opening to the surface. Apparently luminous material can be released voluntarily from the gland. Similar glands are known in the ceratioids. This type of luminescence is called extracellular, contrasting with the intracellular type in which the luminous material is confined to the cells (photocytes) that produce the light.

Control of the display of light can be indirect, by concealing or screening the luminous tissue, as is seen in the anomalopids, some stomiatioids, or in pinecone fishes (Monocentridae). In most self-luminous bony fishes the photophores are supplied with nerve fibers and appear to be under direct nervous control. However, injections of adrenaline usually cause activity of luminous tissue in most species tested, and the exact relationships of the nervous and possible hormonal controls have not yet been discovered.

Considering that about two thirds of deep oceanic fishes produce light, this ability must bestow some adaptive advantage upon the luminous species. Identification of the advantages is difficult for most species because of their habitats and habits, except for some of the shallow-water forms, so much speculation is usually necessary in discussing the function of bioluminescence in fishes. Naturally, the functions of light are thought to parallel in part the functions of color and, of course, color is involved in bioluminescence. Many species have color filters built into the light organs so that, over a range of species, lights shine in many hues. Possible functions of luminescence are as follows.

Concealment. Placement of photophores and other luminous tissue in a ventral position in most luminous species has led to the theory that a midwater luminous fish can match the background of light coming from above, so that predators hunting from below will have less chance of seeing the silhouette. Other theories concerning the ventral placement of photophores are that the reflectors of the organs would cause the bearer to be conspicuous from above even in dim daylight if they were placed dorsally, and that there are more predators living above the luminous species than below, so that selective pressure favors the ventral photophore.

Advertisement. Luminescence may serve several purposes in reproduction. Luminous organs of males and females are known to be disposed in different patterns in many species, so that mates of the proper species and sex may be attracted by a display of the pattern. Midshipmen, *Porichthys*, are known to display the light from photophores during courtship, and other fishes might do the same. In species which have well lighted males and less conspicuous females, the impact of predators on the females might be lessened, but this would be an advantage mainly if the species were polygamous or if males naturally outnumbered females.

There are several other advantages that might accrue to fishes that advertise themselves with lights. Species recognition could aid in keeping schools together or, on the other hand, could aid individuals of nonschooling species in maintaining territories. Sudden displays of light by a single fish or by an aggregation of small fish might serve to startle or confuse a predator. The Atlantic midshipman, *Porichthys porosissimus*, possessor of a venomous spine, is known to flash its lights upon approach of predators, thus warning them away. Continued luminescence after being swallowed, if the light could shine through the tissues of the predator, would advertise the presence of the predator to prey species or to larger predators. Dr. D. E. McAllister has theorized that the dense pigment of the peritoneums or stomachs of deepsea predators serve mainly to guard against the lights of prey shining through. Several predators, including stomiatioids and anglers, show luminous organs or tissue close to the mouth, in the mouth, or on barbels or illicia. In some or all of these cases prey could be attracted to the mouth by a show of light.

Disguise. If the baits mentioned above closely resemble some natural food of prospective prey of the angler, this might be considered some sort of disguise. Some investigators have suggested that the luminous tissue of certain fishes mimics various luminous invertebrates. Perhaps a school of small luminous fish might take on the appearance of a much larger individual.

Special Considerations. Placement of light organs on the head so that objects in the most effective visual field of the fish can be illuminated should make possible the floodlighting of prey, and the observation of feeding stomiatioids appears to bear this out. If an individual of a prey species illuminates a nearby predator, others of the species could possibly avoid it. Or, certain types of signals might alert prey species to the presence of a predator. The term "burglar alarm" has been applied to these hypothetical cases. Studies on a species of *Photoblepharon* in the Red Sea have confirmed multiple uses of the photophores in predation, intraspecific communication, and avoidance of predators.

Function of photophores placed so that the light shines into the eye, as in many stomiatioids and some opisthoproctids, is not known, but different theories have been presented. One is that the light entering the eye adapts it so that the sudden flash of brighter organs will not cause temporary blindness.

Production of Sound

That fishes and other animals living in the sea produce noises has been recognized from ancient times, and in some areas fishermen have long since located certain species by listening for their characteristic sounds. Most detailed information on the production of sound by fishes has been gathered in the past few decades as the importance of underwater sound to naval operations and fisheries has become apparent. Also, the instrumentation necessary for research on sound has become more available in recent years.

Water is a good medium for the transmission of sound. The speed of sound in water is about 1500 meters/second and varies with temperature and salinity, so that the speed in marine water can approach 1540 meters/second. Sounds can be carried long distances in water, being reflected off the bottom, the surface and density layers caused by temperature or salinity differences.

Fishes have evolved excellent sound-detecting organs over their long history, but do not appear to have developed any special organ for sound production, although many structures have been modified for the purpose. Sound-producing structures include skeletal parts, muscles, and the gas bladder. First, attention should be drawn to noises that fishes make incidentally to other activities. The very act of swimming can cause sound to be produced because of the movement of skeletal parts on each other and the turbulence caused by movement through the water. Large schools of fish have been noted to cause sounds as they veer and turn. Feeding activities are often noisy. Parrotfishes (Scaridae), wrasses (Labridae), surfperches (Embiotocidae), and many others that crush molluscs or crustaceans with their strong teeth in the jaws or throat make much incidental sound. Fishes that break the surface of the water, whether in feeding or for other purposes, set sounds in motion. Many fishermen are attracted to the exact spot where the quarry is feeding by hearing the smacking sound of the bluegill or the splash of the trout. Many know the double splash of the leaping carp. Eagle rays are known to make prodigious leaps which result in loud noises as they fall back to the water. Opening and closing of oral and opercular valves in large fishes can be audible, as can the release of air from the mouth or anus of air breathers. *Misgurnus fossilis*, which utilizes the alimentary canal as a respiratory surface, is known as "squeaker" because of the sounds made as respiratory gas is expelled. Release of bubbles because of reduced pressure during rapid ascents probably causes sounds.

Most sounds made "on purpose" by fishes are the result of stridulation or the vibration of the gas bladder. Of course, the gas bladder can be involved in other ways, and may be brought into play to amplify or otherwise modify the stridulatory sounds.

Some fishes have been noted to cause sounds by closing the opercula sharply. Others, including electric rays, shiners, sturgeons, gobies, and blennies produce noises of unknown origin.

Stridulation. Grinding, snapping, or rubbing teeth together is the most common type of stridulation among fishes. Pharyngeal teeth appear to be important noisemakers, especially because they are close enough to the gas bladder so that the sounds can be amplified. One study of the cichlid, *Tilapia mossambica*, in which special muscles run from the occipital region and the first vertebra to the upper pharyngeals, showed that sounds made in the absence of the gas bladder were

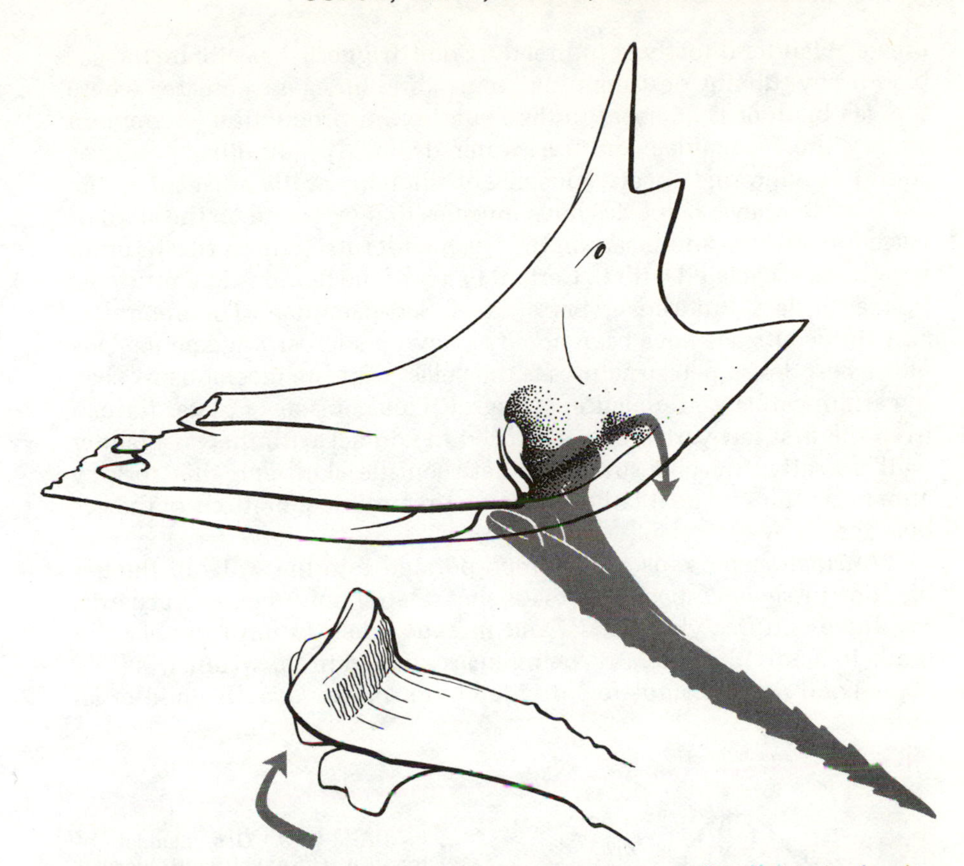

FIGURE 10–7. Right cleithrum and pectoral spine of sea catfish (Ariidae), showing roughened flange with which stridulatory sounds are made.

essentially the same as those produced by the intact animal. Jaw teeth often are used in sound production; some filefishes (Aluteridae) have ridges apparently effective in stridulation, on the backs of the front teeth, and various perciform fishes have been noted to snap their teeth together, making a sharp sound. Other mechanisms of stridulation include: movement of fin spines against their sockets in catfishes (Fig. 10–7), triggerfishes, filefishes, sticklebacks, surgeonfishes, and others; contact between the first dorsal interspinous bone and modified neural spines in some sisorid catfishes; friction between other skeletal parts in triggerfishes, seahorses, and clownfishes.

Sound Production Involving the Gas Bladder. Mention has been made of incidental sounds made by release of air from the lungs or other cavities used in air breathing. Sounds with greater significance than a simple burp may be caused by release of gas by the Atlantic eel and some catfishes. Such sounds are common but their importance is unknown.

Most gas bladder sounds arise because of vibrations set up by some special means. Stridulation can be enhanced by the resonance of the gas bladder, but many noisemakers are equipped with muscles that vibrate the gas bladder directly or indirectly. Some triggerfishes vibrate the gas bladder by rubbing or drumming the pectoral fin against an area where the gas bladder is adjacent to the skin. Sound production is common among the Sciaenidae; the freshwater drum, *Aplodinotus,* produces sound by vibrating special muscles of the body walls adjacent to the gas bladder. Several species have muscles that originate on the skull or vertebral column and insert on the gas bladder itself or on ribs or other structures associated with it. Certain fishes of the families Macrouridae, Priacanthidae, Brotulidae, Serranidae, Scorpaenidae, Theraponidae, and Holocentridae have been noted as having such arrangements. One of the best developed structures is the "elastic spring mechanism" seen in certain catfishes (*Galeichthys, Bagre*). This consists of plates formed from the first few vertebrae and placed in contact with the gas bladder wall dorsally. Muscles stretching between the skull and this springy apparatus can vibrate it rapidly, setting up audible vibrations of the gas bladder.

Intrinsic sonic muscles are incorporated into the walls of the gas bladder in several species of toadfishes, Batrachoididae, in gurnards, Triglidae and Dactylopteridae, and in *Zeus faber.* In most the bladder tends to be divided laterally, being heart-shaped in the toadfishes (Fig. 10—8) and nearly completely divided in gurnards. Usually an internal

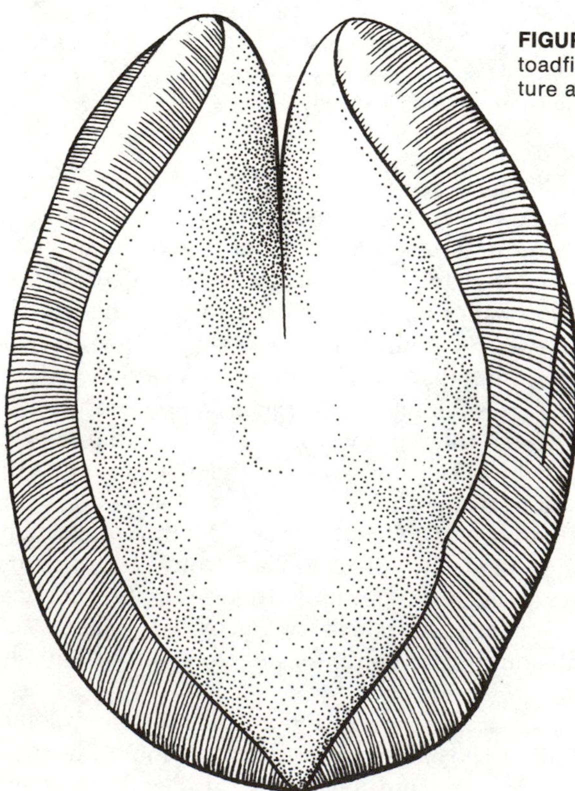

FIGURE 10–8. Gas bladder of toadfish *(Opsanus)* with musculature along outer edges.

diaphragm divides the bladder into anterior and posterior sections. The diaphragm is typically perforated by an opening surrounded by a sphincter muscle. The relationship of the diaphragm to the sounds produced is not understood, experimentation having produced no clear picture of its function.

The sonic muscles of fishes are red in color, except in Therapon-idae. They are very vascular, well supplied with myoglobin, and are resistant to fatigue. They are among the most fast-acting of vertebrate muscles, and can be artificially stimulated to act at more than 100 contractions per second.

Nature of Sounds. Noises made by fishes have been described by a great variety of terms. Sounds of schools of fish swimming have been called rustles or roars. Stridulation produces sounds reminiscent of clicks, rasps, scratches, etc., when not aided by the gas bladder, and croaks, grunts, and knocks when the bladder acts as a resonator. Stridulatory sounds can range from 100 to over 8000 cycles per second (cps), although the gas bladder-aided sounds are generally below 1000 cps and the unaided stridulations produce frequencies usually in the 1000 to 4000 cps range.

Sounds made by vibration of the gas bladder have been described as hoots, boops, grunts, yelps, and croaks, with the toadfishes, *Opsanus* spp., being known for their boat whistle sounds. Gas bladder sounds are harmonic and usually of low frequency, from 40 to 250 cps, with the great majority in the 75 to 100 cps range.

Significance of Sounds. Observation and experimentation have established that certain fish sounds may accompany specific behavior. Several species are known to produce apparent warning, alarm, or escape sounds when prodded or attacked. Some are known to make grunting noises in territorial defense. Others sound off when migrating, feeding, or resting following feeding. Reproductive behavior is accompanied by sounds in several species in both fresh and salt water. In most instances the males seem to make most of the noise and are usually better equipped to do so than the females. Among the sounds made by male toadfishes at the nest are the boat whistle, grunts, and growls. Courtship is accompanied by sounds in the shiners (*Notropis analostanus*), certain cods, seahorses, gobies, and others. Some sounds are thought to attract the female to the spawning area.

Production of Electricity

Three kinds of fishes that generate strong electrical shocks have long been known, although the nature of this power was a mystery to the ancients. The electric rays of the family Torpedinidae are said to have been used by ancient Roman physicians in an early form of electrotherapy. The electric catfish, *Malapterurus electricus*, is featured in Egyptian hieroglyphics, and has been noted as having an Arabic name translating as "father of thunder." In South America the electric eel, *Electrophorus electricus*, the most powerful of all, was apparently well known to the natives before its discovery by European explorers. Other groups with weak electrical powers were confirmed as being electric only after instrumental means of studying electricity were developed, and the strong electrical capability of the electric stargazers was not recognized until the twentieth century.

In all, ten families of fishes representing six orders are recognized as having electric organs and another family, the Petromyzontidae, has been reported as setting up an electric field around the head, but the origin of the field is not yet known. Strongly electric families are Torpedinidae, Malapteruridae, Electrophoridae, and Uranoscopidae. The electric rays, or torpedoes, are widespread in marine waters, some living at considerable depths. They are benthic and slow and some of the larger species, such as *Torpedo nobiliana,* are capable of delivering a shock of 220 volts. Electric catfish live in the murky water of African rivers. They are known to reach a length of about 1 meter and to produce shocks of 350 volts. The electric eel, an Amazonian species, is a sluggish fish like the latter two, and lives in water of low visibility. This is a relatively large fish, with at least one specimen measured at nearly 3 meters in length. Shocks of up to 650 volts have been measured from this fish, but 350 volts is a more usual maximum. Electric stargazers are marine fishes of the western Atlantic that have a habit of burrowing in sand, and can deliver 50 volts.

Weak electric fishes are in families mostly found in tropical fresh waters, but one is marine. All are either benthic or semibenthic, and rather sluggish. The skates (Rajidae), well known and nearly cosmopolitan in the oceans, include numerous species of small to moderate size. Mormyridae, the elephantfishes and relatives, are freshwater fishes of Africa, as are their close relatives, the Gymnarchidae. Many of these fishes are nocturnal in habit, and most live in waters of relatively low visibility. In the fresh waters of South America the gymnotoid fishes form a group that in part is the ecological parallel of the African mormyriforms. Electric families included are the knifefishes, Gymnotidae, Apteronotidae, and Rhamphichthyidae.

Electric organs are made up of specialized cells, called electrocytes, that have evolved from muscle cells. Electrocytes are typically thin and wafer-like and are arranged in bundles or stacks, depending upon the orientation of the organ (Fig. 10–9). One surface of a typical electrocyte is heavily innervated and the opposite face is irregular with numerous papilla-like projections. Gelatinous material surrounds the bundles or columns of electric cells, and the electric organs are rich in blood vessels, nerves, and connective tissue. Electrocytes of skates are of two types, flat and cup-shaped. They are not packed tightly and in some species they retain the striations of muscle cells. One family, Apteronotidae (= Sternarchidae), differs from the other electric fishes in that the electrocytes are modified spinal neurons, the electric cells of muscle origin having been lost. These enlarged neurons pass forward after entering the electric organ and then loop back; they can reach more than 100 microns in diameter in both the forward- and backward-running sections.

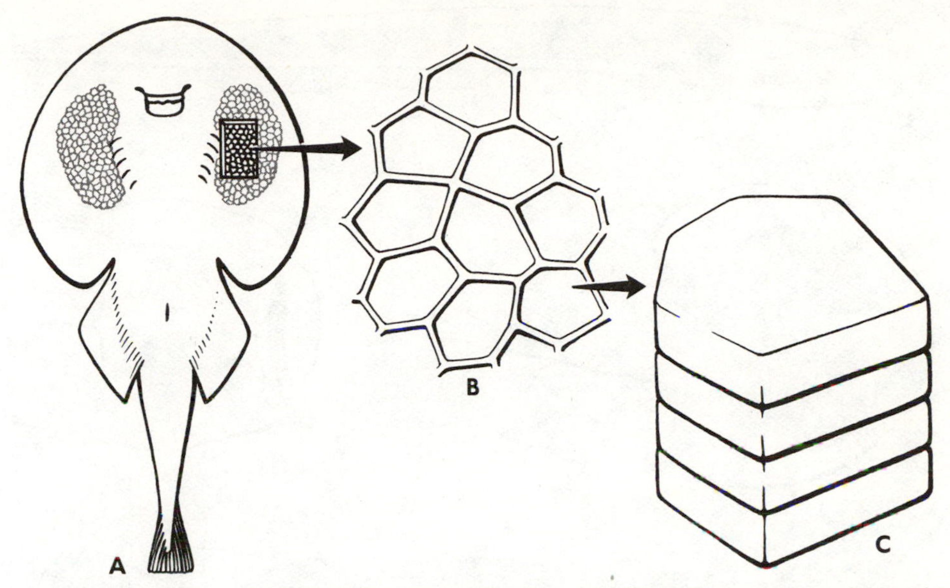

FIGURE 10–9. *A,* Ventral view of electric ray *(Torpedo),* showing position of electric organs; *B,* shape of electrocytes as viewed ventrally; *C,* diagram of arrangement of one "stack" of electrocytes.

Among the strong electric fishes (Fig. 10–10) the electric eel has three separate electric organs forming a large part of its bulk. The hypaxial part of the long caudal region is made up mostly of the main electric organ with a smaller one, the organ of Hunter, running along the ventral aspect. The organ of Sachs is posterior to the main organ. These organs have formed from axial musculature, and the electrocytes are ribbon-like, flattened anteroposteriorly, and extend from the medial septum out toward the skin. A large adult may have over 100,000 electrocytes on each side in the main organ, as there can be up to 6000 vertical arrays of up to 25 of the ribbon-like cells. The posterior surfaces of the electrocytes are innervated from spinal nerves. Current flow in the organ is from back to front, with reverse flow in the surrounding water.

In the electric catfish, the electric organ lies in the skin and covers most of the body musculature. The electrocytes are disc-like, about 1 mm in diameter, with a short stalk on the posterior, innervated face. Each side of the organ is innervated by branches from a large neuron in the anterior part of the spinal cord. The embryonic origin of the organ is pectoral musculature. Current flow is from front to back.

Electric rays have a large kidney-shaped electric organ in each side of the disc, next to the head and branchial region. The columns of hexagonal to roughly circular electrocytes that make up the organ are oriented vertically, and reach from the dorsal to the ventral surface. The electric cells, which reach 7 mm in diameter in some species, are

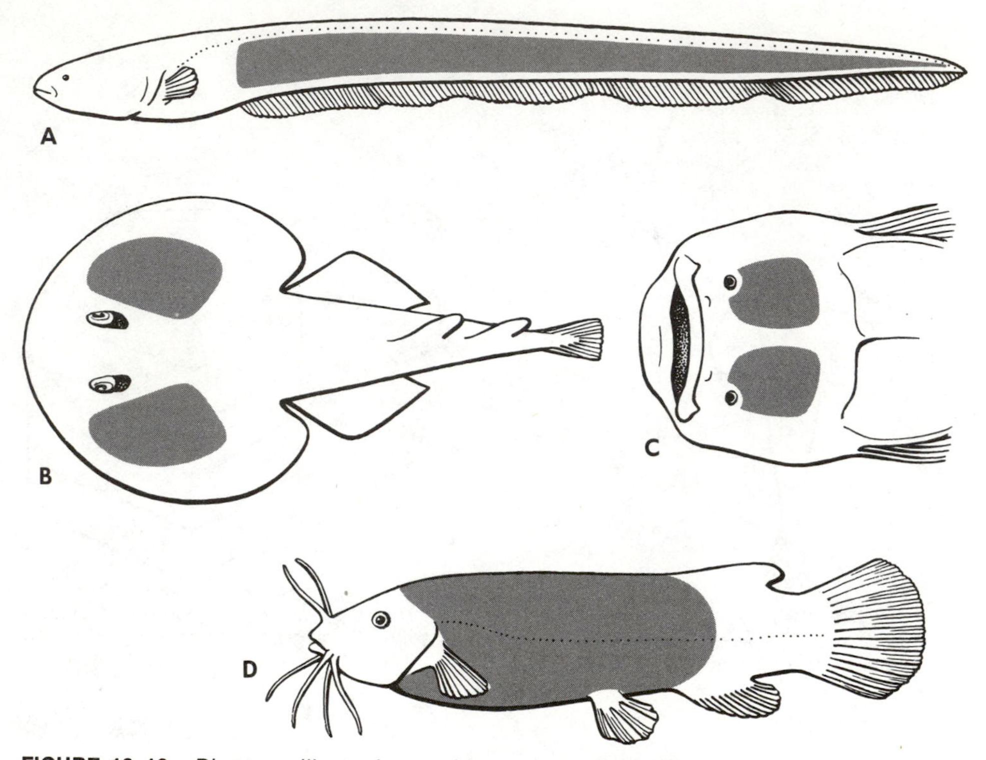

FIGURE 10–10. Diagrams illustrating positions of electrical organs in strong electric fishes. *A*, electric eel *(Electrophorus); B*, electric ray *(Torpedo); C*, stargazer *(Uranoscopus); D*, electric catfish *(Malapterurus).*

derived from branchial musculature. Current flow is from ventral to dorsal. The genus *Narcine* has an accessory organ at the posterior part of each main organ. Like the torpedoes, the stargazers have dorsoventrally flattened electric organs in the head region. The relatively small organs are situated posterior to the eye and are derived from eye musculature. The flattened electrocytes may be as much as 5 mm in diameter. Current direction is dorsal to ventral.

Weakly electric fishes tend to have one or more elongate electric organs along each side in the caudal region (Fig. 10–11). Mormyrids have two columns of cells in each side of the caudal peduncle. *Gymnarchus* has four thin columns per side in the posterior half of the caudal region, and skates are equipped with one pair of organs running most of the length to the tail. In *Gymnotus* the electric organ extends from below the head to the tip of the tail, coursing along the ventral aspect of the body. In addition to elongate, ventral main organs, some gymnotoids (*Steatogenys, Gymnorhamphichthys*) have small organs under the skin of the chin region. *Steatogenys* has another small organ in the pectoral region. Apteronotids have lateral electric organs reaching from above the pectoral fin to the base of the caudal fin. Current

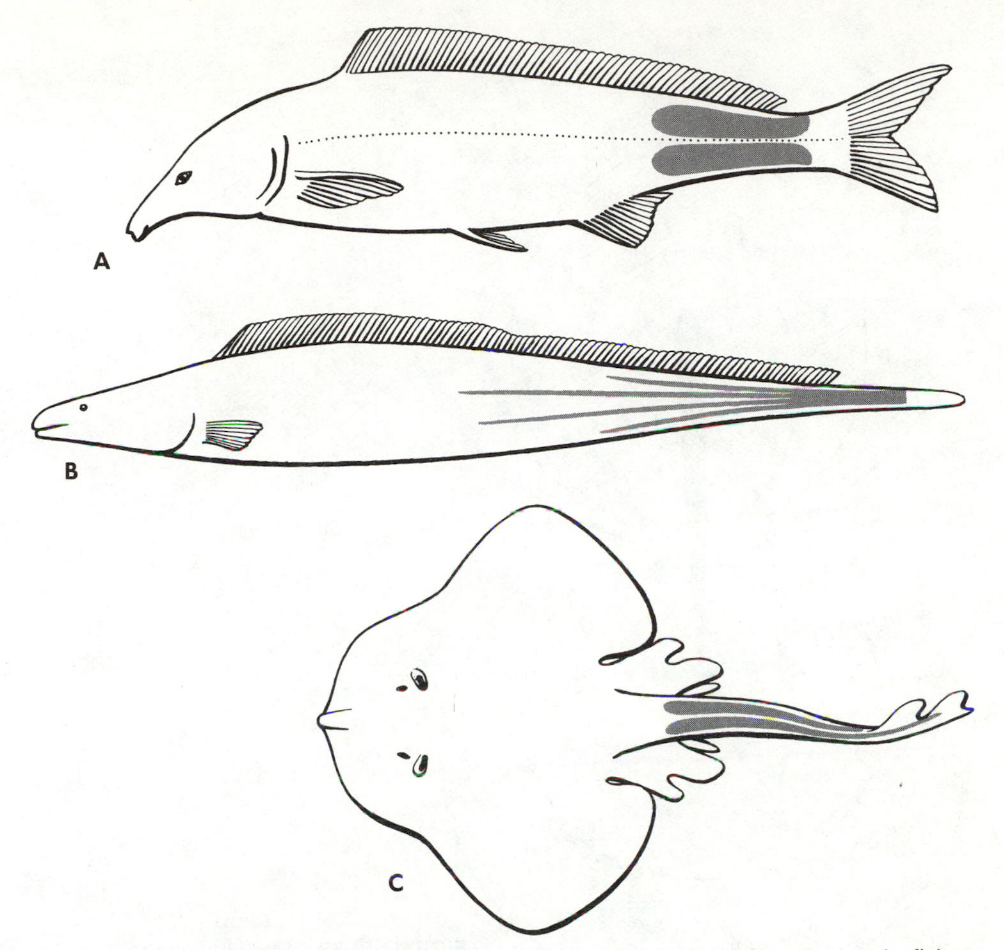

FIGURE 10–11. Diagrams illustrating positions of electrical organs in weak electric fishes. *A*, Mormyridae; *B*, Gymnarchidae; *C*, Rajidae.

flow is forward in *Gymnarchus*, backwards in *Raja*, and both directions are possible in the gymnotoids.

The function of strong electric organs appears to be in the stunning of prey and discouragement of intruders or predators. Use of electricity to obtain prey has been observed in the torpedoes, and its use for this purpose in others seems probable, considering the circumstances under which the electric species live. All are secretive, living near the bottom in situations that probably allow prey to approach closely without alarm. The stargazer can be especially well concealed, allowing small crustaceans or fish to move onto the sand under which the predator is buried.

Electricity in the weakly electric fishes is involved in the fascinating function of electrolocation of nearby objects, except in the skates, in which the function is unknown. The mormyrids, gymnarchids, and gymnotoids all live in turbid waters and some are nocturnal in habit,

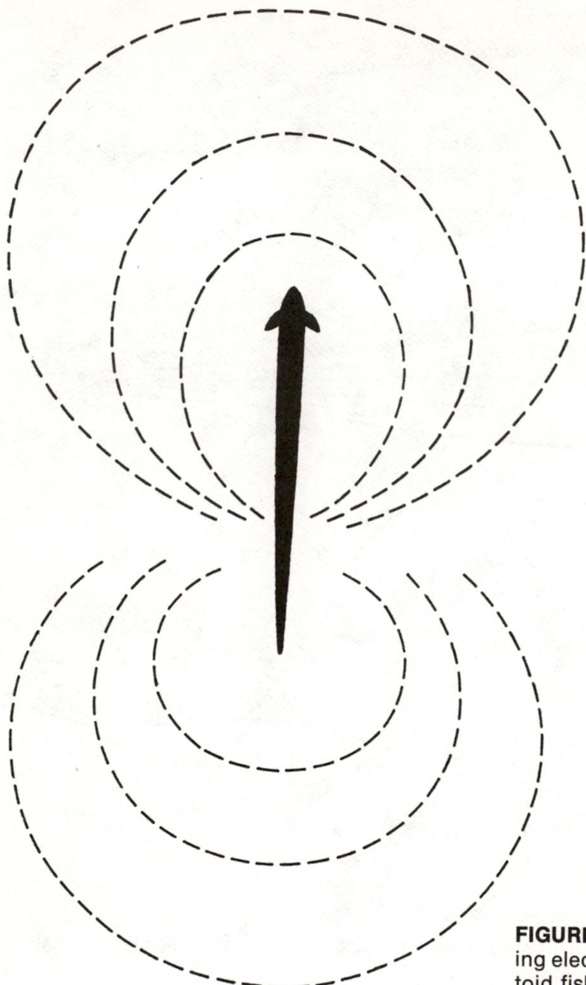

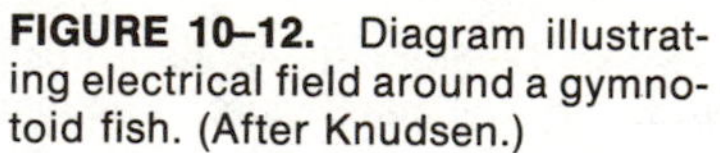

FIGURE 10–12. Diagram illustrating electrical field around a gymnotoid fish. (After Knudsen.)

causing their eyesight to be of limited use. Other senses — hearing, olfaction, and the mechanoreception of the lateralis system — must aid their orientation to their surroundings, as in other nocturnal species, but the possession of a system to generate and detect electrical fields sets these animals apart. For the most part, these electric fishes hold their bodies rigid and straight, depending upon undulations of fins for propulsion. The straight posture ensures symmetry in the electrical field (Fig. 10–12). Interference with or distortion of the field by nearby objects can be detected by special sensory organs. In addition, the impulses can be sensed by other fishes with the proper receivers in a special kind of communication.

Discharges of electric organs are characteristic of individual species, with frequencies and other features of the pulses or waves differing from one species to another. Although many electric fishes have the ability to vary the discharges to some extent, oscillograph tracings show characteristic shapes and amplitudes. The signals of mormyrids and most gymnotoids are pulsed. *Gymnarchus niloticus* and some gymnotoids, such as *Apteronotus*, *Eigenmannia*, and *Sternopygus*, discharge a wave or tone signal.

Various mormyrids discharge from one to six pulses per second as a rule, but can accelerate up to about 130 pulses per second. Voltage in mormyrids generally ranges from about 9 to about 16. *Gymnarchus* is reported to operate at about 4 to 7 volts, discharging at about 300 pulses per second. Investigators have demonstrated a wide range of pulse frequencies in gymnotoids — from 2 up to about 1000 per second.

Electric fishes, although members of diverse systematic groups, share certain attributes. They are generally slow-moving or sedentary, are active at night or in murky waters of low visibility, and have thickened skin that acts as a good insulator. Most have reduced eyes, and some torpedoes are blind. Remarkable adaptations in their brains set fishes with electrosensory systems apart from the others. Generally the cerebellum is enlarged, greatly so in the mormyrids.

References

Color

Abbot, F. S. 1973. Endocrine regulation of pigmentation in fish. Am. Zool., 13: 885–894.

Barlow, G. W. 1972. The attitude of fish eyelines in relation to body shape and to stripes and bars. Copeia, 1972(1):4–12.

Breder, C. H. Jr. 1946. An analysis of the deceptive resemblances of fishes to plant parts with critical remarks on protective coloration. Bull. Bingham Oceanographic Coll., 10(2):1–49.

———. 1972. On the relationship of teleost scales to pigment patterns. Contrib. Mote Mar. Lab., 1:1–79.

———, and Bird, P. M. 1975. Cave entry by schools and associated pigmentary changes of the marine clupeid, *Jenkinsia*. Bull. Mar. Sci., 25(3):377–386.

Cott, Hugh B. 1957. Adaptive Coloration in Animals. London, Methuen.

Denton, E. J., and Nicol, J. A. C. 1966. A survey of reflectivity in silvery teleosts. J. Mar. Biol. Assoc. U.K., 46:655–722.

Fernando, M. M., and Grove, D. J. 1974. Melanophore aggregation in the plaice (*Pleuronectes platessa* L.). I. Changes in *in vivo* sensitivity to sympathomimetic amines. Comp. Biochem. Physiol., 48A(4):711–721.

Fingerman, M. 1965. Chromatophores. Physiol. Rev., 45:296–339.

Fox, D. L. 1957. The pigments of fishes. *In*: Brown, M. E. (ed.) The Physiology of Fishes. New York, Academic Press, pp. 367–385.

Fox, H. M., and Vevers, H. G. 1960. The Nature of Animal Colours. New York, Macmillan.

Fujii, R. 1969. Chromatophores and pigments. *In*: Hoar, W. S., and Randall, D. J. (eds.), Fish Physiology, Vol. 3. New York, Academic Press, pp. 307–353.

Hamilton, W. J. III, and Peterman, R. M. 1971. Countershaded colorful reef fishes, *Chaetodon lunula*: concealment, communication, or both. Anim. Behav., 19(2):357–364.

Hawkes, J. W. 1974. The structure of fish skin. 2. The chromatophore unit. Cell. Tissue Res., 149(2):159–172.

Herring, P. J. 1967. The pigments of plankton at the sea surface. Aspects of Marine Zoology, Symp. Zool. Soc. Lond., (19):215–235.

Iwata, K. S. and H. Fukuda. 1973. Central control of color changes in fish. *In*: Chavin, W. (ed.), Responses of Fish to Environmental Changes. Proceedings. U.S.-Japan Seminar, Tokyo, 1970. Springfield, Ill., C. C Thomas, pp. 316–341.

Kishimoto, H. 1974. Changing color patterns in a labrid fish, *Stethojulis interrupta*. Jap. J. Ichthyol., 21(2):101–107.

Lagler, K., Bardach, J., and Miller, R. 1962. Ichthyology. New York, John Wiley & Sons, pp. 120–130.

Lorenz, K. 1962. The function of colour in coral reef fishes. Proc. R. Inst., 39: 282–296.

Odiorne, J. M. 1957. Color changes. *In*: Brown, M. E. (ed.), The Physiology of Fishes, Vol. 2. New York, London, Academic Press, pp. 387–401.

Parker, G. H. 1948. Animal Colours and their Neurohumours. London, Cambridge University Press.

Peterman, R. M. 1971. A possible function of coloration in coral reef fishes. Copeia, 1971(2):330–331.

Randall, J. E., and Randall, H. E. 1960. Examples of mimicry and protective resemblance in tropical marine fishes. Bull. Mar. Sci. Gulf Carib., 10:444–480.

Rasquin, P. 1958. Studies in the control of pigment cells and light reactions in recent teleost fishes. Bull. Am. Mus. Nat. Hist., 115:1–68.

Simon, H. 1971. The Splendor of Iridescence; Structural Colors in the Animal World. New York, Dodd & Mead.

Springer, V. G., and Smith-Vaniz, W. F. 1972. Mimetic relationships involving fishes of the family Blenniidae. Smithsonian Contrib. Zool., 112:1–36.

Topp, R. W. 1970. Behavior and color change of the rudderfish, *Kyphosus elegans*, in the Gulf of Panama. Copeia, 1970(4):763–765.

Waring, H. 1963. Color Change Mechanisms of Cold-Blooded Vertebrates. New York, Academic Press.

Wickler, W. 1968. Mimicry in Plants and Animals. London, Weidenfeld and Nicolson.

Luminescence

Anctil, M. 1972. Stimulation of bioluminescence in lanternfishes (Myctophidae). II. Can. J. Zool., 50(2):233–237.

Bassot, J. M. 1966. On the comparative morphology of some luminous organs. *In*: Johnson, F. H., and Haneda, Y. (eds.), Bioluminescence in Progress. Princeton, N.J., Princeton University Press, pp. 557–610.

Cohen, D. M. 1964. Bioluminescence in the Gulf of Mexico anacanthine fish *Steindachneria argentea*. Copeia, 1964(2):406–409.

Cormier, M. J., Crane, J. M. Jr., and Nakano, Y. 1967. Evidence for the identity of the luminescent systems of *Porichthys porossimus* (Fish) and *Cypridina hilgendorfii* (Crustacean). Biochem. Biophys. Res. Comm., 29(5):747–752.

Crane, J. M. Jr. 1965. Bioluminescent courtship display in the teleost *Porichthys notatus*. Copeia, 1965(2):239–241.

———. 1968. Bioluminescence in the batfish *Dibranchus atlanticus*. Copeia, 1968(2):410–411.

Haneda, Y., and Tsuji, F. I. 1971. The source of light in luminous fishes, *Anomalops* and *Photoblepharon*, from the Banda Islands. Sci. Rep. Yokosuka City Mus., (18):18–28.

Harvey, E. N. 1931. Stimulation of adrenalin of the luminescence of deep-sea fish. Zoologica, 12:67–69.

———. 1957. The luminous organs of fishes. *In*: Brown, M.E. (ed.), The Physiology of Fishes, Vol. 2. New York, Academic Press, pp. 345–366.

Hastings, J. W. 1971. Light to hide by: ventral luminescence to camouflage the silhouette. Science, 173:1016–1017.

Idyll, C. P. 1964. Abyss. New York, T. Y. Crowell.

Johnson, F. H., and Haneda, Y. 1966. Bioluminescence in Progress. Princeton, N.J., Princeton University Press.

Marshall, N. B. 1954. Aspects of Deep Sea Biology. New York, Philosophical Library.

———. 1965. The Life of Fishes. London, Weidenfeld and Nicholson, pp. 173–180.

McAllister, D. E. 1967. The significance of ventral bioluminescence in fishes. J. Fish. Res. Bd. Can., 24(3):537–554.

McElroy, W. D., and Seliger, H. H. 1962. Biological luminescence. Sci. Am., Dec. 1962, pp. 76–89.

Morin, J. G., Harrington, A., Nealson, K., Krieger, N., Baldwin, T. O., and Hastings, J. W. 1975. Light for all reasons: versatility in the behavioral repertoire of the flashlight fish. Science, 190:74–76.

Nicol, J. A. C. 1957. Observations on photophores and luminescence in the teleost *Porichthys*. Q. J. Microsc. Sci., 98:179–188.

———. 1969. Bioluminescence. *In*: Hoar, W. S., and Randall, D. J. (eds.), Fish Physiology. Vol. III. New York, Academic Press, pp. 355–400.

O'Day, W. T. 1973. Luminescent silhouetting in stomiatoid fishes. L.A. Nat. Hist. Mus., Contrib. Sci., 246:1–8.

———. 1974. Bacterial luminescence in the deep-sea anglerfish. *Oneirodes acanthias* (Gilbert 1915). L.A. Nat. Hist. Mus., Contrib. Sci., 255:1–12.

Strum, J. 1969. Photophores of *Porichthys notatus*: ultrastructure of innervation. Anat. Rec., 164:463–478.

Tett, P. B., and Kelly, M. G. 1973. Marine bioluminescence. Oceanogr. Mar. Biol. Ann. Rev., 11:89–175.

Sound Production and Reception

Demski, L. S., Gerald, J. W., and Popper, A. N. 1973. Central and peripheral mechanisms of teleost sound production. Am. Zool., 13:1141–1167.

Harris, G. G. 1964. Considerations on the physics of sound production by fishes. *In*: Tavolga, W. N. (ed.), Marine Bio-acoustics. New York, MacMillan, pp. 233–252.

Lanyon, W. E., and Tavolga, W. N. 1960. Animal Sounds and Communication. Washington, D.C., Intelligencer Printing Co., Publ. No. 7, AIBS.

Lanzing, W. J. R. 1974. Sound production in the cichlid *Tilapia mossambica* Peters. J. Fish. Biol., 6:341–347.

Marshall, J. A. 1976. Sound production by freshwater fishes—a bibliography. W. Va. Acad. Sci., 48(1):1–67.

Marshall, N. B. 1967. Sound-producing mechanisms and the biology of deep-sea fishes. *In*: Tavolga, W. N. (ed.). Marine Bio-acoustics, Vol. 2. Oxford, Pergamon Press, pp. 123–133.

Moulton, J. M. 1963. Acoustic behaviour of fishes. *In*: Busnel, R.-G. (ed.), Acoustic Behaviour of Animals. Amsterdam, Elsevier, pp. 665–693.

Schneider, H. 1967. Morphology and physiology of sound-producing mechanisms in teleost fishes. *In*: Tavolga, W. N. (ed.), Marine Bio-acoustics, Vol. 2. Oxford, Pergamon Press, p. 135–158.

Schuijf, A., and Hawkins, A. D. (eds.) 1976. Sound Reception in Fish. Amsterdam, Elsevier.

Tavolga, W. N. (ed.) 1964. Sonic characteristics and mechanisms in marine fishes. *In*: Marine Bio-acoustics, Vol. 1. New York, Macmillan, pp. 195–211.

———. 1971. Sound production and detection. *In*: Hoar, W. S., and Randall, D. J. (eds.), Fish Physiology, Vol. 5. New York, Academic Press, pp. 135–205.

———. (ed.) 1976. Sound reception in fishes. Stroudsburg, Pa. Dowden, Hutchinson, & Ross (distrib. Halstead Press).

Winn, H. E. 1964. The biological significance of fish sounds. *In*: Tavolga, W. N. (ed.), Marine Bio-acoustics, Vol. 1. New York, Macmillan, pp. 213–231.

Electricity

Bennett, M. L. V. 1967. Mechanisms of electroreception. *In*: Cahn, P. (ed.), Lateral Line Detectors. Bloomington, Indiana University Press, pp. 313–393.

———. 1970. Comparative physiology: Electric organs. Ann. Rev. Physiol., 32: 471–528.

———. 1971. Electrolocation in fish. Ann. N.Y. Acad. Sci., 188:242–269.

———. 1971b. Electroreceptors. *In*: Hoar, W. H., and Randall, D. J. (eds.), Fish Physiology, Vol. 5, New York, Academic Press, pp. 347–491.

———, Nakajima, Y., and Pappas, G. D. 1967. Physiology and ultrastructure of electrotonic junctions. III. Giant electromotor neurons of Malapterurus electricus. J. Neurophysiol., 30:209–235.

Black-Cleworth, P. 1970. The role of electric discharges in the nonreproductive social behavior of *Gymnotus carapo*. Anim. Behav. Monogr., 3:1–77.

Brown, M. V., and Coates, C. W. 1952. Further comparisons of length and voltage in the electric eel *Electrophorus electricus*. Zoologica, 37:191–197.

Bullock, T. H. 1973. Seeing the world through a new sense: Electroreception in fish. Am. Sci., 61(3):316–325.

———. Hamstra, R. H. Jr., and Scheich, H. 1972. The jamming-avoidance response of high-frequency electric fish. I. General features. J. Comp. Physiol., 77:1–22.

Grundfest, H. 1967. Comparative physiology of electric organs of elasmobranch fishes. *In*: Gilbert, P. W., Mathewson, R. F., and Rall, D. P. (eds.), Sharks, Skates and Rays. Baltimore, Johns Hopkins, pp. 399–432.

Heiligenberg, W. 1973. Electrolocation of objects in the electric fish, *Eigenmannia* (Rhamphichthyidae, Gymnotoidei). J. Comp. Physiol., 87:137–164.

Hoar, W. S., and Randall, D. J. (eds.). 1971. Fish Physiology, Vol. 5. Sensory Systems and Electric Organs. New York, Academic Press.

Hopkins, C. D. 1972. Sex differences in electric signalling in an electric fish. Science, 176:1035–1037.

———. 1974. Electric communication in fish. Am. Sci., 62(4):426–437.

Kalmijn, A. J. 1971. The electric sense of sharks and rays. J. Exp. Biol., 55: 371–383.

———. 1974. The detection of electric fields from inanimate and animate sources other than electric organs. *In*: Fessard, A. (ed.), Handbook of Sensory Physiology, 3(3). Berlin and New York, Springer Verlag, pp. 147–200.

Knudsen, E. I. 1975. Spatial aspects of the electric fields generated by weakly electric fish. J. Comp. Physiol., 99(2):103–118.

Lissman, H. W. 1958. On the function and evolution of electric organs in fish. J. Exp. Biol., 35:156–191.

———, and Machin, K. E. 1958. The mechanism of object location in *Gymnarchus niloticus* and similar fish. J. Exp. Biol., 35:451–486.

———, and Mullinger, A. M. 1968. Organization of ampullary electric receptors in Gymnotidae (Pisces). Proc. R. Soc., B169:345–378.

McCleave, J. D., Rommel, S. A., and Cathcart, C. L. 1971. Weak electric and magnetic fields in fish orientation. Ann. N.Y. Acad. Sci., 188:270–282.

Moffler, M. D. 1972. Plasmonics: communication by radiowaves as found in Elasmobranchii and Teleostii fishes. Hydrobiologia, 40(1):131–143.

Moller, P., and Bauer, R. 1973. "Communication" in weakly electric fish, *Gnathonemus petersii* (Mormyridae). II. Interaction of electric organ discharge activities of two fish. Anim. Behav., 21:501–512.

Mullinger, A. M. 1964. The fine structure of ampullary electric receptors in *Amiurus*. Proc. R. Soc., B160:345–359.
Murray, R. W. 1967. The function of the ampullae of Lorenzini of elasmobranchs. *In*: Cahn, P. (ed.), Lateral Line Detectors. Bloomington, Indiana University Press, pp. 277–293.
Rommel, S. A., Jr., and McCleave, J. D. 1972. Oceanic electric fields: perception by American eels? Science, 176:1233–1235.
Scheich, H., and Bullock, T. H. 1974. The detection of electric fields from electric organs. *In*: Fessard, A. (ed.), Handbook of Sensory Physiology, 3(3). Berlin and New York, Springer Verlag, pp. 201–256.
Srivastava, C. B. L., and Szabo, T. 1972. Development of electric organs of *Gymnarchus niloticus* (Fam. Gymnarchidae): I. Origin and histogenesis of electroplates. J. Morphol., 138(3):375–385.
———. 1973. Development of electric organs of *Gymnarchus niloticus* (Fam. Gymnarchidae): II. Formation of spindle. J. Morphol., 140(4):461–465.
Suga, N. 1967. Electrosensitivity of specialized and ordinary lateral line organs of the electric fish, *Gymnotus carapo*. *In*: Cahn, P. (ed.), Lateral Line Detectors. Bloomington, Indiana University Press, pp. 395–409.
Szabo, T., Kalmijn, A. J., Enger, P. S., and Bullock, T. H. 1972. Microampullary organs and a submandibular sense organ in the freshwater ray, *Potamotrygon*. J. Comp. Physiol., 79:15–27.
Szamier, R. B., and Wachtel, A. W. 1969. Special cutaneous receptor organs of fish: III. The ampullary organs of *Eigenmannia*. J. Morphol., 128:261–289.
———. 1970. Special cutaneous receptor organs of fish. VI. The tuberous and ampullary organs of *Hypopomus*. J. Ultrastruct. Res., 30:450–471.
Wachtel, A. W., and Szamier, R. B. 1966. Special cutaneous receptor organs of fish: The tuberous organs of *Eigenmannia*. J. Morphol., 119:51–80.

11
SENSORY FUNCTION IN THE LIFE OF FISHES

Vision

Although the eyes of fishes (Fig. 4–2) are built on the same basic plan as those of other vertebrates, many structural adaptations suit them for function in the range of ecological situations exploited by fishes. Remember that the fish lives in a medium having greatly different optical properties from those of the atmosphere. Depending on the angle of incidence of light, a calm water surface can reflect up to 80% of light striking it. If the water is rough there is great variation in the transmission of light regardless of the angle of incidence. The bending of light rays entering water is such (approx. 48.6°) that a fish in water with a perfectly smooth surface views objects above the water through a circle subtended by a 97.2° cone above each eye (Fig. 11–1). Nearly all objects from horizon to horizon appear in the circle, which is surrounded by the reflective undersurface seen beyond the limits of the cone. In rough water the circular window in the surface is broken up and light transmitted through ever-changing patterns. Water absorbs light rapidly and differentially with red, for instance, attenuating rapidly, and blue penetrating to relatively great depths. Furthermore, various species of fish can cope with life at the bright surface of the waters, in brilliant coral reefs, in dimly lit caves, sheltered forest streams, dark bogs, and ocean depths where sunlight disappears and the only light is that of bioluminescent creatures. Some even invade the land or commonly peer above the water's surface and encounter the problems that aerial vision brings. Some, of course, subjugate sight to other means of sensing the environment, and have reduced powers of vision.

Because streamlining is a major consideration in the functional morphology of fishes, eyelids, except for the nictitating membranes of some elasmobranchs, are not developed, and the eyes of most are placed on the sides of the head. That placement, as shall be seen, does not preclude binocular vision in certain segments of the visual field, for adaptations of various parts of fish eyes allow for it. Some species specialize in the inspection of more restricted parts of their surroundings and, for example, have eyes set forward for binocular front vision (*Gigantura* and others) or upward for binocular vision (*Argyropelecus*

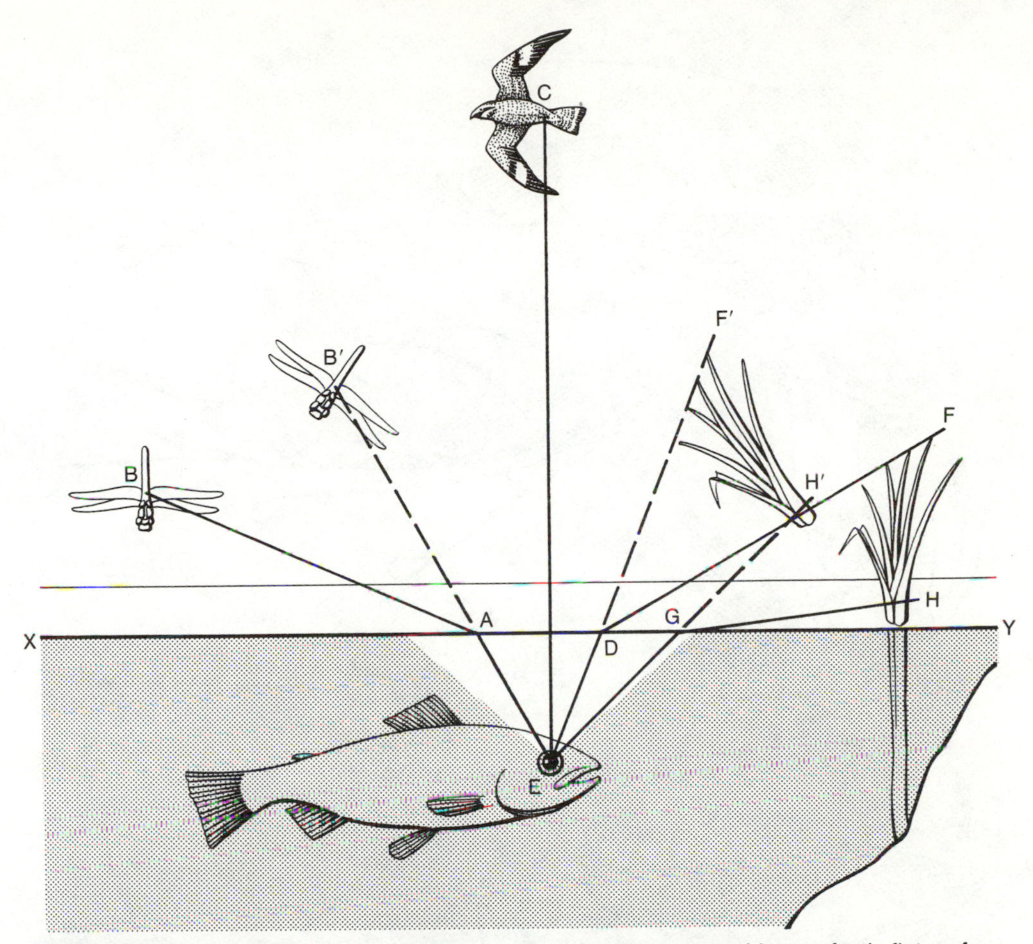

FIGURE 11–1. Diagram showing refraction of light entering water with a perfectly flat surface (*XY*). Because of the bending of the light rays, the fish's eye (*E*) does not receive light striking the surface above the shaded area. The bird at *C*, directly above *E*, is seen in its actual position. The insect at *B* is perceived as if at *B'*, and the angles *EDF* and *EGH* cause the plant to be seen as if the top were at *F'* and the bottom at *H'*.

and others). Some species have the eyes positioned for a wider field of vision below (*Hypophthalmichthys*) or above, as in many topminnows that feed from the surface or many bottom fishes that have no need of vision below. The four-eyed fish of the ocean depths, *Bathylychnops*, has small auxiliary eyes that view the lower scene while the main section of the eye looks obliquely upward. Some bottom-living stargazers have eyes on short stalks (Fig. 11–3).

In most fishes appreciation of the maximum field of view is achieved by the placement of the spherical lens so that it bulges through the opening of the pupil and nearly touches the cornea. The lens can thus gather light practically from an entire hemisphere. This condition is parallelled by mammals in meadow mice but, whereas the spherical lens of fish is adapted to the 1.33 refractive index of water,

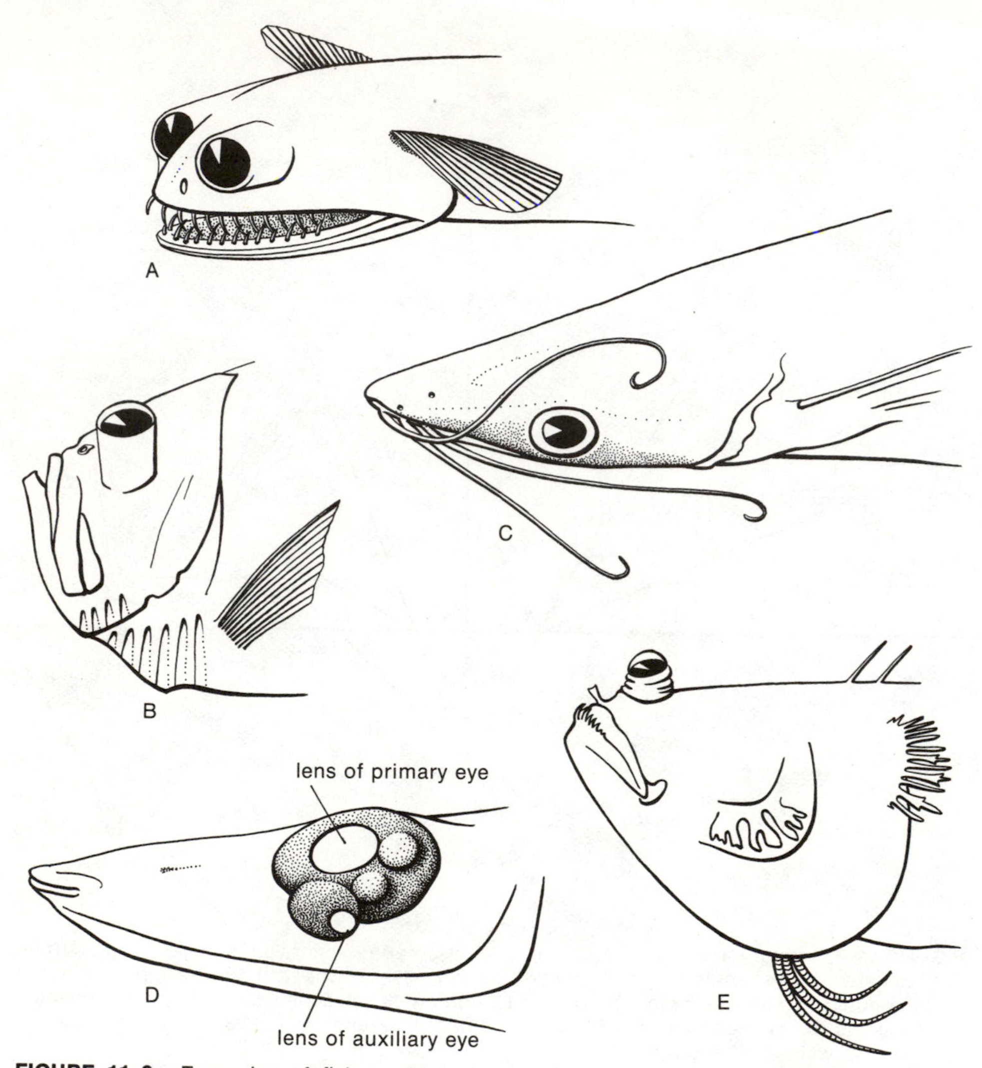

FIGURE 11–2. Examples of fishes with unusual eyes. *A, Gigantura; B, Argyropelecus; C, Hypophthalmus* (eyes directed obliquely downward); *D, Bathylychnops; E, Dactyloscopus.*

and visual acuity can be great at close range, the mouse sacrifices acuity for the detection of movement throughout the greatest possible field. The fish cornea is usually of a single layer with a refractive index approximating that of water so it does not bend the light, whereas the lens is constructed so as to have a high refractive index of about 1.67.

Although the lens of the typical teleost eye is spherical, or nearly so, the vitreous chamber is not, giving the retina the shape of an ellipsoid (Fig. 11–3*A*). The effect achieved is that relatively distant objects lateral to the fish are in focus and close objects are not. Anteriorly, near objects in the binocular field are in better focus than more

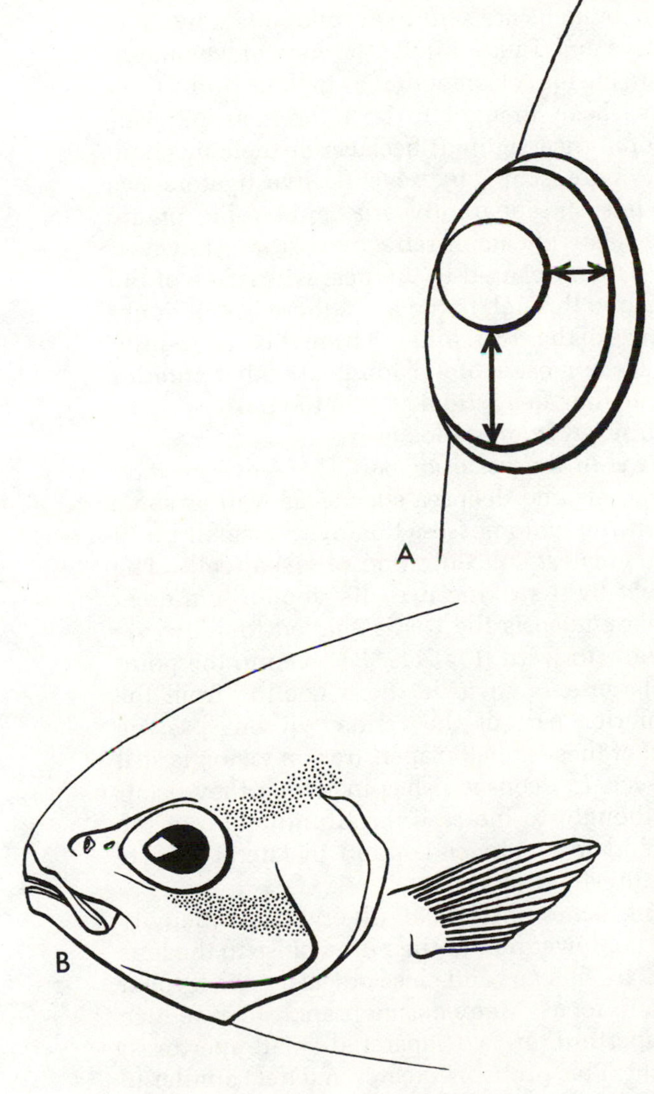

FIGURE 11–3. *A*, Diagram of eyeball shape and placement of lens in relation to retina in many teleosts. Close objects in the anterior field can be in sharp focus, whereas the lateral field is adapted to more distant vision. *B*, Diagram of elliptical eye shape (Girellidae) that allows for a large anterior field of vision by a laterally placed eye. Directly in front of the eye is a groove that facilitates forward vision.

distant objects. Accommodation to distant vision in the anterior field is accomplished by moving the lens posteriorly by means of the retractor lentis muscle. Fishes do not have equal powers of accommodation. Most marine teleosts studied by Drs. Somiya and Tamura were found to have well developed, triangular lens muscles, and accommodated very well. A few freshwater species in their study (largemouth bass, bluegill, and snakehead) shared the well developed muscle and good accommodation, but most freshwater fishes had less well developed muscles and moderate powers of accommodation. The lens muscle appears to be rod-like in many freshwater species, and in some with poor accommodation (*Anguilla*, some catfishes) the muscle is fine and feeble. Among marine species, *Mugil cephalus* was noted as having a nonfunctional lens muscle, and a few others showed little or no accommodation. Although elasmobranchs are said to accommodate by moving the lens forward, Somiya and Tamura noted no lens movement or deformation of the eyeball in the four elasmobranchs in their study. Generally, elasmobranchs have been thought to be hypermetropic (farsighted), and teleosts myopic (nearsighted) because of their mechanisms for accommodation. Retinoscopy by several investigators has demonstrated that many teleosts apparently are emmetropic or are slightly farsighted, having relatively small refractive errors. However, because some researchers have measured to the nearest surface of the retina and some have measured through to the back, there is still doubt as to the degree of significance of the error. Judging from the distribution of visual cells in the retina, the most acute vision is aimed at anterior objects. Cones are most numerous in a retinal area in the posterior part of the eye, where images from in front are focused.

If a fovea is present, it is in the posterior part of the eye. Several species of teleosts, both shallow and deepsea species as well as some freshwater fishes, have been shown to possess the fovea, a small pit in the retina at the site of the greatest concentration of visual cells. The shape of the pit is such that light striking it is distributed to a great number of receptors. In many teleosts the iris is shaped to allow the greatest possible oblique view forward (Fig. 11–3*B*), even to the point of providing an opening large enough to let light coming from the lateral field strike the anterior part of the retina without passing through the lens. The effect of these aphakic apertures on vision is still under investigation. However, in deepsea fishes in which they occur around the lens, they are thought to increase the illumination of the central part of the retina and to aid in perception of lateral objects under conditions near the threshold of vision.

In some fishes, including skates, the eyeball diverges considerably from the spherical, placing the lower part of the retina close to the lens so that distant objects above are in focus and close objects at the level of the eye or below it are also in focus. Many elasmobranchs can reduce the pupil to a very small aperture (or two separated small apertures) when adapting to bright light. This probably creates an effect similar to a pinhole camera, giving reasonably good focus to both close and far objects. In the barrel-shaped eyes of certain deepsea species the immobile lens allows for no accommodation, but the retina is specialized by division into two parts so that objects at two different distances can be seen in good focus.

Regulation of light as it enters or after it enters the eye and adaptation to light or dark are accomplished by several means. One means of regulating light entering the eye is to swim to or away from the source of illumination. Some fishes have pigment in the cornea or lens that acts as a filter and eliminates certain wavelengths. Many benthic species have specially constructed corneas containing, especially dorsally, a layer of lamellae, which, because of their regular arrangement and thickness, cause iridescence when illuminated from above; this reduces the light that enters the eye. Most elasmobranchs and a few teleosts have contractile irises that can control the amount of light entering the eye. Contraction of the iris is usually more rapid than dilation. Some teleost species have mobile pupils under nervous control, but the eels of the genus *Anguilla*, like elasmobranchs, appear to have irises that respond directly to light. Most rays, many flatfishes, especially Bothidae, stargazers, and the armored catfishes (*Plecostomus*) have a pupillary operculum that can expand and cut off most of the light reaching the pupil. Some sharks have nictitating membranes that can be drawn across the eye to reduce excessive illumination.

A choroidal tapetum lucidum is present in most elasmobranchs and a retinal tapetum is present in many freshwater teleosts. These tapeta involve guanine crystals as reflectors, which are as efficient as a good mirror. A different type of choroidal tapetum consisting of reflective fibers is seen in some marine fishes. The tapetum acts to reflect light that has passed through the visual cells back into the system, and is an effective adaptation for sight in dim light (scotopic vision). In sharks the cells in which guanine provides a reflective surface are disposed parallel to the plane of the retina throughout the back, or fundus, of the eye, but are oblique peripherally, so that they are always perpendicular to the light entering the eye. Associated with the guanine reflectors is the pigment melanin, which can migrate along the reflectors and mask the reflectivity. This has two possible functions; one is to adapt the eye for vision in bright light, while the other may be to reduce eye shine and make the fish (especially sharks) less conspicuous. Extraneous light is prevented from entering the eyeball through its walls by the stratum argenteum, which is a reflective layer outside the choroid. This is especially well developed in fishes with translucent tissue around the orbits. Reflective layers are common in the irises of teleosts.

Light and dark adaptation in teleosts is largely accomplished by retinomotor movements of pigment and visual cells (Fig. 11–4). Pigment cells in the outer layer of the retina contain processes through which melanin can move to or from the outer parts of the visual cells. Under bright illumination the eye adapts by movement of melanin toward the visual cells, and by movement of the outer segments of the rods into the pigmented area where they are shielded from the light. In dim light the pigment is drawn back and the contractile or myoid part of the rods pulls them away, exposing them again. Movement of the cones is opposite that of the rods, but the cones are not usually hidden by the pigment. In wrasses (Labridae) a red pigment is prominent in the pigment cells, which also contain melanin. Cylinders of pigment are extended to screen rods and long cones, and permit only red light to reach these visual cells during light adaptation (photopic vision). In dark adaptation the rods shorten and are not screened by the red

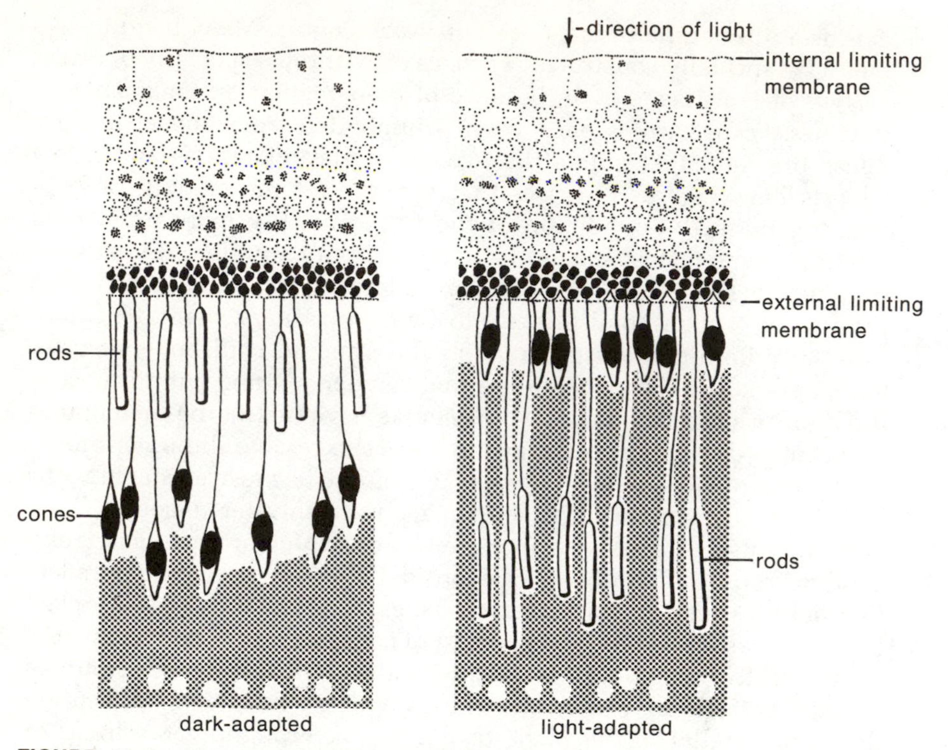

FIGURE 11–4. Diagram illustrating movement of rods, cones, and pigment in the retina of teleosts. In light adaptation (*right*) the rods are moved away from the light and are protected by forward movement of pigment.

cylinders. The time required for pigment movement in the tapetum of sharks or the shifting of pigment and visual cells in teleosts is considerable—about 2 hours for advancing or receding in the shark tapetum, and about 30 minutes for light adaptation and 1 hour or more for dark adaptation in teleosts (*Oncorhynchus*).

Visual cells of fishes include rods, cones, and double cones, with subtypes in various species, some having both long and short single cones. Furthermore, the cones within some species can carry blue, green, or red pigments. This variety leads to the supposition that certain cells are specialized as to the wavelengths and intensities of light to which they best respond. Rods function in dim light and outnumber the cones in those fishes with a duplex retina. Most elasmobranchs and deepsea teleosts have either pure rod retinas or retinas with only a few cones. Cones are set in regular patterns in many species, and are often arranged in a mosaic of squares throughout the central part of the retina. This regular arrangement ensures that the most important parts of the retina are supplied with each type of specialized cell.

In species that commonly feed in dim light numerous rods may connect with only a few bipolar cells, which in turn connect to a single ganglion cell. This adaptation aids in summing the impulses and increases sensitivity in subdued light. The numbers of rods in nocturnal or deepsea species can reach up to 20 million per mm^2. Some of these fishes have retinas that absorb 90% or more of the light striking them, so the thresholds at which they can detect light are probably lower than that of man, in which the retina absorbs about 30% of the blue-green light striking it.

The photosensitive pigments in visual cells are purplish rhodopsins and rose-colored porphyropsins. These complex chemicals consist of an aldehyde of vitamin A (a retinene) and a protein (an opsin). When exposed to light, changes in these excite nerve cells and result in the sensation of vision. A broad generalization is that freshwater fishes have mainly porphyropsin in their retinas and marine fishes have mainly rhodopsins. Actually, many freshwater species have mixtures of the two pigments. Many cyprinids and a few others have rhodopsin only. There is evidence of seasonal changes in the proportions of the two types of pigments in certain minnows. Some species of anadromous fishes (*Petromyzon, Oncorhynchus*) may have a preponderance of one pigment over the other at certain life history stages. Pacific salmon are known to switch from rhodopsin to porphyropsin on the spawning migration. Among freshwater fishes, benthic species tend to have more rhodopsin and surface species more porphyropsin. Some deepsea fishes have a pigment called "visual gold" or chrysopsin in the retina.

Dependence upon sight varies greatly among fishes, and a general clue to the use of the eye is its relative size. Sight-feeding diurnal fishes such as trout, bass, and sunfishes have prominent eyes with a diameter equal to about one fifth or one sixth the length of the head. The pike is a daytime predator but has an extremely long head, not an especially small eye. Crepuscular or nocturnal fishes that hunt by sight, and at least partly sight-dependent deepsea fishes, tend to have eyes that are one third to one half the head length. Those night-active or deepwater fishes that depend largely on other senses such as olfaction, taste, the acoustico-lateralis sense, or electroreception tend to have very small eyes. Obviously, the quality of eyes of the same size can differ markedly from species to species. The number, disposition, and types of visual cells, connections of the cells to the optic neurons, mechanisms for accommodation, effectiveness of the tapetum lucidum, and so on determine the efficiency of the eye.

Some of the most remarkable adaptations of the eye are seen in abyssal fishes. In some, the parts of the eye retain approximately the same proportions as those of the eyes of shallow-water species, and are increased in size. In others the lens becomes greatly enlarged, so that an eye in proportion to the lens size would be as large as the head. In these the eyes are barrel-like or tubular, and are equipped to gather the maximum amount of light in the huge lens and focus it on a small part of the retina. These eyes are fixed, lack any mechanism for accommodation, and therefore receive clear images from one direction only. Some have two effective parts of the retina, and can form images from two directions. One of the strangest eye modifications of deepsea fish is that

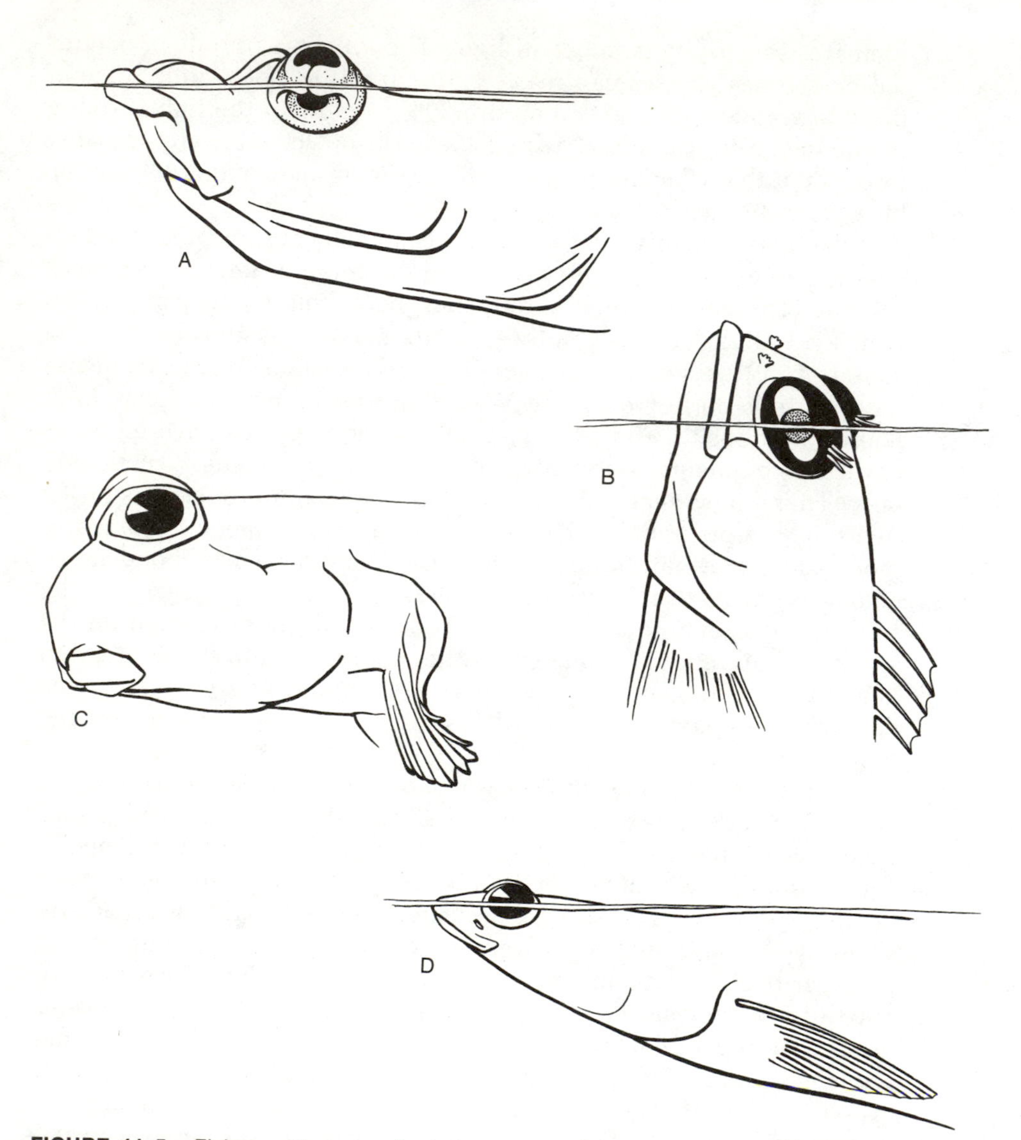

FIGURE 11–5. Fishes with eyes adapted to aerial vision. *A, Anableps,* in which the eye is modified to have an aerial *(above)* and an aquatic *(below)* aperture; *B, Dialommus,* in which the anterior part of the eye is held above the water surface; *C, Periophthalmus,* which spends much time completely out of water; *D, Rhinomugil corsula*, which often swims with the eyes above the water surface.

of *Ipnops*, which has no eyeball or lens, but has visual cells disposed in two large areas on top of the flattened head.

A few fishes are adapted to aerial vision, and at least two have eyes greatly modified so that both aerial and aquatic vision is good (Fig. 11–5). The four-eyed fish of C. America *(Anableps)* swims at the surface with the upper half of the eye exposed to the air. The iris is modified so that two flaps divide the pupil of the eye at the level of the water surface, and the lens is egg-shaped so that light entering from above the surface passes through a short axis, and that entering from the water passes through a long axis. The upper part of the cornea is thicker and is curved more than the lower, and the lower part of the retina has more visual cells than the upper. *Anableps* must submerge its eyes frequently to prevent drying. A blenny (*Dialommus*) of the Galapagos Islands has a similar adaptation, except that the eye is elongate from front to back, the dividing septum of the iris is vertical, and the anterior part of the eye is held above water as the fish clings vertically to the edge of tide pools.

Mudskippers (*Periophthalmus*), which spend much of their time on mud flats or among mangrove roots, have prominent eyes set high on the head. The lens of the eye is flattened more than in most fishes, so that aerial vision is good. A pocket below the eye carries moisture into which the eye can be retracted to prevent drying. A mullet of India, *Rhinomugil corsula*, shows a measure of evolutionary convergence with *Anableps* in the shape of body and head and the placement of the eyes. *R. corsula* can hold its eyes above water as it swims at the water surface. Aerial vision appears to be good, for the slightest movement of an observer is sufficient to send this nimble fish below the surface to pop up again at a safe distance. The archerfish, *Toxotes jaculator*, although it keeps its eyes below the surface, has excellent aerial vision and can squirt water at insects and other small prey with great accuracy.

Although many fishes in several taxonomic groups are blind, most retain vestigial eyes beneath the skin. A few are completely eyeless, such as some of the whalefishes (Cetomimidae) and *Phreatichthys andruzzii* from the Somali Republic. The latter lacks optic nerves and has markedly reduced optic lobes. "Blindness" does not necessarily mean a complete lack of sensitivity to light, for light sensors exist in the skin of some blind species and are associated with the central nervous system of many. Photoreceptors in the caudal region of lamprey ammocoetes appear to be part of a system that triggers burrowing movements when the animal is exposed to light. Blind cavefishes possess photoreceptors in the skin, as do hagfishes.

Probably the most important site of extraocular photoreception is the pineal organ and associated structures. Electrophysiological studies have shown receptor potentials in the pineal nerve following stimulation by light. The role of the pineal complex in regulating chromatophores has been demonstrated in many experiments. Cave-dwelling species initiate swimming movements away from a light source stimulating the pineal region. Most fishes fall into three categories regarding illumination of the pineal region, as shown by Drs. Charles M. Breder and P. Rasquin. Fishes of group 1 have transparent or translucent tissue covering the pineal complex and usually react positively to light. In group 2 the fishes have an opaque covering over the pineal and usually are light-negative. Group 3 fishes can control entry of light to some extent by the action of chromatophores above the pineal complex; these vary in their reaction to light.

The pineal window in some sharks can allow the transmission of up to seven times as much light as adjacent parts of the head. In some species the threshold for detection of light at this site is below the energy level of moonlight. Tunas have, in addition to a window, a tube-like translucent structure that directs light to the dorsal part of the brain. About 25% of the incident light can be transmitted to the vicinity of the apparently photosensitive pineal. Experiments with the eyeless fish, *Phreatichythys*, show that its central nervous system is sensitive to light.

Experimental study has credited many fish species with exceptionally acute vision and a good ability to recognize a variety of shapes. In conditioning programs, fishes have been taught to respond to block figures and to such closed figures as squares and triangles, as well as to letters of the alphabet (sometimes forming words). Fishes also have been found to respond to open figures such as right, obtuse, or acute angles, and to colors.

Acoustico-Lateralis System

The vestibular system or ear of the fish and the lateral line system are generally considered together because of their structural similarity and obvious close relationships of the end organs, the neuromasts. The functions of the two systems are similar but differ in the type of stimuli to which they respond. The ear responds to displacement of the head, angular acceleration, and sound, whereas the lateral line responds mainly to movements of water, both strong and slight.

Because water is so much denser than air (about 100 times as dense) a greater amount of energy is required to cause sound in water, but when sound is propagated it travels at about five times the speed of sound in air, and is not rapidly attenuated. Sound energy in water consists of both particle displacement and compression waves, the former being important close to the source of the sound. Low frequency sound propagation is relatively easier in water than propagation of high frequencies.

Many natural physical processes cause sound in water. Such factors as earthquakes, volcanism, winds and the water movements caused by them, precipitation, movement of bottom materials caused by currents or waves, and action of ice provide background noise. Man-caused sound may be evident in certain places, and may consist of noise from ships and submarines and from industrial operations both on shore and sea-based. Biological sound may result from interaction of animals with the physical environment or from sound production, incidental or deliberate, by the animals. Marine mammals, molluscs, crustaceans, and fishes are all sound producers.

One of the most interesting phenomena concerning sound in the ocean is the "sound-scattering layer." Apparently it consists of great numbers of mesopelagic animals from which sound impulses sent by sonic devices of ships and submarines are reflected.

Function of the Membranous Labyrinth. The inner ear of fishes is active in the detection of changes in position due to acceleration in any plane, of position changes in respect to gravity (equilibration), and of the pressure waves or water displacements called sound. In addition, the organ appears to be involved in the maintenance of muscle tone. Angular acceleration, also referred to as "dynamic equilibrium," is detected by the ampullar neuromasts of the semicircular canals. The hair cells in these neuromasts, like the hair cells in other parts of the acoustico-lateralis system, are cylindrical, with two types of cilia-like structures projecting from them (Fig. 11–6). Each has a single kinocilium, which is longer, larger, and more complex than the numerous, shorter stereocilia that form a sloping series with the longest next to the kinocilium. Hair cells are innervated by afferent and efferent nerve endings. Because the cells in a crista or macula are usually definitely oriented with the slope of stereocilia in the same direction, researchers have assumed that deformation of the hairs in one direction may inhibit activity of the cell and movement in the opposite direction may be excitatory.

The gelatinous cupula into which the hairs of an ampullary neuromast protrude fills much of the ampulla, and is deformed by the inertia of the endolymph as the head is rotated. Rotation on a vertical axis activates the hair cells of the horizontal semicircular canals. The vertical canals respond to rotation in any direction, but there appears to be scope for greater stimulation by movement on horizontal axes.

Detection of gravity, or "static equilibrium," involves the utriculus

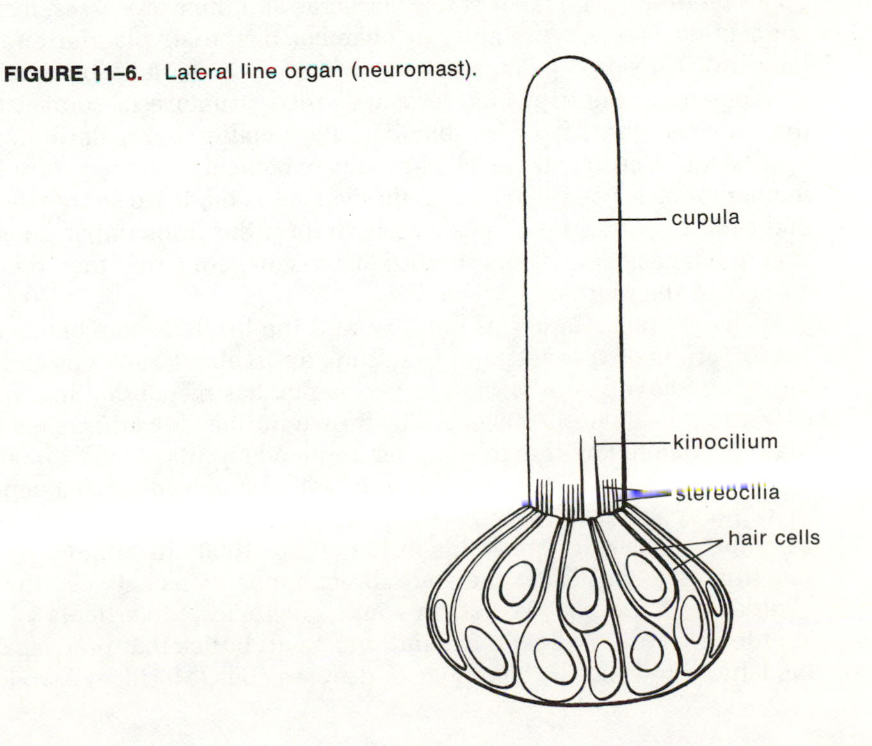

FIGURE 11–6. Lateral line organ (neuromast).

in most fishes, although there is evidence that the utriculus is active in sound perception in some species and that in others the lagena or part of the sacculus is involved in equilibration. Impulses from the utriculus control a series of reflexes governing posture. Some that are easiest to observe are the curling of fins and rolling of eyes as a live specimen is held at an angle. Change of position in relation to gravity is apparently detected by means of deformation of sensory hairs as the utricular otolith (lapillus) shifts. If a fish is rotated to an angular position, there will be initial stimulation of semicircular canal cristae as well as of the utricular maculae, but when the endolymph of the canals has stabilized the otoliths will continue to deform the sensory hairs.

Sound detection by fishes has engendered much interest on the part of scientists, and at one time there was a prevalent belief that fishes were deaf. However, careful study over the years has shown that fishes are sensitive to sounds, and that in most species the energy is received by the sacculus and lagena. Compression waves and the water displacement of near-field sound move or vibrate the otoliths on the beds of sensory cells. Response to a wide range of frequencies has been measured in fishes but the range of response is directly related to the auditory equipment of the various species. Species with no connection of any kind between the gas bladder and the ear are not as sensitive to sound as those having this connection, nor can they respond to as great a range of frequencies. Among fishes with the otophysic connection the cypriniform and siluriform fishes (Ostariophysi) appear to have the best powers of hearing. In these the Weberian apparatus, consisting of bones modified from the first few vertebrae and their processes, forms a connection between the anterior chamber of the air bladder and the labyrinth (Fig. 11–7). The ossicles through which the vibrations are conducted are the tripus, a crescent-shaped structure in contact with the anterior part of the air bladder, the smaller intercalarium, scaphium, and claustrum. The claustrum is in contact with the walls of the membranous labyrinth, which in these fishes is modified so that the left and right organs coalesce posteriorly to form the sinus impar. In addition, the lagena and lagenar otolith (asteriscus) are larger than the sacculus and the sagitta.

The goldfish, *Carassius auratus*, and the loach, *Nemacheilus barbatula*, are known to respond to sounds up to about 3500 cps and the minnow, *Phoxinus*, in various experiments, has responded to a range of from 20 to 5000 to 7000 cps. The brown bullhead, *Ictalurus nebulosus*, has been reported to have upper frequency limits of from 10,000 to 13,000 cps. Sensitivity is usually greatest in much lower frequencies than the upper limits. For instance, the goldfish and brown bullheads have lowest hearing thresholds in the 200 to 1000 cps range.

In a variety of fishes there are direct connections between the gas bladder and the ear. In herrings and anchovies, diverticula of the bladder enter the skull and expand into small bullae that press against the labyrinth wall. The hakelings or deepsea cods (Moridae) have large

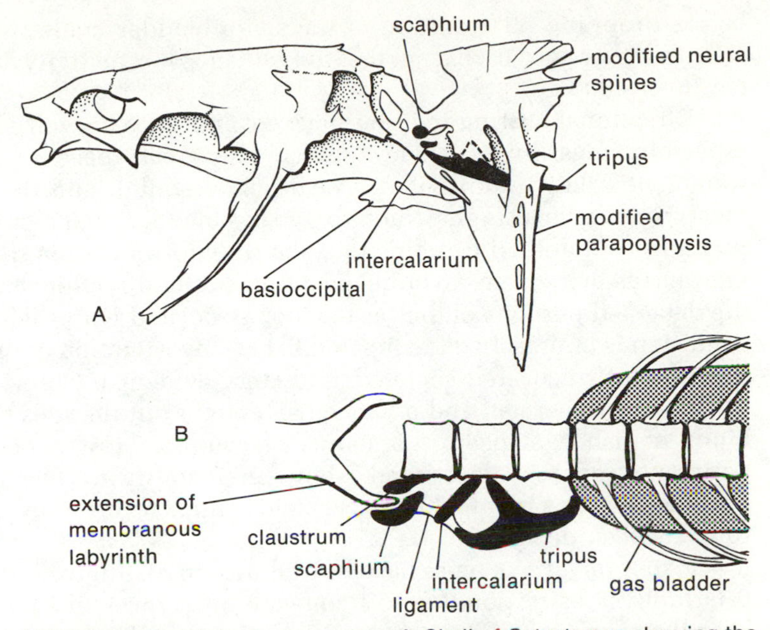

FIGURE 11–7. Weberian apparatus. *A,* Skull of *Catostomus* showing the modified first three vertebrae and the relationship of the left Weberian ossicles (*in black,* claustrum not shown) to them; *B,* diagram showing the relationship of the left Weberian ossicles *(in black)* to gas bladder, vertebrae (here shown unmodified and unfused), and the extension of the sinus impar of the membranous labyrinth.

branches of the gas bladder in contact with the skull and, in the mormyriform fishes, small portions of the gas bladder become separated during early development and are enclosed within the skull in contact with the ear. Featherbacks (Notopteridae), tarpons (Elopidae), porgies (Sparidae), and squirrelfishes (Holocentridae) also have the gas bladder in contact with the ear. Others with the bladder in close relationship to the skull are drums (Sciaenidae) and triggerfishes (Balistidae). Fish with air-breathing chambers in the head, such as climbing perch and snakeheads, maintain some of the atmosphere close to the auditory region.

Fishes without a close connection between the gas bladder and the ear usually respond at a much lower range of sound than those with this otophysic linkage. Responses are elicited either by conditioning the fishes to show a specific behavior upon receiving a sound, or by inserting an electrode into the auditory nerve and measuring microphonic potentials resulting from stimulation of the ear by sound. The highest frequencies causing response in these fishes are mostly below 1000 cps, often in the 500 to 800 cps range. Species without gas bladders appear to be less responsive to sound than those having this organ. In one study that involved implanting electrodes in the ears of a *Tilapia* and a channel catfish, both species had similar sensitivities to vibrations up to 600 cps. The response of the *Tilapia* dropped off above that frequency, but the sensitivity of the catfish increased to 4000 cps before dropping off. Removal of the swim bladder made little difference in either species, except that the catfish lost sensitivity at frequencies over 4000 cps.

Directional hearing in fishes does not appear to be well developed, especially regarding far-field sounds. The great speed of sound in water, the relative transparency of fishes to sound, and the fact that most fishes have a single structure, the gas bladder, that can serve as a pressure receptor, all combine to make sound localization difficult. In mormyrids there seems to be good equipment for directional hearing in the detached portions of the gas bladder associated with each ear. This condition is approached in clupeoid fishes. Investigation of directional hearing has uncovered some contradictory evidence. Observations in the field by fishermen and others have brought out the idea that many kinds of fish can locate the source of sounds. Most controlled experiments, on the other hand, show an inability or low ability to localize sounds. Orientation to near-field sounds received by the lateral line has been demonstrated.

Fishes of various types have been shown to distinguish tones, with ostariophysines responding to frequency differences of from 3 to 5% over the range of 200 to 500 cps. Nonostariophysines are less able to discriminate frequencies with some, such as sharks and eels, responding only to frequency differences of 50% or more. Fishes can also distinguish intensity of sound.

Function of the Lateral Line in Reception of Mechanical Stimuli. Typical neuromast organs of the lateral line system are found not only in canals (Fig. 11–8), but also in superficial positions on the skin. These organs may differ in size and in the number of stereocilia per hair cell, but their function appears to be the same. They respond to deformation by mechanical stimuli, particularly water movement in relation to the fish. Thus, currents from whatever source, the swimming of the fish, water displacements caused by other organisms, and the small displacements caused by sources of sound can be detected.

The free neuromasts are usually arranged in series oriented with the long axis of the fish, and the hair cells are further oriented so that they respond mainly to currents moving along the axis, either backward or forward. Numerous hair cells are covered by each of the cupulae in the lateral line canals. Adjacent hair cells usually show opposing orientation, with the kinocilia oriented headward in some and tailward in the others. Water movement striking the side of the fish sets up an impulse in the mucus of the lateral canal. This causes slight deformation of the cupulae, bringing on a shearing motion over the hair cells and thus stimulating them.

The significance of the lateralis sense is great. This sense allows the fish to detect the presence of predators or prey, to orient properly to the current, to maintain position in a school, and to avoid obstacles. These abilities would seem to be especially important among species of nocturnal habit or those living in caves or the deep ocean. The latter two groups certainly show great modification of the lateral line organs,

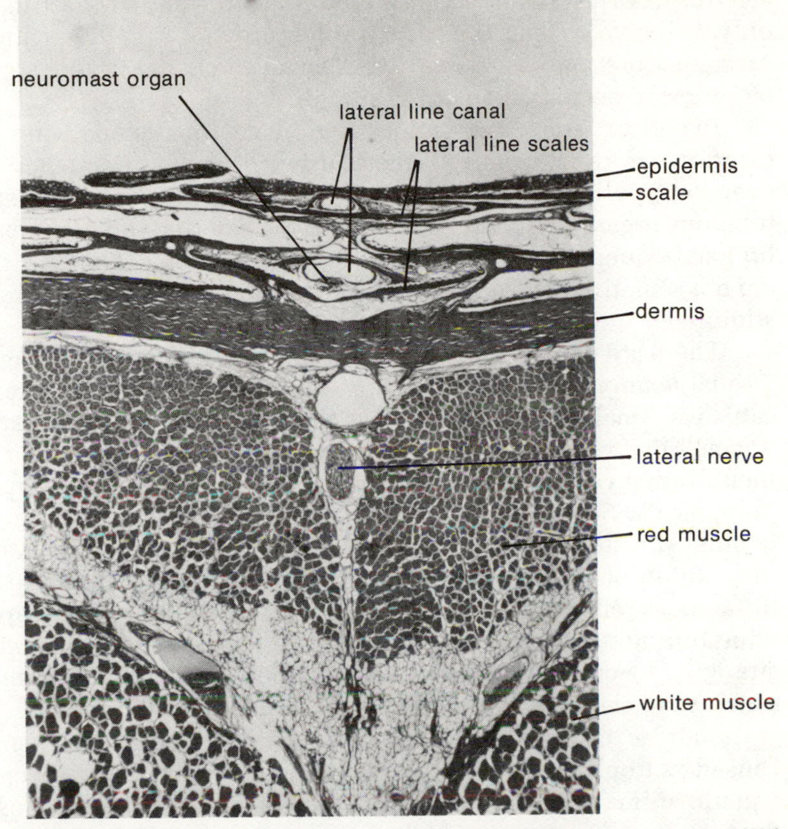

FIGURE 11–8. Photomicrograph showing relationship of lateral line canal to scales, skin, and muscles. (Photo courtesy of Prof. Joseph Wales.)

some even having them placed on ridges or papillae instead of in canals.

Electroreception

The weak electric organs of mormyroids and gymnotoids were once referred to as "pseudoelectric," probably because of lack of sensitive instruments to measure the discharges, although as early as 1841 they were assumed to have an electric function. Once the discharges were confirmed, about 1880, the search began for the function of the organs. Matching the production of electrical fields with the presence of receptors responsive to the fields required clever experimentation and solid scientific reasoning. Because some electroreceptors were found to be sensitive to changes in temperature, they were first considered to be thermoreceptors. Some apparently nonelectric fishes (sharks, rays, catfishes, and others) possess the electrosensory ampullae of

Lorenzini or similar organs. These fishes are said to have passive electrosensory systems because they react to external electric stimuli only. In active systems the electric fishes provide stimuli to which their own electroreceptors are sensitive. The receptors in active systems can, of course, react also to external stimuli.

In general, ampullary organs, equipped with a canal opening to the exterior, are responsive to direct current or to low frequency stimuli over long periods. These are called tonic receptors in contrast to the tuberous organs, which do not open to the exterior and are sensitive to higher frequencies. The tuberous organs are called phasic receptors, are not sensitive to direct current, and become insensitive to prolonged stimuli.

There are various sizes and types of electroreceptors among fishes. The large ampullae of Lorenzini are found in marine sharks, rays, and catfishes. Smaller ampullary organs, called by such names as "small pit organs" or "microampullae," are present in many freshwater fishes, including rays, gymnotoids, mormyroids, catfishes, lungfishes, and polypterids. Some species have as many as three types of ampullary organs with several subtypes, and there are differences in structure and disposition of electroreceptors among species. Presumably, structural differences reflect differences in function, and fishes may be capable of "fine-tuning," having receptors sensitive to different frequencies and strengths of current. Ordinary lateral line organs are not sufficiently sensitive to electricity, so they are not considered to be electroreceptors.

Sources of electrical stimuli are both animate and inanimate, caused not only by activity of electric organs or by other processes of aquatic animals but also by movements of water masses, atmospheric processes, and various geological and electrochemical processes. Volcanoes, earthquakes, lightning, and man's electric and electronic activities all provide stimuli to which electroreceptors can respond.

Functions of electroreception include, but may not be restricted to, location of objects, communication, and navigation. Fishes with active electric systems hold their bodies straight and swim by undulating the dorsal, anal, or pectoral fins. Some species move backward equally as well as forward, the gymnarchids especially tending to test new situations tail-first. The tail-first approach is probably advantageous in that the back ends of the electric organs are somewhat electrically isolated from the rest of the body so that maximum current density is set up around the tip of the tail. Distortions of the field by objects are thus maximized. In addition, only the tail is exposed to the possible dangers of a new situation. The straight posture allows them to set up a symmetrical electric field around themselves, the extent of the field governed by the resistance of the water and the nature of the electric discharge. The field approximates a dipole field in some species and in the young of others. In apteronotids and rhamphichthyids, the anterior three quarters of the fish acts as a distributed source of current while the tail, where the ends of the electric organs are close to the surface and are not surrounded by body fluids, acts more like a point source. This causes the equipotential lines to be somewhat parallel to the body

(Fig. 10–12). Objects encountered in the field, whether good or poor conductors, distort the field. The high frequency receptors sense the change in impedance and the fish reacts appropriately to the information received. Although the range of this system is not great, probably operating best within only a few centimeters of the fish, it is apparently of significance to nocturnal fishes and to those living in turbid waters.

Communication between electric fishes may have significance in reproduction, spacing of individuals, and so on. Range of communication may be from less than 50 cm to nearly 7 meters, depending upon water conditions and the species involved. Some species can quickly change the frequency of their pulses to avoid "jamming" by interfering frequencies.

Passive electrolocation allows fishes to react to the fields that emanate from living organisms in the water, as well as to those from inanimate sources. Activity of muscles sets up very small A.C. fields, and electric organs set up larger fields. D.C. fields from organisms originate from potentials involving the body fluids and the surrounding medium. Wounds are reported to strengthen the D.C. fields, so that a wounded organism can be detected electrically at a greater distance than can an entire organism. Certain sharks have been shown to be able to locate flatfish concealed by covers that allowed passage of an electrical field, but prevented the passage of odors. Many electrosensory species react to magnets or to other inanimate sources of electrical fields, and sharks will readily seek out a dipole field from electrodes set out to imitate the field around a flatfish.

Use of sensitivity to electrical fields for orientation during long distance navigation is suspected in salmon and eels, but most of the evidence is indirect, and the nature of the receiving system is unknown. In an experiment involving tracking of salmon that had electronic devices inserted in their stomachs, some investigators found that the fish would continue or reverse their direction of migration depending upon the polarity of the devices, so that care had to be taken to ensure that they were inserted with the right pole forward.

The possibility of radio communication in fishes has been reported by M. D. Moffler, who found that fishes emit hydronic radio waves that can be detected by a dipole antenna and analyzed by appropriate instrumentation. He tested over 100 species, all of which emitted signals, and found that predators emitted stronger search signals than nonpredators. Poorly-sighted or blind fishes sent strong signals, as did nocturnal species, with signal activity in schools suggesting interspecific communication.

Chemical Senses

Olfaction. The excellent solvent properties of water allow myriads of organic and inorganic substances to be carried in solution and even though many are only sparingly soluble, they may still be detected by the acute olfactory organs of fishes. Biologically significant odors arise from many sources in the environment of fishes, some of which are known to detect the smell of prey, predators, individuals of their own species, and even of specific small streams. Odor-laden water is carried to the olfactory epithelium by currents moving past the motionless fish, by the movement of the fish through the water, or by a pumping or ciliary action in the olfactory sac. Reaction of the fish to the odors depends not only upon the species involved, but upon the previous learning experiences of the individual.

Detection of food is a major function of olfaction in many species that feed in dim light or search through bottom materials or vegetation for edible objects. Most active predators among the bony fishes are primarily sight-oriented hunters but, among the sharks, olfaction is of great importance, along with other chemical senses. Obvious attraction of sharks to baits or to wounded fish upstream from the sharks even in gentle currents has been observed by both scientists and laymen, and the knowledge has been refined by experimentation. Some sharks lost the ability to find food placed into a tank with them when both nostrils were plugged with cotton, but sought and found it when only one nostril was occluded, even though the pattern of search differed from that followed when both nostrils were clear. In many of these experiments, the possibility of sighting the food was removed by concealing it in cheesecloth or some other substance through which odor could penetrate. In others, sharks were blinded but would still perform the typical figure-eight search pattern in locating food.

Some of the most interesting studies have involved the introduction of extracts, dilutions, or washes of various substances into experimental tanks with sharks. Extracts of fish flesh, especially those with a considerable oil content, caused search and feeding activity even when presented in very dilute concentrations. Flesh rotted to foulness appeared to repel sharks to some extent. In studying the response of sharks to living fish, Dr. Albert Tester found that water flowing over intact but distressed or excited fish caused a greater response in the test animals than did water flowing over quiescent fish. Starved sharks were more responsive than well-fed specimens, some detecting the odor of food at a concentration of 0.0001 ppm, a dilution that may be stronger than the undetermined threshold level.

Because of the occasional attacks by sharks upon humans, there has been interest in the attractiveness of human odors to sharks. Human urine appears to be detected by some sharks but causes no particular activity, whereas human blood attracts some of the species tested. In one series of experiments, human sweat caused what Dr. Tester termed "aversion" in sharks at about 1 ppm. Repulsion of sharks by odors has been studied for many years in the hope that a suitable repellent can be found. Certain dyestuffs and copper acetate have shown limited promise, but a repellent based on the poison found in a tropical flatfish appears to be the most effective.

Fishes other than sharks and rays that are known, because of experimentation, to orient to food by olfaction include lampreys, hagfishes, African lungfishes, and many teleosts. Among the latter are numerous minnows, catfishes, eels, perches, wrasse, and cods. Field observations and examination of the relative sizes and degree of development of olfactory organs and olfactory centers have led investigators to believe that certain plankton feeders, of both shallow and deep waters, and some swift predators orient toward food at least partially by olfaction.

Eels of the genus *Anguilla* have especially acute olfaction and have been used successfully in experiments involving conditioned responses to extremely dilute solutions of food extracts or other substances. For example, only a few molecules of p-phenylethyl alcohol in the olfactory sac have been reported to cause a conditioned response in a trained eel. The olfactory organ of the eel is relatively large and is equipped with many folds in the epithelium, which, unlike most fishes, is pigmented.

Orientation to the environment is probably achieved partly through the sense of smell in many fishes, possibly in more than are presently recognized. The experiments of Dr. A. D. Hasler and others have shown that some minnows can be taught to distinguish between the odors of different species of aquatic plants in very dilute solutions, indicating that they might be able to recognize localities by small olfactory cues. The ability to do so has been proven for Pacific salmon (*Oncorhynchus*), which can return to the stream locality from which they migrated months or years earlier. Olfaction seems to play a prominent part in the salmon's homing behavior during the spawning migration. Field experiments have shown that blinded salmon taken from a spawning stream and displaced downstream can make the correct choices of tributaries and return if allowed full use of the olfactory organs, but distribute in a random manner if the nostrils are plugged. Electroencephalographic studies carried out at the University of Washington have shown that strong impulses can be recorded from the olfactory bulb of salmon stimulated by home stream waters, but some react to other natural waters as well. Other fishes, including sunfishes (*Lepomis*), have been shown to be able to locate their home areas by means of the olfactory sense.

Olfaction has been implicated or suspected to be of importance in many aspects of fish behavior. In reproductive behavior, in addition to homing to the spawning stream, the sense of smell is important in location of mates, triggering certain phases of the spawning act, recognition of young, and territorial defense. Olfaction is involved in the social behavior of fishes in many ways, including recognition of members of the same species, collectively in schools or individually. One of the most remarkable implications is in the sensing of alarm substances released from injured individuals of minnows and related fishes. Minute amounts of the substance (*Schreckstoff* in German) issuing from the epidermis of a wounded fish cause almost immediate fright reaction and retreat in members of the same species. A dilution of skin extract of about 2×10^{-11} can be sensed.

Fright reaction is caused in Pacific salmon by rinses or extracts of mammalian skin containing L-serine. This may represent recognition of the odors of potential predators. There is evidence that minnows surviving attacks by predators show fright reactions when exposed to the odor of the predator species.

Taste. Taste receptors in fishes are by no means localized in the oral cavity. They are found there but are also distributed on the lips, in the branchial cavity, on the exterior surfaces of the head and body, and especially on the barbels and certain modified fin rays. As pointed out earlier, there are three kinds of receptors sensitive to taste or a similar chemical sense — taste buds, certain free nerve endings, and the spindle cells that resemble the receptor cells of the taste buds.

Although taste buds may be widespread on the body, there are usually concentrations of them on palatal organs, lips, barbels, and the lower parts of the head. Bullhead catfishes (*Ictalurus*) are good examples of fishes that have taste buds over most of the body and fins but have great concentrations on the barbels. Regardless of location, taste buds are innervated by cranial nerves; the facial nerve (VII) serves the mouth, barbels, head and most of the body, while the glossopharyngeal (IX) and vagus (X) innervate the buds in the pharyngeal region, with the vagus nerve serving some of the body taste buds. The single sensory cells (so-called spindle cells) on the body and fins are innervated by spinal nerves. The searobins (Triglidae), which show a well developed gustatory sense located in the modified lower pectoral rays, have only the special cells in the skin of those rays. These are innervated by the third spinal nerve, as shown by Dr. John Bardach and colleagues. Other fishes with modified fin rays that bear taste buds innervated by cranial nerves are the hakes of the genus *Urophycis*, the rocklings (*Ciliata*), and the gouramies (Belontiidae).

Some sharks are known to react to bitter and sour substances and some react to salt, seemingly through the gustatory sense. Others have been observed to prefer certain food fishes in taste tests, but comparatively little research has been carried out on their sense of taste. Much more is known of the taste capabilities of bony fishes, various species of which are known to react to bitter, sweet, salt, and sour tastes as well as to amino acids. Through electrophysiological studies and training for conditioned responses, various minnows (especially *Cyprinus* and *Phoxinus*) have been shown to respond to numerous substances placed into contact with the taste receptors. Sugar and other sweet substances, acids, amino acids, many salts, quinine and related bitter materials, as well as saliva, milk, and extracts of earthworms, silkworm pupae, and other food items all cause gustatory response. The palatal organ of the carp responds strongly to carbon dioxide. Experimental evidence has shown that taste receptors on various parts of the body have different capabilities and thresholds, and that there are notable specific differences among fishes. Even within the same species the responses can differ between strains, as shown by Drs. J. Konishi and Y. Zotterman for

Japanese and Swedish carp. Japanese carp reacted more strongly than did the Swedish strain to quinine and extracts of worms and silkworm pupae, whereas the Swedish strain responded only weakly to quinine but much more strongly to sucrose than did the Oriental strain.

In testing responses of single nerve fibers, investigators have noted some specialization among the receptors; some react to all four classes of taste, but others only to specific tastes or combinations. Carp appear to have seven, and puffers at least three, kinds of receptors.

The thresholds at which fish can respond to tastes vary from species to species but, generally speaking, fishes react to smaller quantities of sapid substances than man. In minnows, for instance, thresholds for sucrose are from 500 to 900 times lower than for man. The ability of minnows to detect fructose is about 2500 times greater than that of man, and the ability to taste sodium chloride is about 200 times better. The Mexican blind cavefish, *Anoptichthys*, is reported to have a much greater taste sensitivity than minnows — up to thousands of times better. *Phoxinus* can detect sucrose at a concentration of 2×10^{-5} molar and sodium chloride at 4×10^{-5} molar, while *Ictalurus* can taste quinine at about 1×10^{-4} molar.

Because olfaction is generally more acute than taste in fishes it may be of greater importance in food location and various reproductive activities, but taste can be involved in these as well as in the eventual selection and ingestion of food. Bullheads (*Ictalurus*) are known to rely heavily upon the gustatory sense in finding food, and the same appears to be at least partially true in other fishes with numerous taste buds over the body and fins. In most fishes, the taste receptors on the lips, in the mouth, and on the branchial arches are instrumental in the final detection of food items and in initiating the reflexes involved in seizing and swallowing, as well as in the rejection of unwanted items. Some function of taste is suspected in the courtship of certain fishes — cichlids and gouramies, for instance — because of the mouth and fin contact during mate selection and other processes. Recognition of young may be in part dependent upon taste, although olfaction may be of greatest importance.

Although olfaction and taste are probably the chemical senses of greatest importance to fishes, there are others that might be of basic importance. A general chemical sense, attributed to free nerve endings in the skin, has been suggested as a primitive sensor of salts. The sense, if truly present, has been studied very little and its true significance is not well known. On the other hand, the function of pit organs and free neuromasts of the lateral line system as chemoreceptors has been demonstrated by Drs. Katsuki and Onada. Sharks and freshwater and marine fishes have been shown through electrophysiological studies to respond to monovalent cations applied to the appropriate neuromasts. All responded strongly to K^+, but the shark and marine teleosts showed only a weak response to Na^+. Ca^{2+} appears to have a suppressive effect. This salt detection sense may be of significance in habitat selection.

References

Vision

Ali, M. A. (ed.) 1975. Vision in Fishes: New Approaches in Research. Plenum Press, New York.

———, and Anctil, M. 1976. Retinas of Fishes. Berlin, Springer-Verlag.

Arnott, H. J., Maciolek, N. J. and Nicol, J. A. C. 1970. Retinal tapetum lucidum: a novel reflecting system in the eye of teleosts. Science, 169:478–480.

Borwein, B., and Hollenberg, M. J. 1973. The photoreceptors of the "four-eyed" fish, *Anableps anableps*. J. Morphol., 140:405–439.

Braekevelt, C. R. 1974. Fine structure of the retinal pigment epithelium, Bruch's membrane, and choriocapillaris in the northern pike (*Esox lucius*). J. Fish. Res. Bd. Can., 31(10):1601–1605.

Breder, Charles M., Jr., and Rasquin, P. 1950. A preliminary report on the role of the pineal organ in the control of pigment cells and light reactions in recent teleost fishes. Science, 111:10–12.

Brett, J. R. 1957. The eye. *In*: Brown, M. E. (ed.), The Physiology of Fishes, Vol. 2. New York, Academic Press, pp. 121–154.

Denton, E. J., and Nicol, J. A. C. 1964. The choroidal tapeta of some cartilaginous fishes (Chondrichthyes). J. Mar. Biol. Assoc. U.K., 44:219–258.

Engstrom, K. 1963. Cone types and cone arrangements in teleost retinae. Acta Zool., 44(1–2):179–243.

Ercolini, A., and Berti, R. 1975. Light sensitivity experiments and morphology studies of the blind phreatic fish *Phreatichthys andruzzii* Vinciguerra from Somalia. Monit. Zool. Ital., Suppl. 6(2):29–43.

Fenwick, J. C. 1970. The pineal organ. *In*: Hoar, W. S., and Randall, D. J. (eds.), Fish Physiology, Vol. 3. New York, Academic Press, pp. 91–108.

Fineran, B. A., and Nicol, J. A. C. 1974. Studies on the eyes of New Zealand parrot fishes (Labridae). Proc. R. Soc. Lond., B186 (1084):217–247.

Gilbert, P. W. 1963. The visual apparatus of sharks. *In*: Gilbert, P. W. (ed.), Sharks and Survival. Boston, D. C. Heath, pp. 283–326.

Gruber, S. H., Hamasaki, D. K., and Davis, B. L. 1975. Window to the epiphysis in sharks. Copeia, 1975(2):378–380.

Hafeez, M. 1971. Light microscopic studies on the pineal organ in teleost fishes with special regard to its function. J. Morphol., 134:281–313.

Locket, N. A. 1974. The choroidal tapetum lucidum of *Latimeria chalumnae*. Proc. R. Soc. Lond., B186(1084):281–290.

Lythgoe, J. N. 1975. The iridescent cornea of the sand goby *Pomatoschistus minutus* (Pallas). *In*: Ali, M. A. (ed.), Vision in Fishes: New Approaches in Research. New York, Plenum Press, pp. 263–277.

Munk, D., and Frederickson, R. D. 1974. On the function of aphakic apertures in teleosts. Vidensk. Medd. Dan. Naturhist. Foren. Khb., 137:65–94.

Munz, F. W. 1971. Vision: visual pigments. *In*: Hoar, W. S., and Randall, D. J. (eds.), Fish Physiology, Vol. 5. New York, Academic Press, pp. 1–32.

Nicol, J. A. C. 1963. Some aspects of photoreception and vision in fishes. Adv. Mar. Biol., 1:171–208.

Prazdnikova, N. V. 1967. Peculiarities of the distinction of visual images by fish. *In*: Karzinkin, G. S., and Maliukina, G. A. (eds.), Behavior and Reception in Fish. Moscow, "Nauka," Akademiya Nauk SSR, Ministerstvo Rybnogo Khozyaistva SSSR, Ikhtiologicheskaya Komissiya, transl. by Robert M. Howland, pp. 79–86.

Rivas, L. R. 1953. The pineal apparatus of tunas and relative scombroid fishes as a possible light receptor controlling phototactic movements. Bull. Mar. Sci. Gulf Carib., 3:168–180.

Scholes, J. H. 1975. Colour receptors, and their synaptic connexions in the retina of a cyprinid fish. Philos. Trans. R. Soc. Lond. [Biol. Sci.], 270(902):61–118.

Schwanzara, S. A. 1967. The visual pigments of freshwater fishes. Vision Res., 7:121–148.

Somiya, H., and Tamura, T. 1973. Studies on the visual accommodation in fishes. Jap. J. Ichthyol., 20(4):193–206.

Tabata, M., Tamura, T., and Niwa, H. 1975. Origin of the slow potentials in the pineal organ of the rainbow trout. Vision Res., 15(6):737–740.

Walls, G. L. 1963. The Vertebrate Eye and Its Adaptive Radiation. New York, Hafner.

Walters, L. H., and Walters, V. 1965. Laboratory observations on the cavernicolous poeciliid from Tabasco, Mexico. Copeia, 1965(2):214–223.

Wang, R. J., and Nicol, J. A. C. 1974. The tapetum lucidum of gars (Lepisosteidae) and its role as a reflector. Can. J. Zool., 52(12):1523–1530.

Acoustico-Lateralis System

Allis, E. P. 1899. The anatomy and development of the lateral line system in *Amia calva*. J. Morphol., 2:463–540.

Backus, R. H. 1963. Hearing in elasmobranchs. *In*: Gilbert, P. (ed.), Sharks and Survival. Boston, D.C. Heath.

Bamford, T. W. 1941. The lateral line and related bones of the herring (*Clupea harengus* L.). Ann. Mag. Nat. Hist., 8:414–438.

Bergeijk, W. A. van. 1964. Directional and non-directional hearing in fish. *In*: Tavolga, W. N. (ed.), Marine Bioacoustics. New York, Macmillan, pp. 281–299.

Bishop, S. C. 1950. Sound perception apparatus in fish. Copeia, 1950(4):315–317.

Blacker, R. W. 1974. Recent advances in otolith studies. Harden-Jones, F. R. (ed.), Sea Fisheries Research. New York, John Wiley & Sons, pp. 67–90.

Branson, B. A. 1961. The lateral-line system in the Rio Grande perch, *Cichlasoma cyanoguttatum* (Baird and Girard). Am. Midl. Nat., 65:446–458.

———, and Moore, G. A. 1962. The lateralis components of the acoustico-lateralis system in the sunfish family Centrarchidae. Copeia, 1962(1):1–108.

Cahn, P. H. (ed.). 1967. Lateral Line Detectors. Bloomington, Indiana University Press.

Chapman, C. J., and Johnstone, A. D. F. 1974. Some auditory discrimination experiments on marine fish. J. Exp. Biol., 61(2):521–528.

Chapman, M. W. 1957. The ear of a fish. Aquarium J., 18(1):8–18.

Denison, R. H. 1966. The origin of the lateral line sensory system. Am. Zool., 6:369–370.

Denny, M. 1937. The lateral-line system of the teleost *Fundulus heteroclitus*. J. Comp. Neurol., 68:4965.

Dijkgraaf, S. 1963. The functioning and significance of the lateral-line organs. Biol. Rev., 38:51–105.

Disler, N. N. 1971. Lateral Line Sense Organs and Their Importance in Fish Behavior. Jerusalem, Israel Program for Scientific Translations.

Evans, H. M. 1925. A contribution to the anatomy and physiology of the air-bladder and Weberian ossicles in Cyprinidae. Proc. R. Soc. Lond., B97(686):545–576.

Fay, R. R., and Popper, A. N. 1975. Modes of stimulation of the teleost ear. J. Exp. Biol., 62(2):379–387.

Gosline, W. A. 1949. The sensory canals on the head in some cyprinodont fishes, with particular reference to the genus *Fundulus*. Occ. Pap. Mus. Zool. Univ. Mich., 519:1–17.

Hawkins, A. D. 1973. The sensitivity of fish to sounds. Oceanogr. Mar. Biol. Ann. Rev., 11:291–340.

Howland, H. C. 1971. The role of the semicircular canals in the angular orientation of fish. Ann. N.Y. Acad. Sci., 188:202–216.

Illick, H. J. 1954. A comparative study of the cephalic lateral-line system of North American Cyprinidae. Am Mid. Nat., 56:204–223.

Katsuki, Y., Hashimoto, T., and Kendall, J. I. 1971. The chemoreception in the lateral-line organs of teleosts. Jap. J. Physiol., 21(1):99–118.

———, and Onada, N. 1973. The lateral-line organ of fish as a chemoreceptor. *In*: Chavin, W. (ed.), Responses of Fish to Environmental Changes, Springfield, Ill., Charles C Thomas, pp. 389–411.

Krumholz, L. A. 1943. A comparative study of the Weberian ossicles of North American ostariophysine fishes. Copeia, 1943 (1):33–40.

Lowenstein, O. 1971. The labyrinth. *In*: Hoar, W. S., and Randall, D. J. (eds.), Fish Physiology, Vol. 5. New York, Academic Press, pp. 207–240.

Merrilees, M. J., and Crossman, E. J. 1973. Surface pits in the family Esocidae: I. Structure and types. J. Morphol., 141:307–320.

Moffler, M. D. 1972. Plasmonics: communication by radio waves as found in Elasmobranchii and Teleostii fishes. Hydrobiologia, 40(1):131–143.

Moore, G. A. 1956. The cephalic lateral-line system in some sunfishes (*Lepomis*). J. Comp. Neurol., 104:49–56.

———, and Burris, W. F. 1956. Description of the lateral-line system of the pirate perch, *Aphredoderus sayanus*. Copeia, 1956 (1):18–20.

Moulton, J. M., and Dixon, R. H. 1967. Directional hearing in fishes. *In*: Marine Bio-acoustics, Vol. 2. Oxford, Pergamon Press, pp. 187–202.

Neave, F. 1946. Development of the lateral line organs in salmonids. Trans. R. Soc. Can., 5:113–118.

Norris, H. W. 1924. The lateral line organs of the shovelnose sturgeon, distribution and innervation. Proc. Iowa Acad. Sci., 31:443–444.

Parrington, F. R. 1948. A theory of the relations of lateral lines to dermal bones. Proc. Zool. Soc. Lond., 119:65–78.

Peters, R. C., Loos, W. J. G., and Gerritsen, A. 1974. Distribution of electroreceptors, bioelectric field patterns and skin resistance in the catfish *Ictalurus nebulosus* (Le S.). J. Comp. Physiol., 92(1):11–22.

Popper, A. N., and Fay, R. R. 1973. Sound detection and processing in fish: a critical review. J. Acoust. Soc. Am., 53:1515–1529.

———, Salmon, M., and Parvulescu, A. 1973. Sound localization by the Hawaiian squirrel fishes, *Myripristis berndti* and *M. argyrosomus*. Anim. Behav., 21(1):86–97.

Reno, H. W. 1971. The lateral-line system of the silverjaw minnow, *Ericymba buccata* Cope. Southwest Nat., 15(3):347–358.

Sand, A. 1937. The mechanism of the lateral sense organs of fishes. Proc. R. Soc. Lond., 123:472–493.

Tavolga, W. N. (ed.). 1976. Sound Reception in Fishes. Stroudsburg, Pa., Dowden, Hutchinson and Ross.

———, and Wodinsky, J. 1963. Auditory capacities in fishes. Bull. Am. Mus. Nat. Hist., 126(2):177–240.

Trumarkin, A. 1955. Evolution of the auditory conducting apparatus. J. of Evol., (2):313–356.

Wisby, W. J., Richard, J. D., and Nelson, D. R. 1964. Sound perception in elasmobranchs. *In*: Tavolga, W. N. (ed.), Marine Bio-acoustics. New York, Macmillan, pp. 255–268.

Wodinsky, J., and Tavolga, W. N. 1964. Sound detection in teleost fishes. *In*: Tavolga, W. N. (ed.), Marine Bio-acoustics. New York, Macmillan, pp. 269–280.

Olfaction and Taste

Bannister, L. H. 1965. The fine structure of the olfactory surface of teleostean fishes. Q. J. Microsc. Sci., 106:333–342.

Bardach, John E., and Atema, J. 1971. The sense of taste in fishes. *In*: Beidler, L. M. (ed.), Handbook of Sensory Physiology, Vol. IV, Part 2: Taste. Berlin, Springer-Verlag, pp. 293–336.

———, Fujiya, M., and Holl, A. 1967. Investigations of external chemoreceptors of fishes. *In*: Hayashi, T. (ed.), Olfaction and Taste. II. Proc. 2nd Int. Symp., Oxford, Pergamon Press, pp. 647–665.

———, and Villars, T. 1974. The chemical senses of fishes. *In*: Grant, P. T., and Mackie, A. M. (eds.), Chemoreception in Marine Organisms. New York, Academic Press, pp. 49–104.

Crisp, M., Lowe, G. A., and Laverack, M. S. 1975. On the ultrastructure and permeability of taste buds in the marine teleost *Ciliata mustela*. Tiss. Coll., 7(1):191—202.

Grant, P. T., and Mackie, A. M. (eds.). 1974. Chemoreception in Marine Organisms. New York, Academic Press.

Hara, T. J. 1971. Chemoreception. In Hoar, W. S., and Randall, D. J. (eds.), Fish Physiology, Vol. 5. New York, Academic Press, pp. 79–120.

———. 1975. Olfaction in fish. Progr. Neurobiol. 5(5):271–335.

Hasler, A. D. 1957. The sense organs: olfactory and gustatory senses of fishes. *In*: Brown, M. E. (ed.), The Physiology of Fishes, Vol. 2. New York, Academic Press, pp. 187–209.

Hidaka, I., Kiyohara, S., Tabata, M., and Yonezawa, K. 1975. Gustatory responses in the puffer. Bull. Jap. Soc. Sci. Fish., 41(3):275–281.

Hodgson, E. S., and Mathewson, R. F. 1971. Chemosensory orientation in sharks. Ann. N.Y. Acad. Sci., 188:175—182.

Idler, D. R., Fagerlund, U. H. M., Mayoh, H., Brett, J. R., and Alderdice, D. F. 1956. Olfactory perception in migrating salmon, I. L-Serine, a salmon repellent in mammalian skin. J. Gen. Physiol., 39:889–892.

Johnston, J. W., Moulton, D. G., and Turk, A. 1970. Advances in Chemoreception, Vol. 1. Communication by Chemical Signals. New York, Appleton-Century-Crofts.

Kapoor, B. G., Evans, H. E., and Pevzner, R. A. 1975. The gustatory system in fish. Adv. Mar. Biol., 13:53–108.

Katsuki, Y., and Onada, N. 1973. The lateral-line organ of fish as a chemoreceptor. *In*: Chavin, W. (ed.), Responses of Fish to Environmental Changes. Springfield, Ill., Charles C Thomas, pp. 389–411.

Kiyohara, S., Hidaka, I., and Tamura, T. 1975. Gustatory response in the puffer. 2. Single fiber analyses. Bull. Jap. Soc. Sci. Fish, 41(4):383–391.

Kleerekoper, H. 1969. Olfaction in Fishes. Bloomington, Indiana University Press.

Konishi, J., and Zotterman, Y. 1963. Taste functions in fish. *In*: Zotterman, Y. (ed.), Olfaction and Taste. Proc. 1st Intern. Symp., New York, Macmillan, pp. 215–233.

Lowe, G. A. 1974. The occurrence of ciliary aggregations on the olfactory epithelium of two species of gadoid fish. J. Fish Biol., 6(4):537–539.

———, and MacLeod, N. K. 1975. The ultrastructural organization of olfactory epithelium in two species of gadoid fish. J. Fish Biol., 7(4):529–532.

Marshall, N. B. 1967. The olfactory organs of bathypelagic fishes. *In*: Marshall, N. B. (ed.), Aspects of Marine Zoology. New York, Academic Press, pp. 57–70.

Oguri, M., and Omura, Y. 1973. Ultrastructure and functional significance of the pineal organ of teleost. *In*: Chavin, W. (ed.), Responses of Fish to Environmental Changes. Springfield, Ill., Charles C Thomas, pp. 412–434.

Ojha, P. O., and Kapoor, A. S. 1973. Structure and function of the olfactory apparatus in the fresh-water carp, *Labeo rohita* Ham. Buch. J. Morphol., 140(1): 77–85.

Pfaffman, Carl (ed.). 1960. Olfaction and taste. Proc. 3rd Intern. Symp. New York, Rockefeller University Press.

Tester, Albert L. 1963. Olfaction, gustation, and the common chemical sense in sharks. *In*: Gilbert, P. W. (ed.), Sharks and Survival. Boston, D. C. Heath, pp. 255–282.

Thornhill, R. A. 1967. The ultrastructure of the olfactory epithelium of the lamprey *Lampetra fluviatilis*. J. Cell. Sci., 2:591–602.

Wolstenholme, G. E. W., and Knight, J. (eds.). 1970. Taste and Smell in Vertebrates. London, J. and A. Churchill.

12 CIRCULATION, RESPIRATION, AND THE GAS BLADDER

Because fishes live in an environment that is oxygen-poor compared to the atmosphere we breathe, and because their simpler hearts must force blood past an oxygenating surface before its distribution to the tissues, we have some particular interests in the solutions of their problems of circulation and respiration. Special adaptations to many situations have involved the composition of the blood, the circulatory apparatus, behavioral responses, and the structure and function of the gills and other respiratory surfaces. Some of the more interesting adaptations involve the direct utilization of the atmosphere as a source of oxygen. The lungfishes differ markedly from the other bony fishes in this aspect. The following discussion will cover blood, gill function, air breathing and the gas bladder in relation to buoyancy.

The Blood of Fishes

Fishes generally have less blood volume than other vertebrates, the volume usually ranging between about 2 and 4 ml/100 g in bony fishes. Lampreys appear to have greater volumes than these, about 8.5 ml/100 g, hagfishes even more, 17 ml/100 g, and elasmobranchs are reported to have blood volumes of 6 to 8 ml/100 g. Salmonids have been reported in earlier studies as having about 3 to 3.5 ml/100 g, but recent physiological work has concluded that they approach the blood volumes of elasmobranchs, having from 5 to more than 7 ml/100 g. Tunas have a high blood volume, ranging from about 8 ml/100 g in 9-kg fish to 13 ml/100 g in 4.5-kg fish; smaller individuals are reported to have even greater volumes. Some investigators have pointed out that there may be a phylogenetic decrease in blood volume throughout the fishes. The higher bony fishes possess a more perfected vascular system, and thus need less blood for transport of oxygen and other materials.

The blood transports a variety of materials, including inorganic ions and a number of organic constituents such as hormones, vitamins, and several plasma proteins that may make up from 2 to 6 g/100 ml. These proteins may include two forms of alpha globulin, two of beta globulin, and gamma globulin, as well as albumin, transferrin, and others. They are involved in certain immune responses, buffering against pH changes, and maintenance of osmotic pressure important to the movement of water through capillary walls.

The osmotic concentration of fish blood varies according to the habitat and means of osmoregulation developed in the species involved. Osmoregulation will be covered later, but mention of osmoconcentration will be mentioned here briefly. The osmoconcentration of bony fish blood ranges from somewhat below 200 milliosmoles (mOsm) in freshwater fish to more than 400 mOsm in marine species. Sodium and chloride are the main contributors to the concentration, with lesser contribution from potassium, calcium, and magnesium, plus urea and free amino acids. The freezing point depression of bony fish blood ranges generally from about $\Delta_{fp} = -0.6$ in freshwater fishes to about -0.75 in marine species without special antifreeze protection. Some arctic fishes live in waters of $-1.7°$ C; Antarctic waters are as cold as $-1.86°$ C. Fishes that live in such habitats are protected from freezing by blood glycoproteins that may account for one half or more of the osmolality of the blood. In arctic fishes the Δ_{fp} ranges down to $-1.0°$ C, so those living in the coldest water have supercooled body fluids and, with few exceptions, will freeze if brought into contact with ice. A similar situation occurs in the Antarctic, where most species have freezing points of $-0.9°$ C to $-1.54°$ C. Certain nototheniids freeze only at temperatures as low as $-2.97°$ C. Many temperate and polar fishes are known to adjust osmolality of their blood with the seasons. Some increase the concentration of NaCl when adapting to cold; others increase the concentration of organic compounds.

Cellular constituents of the blood are the red blood cells, or erythrocytes, and the white cells, or leucocytes. Erythrocytes obtain their characteristic color from hemoglobin, made up of the colorless protein globin, and the red-yellow pigment heme, which contains iron. Hemoglobin molecules of elasmobranchs and bony fishes are made up of four complex chains (tetrameric) and have molecular weights of about 61,000 to 70,000. Hemoglobin of lampreys is much like myoglobin, a hemoglobin found in muscle tissue, in that it is monomeric, with a molecular weight of 18,200. More than one type of hemoglobin can be found in some species. Hemoglobin transports oxygen in combination with the ferrous iron of the heme. This combination is reversible, depending upon the partial pressure of the oxygen. Only a few fishes lack hemoglobin; for instance, some chaenichthyids of the Antarctic and leptocephalus larvae of eels have colorless blood.

Erythrocytes of fishes are nucleated and usually oval in shape, with relatively few species having cells with a nearly round shape. Lampreys have round red cells about 9 microns in diameter. Elasmobranchs have large erythrocytes, the length ranging from 20 to 27 microns and width from around 14 to 20 microns. Erythrocytes of bony fishes generally range from 12 to 14 microns in length and 8.5 to 9.5 in width, but lungfishes have large red cells, about 36 microns long.

Generally there is an inverse relationship between size and number of red blood cells, with sharks and rays having fewer than half a million cells per cubic mm. However, some gobies may have similar counts — for instance *Gobius exanthematus* is reported to have a count of 0.425 $\times$ 10^6/mm^3. An Antarctic fish, *Trematomus*, has from 0.66 to 0.80 $\times$ 10^6/mm^3. Most bony fishes have red cell counts of 1 to 3 $\times$ 10^6/mm^3, with a majority under 2 $\times$ 10^6/mm^3, but some active marine fishes have higher numbers, ranging from 4 to 6 $\times$ 10^6/mm^3.

The percentage of the blood made up of red cells (the percentage of packed red cells) is called the hematocrit, and is correlated with the red cell count. Elasmobranchs usually have hematocrits under 25%; most teleosts are in the 20 to 30% range, with some marine species, such as *Mugil cephalus* and *Thunnus thynnus*, having hematocrits of about 42%. Hemoglobin concentration in fish blood, expressed as g/100 ml, is usually 7 to 10. Red blood cells, and consequently the hematocrit and hemoglobin concentration, can vary with season, temperature, and nutritional state and health of the fish. Circadian changes have been noted in some species.

White blood cells (leucocytes) are not as numerous as red cells, usually numbering fewer than 150,000 per cubic mm (0.15 $\times$ 10^6/mm^3) in most fishes. The range within a single species may be great; counts for *Cyprinus carpio*, for example, have been reported as ranging from about 0.032 $\times$ 10^6/mm^3 to 0.146 $\times$ 10^6/mm^3. The white cells include four kinds — granulocytes, thrombocytes, lymphocytes, and monocytes — although the presence of the latter in some fishes has been questioned. Thrombocytes are involved in blood clotting; they carry a chemical that promotes the conversion of prothrombin to thrombin. They are more numerous than the other white cells in many marine fishes, constituting about half the total. Granulocytes include three types of cells, named for their staining properties; these are neutrophils, eosinophils, and basophils, the latter seldom found in fishes except for a few marine species. Eosinophils are not always present, but neutrophils are common in most species. Granulocytes are phagocytic, involved in combatting disease, and may increase in number when the fish is infected by bacteria. Lymphocytes include the phagocytic macrophages, plasma cells, and small lymphocytes that may be active in protein production. Lymphocytes can make up more than 90% of the leucocytes in carp and trout.

Blood cell formation occurs at several sites in fishes. Erythrocytes are formed mainly in the kidney and spleen in most fishes; the head kidney is usually the most important site. Leucocytes are formed in the kidneys of many teleosts, but in elasmobranchs there is a special organ (organ of Leydig) that assumes this function. This is usually associated with the wall of the alimentary canal, commonly found along the esophagus. Similar tissue may occur in various places in other fishes — in mesenteries, the orbit, the meninges, and the base of the cranium. The spleen is the site of formation of thrombocytes in some species.

The oxygen capacity of fish blood includes oxygen carried in solution as well as that carried in combination with hemoglobin in the erythrocytes. *Chaenocephalus*, the Antarctic icefish, which has no hemoglobin, has a reported blood oxygen capacity of 0.45 to 1.08 ml/100 ml, whereas most teleosts have capacities in the 8 to 12 volume % range. Very active fishes, such as the pelagic scombroids and species adjusted to oxygen-poor waters, have blood oxygen capacities up to 20 volume %. The oxygen capacity of the blood of sharks and rays is typically less than that of teleosts, usually ranging from 3.5 to 6 volume %.

The actual oxygen content of the blood depends on many factors, including the partial pressure of oxygen in the medium, the partial pressure of carbon dioxide, pH, temperature, and the activity of the fish. Normally blood from the dorsal aorta is at 85 to 95% of saturation, while venous blood usually carries oxygen at 30 to 60% saturation. Trout undergoing strenuous exercise have been reported to carry no oxygen in the venous blood as it returns to the heart.

The relationships among CO_2, pH, and the affinity of the blood for oxygen is of special interest. One of these relationships, the Bohr effect, involves a decreased affinity of hemoglobin for oxygen at low pH due to altered configuration of the hemoglobin molecule by the increased hydrogen ion concentration. This effect is prominent in fishes that are adapted to habitats with high oxygen content and low CO_2. A distinct advantage exists for those species in that, at the gills, in the presence of a low partial pressure of CO_2, the blood can easily load oxygen even at low partial pressure; then, in the tissues, at higher partial pressure of CO_2, oxygen can be released independently of its partial pressure. A disadvantage of the Bohr effect in fishes adapted to low CO_2 and high oxygen is that if the CO_2 content of the medium rises, a greater and greater dissolved oxygen content becomes necessary to facilitate loading of the hemoglobin. Fishes adapted to swamp life and other slow-water habitats, such as bullheads, carp, bowfin, and lungfishes, where low pH and low oxygen content is normal have blood with a very weak Bohr effect.

Related to the Bohr effect is the Root effect, which involves a decrease in the oxygen capacity of the blood with rising partial pressure of CO_2. A decrease in pH will render fish blood incapable of becoming 100% saturated with oxygen regardless of the pressure of oxygen. This has been demonstrated in experiments in which oxygen pressures up to 140 atmospheres were used.

Branchial Circulation and Function

Effective exchange of gases in a gill-breathing fish depends on bringing the blood and respiratory water into close apposition on either side of a membrane, usually one or two cells thick, through which the gases can diffuse. Such a system works best if the blood and water flow opposite to each other, and fishes have developed this countercurrent system. The system requires a force to move the water and one to move the blood, and finely divided channels through which these fluids can flow. The channels are provided by the structure of the gills.

Each gill arch bears numerous filaments, so that the total in the entire apparatus is several hundred; the actual number varies with factors such as size and surface area of the fish and the general habits of the species. Active fishes generally have greater numbers than sluggish species. Small bottom-living darters have about 300 filaments in total. Mackerel weighing 800 g have about 2400, and perch of 30 g have about 1500. Each gill filament carries an abundance of secondary lamellae at right angles to the long axis. The lamellae, which are distributed on both the upper and lower surfaces of the filaments, are fragile ridges with thin walls. The thickness of these walls, which constitute the respiratory membrane through which gases must diffuse, varies with mode of life of the different species. Tunas and their close relatives have very thin lamellar walls of about 0.53 to 1.0 micron; most bony fishes have walls of 2 to 4 microns, and some demersal species have respiratory membranes 5 to 6 microns thick. Most elasmobranchs for which data are available have lamellar thicknesses of 5 to 11 microns.

The total number of lamellae in marine fishes studied by Dr. G. M. Hughes ranged from 52,000 to 689,000. Of course, the number varies with the number of filaments and the count of lamellae per unit length of filament as well as with size of the fish. Slow-moving fishes usually have from 10 to 20 lamellae per mm, whereas active fishes have 30 to 40 per mm. Most bony fishes are in the 15 to 30/mm range. Some air-breathing species have fewer lamellae, with walls up to 20 microns thick. These appear to be of little use in aquatic respiration.

The product of numerous filaments bearing the small but numerous lamellae is a great surface area for exchange of respiratory gases. As with other respiratory features, active and sluggish fishes differ greatly in respect to gill area. Some examples of estimates of gill area are: whitefish *(Coregonus)* of 1000 g, 290 mm^2/g; bullhead *(Ictalurus)* of 50 g, 158 mm^2/g; herring *(Clupea)* of 11 g, 636 mm^2/g; mackerel *(Scomber)* of 800 g, 533 mm^2/g. Drs. S. de Jager and W. J. Dekkers, converting published gill measurements to that expected in 200-g fish, have shown that, for fish of that size, scombrids generally have gill areas of more than 1000 mm^2/g, with various tunas having 1500 to 3500 mm^2/g. Other active pelagic fishes have 500 to 1000 mm^2/g, and most bony fishes for which measurements were available were found to be in the range of 150 to 350 mm^2/g.

The effectiveness of the gill area in exchange of respiratory gases depends upon the contact made with water being pumped through the system. In each hemibranch, lamellae of adjacent filaments meet to form tiny channels through which water is forced (Fig. 12–1). The filaments are equipped with muscles that hold the tips of the filaments of posterior hemibranchs of each arch against the tips of the filaments of the anterior hemibranch on the following arch, so that all water must pass through the lamellar channels (Fig. 12–2). As the space in the oral

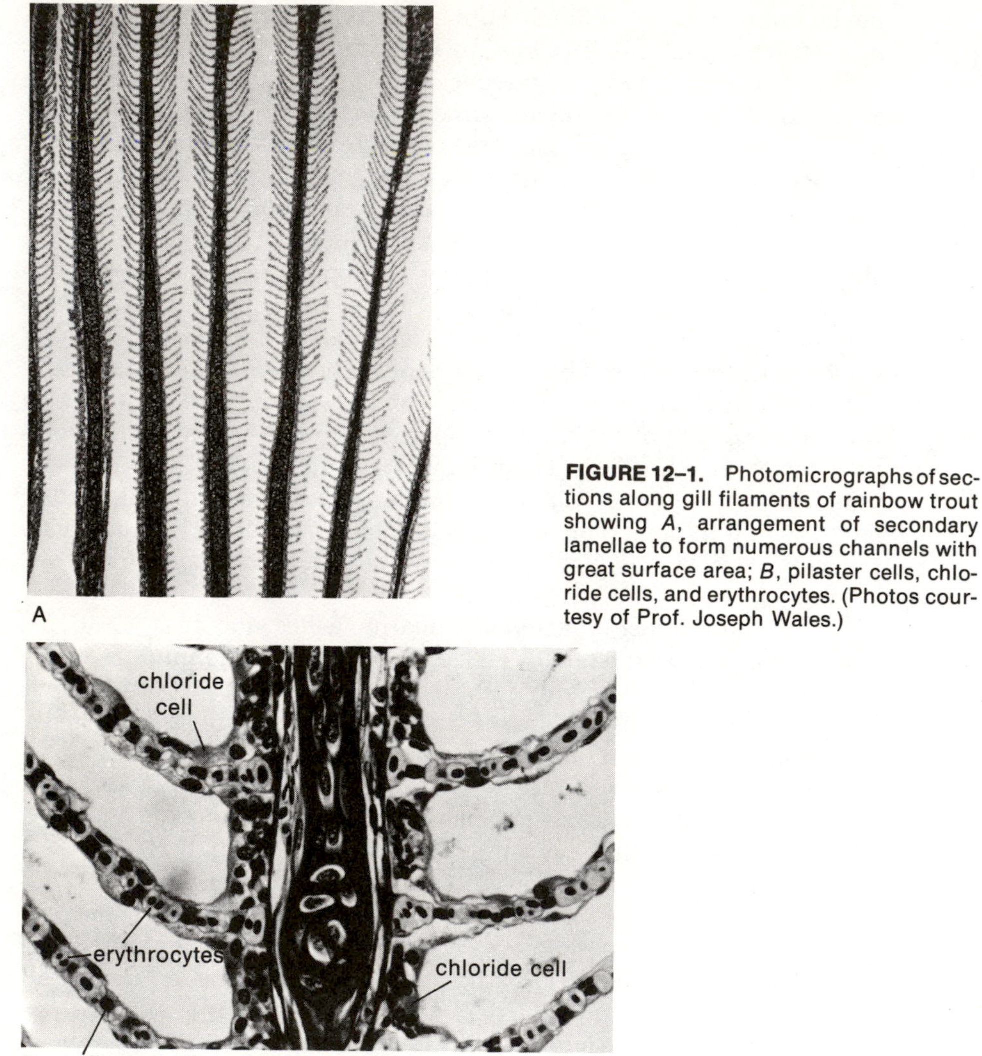

FIGURE 12–1. Photomicrographs of sections along gill filaments of rainbow trout showing *A*, arrangement of secondary lamellae to form numerous channels with great surface area; *B*, pilaster cells, chloride cells, and erythrocytes. (Photos courtesy of Prof. Joseph Wales.)

and branchial cavities changes with respiratory movements, there is compensatory change in the space occupied by the gill mass due to the action of gill arch and gill filament musculature. The tips of the filaments from adjacent arches remain in contact most of the time, being separated briefly, at least in some species, during part of each opercular cycle. The tips are separated during coughing, and during bypassing of excessive water flow.

In agnaths and elasmobranchs the filaments are bound to the gill septa, but the secondary lamellae stop short of the septa (Fig. 12–3) so that passages or canals are formed next to them. Water passes through the interlamellar channels to these canals along the septa and then toward the gill slit. Countercurrent flow of blood and water is imperfect

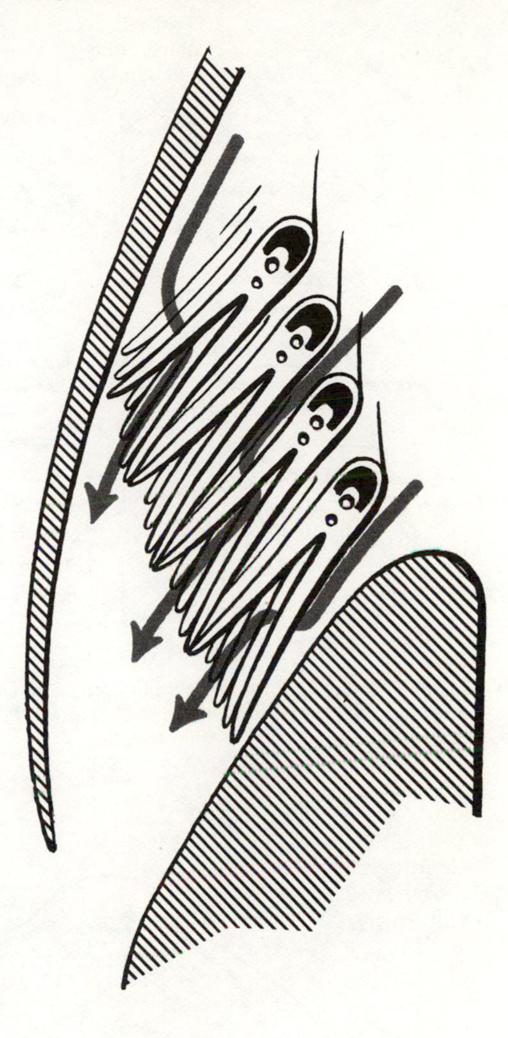

FIGURE 12–2. Diagram of frontal section through gill area of teleost, showing tips of filaments from adjacent arches held together so water flow must cross filaments through lamellar channels. (Lamellae, which are at right angles to the axes of the filaments, are not shown.)

or lacking in certain elasmobranchs, according to some investigators. Instead, a serial multicapillary arrangement brings small blood vessels in contact with respiratory water, each for a short distance.

Internally, the secondary lamellae of the gill filaments are divided into numerous capillary-sized channels by pillar or pilaster cells that are disposed in more or less regular rows. A marginal channel rims each lamella (Fig. 12–4). These small spaces receive blood from capillaries branching from the afferent arterioles of the filaments and pass the blood, counter to the flow of water outside the lamellae, to the efferent arterioles. There is a central sinus in the filament through which some investigators believe the blood can be bypassed without passing through the lamellae. Shunting of blood through the central cavity or around the tips of the filaments under conditions of abundant dissolved oxygen has been suggested by certain investigators, but others believe that no shunting occurs and that the only control of

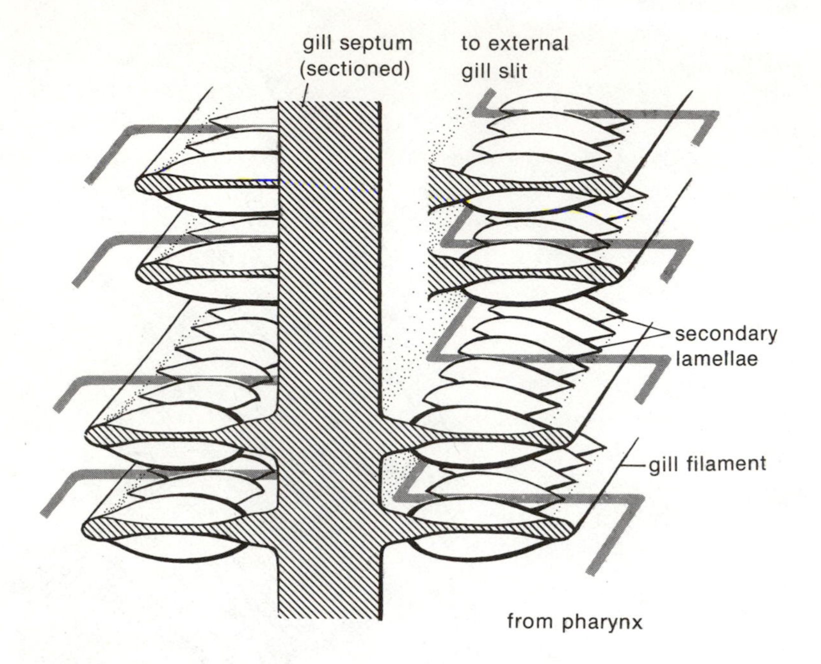

FIGURE 12–3. Diagram of gill filaments and secondary lamellae of dogfish *(Squalus acanthias),* showing path of respiratory water between lamellae, then through channels next to gill septum.

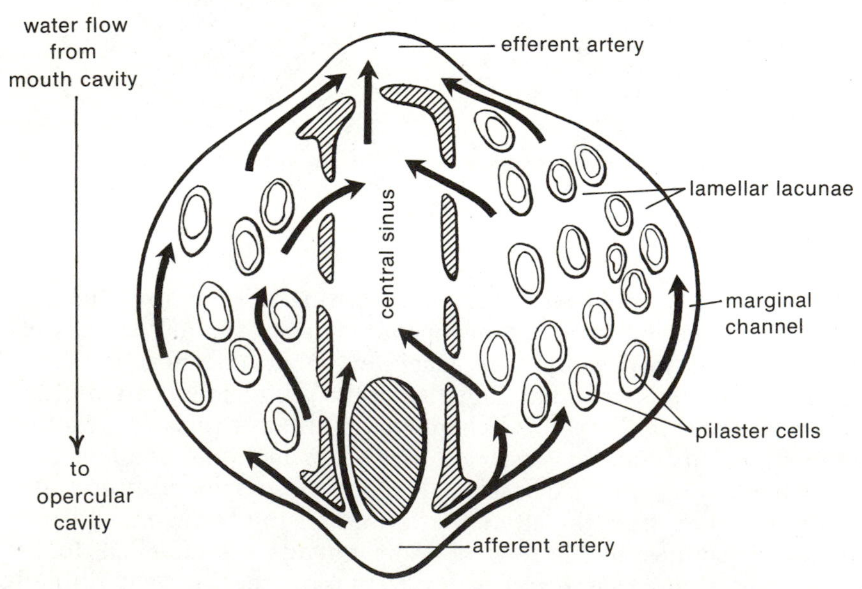

FIGURE 12–4. Diagram of section through gill filament of teleost at level of secondary lamellae. Arrows depict flow of blood through spaces (lacunae) among pilaster cells.

blood flow through the filament is by the muscles of the arterioles. Shunting of blood around the gill tissue is necessary in air-breathing fishes while in oxygen-poor water, or else they would lose oxygen from the blood to the water. In these cases only enough blood to take care of ammonia excretion and ion exchange must go through the relatively thick-walled lamellae.

The importance of gill dimensions can be seen in the expression of Fick's Law of Diffusion:

$$m = \frac{k \cdot a \cdot dP}{D}$$

in which m = oxygen uptake in mg/hr; k = diffusion constant for tissue involved (number of ml of oxygen that diffuses through 1 mm^2 in 1 hour at a pressure gradient of 1 atm/micron. A figure of 0.12 has been used for calculations pertaining to fish): a = diffusion surface in mm^2 (gill area); D = length of diffusion path (thickness of lamellar wall); and dP = difference in oxygen pressure (in atm) on either side of lamellar wall.

The respiratory pump that forces the water across the gills consists of the buccopharyngeal cavity plus all the mechanisms for opening, enlarging, and constricting it, and the parabranchial cavity, which can be enlarged or constricted by action of the operculum and the branchiostegals. Coordinated action of these two cavities can produce a continuous flow of water across the gills. As the mouth is opened and water is sucked in by the enlarging of the buccal cavity, the parabranchial cavity is rapidly enlarged, but with its opening closed. This causes a negative pressure that draws water across the gills. When the mouth is closed, the oral valves prevent escape of the water past the lips, and the water is forced across the gills, through the parabranchial cavity, which is now constricting, and through the opercular opening.

In hagfishes, respiratory water is pumped to the gills by a velar pump situated in a chamber posterior to the buccal cavity. Water enters the chamber through the nostril and is forced toward the gill pouches by the action of paired scroll-like structures that unroll into the velar chamber from the top, and then force the water out of the chamber as they rapidly roll up again with the edge of the roll in contact with the lateral walls of the chamber. This action also sucks water into the chamber from the nostril. Adult lampreys apparently can draw a respiratory current through the mouth when free-swimming, but usually depend on a tidal action and move water in and out of the external branchial apertures. This mode is necessary when attached to prey or other objects. The branchial region can be constricted to force water out of the gills, and the elasticity of the branchial skeleton then greatly aids in expanding the gill chambers to draw water in. A valvular arrangement carries incoming water to the mesial side of the gill pouch so that it can pass through the gill lamellae counter to the flow of blood. Larval lampreys utilize a velar pump in moving respiratory water, in addition to contractions of the branchial basket.

The relationships between the habits of fishes and their respiratory apparatus have long been recognized, and a classification of branchial pumps of teleosts was proposed by S. Baglioni in 1908. Most teleosts fit into the system, which points out the increasing use of the branchiostegal apparatus in species with more and more sedentary habits. Pelagic species tend to depend mainly upon opercular movements in pumping respiratory water. Less active species that spend some time resting on the bottom tend to have the branchiostegal part of the branchial pump better developed, and combine opercular and branchiostegal movements in gill irrigation. Demersal species depend greatly on the branchiostegal apparatus, and may have the opercular elements reduced. Some fishes, such as eels, pipefishes, and others with unusual branchiostegal configuration, were placed into a miscellaneous group by Baglioni.

Swift species such as tunas and their relatives can depend on their speed to force water over the gills, and some scombrids have lost the facility of irrigating the gills by pumping. This ram irrigation takes place at speeds over 1 km/hr and has been calculated to take less than 1% of the total energy expended by a swimming skipjack. Several marine fishes other than scombrids retain the ability to pump water across the gills, but can ram irrigate when forced to swim at speeds of 1.5 to over 2 km/hr. Some stream fishes are known to hold the mouth and opercula open while maintaining position in swift water, and irrigate the gills passively. Demersal marine fishes are not known to ram irrigate, but some stream fishes, such as Gyrinocheilidae, cling to the substrate while allowing the swift current to force water over the gills.

The volume of water pumped over the gills (respiratory volume) varies with many factors such as species, size, temperature, carbon dioxide content of the water, oxygen content, activity, and other factors. Fishes respond to stresses of activity, lowered oxygen, and so on by increasing the number and amplitude of respiratory movements, so that the respiratory volume can be increased greatly. Experiments performed with the sucker, *C. catostomus,* which showed a respiratory volume of about 50 ml/min/kg under certain conditions, obtained volume of up to 6000 ml/min/kg under exercise and 12,900 ml/min/kg when the dissolved oxygen in the medium was decreased. Relationships of breathing rate, amplitude of movement, and volume can be seen for the rainbow trout, in which the normal breathing rate is about 80/min. When exercised, the rate was found to increase to about 100/min, but the respiratory volume increased from 594 to 3042 ml/min/kg. Efficiency of oxygen removal from the water is usually impaired at high irrigation rates because not all the water pumped through the gills comes in close enough contact with the lamellae for a sufficient time. However, the blood passing through the gills is usually 85% to 95% saturated with oxygen.

The heart provides the force to move the blood through the ventral aorta, up the afferent branchial arteries, and into the gill filaments. Contractions of the ventricle generate a considerable pressure that is attenuated as the blood is forced through the interlamellar channels and into the dorsal aorta via the efferent branchial arteries. Examples of systolic pressures recorded from the ventral aorta in various teleosts (in mm Hg) are about 30 in the cod (*Gadus morhua*), 40 in the eel (*Anguilla*), 44 in the carp, 48 in the lingcod (*Ophiodon*), and 75 to 82 in the chinook salmon. The bulbus arteriosus, because of its elasticity, maintains a positive pressure on the blood even though the diastolic ventricular pressure may drop to zero. Blood pressure in the ventral aorta of elasmobranchs is generally slightly less than that of teleosts, having been measured at 30 to 70 mm Hg in various species. Pressure in the ventral aorta of the hagfish, *Myxine*, ranges to a maximum of 15 mm Hg.

The pressure drops markedly as the blood passes through the gills. Pressure in the dorsal aorta is usually 40% to 50% that in the ventral aorta in bony fishes, but some, such as the carp, maintain from 50% to 70% of the ventral aortic pressure in the dorsal aorta. The recorded dorsal aortic pressures of the carp, 22 to 32 mm Hg, contrast with the low figures of 8 to 12 for the Antarctic icefish. Systolic pressure in the dorsal aorta of various elasmobranchs has been measured at 25% to 40% of that of the ventral aorta. Pressure in the dorsal aorta of the hagfish, *Myxine*, ranges from about 2 to 14 mm Hg, the higher pressures apparently being due to the contraction of the gillpouches. The gill contractions superimpose a strong pulse on the pulse caused by the contractions of the heart, but that does not seem to hold true for the genus *Eptatretus*.

Volume of blood flow through the gills is variable, depending upon heart rate and gill resistance. Typical bony fishes have cardiac outputs in the range of 15 to 20 ml/kg/min but outputs from 5 to 100 ml/kg/min have been reported. Elasmobranchs have cardiac outputs in the range of 20 to 25 ml/kg/min.

Although the heart supplies the major force that maintains blood flow in the dorsal aorta, systemic arteries, and veins, other structures may assist in the flow, such as the gill contractions mentioned earlier in regard to *Myxine*. Serial contractions of body musculature aid the blood flow, and one of the consequences of heart contraction inside a nonelastic pericardium is a sucking action that hastens venous blood flow. Among the special structures that help maintain flow of body fluids are lymph hearts, the caudal and hepatic hearts of hagfishes, and possibly the elastic ligament in the lumen of the dorsal aorta of trout. The ligament is so constructed that it is believed to pump blood as the body flexes and forces the ligament against the walls of the aorta.

Under favorable conditions, fishes can remove about 85% to 90% of the dissolved oxygen from the water passing over the gills. Such efficient removal rates usually occur when dissolved oxygen is high and respiratory volume low. More typically the range is from 50 to 60%, and under conditions of low dissolved oxygen, high tempera-

Table 12–1. REPRESENTATIVE RESTING OXYGEN CONSUMPTION IN SELECTED FRESHWATER FISHES

Species	Temperature (°C)	O_2 Consumption (mg/kg/hr)	Source
Salmo trutta	10	81	Beamish
	10	128	Beamish
	20	282	Beamish
Carassius auratus	10	15.7	Beamish and Mookherjii
	20–22	30–160	Beamish and Mookherjii
	32–35	127–262	Beamish and Mookherjii
Cyprinus carpio	10	17	Beamish
	20	48	Beamish
	30	104	Beamish
Cottus spp.	15	92–157	Original
	25	150–264	Original

tures, and increased respiratory volume, the utilization may fall to 10 to 20% or even lower. Actual oxygen consumption in fishes depends on many factors, including size, temperature, activity, the standard metabolism of the species, oxygen pressure, carbon dioxide pressure, pH of the medium, and salinity. The immediate past history of the individual in regard to acclimation or acclimatization to some of the above factors is important to oxygen consumption, as are circadian or seasonal cycles. Resting oxygen consumption for some familiar freshwater fishes is shown in Table 12–1.

The importance of maintaining optimum water conditions for valuable fish life can be appreciated when considering the effect that can be achieved by combining some of the factors mentioned. When dissolved oxygen content of the water is lowered, for instance, the fish responds by increasing the rate of gill ventilation. This added activity requires a greater consumption of oxygen, so the individual can be placed in the position of trying to extract a greater amount of oxygen from a smaller supply. A rise in temperature, or increased carbon dioxide content, could combine with the lowered oxygen and increased activity to make the situation intolerable.

Skin as a Respiratory Surface. In eels and other fishes without strong squamation and with a favorable ratio of body surface to volume, significant amounts of oxygen can be absorbed through the skin. This is of relative importance during periods of low activity or of low temperature. Larval fish obtain oxygen via diffusion through the skin, and some larvae have special cutaneous gills. These feathery structures are seen in the larvae of *Polypterus* and lungfishes.

Air Breathing in Fishes. The capability of extracting oxygen from the atmosphere has a surprising distribution among the fishes, from the standpoints of both phylogeny and geography. The habit is encountered from the tropical swamps and beaches to the freezing arctic bogs where the blackfish live, and members of many orders, from primitive to derived, are involved. Overall, there are many more air breathers among warm-water fishes than temperate or cold water types, with the habit most common among the tropical swamp dwellers.

Air breathing by means of lungs is ancient among fishes, having probably originated in oxygen-poor environments during the late Silurian and early Devonian periods. Although lungs may have arisen in hypersaline Silurian seas, they were common among freshwater fishes of the Devonian swamps, and the heritage has been passed on to our present-day lungfishes and perhaps is reflected in the living holosteans. In contemporary fishes, not only the lung or gas bladder is used by air breathers, but several other structures have been modified for the purpose. Environmental pressures that may have caused evolution of the ability to take advantage of the great atmospheric reservoir of oxygen can be deduced from the living conditions of several of our present species. In others, utilization of atmospheric oxygen must be involved in aspects of life history or behavior not necessarily connected to the lack of oxygen in the aquatic environment.

Obviously, waters permanently low or lacking in oxygen can be inhabited by fishes only if those fishes can derive their oxygen from an alternate source. Throughout the tropics, swamps of high organic content and heavy vegetative cover support year-round populations of fishes, some of which are obligate air breathers. Swamps and streams that provide good dissolved oxygen supplies during part of the year, but which stagnate and even dry up at other times, may maintain a complement of specialized fishes that are facultative or obligate air breathers and can cope with the drying of the water either by burrowing and aestivating or by moving overland to more permanent bodies of water. Some mountain streams of the tropics support species that can withstand the torrents of the rainy season and can later adapt to the oxygen-poor pools of the dry season.

There are some fishes that expose themselves to air even though the surrounding water contains sufficient dissolved oxygen. Certain blennioids habitually remain in place under rocky cover as the tide recedes and returns, living in a dewatered but damp habitat for 2 or 3 hours every tidal cycle. Eels (*Anguilla*) increase the living space available to them by moving overland through wet vegetation to enclosed ponds and lakes. Species of walking catfishes owe part of their geographical distribution to overland forays that place them in new bodies of water. The spread of the exotic *Clarias batrachus* in Florida is an example. Several species of air breathers, including synbranchoids and mudskippers, actively seek food while out of the water. The mudskippers are contentedly at home on sunny mud flats, where they move about freely, engaging in aggressive displays and other social behavior.

Adaptation to the aerial mode of respiration demands that air be brought into contact with highly vascular tissue of considerable area.

In a few species the skin itself may be modified to aid considerably in oxygen uptake, but usually some cavity in the head or body is modified for the purpose. In some instances an existing cavity is modified, in others a new cavity is formed, and in others existing structures are modified to provide the requisite surface area. In addition, air-breathing species must be equipped to carry on the functions of osmoregulation, releasing carbon dioxide, and the excretion of ammonia. The pumping of blood through very thin gill tissue to facilitate one or more of these functions could lead to loss of oxygen from relatively oxygen-rich blood to the oxygen-poor water. Development of thicker gill lamellae, reduced branchial irrigation, and other modifications have aided in overcoming the problems.

A question that naturally arises is why the gills, with their great surface area and blood supply, will not suffice for aerial respiration. If gills could be kept moist, and all the filaments and lamellae spread out, they would be of use in breathing air, but the soft nature of these structures allows the fine tissues to collapse together without the support of the water and thus reduce the surface area. The gill filaments are stiffened to a small extent by thin rays of cartilage with tiny side branches that correspond to the lamellae, but these are not strong enough to keep the gills spread. Perhaps a few species can rely on the unmodified gills for some aerial respiration; many fishermen have noted the ability of bullheads *(Ictalurus)* to continue respiratory movements of the gills and to remain alive for hours in spite of being held in a dry sack. The gills of *Synbranchus marmoratus*, a swamp eel of South America, and of *Hypopomus*, a South American knifefish, are modified to remain spread while unsupported by water and constitute part of the air-breathing surface.

The skin of some fishes, especially those without scales or those with small embedded scales, functions as a respiratory surface. The eels, *Anguilla*, exemplify the latter — while in the water, about 10% of their oxygen intake is through the skin, and while in air about 66% of their oxygen is taken in through the skin, the remainder coming via the gills. Members of the goby genera *Periophthalmus* and *Boleophthalmus*, while out of the water, acquire part of their oxygen through the skin. Another goby, *Gillichthys mirabilis*, respires readily through the skin.

Most air-breathing species rhythmically or periodically empty and fill a specialized cavity with air, so the atmospheric oxygen can come in contact with well vasculated tissue in that cavity. In the case of the electric eel *(Electrophorus electricus)* the cavity is the mouth and pharynx, where the lining is folded and otherwise modified to provide a large surface rich in blood vessels. The mouth lining of some of the synbranchoids, including *Synbranchus* and *Monopterus*, plays a significant part in respiration, although it is not much modified for an increase of surface area. The common carp brings a bubble of air into contact with a specialized part of the palate when in oxygen-poor water. The pharyngeal walls of *Periophthalmus* are reported to be a respiratory surface. The walls of the pharynx are enlarged by diverticula in several species including all of the snakeheads, Channidae, and the synbranchoid, *Amphipnous cuchia.* The branchial chamber is the site of respiratory epithelium in many air breathers. Some, such as the so-called labyrinth fishes (climbing perch, *Anabas,* and the gouramies and allies) and the walking catfishes *(Clarias)* have parts of gill arches modified into firm structures of large surface area bearing respiratory epithelium. In the labyrinth fishes the respiratory organ is in an enlarged cavity above the gills and consists of a number of folded and crenulated plates. The corresponding organ of the catfish develops from the second and fourth gill arches and is arborescent in nature, taking the name "gill tree." *Saccobranchus fossilis,* an Asian catfish, has a pair of sac-like diverticula leading from the branchial chamber into the lateral musculature.

A number of physostomes use the gas bladder for aerial respiration. Included among these are *Polypterus, Lepisosteus, Amia, Arapaima,* and several other osteoglossoids, *Umbra, Megalops* and *Erythrinus.* The latter is a characoid of South American swamps, and is notable for having a secondarily acquired gas bladder structure very similar to a lung. The best development of the gas bladder as a lung is in the lungfishes, especially in the African and South American species, all of which have the lung divided into right and left sections. The Australian lungfish has an undivided lung. All have a special pulmonary circulation.

One of the most remarkable of adaptations for the use of atmospheric oxygen is the modification of parts of the alimentary canal for respiratory purposes. Some South American armored catfishes, *Plecostomus* and *Ancistrus,* for example, use the stomach as a respiratory organ. Many loaches, Cobitidae, the swamp eel, *Monopterus,* and several of the armored catfishes are intestinal breathers, using a large section of the gut exclusively as a "lung."

Several air breathers have lost the capability to keep themselves supplied with oxygen from the water, even when in well aerated situations, so if restrained from reaching the surface they soon die. Such obligate air breathers include *Lepidosiren paradoxa,* the South American lungfish, *Protopterus* spp., the African lungfishes, *Arapaima gigas, Electrophorus,* the electric eel, *Channa* spp., the snakeheads, and *Hoplosternum,* a South American armored catfish. Many others must rely on air breathing during periods of low dissolved oxygen, or during excursions out of the water, but their gills can maintain their respiratory needs only if the dissolved oxygen content of the water is high enough. Those species that aestivate, spending the dry season buried in the mud, in burrows, or even in mucous cocoons as in the case of the lungfishes, must be able to maintain themselves by breathing air over periods lasting for months. Other than the African and South American lungfishes, fishes known to aestivate include *Amia calva,* the bowfin, *Umbra limi,* a mudminnow, *Clarias* spp., the walking catfishes, *Synbranchus* and *Amphipnous,* swamp eels, Channidae, the snakeheads, and *Anabas scandens,* the climbing perch.

Other than certain species found typically at the ocean's edge, where they may expose themselves during the changes of the tide, there are few marine fishes capable of aerial respiration. Tarpon, the young of which are commonly found in brackish or fresh water, have a lung-like gas bladder. Some sharks of the genus *Chiloscyllium* have been observed to gulp air.

The Gas Bladder and Buoyancy

Although the gas bladder may have had its origin back in the early history of fishes as a lung, its major function among teleosts presently appears to be to help provide buoyancy. Indeed, the gas bladder is the major means of maintaining neutral buoyancy or something close to it. Except for fats, the tissues of a fish have densities greater than water. Scales and bone have a specific gravity of about 2.0, and other tissues are in the 1.05 to 1.1 range. Fats and oils have a specific gravity of about 0.9 to 0.93. The combination of all tissues in a fish without a gas bladder or some other device for maintaining buoyancy causes the specific gravity to be in the range of 1.06 to 1.09. Because fresh water has a specific gravity of 1 and the oceans about 1.026, such fishes are destined to sink or must exert enough continuous effort to prevent sinking.

There are numerous species of demersal fish that rest on the bottom or even burrow into it. These would be, for the most part, hindered by neutral buoyancy and must have no gas bladder or have it greatly reduced. Sculpins, flatfishes, and clingfishes are examples among the teleosts. Gurnards, which are rather heavily built and live mostly on the bottom, gain lift from large pectorals and a small gas bladder when they swim off the substrate. Sharks and rays have no gas bladders, and most are demersal in habit. Those elasmobranchs that are pelagic or that spend considerable time off the bottom must swim constantly to prevent sinking. Large pectoral fins provide lift in many species of sharks, and of course the rays gain both lift and great physical resistance to sinking from their broad discs. Some elasmobranchs are aided in maintaining buoyancy by inclusion of much oil and a special hydrocarbon called squalene in their large livers. Squalene has a specific gravity of 0.86.

Although most bony fishes get only negligible buoyancy from fats and oils, these lipids can be of some significance in exceptionally oily species or in those with special inclusions of lipids in the body cavity, liver, or bones. Lipids may be of relatively greater importance in those species having reduced bones and watery flesh. The pelagic *Mola mola,* which has no gas bladder, is remarkably watery, as are numerous deepsea fishes. The little lift afforded by body fluids in marine fishes is due to the fact that body fluids have about half the salinity of sea water and are therefore a bit lighter. Marine fishes in which water, on a weight basis, is as much as 86% of the tissues, gain considerable buoyancy. Most bathypelagic species found below 1000 meters lack gas bladders, but benthic fishes found as deep as 7000 meters have small gas bladders. At pressures found at the latter depth, the specific gravity of oxygen is 0.7.

Fishes that make their living near the surface, in midwater, or free-swimming close to the bottom gain an advantage from a gas bladder in that they are relieved of the necessity of maintaining a chosen depth by muscular effort, which in the absence of a gas bladder amounts to a force of about 5% of the body weight in sea water or about 7% in fresh water. Gaining and maintaining the advantage of weightlessness is not simple for most species because of the change in pressure with depth, which is about 1 atmosphere per 10 meters. A descending fish with a given quantity of gas in the gas bladder becomes less buoyant as the gas compresses and the fish consequently displaces less water. A fish descending from the surface to 10 meters has the gas bladder compressed to one half its volume at the surface. An ascending fish faces a problem that can be very serious; the release of pressure on the gas bladder allows it to expand, and if no relief valve is provided, the other internal organs can be greatly crowded and the stomach can be forced out into the mouth. These drastic results do not generally occur naturally, but can be seen in fish brought from deep water by fishermen. Sometimes in a forced ascent, if the fish is being brought from a sufficient depth, the buoyancy of the expanding bladder can become great enough so that the fish cannot overcome it to swim back to the accustomed depth. Furthermore, dragging a fish up from a great enough depth can result in a ruptured gas bladder. In natural vertical movements, fishes generally do not move rapidly through a great enough depth range to bring about more than a 25% change in gas bladder volume. At great depths the change in pressure in terms of percentage is small for each 10-meter change in depth. For instance, ascending from 100 to 90 meters increases the volume of the gas bladder only 10%. Those species moving through greater ranges are specially equipped with means of emptying and filling the gas bladder.

Obviously, the presence, absence, size, structural modifications, and placement of gas bladders have relationships to the ecology and habits of fishes. The matter of benthic fishes with small or no gas bladders has already been mentioned. These depend upon friction with the bottom materials, wedging into tight places, digging in, or holding on with special suction surfaces to maintain position. Certain flatfishes are known to be 5% denser than sea water. Those fishes that swim up to capture food above the bottom can derive some benefit from a small gas bladder. Some of these have gas bladders of about 0.3 to 1% of their volume. Marine species that live off the bottom but confine their activities to narrow depth ranges usually have gas bladders of about 5 to 5.6% of their volume, compared to about 7 to 10.6% in freshwater species. These fishes can hold themselves motionless with comparative ease, using slight fin or body movements to counteract weak current. Many of these species may be slightly heavier than water and sink slowly between adjustments of position. Deep-bodied fishes usually have the gas bladder placed above the center of gravity so that no effort is required to hold the body upright. In some slender fishes, the organ is below the center of gravity and the fish must use fin motion to maintain an even keel. These will go "belly-up" when anesthetized.

Species of fish adapted to running water tend to have smaller gas bladders than still-water fishes, as the less buoyant condition is favorable to maintaining a given station in the stream. In related stream fishes, those species habitually living in the swiftest water have less capacious gas bladders than those that live in slower currents. Furthermore, in laboratory experiments, individuals of stream species reared in still water proved to have larger gas bladders than those reared in a current. Within a given species there is a general capability to adjust buoyancy to suit the current encountered. Such adjustment may be made to aid in maintaining a station in swift water (reduction of buoyancy), or to relieve the muscular effort required to maintain position as the current changes to a lower velocity (increase of buoyancy). Over the course of the life history of some species, the relative volume of the gas bladder changes considerably. In the longnose dace *(Rhinichthys cataractae)*, as well as similar fish, the young live in slow water at stream margins and have a relatively large gas bladder. As the fish become older and move into swift current, the gas bladder grows at a much slower rate than the body of the individual. In certain anadromous trout and salmon, the young are adapted to a stream life for periods of up to 3 years or more. During this time the young are less buoyant than they will be when they begin their downstream migration to the sea. Becoming more buoyant in a current serves as a dispersal mechanism.

Vertical Movement. The problem of vertical movement has been solved by a number of species. Many powerful scombroids lack a gas bladder and govern their depth mainly by rapid swimming, using the pectoral fins for lift when necessary. The larger the pectorals, the less speed needed to maintain the lift. Such fishes can make rapid changes in depth. Examples of scombroids without gas bladders are Spanish mackerels and close relatives (*Scomberomorus* spp.), skipjacks (*Euthynnus* spp.), and mackerels (*Scomber* spp.). Larger tunas and their xiphioid relatives (billfishes and swordfish) have gas bladders and consequently can rely on a slower swimming speed than smaller species, although their vertical movements near the surface are restricted unless the gas bladders have extremely tough walls. The yellowfin tuna, *Thunnus albacares*, is interesting in that its gas bladder is not inflated in fish of about 2 kg, and these small fish have a density of about 1.09. Quick growth and inflation of the bladder reduces the density of 10-kg fish to about 1.05.

Many mesopelagic fishes perform daily vertical migrations, so that the greatest concentrations of the species will be at the surface (or near it) during the night, and at depths of from 400 meters to greater than 500 meters in the daytime. Others may move from greater depths to a nocturnal depth of about 150 meters. The lanternfish family (Myctophidae) has a greater number of species that migrate to the surface at night than the other mesopelagic families. Other families with vertical migrants among the members include Chauliodontidae, viperfishes, Stomiatidae, scaly dragonfishes, Astronesthidae, snaggletooths, Melanostomiatidae, scaleless dragonfishes, and Idiacanthidae, stalkeyes. Some families have members that migrate vertically but do not reach the surface. Included are the Melamphaidae, or bigheads, and the Sternoptychidae, hatchetfishes. Certain epipelagic fishes, such as *Clupea harengus* and other herrings, are known to go as deep as 150 meters and return to the surface in the course of a day.

Many of these vertical migrants have gas-filled bladders and face the difficulties inherent in moving captured gas through pressure changes of 15 to 50 atmospheres or more. The problems are greatest for those that move all the way to the surface. Some species have avoided the problems in part by filling the gas bladder partially or completely with fat, as in the case of the black scabbardfish and certain macrourids (rattails), or by reducing the gas bladder and investing it with fat, as is found in some of the lanternfishes and anglemouths. Some lipids in or around the gas bladder of lanternfishes have a specific gravity close to that of squalene. For the most part, however, gas is maintained in the bladder by special structures that can secrete or absorb gas. Of course, some physostomous fishes retain the ability to release gas through the pneumatic duct or, in the herrings, through a posterior duct opening behind the anus, but even these must secrete gas into the lumen of the gas bladder if a reasonable hydrostatic balance is to be maintained. There is little evidence indicating that perfect hydrostatic balance is kept throughout vertical migration; many lanternfish species netted at the surface are heavier than sea water.

Gas Resorption. Gas bladders are nearly impervious to gas because of the structure of the wall. There are usually four layers, the outer one consisting of densely woven but elastic fibers. The next layer is of more loosely organized fibers and the inner two are smooth muscle and epithelium. In many species the wall is gas-proofed by a layer of guanine crystals just below the outer elastic fiber layer. Tiny overlapping platelets may aid in gas-proofing gas bladders of certain poeciliids. Wall thickness is usually in the range of 50 to 300 microns, but exceptions, such as *Aphanopus carbo*, the black scabbardfish, with walls 1.5 mm thick must be noted. The problem of removing gas from a gas-tight sac is overcome by special areas in the wall where a rich bed of capillaries can be exposed to the lumen of the gas bladder. The capillaries are usually disposed in a subcircular or oval area in the dorsal wall of the gas bladder, and can be isolated from the lumen during the nonabsorbent phase of operation by a sphincter. (Because of its shape, the structure is usually called the "oval.") Relaxation of the sphincter and contraction of radial muscles in the oval bring about maximum exposure of the capillaries to the gases in the bladder. There are exceptions to the arrangement described. In some fishes the capillaries cannot be isolated, but resorption can be prevented by constriction of the capillaries or by thickening of the epithelial lining by muscle contractions.

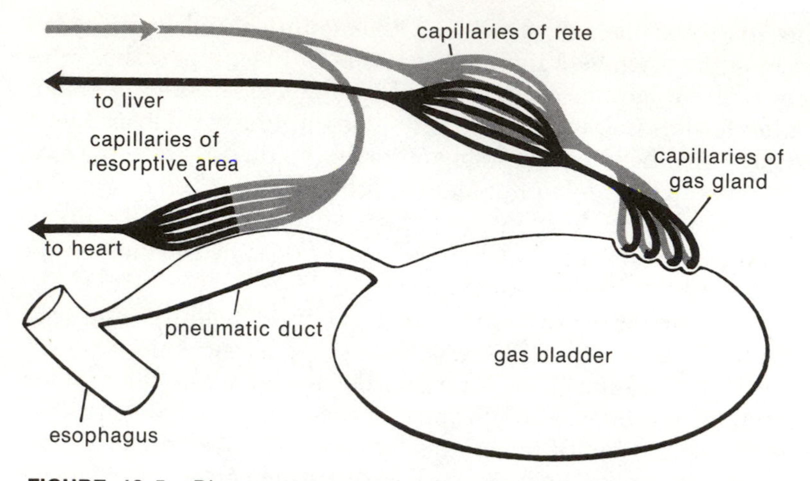

FIGURE 12–5. Diagram representing gas bladder of *Anguilla,* showing relationships of rete and gas gland, and resorptive area on pneumatic duct.

In the eels of the genus *Anguilla,* the pneumatic duct is modified for resorption of gas (Fig. 12–5). In several percoid fishes, pipefishes, sticklebacks and others a diaphragm separates the posterior, or gas-resorbing, part of the gas bladder from the anterior, gas-producing section. Such fishes are called "euphysoclistic," distinguishing them from those "paraphysoclists" in which the gas-secreting and gas-diffusing areas are not well separated (Fig. 12–6). Gas resorption in many euphysoclists is accompanied by contraction of the gas-secreting chamber and may involve anterior displacement of the diaphragm as well as expansion of the resorbent chamber.

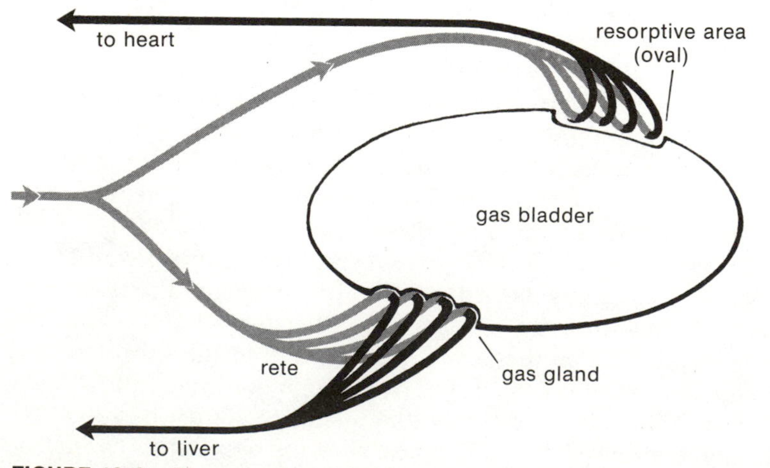

FIGURE 12–6. Diagram of gas bladder, rete, and resorptive oval of physoclistous fish.

The veins leading from the oval or other resorbent structure lead into the cardinal vein in most physoclists. In the eel, blood leaves the resorbent structure via the pneumatic duct vein, which proceeds directly to the heart. Because the blood with its load of gas thus must pass through the gills before being distributed to the systemic circulation, there is an opportunity for excess gas to be diffused into the surrounding water. In considering the rapid ascents made by lanternfishes moving from 500 meters to the surface, the resorption process might seem too slow for proper adjustment of gas bladder volume to be made. Certainly the rates of resorption calculated for freshwater fishes would not allow for rapid ascent. However, anatomical studies of the gas resorption apparatus in lanternfishes reveal large areas of capillaries in relation to the volume of the gas bladder as well as capacious arteries and veins serving the oval. Capillaries of the oval are separated from the lumen of the bladder by tissue 1 micron thick, or less. Furthermore, most vertical migrants are small, usually less than 100 mm, and have a small gas bladder in relation to their size, as buoyancy is often aided by inclusion of lipid.

Gas Secretion. Filling the gas bladder on downward migration cannot be as easily accomplished as the removal of gas. Instead of diffusion from a high gas tension to a lower tension in the blood, at a rate governed by the tensions, temperature, area of the capillary bed, and rate at which the blood flows through the bed, gas must somehow be secreted from the lower pressure in the blood into the higher pressure in the gas bladder. Oxygen, for instance, found essentially at a pressure of 0.2 atmosphere or less in the blood, must be secreted into the gas bladder to pressures up to hundreds of atmospheres. The apparatus by which the secretion is accomplished consists of bundles of arterial and venous capillaries running counter to each other, plus a specialized bed of capillaries from which gases are actually secreted into the gas bladder. The capillary bundles comprise the rete mirabile, or wonder net. The name is somewhat inappropriate, as a network is not formed; the small vessels run parallel to each other, forming an efficient countercurrent structure that serves to concentrate gases at the site of the gas gland. Depending on species, there may be from a few hundred to over 200,000 capillaries in the rete. The process of concentration of gas is aided by the secretion of lactic acid into the blood by the gas gland. This acid increases the partial pressure of carbon dioxide in the blood by releasing CO_2 from bicarbonate, but the greatest effects are on the partial pressure of oxygen. Acid activates the Bohr effect, so that higher partial pressures are required to keep oxygen in combination with hemoglobin. The Root effect is operative as the pH is lowered, so the quantity of oxygen that can combine with hemoglobin, regardless of how high the partial pressure, is decreased. There is, in addition, a small "salting out" effect—the acid ions decrease the amount of gas that can be held in solution.

Because of the effects noted above, the blood leaving the gas gland through the venous capillaries of the rete has greater partial pressures of gases, especially oxygen, than the blood in the arterial capillaries. The two sets of capillaries are intimately associated, so that a cross section of the rete would show them in a checkerboard or mosaic pattern. The capillaries of the rete are very long in relation to other capillaries. In some relatively large deepsea fishes they are up to 25 mm long, but are shorter in fishes from shallow depths. For instance, the rosefishes *(Sebastes marinus)*, 30 to 45 cm long, have retes of 7 to 10 mm long. The rete of *Anguilla* is about 4 mm long. The length and great number of capillaries allow ample opportunity for the oxygen at high partial pressure in the venous capillaries to diffuse to the blood in the arterial capillaries. Eventually, the pressure in the small vessels is higher than that in the lumen of the gas bladder, so that bubbles of gas are released at the gas gland, serving to inflate the bladder.

Structure of gas glands differs among species. The glandular epithelium may be single-celled, multicellular, or folded. Capillaries penetrate between folds or may enter the giant cells found in some species. Formation of bubbles has been observed by investigators who placed plastic film over the gas gland of an opened gas bladder and stimulated secretion of gas by injections of the drug yohimbine.

The gases contained in the gas bladders of shallow-water fish usually consist mainly of nitrogen and oxygen in the proportions found in the atmosphere— about 80% nitrogen and 20% oxygen. Members of the Salmonidae tend to have greater proportions of nitrogen, although in some species gas secreted into the gas bladder may be high in oxygen that is later exchanged for nitrogen by diffusion. Salmonids do not have well developed retia or gas glands, so that gas enters the lumen of the bladder over a wide area of the bladder wall. Gases contained in gas bladders of deepsea fish contain higher proportions of oxygen than nitrogen. A deepsea eel, *Synaphobranchus*, has been noted as having 75.1% oxygen, 20.5% nitrogen, 3.1% carbon dioxide, and 0.4% argon in the gas bladder. Various deepwater species including lanternfishes contain from 76% to 88% oxygen in the gas bladder, but lanternfishes captured at the surface show about 42% oxygen.

Muscle and Choroid Retes. Although not concerned with the gas bladder, there are other retial systems in fishes. There is a connection of a retial system with hydrodynamic equilibrium in the scombroids and the lamnid sharks, groups in which most members must swim constantly and vigorously to stay at a given depth. This rete is associated with the lateral red muscle, the site of nearly constant activity that is favored by a temperature higher than the medium in which the fish swims. The red muscle is surrounded by white muscle, supplied with blood by large cutaneous arteries, and drained by lateral cutaneous veins. A rete is imposed between the blood vessels mentioned and the red muscle, forming a countercurrent system that functions in the exchange of heat (Fig. 12–7). The heat generated by the activity of the red muscle is conserved deep within the fish.

Another retial system is found in the choroid of the eyes of teleosts and *Amia*. The choroid rete forms a horseshoe-shaped body around the

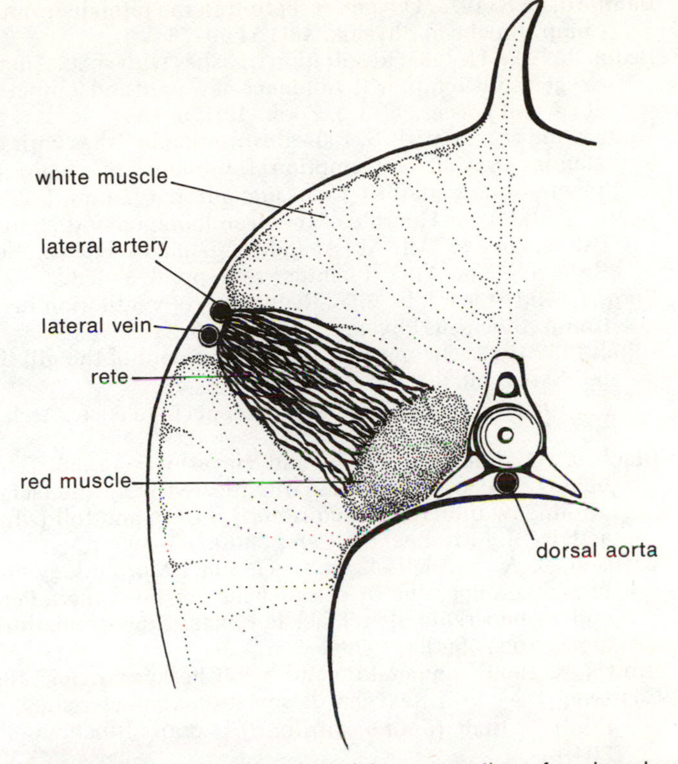

FIGURE 12–7. Diagram of partial cross section of porbeagle shark, showing relationships of lateral blood vessels and rete to deep-seated red musculature.

optic nerve and serves to maintain a high interocular partial pressure of oxygen. Predators that depend heavily on sight have large choroid retes and interocular partial pressures up to more than 1300 mm Hg. More sedentary fish may have partial pressures of only 25% of that figure. Elasmobranchs do not have choroid retes and may have interocular partial pressure of oxygen as low as 30 mm Hg.

References

Circulation and Respiration

Alabaster, J. S., and Welcomme, R. L. 1962. Effect of concentration of dissolved oxygen on survival of trout and roach in lethal temperatures. Nature, 194:107.

Baglioni, S. 1908. Der Attmungmechanismus der Fische. Z. allgem. Physiol., 7:177–282.

Bamford, O. S. 1974. Oxygen reception in the rainbow trout *(Salmo gairdneri)*. Comp. Biochem. Physiol., 48(1A):69–76.

Beamish, F. W. H. 1964. Respiration in fishes with special emphasis on standard oxygen consumption. II. Influence of weight and temperature on respiration of several species. Can. J. Zool., 42:176–188.

———, and Mookerjii, P. S. 1964. Respiration in fishes with special emphasis on standard oxygen consumption. I. Influence of weight and temperature on respiration of goldfish, *Carassius auratus* L. Can. J. Zool., 42:161–175.

Beatty, D. D. 1975. The role of the pseudobranch and choroid rete mirabile in fish vision. *In* Ali, M. A. (ed.), Vision in Fishes; New Approaches in Research. New York, Plenum Press, pp. 673–678.

Berg, T., and Steen, J. B. 1966. Regulation of ventilation in eels exposed to air. Comp. Biochem. Physiol., 18:511–516.

Bijtel, J. A. 1947. The mechanism of movement of the gill filaments in Teleostei. Experientia, 3:158–165.

———. 1949. The structure of gill filaments in teleosts. Arch. Ned. Zool., 8:267–283.

Black, E. C., Manning, G. T., and Hayashi, K. 1966. Changes in levels of hemoglobin, oxygen, carbon dioxide, pyruvate and lactate in venous blood of rainbow trout *(Salmo gairdneri)* during and following severe muscular activity. J. Fish. Res. Bd. Can., 23(6):783–795.

Branson, B. A., and Ulrickson, G. U. 1967. Morphology and histology of the branchial apparatus in percid fishes of the genera *Percina, Etheostoma,* and *Ammocrypta* (Percidae: Percinae: Etheostomatini). Trans. Am. Microsc. Soc., 86(4):371–389.

Brett, J. R. 1969. Temperature and fish. Chesapeake Sci., 10(3 & 4):275–276.

Cameron, J. N. 1971. Oxygen dissociation characteristics of the blood of the rainbow trout *(Salmo gairdneri)*. Comp. Biochem. Physiol., 38A:699–704.

———. 1974. Evidence for the lack of bypass shunting in teleost gills. J. Fish. Res. Bd. Can., 31(2):211–213.

———. 1975. Morphometric and flow indicator studies of the teleost heart. Can. J. Zool., 53(6):691–698.

———, Randall, D. J., and Davis, L. C. 1971. Regulation of the ventilation-perfusion ratio in the gills of *Dasyatis sabina* and *Squalus suckleyi*. Comp. Biochem. Physiol., 39A:505–519.

Carey, F. G. 1973. Fishes with warm bodies. Sci. Am., 228:36–44.

———, Teal, J. M., Kanwisher, J. W., Lawson, K. D., and Beckett, J. 1971. Warm-bodied fish. Am. Zool., 11(1):137–145.

Copeland, D. E. 1974. The anatomy and fine structure of the eye in teleosts: II. The vascular connections of the lentiform body in *Fundulus grandis*. Exp. Eye Res., 19(6):583–590.

DeVries, A. L. 1971. Freezing resistance in fishes. *In*: Hoar, W. S., and Randall, D. J. (eds.), Fish Physiology. Vol. 6. New York, Academic Press, pp. 157–190.

Duman, J. G., and DeVries, A. L. 1974. The effects of temperature and photoperiod on antifreeze production in cold water fishes. J. Exp. Zool., 190(1):89–98.

Feeney, R. E., and Hofman, R. 1973. Depression of freezing point by glycoproteins from an antarctic fish. Nature, 243:357–359.

Fry, F. E. J. 1947. Effect of the environment on animal activity. University of Toronto, Biology Series, SS/Publ., Ontario Fish. Res. Lab., 68:1–62.

———. 1971. The effect of environmental factors on the physiology of fish. *In*: Hoar, W. S., and Randall, D. J. (eds.), Fish Physiology, Vol. 6. New York, Academic Press, pp. 1–98.

Hawkins, R. I., and Mawdesley-Thomas, L. E. 1972. Fish haematology — a bibliography. J. Fish Biol., 4:193–232.

Hayden, J. B., Lech, J. J., Jr., and Bridges, D. W. 1975. Blood oxygen dissociation characteristics of the winter flounder, *Pseudopleuronectes americanus*. J. Fish Res. Bd. Can., 32(9):1539–1544.

Hemmingson, E. A., and Douglas, E. L. 1972. Respiratory and circulatory responses in a hemoglobin-free fish, *Chaenocephalus aceratus*, to changes in temperature and oxygen tension. Comp. Biochem. Physiol., 43A(4A):1031–1043.

Hughes, G. M. 1966. The dimensions of fish gills in relation to their functions. J. Exp. Biol., 45:177–195.

———. 1972. The relationship between cardiac and respiratory rhythms in the dogfish, *Scyliorhinus canicula* L. J. Exp. Biol., 57:415–434.

———, and Grimstone, A. V. 1965. The fine structure of the secondary lamellae of the gills of *Gadus pollachius*. Q. J. Microsc. Sci., 106:343–353.

———, and Shelton, G. 1962. Respiratory mechanisms and their nervous control in fish. Adv. Comp. Physiol. Biochem., 1:275–364.

———, Singh, B. R., Thakur, R. N., and Munshi, J. S. D. 1974. Areas of the air breathing surfaces of *Amphipnous cuchia* (Ham.). Proc. Ind. Nat. Sci. Acad., Part B, 40(4):379–392.

Hunn, J. B. 1967. Bibliography on the blood chemistry of fishes. U.S. Fish and Wildlife Service, Res. Rep., 72:1–32.

Jager, S. de, and Dekkers, W. J. 1975. Relations between gill structure and activity in fish. Netherl. J. Zool., 25(3):276–308.

Johansen, K. 1966. Air breathing in the teleost *Synbranchus marmoratus*. Comp. Biochem. Physiol., 18:383–395.

———. 1970. Air breathing in fishes. *In:* Hoar, W. S., and Randall, D. J. (eds.), Fish Physiology, Vol. 4. New York, Academic Press, pp. 361–409.

Johnston, I. A. 1975. Anaerobic metabolism in the carp (*Carassius carassius* L.). Comp. Biochem. Physiol., 51B(2):235–241.

Kashiwagi, M., Yoshida, F., and Sato, R. 1968. Blood cell constituents of chum salmon, *Oncorhynchus keta* (Walbaum), with special reference to erythrocyte series according to growth of the fish from hatching. Tohoku J. Agr. Res., 19(3):188–194.

Katz, M. 1951. The number of erythrocytes in the blood of the silver salmon. Trans. Am. Fish. Soc., 80:184–193.

Landolt, J. C., and Hill, L. G. 1975. Observations of the gross structure and dimensions of the gills of three species of gars (Lepisosteidae). Copeia, 1975(3):471–475.

Lykkeboe, G., Johansen, K., and Maloiy, G. 1975. Functional properties of hemoglobins in the teleost *Tilapia grahami*. J. Comp. Physiol., 104(1):1–11.

Lysak, A. 1959. Hematologic observations on the small whitefish (*Coregonus albula* L.) and on the hybrids of small whitefish × whitefish (♂ *C. albula* L. × ♀ *C. lavaretus maraenoides* Poljakov). Acta Hydrobiol., 1,2:139–147.

———. 1961. Further investigations on the influence of blood sampling in carp on their blood picture and rate of growth. Acta Hydrobiol., 3,4:261–279.

Magid, A. M. A., and Babiker, M. M. 1975. Oxygen consumption and respiratory behavior of three Nile fishes. Hydrobiologia, 46(4):359–367.

Meijer, N. W. 1975. Cranial motor nerves innervating superficial respiratory muscles in carp (*Cyprinus carpio* L.). Netherl. J. Zool., 25(1):103–113.

Nash, A. R., and Thompson, E. O. P. 1974. Haemoglobins of the shark, *Heterodontus portusjacksoni*. Aust. J. Biol. Sci., 27(6):607–615

Ngan, P. V., Hamamori, K., Hanyu, I., and Hibiya, T. 1974. Measurement of blood pressure of carp. Jap. J. Ichthyol., 21(1):1–8.

Pasztor, V. M., and Kleerekoper, H. 1962. The role of the gill filament musculature in teleosts. Can. J. Zool., 40(5):785–802.

Peterson, G. L., and Shehadeh, Z. H. 1971. Changes in blood components of the mullet, *Mugil cephalus* L. following treatment with salmon gonadotropin and methyltestosterone. Comp. Biochem. Physiol., 38B:451–457.

Piiper, J., and Baumgarten-Schumann, D. 1967. Efficiency of oxygen exchange in the gills of the dogfish *(Scyliorhinus stellaris)*. Resp. Physiol., 2:135–148.
———. 1968. Effectiveness of O_2 and CO_2 exchange in the gills of the dogfish *(Scyliorhinus stellaris)*. Resp. Physiol., 5:338–349.
Priede, I. G. 1975. The blood circulatory function of the dorsal aorta ligament in rainbow trout *(Salmo gairdneri)*. J. Zool. Lond., 175(1):39–52.
Randall, D. J. 1970. The circulatory system. *In:* Hoar, W. S., and Randall, D. J. (eds.), Fish Physiology, Vol. 4. New York, Academic Press, pp. 133–172.
———. 1970. Gas exchange in fish. *In:* Hoar, W. S., and Randall, D. J. (eds.), Fish Physiology, Vol. 4. New York, Academic Press, pp. 253–292.
Richards, B. D., and Fromm, P. O. 1969. Patterns of blood flow through filaments and lamellae of isolated-perfused rainbow trout *(Salmo gairdneri)* gills. Comp. Biochem. Physiol., 29:1063–1070.
Riggs, A. 1970. Properties of fish hemoglobins. *In:* Hoar, W. S., and Randall, D. J. (eds.), Fish Physiology, Vol. 4. New York, Academic Press, pp. 209–252.
Roberts, J. L. 1975. Active branchial and ram gill ventilation in fishes. Biol. Bull., 148(1):85–105.
Satchell, G. H. 1971. Circulation in Fishes. Cambridge, Cambridge University Press.
Saunders, D. C. 1968. Differential blood cell counts of 50 species of fishes from the Red Sea. Copeia, 1968(3):491–498.
Saunders, R. L. 1961. The irrigation of the gills in fishes. I. Studies of the mechanism of branchial irrigation. Can. J. Zool., 39(5):637–653.
———. 1962. The irrigation of the gills in fishes. II. Efficiency of oxygen uptake in relation to respiratory flow activity and concentrations of oxygen and carbon dioxide. Can. J. Zool., 40(5):817–862.
Shelton, G. 1970. The regulation of breathing. *In:* Hoar, W. S., and Randall, D. J. (eds.), Fish Physiology, Vol. 4. Academic Press, pp. 293–359.
Smith, L. S. 1966. Blood volumes of three salmonids. J. Fish. Res. Bd. Can., 23(9):1439–1446.
———, Brett, J. R., and Davis, J. C. 1967. Cardiovascular dynamics in swimming adult sockeye salmon. J. Fish. Res. Bd. Can., 24(8):1775–1790.
Smith, H., Vandenberg, R. J., and Kijne-den Hartog, I. 1974. Some experiments on thermal acclimation in the goldfish (*Carassius auratus* L.). Netherl. J. Zool., 24(1):32–49.
Steen, J. B., and Berg, T. 1966. The gills of two species of haemoglobin-free fishes compared to those of other teleosts — with a note on severe anemia in an eel. Comp. Biochem. Physiol., 18:517–526.
———, and Kruysse, A. 1964. The respiratory function of teleostean gills. Comp. Biochem. Physiol., 12:127–142.
Stevens, E. D., Bennion, G. R., Randall, D. J., and Shelton, G. 1972. Factors affecting arterial pressures and blood flow from the heart in intact, unrestrained lingcod, *Ophiodon elongatus*. Comp. Biochem. Physiol., 43:681–696.
———, Lam, H. M., and Kendall, J. 1974. Vascular anatomy of the countercurrent heat exchanger of skipjack tuna. J. Exp. Biol., 61(1):145–153.
Wittenberg, J. B., and Haedrich, R. L. 1974. The choroid rete mirabile of the fish eye. II. Distribution and relation to the pseudobranch and to the swimbladder rete mirabile. Biol. Bull., 146:137–156.
———, and Wittenberg, B. A. 1962. Active secretion of oxygen into the eye of fish. Nature, 194:106–107.

The Gas Bladder and Buoyancy Control

Alexander, R. M. 1961. The physical properties of the swim bladders of some South American Cypriniformes. J. Exp. Biol., 36:347–355.

———. 1966. Physical aspects of swimbladder function. Biol. Rev., 41:141–176.
———. 1970. Functional Design in Fishes, 2nd ed. London, Hutchinson.
———. 1975. The Chordates. London, Cambridge Press.
Bass, A. J., and Ballard, J. A. 1972. Buoyancy control in the shark *Odontaspis taurus* (Rafinesque). Copeia, 1972(3):594–595.
Beldridge, H. D., Jr. 1972. Accumulation and function of liver oil in Florida sharks. Copeia, 1972(2):306–325.
Bone, Q. 1971. On the scabbard fish *Aphanopus carbo.* J. Mar. Biol. Assoc. U.K. 51:219–225.
———. 1973. A note on the buoyancy of some lanternfishes (Myctophoidei). J. Mar. Biol. Assoc. U.K., 53:619–633.
Borjeson, H., and Hoglund, L. B. 1976. Swimbladder gas and Root effect in young salmon during hypercapnia. Comp. Biochem. Physiol., 54A(3):335–339.
Butler, J. L., and Pearcy, W. G. 1971. Swimbladder morphology and specific gravity of myctophids off Oregon. J. Fish. Res. Bd. Can., 29(8):1145–1150.
Crawford, R. H. 1974. Structure of an air-breathing organ and the swimbladder of the Alaska blackfish *Dallia pectoralis* Bean. Can. J. Zool., 52(10):1221–1225.
D'Aoust, B. A. 1969. Hyperbaric oxygen: toxicity to fish at pressures present in their swimbladders. Science, 163(3867):576–578.
———. 1970. The role of lactic acid in gas secretion in the teleost swimbladder. Comp. Biochem. Physiol., 32(4):637–668.
Denton, E. J. 1961. The buoyancy of fish and cephalopods. Progr. Biophys. Chem., 11:178–234.
———. 1963. Buoyancy mechanisms of sea creatures. Endeavour, 22:3–8.
———. 1968. The buoyancy of marine animals. Sci. Am., 203:119–128.
Fange, R. 1966. Physiology of the swimbladder. Physiol. Rev., 46:299–322.
———, Holmgren, S., and Nilsson, S. 1975. Drug effects on gas production and smooth muscles in a fish swimbladder. Acta. Physiol. Scand., 95(2):38A–39A.
Gee, J. H. 1968. Adjustment of buoyancy by longnose dace (*Rhinichthys cataractae)* in relation to velocity of water. J. Fish. Res. Bd. Can., 25:1485–1496.
———. 1970. Adjustment of buoyancy in blacknose dace, *Rhinichthys atratulus.* J. Fish. Res. Bd. Can., 27:1855–1859.
———. 1972. Adaptive variation in swimbladder length and volume in dace, genus *Rhinichthys.* J. Fish. Res. Bd. Can., 29:119–127.
———. 1974. Behavioral and developmental plasticity of buoyancy in the longnose, *Rhinichthys cataractae,* and blacknose, *R. atratulus* (Cyprinidae) dace. J. Fish. Res. Bd. Can., 31:35–41.
———, and Gee, P. A. 1976. Alteration of buoyancy by some Central American streamfishes, and a comparison with North American species. Can. J. Zool., 54(3):386–391.
———, Machniak, K., and Chalanchuk, S. M. 1974. Adjustment of buoyancy and excess internal pressure of swimbladder gases in some North American freshwater fishes. J. Fish. Res. Bd. Can., 31:1139–1141.
———, and Northcote, T. G. 1963. Comparative ecology of two sympatric species of dace *(Rhinichthys)* in the Fraser River system, British Columbia. J. Fish. Res. Bd. Can., 20(1):105–118.
Hemmingsen, E. A. 1975. Clathrate hydrate of oxygen: does it occur in deep-sea fish? Deep-Sea Res., 22:145–149.
Horn, M. H. 1975. Swim bladder state and structure in relation to behavior and mode of life in stromateoid fishes. Fish. Bull. U.S., 73(1):95–109.
Jones, F. R. H., and Marshall, N. B. 1953. The structure and function of the teleostean swimbladder. Biol. Rev., 28:16–83.

Lee, R. F., Fuller, C. F., and Horn, M. H. 1975. Composition of oil in fish bones: possible function in neutral buoyancy. Comp. Biochem. Physiol., 50B:13–16.

Machniak, K., and Gee, J. H. 1975. Adjustment of buoyancy by tadpole madtom, *Noturus gyrinus,* and black bullhead, *Ictalurus melas,* in response to a change in water velocity. J. Fish. Res. Bd. Can., 32(2):303–307.

Magnuson, J. J. 1970. Hydrostatic equilibrium of *Euthynnus affinis,* a pelagic teleost without a gas bladder. Copeia, 1970 (1):56–85.

———. 1973. Comparative study of adaptations for continuous swimming and hydrostatic equilibrium of scombroid and xiphoid fishes. Fish. Bull. U.S., 71(2):337–356.

Marshall, N. B. 1960. Swimbladder structure of deep-sea fishes in relation to their systematics and biology. Disc. Rep., 31:1–122.

McCutcheon, F. H. 1958. Swimbladder volume, buoyancy and behaviour in the pinfish, *Lagodon rhomboides* (Linn.). J. Cell. Comp. Physiol., 52:453–480.

———. 1962. Swimbladder volume control in the pinfish, *Lagodon rhomboides* (Linn.). J. Cell. Comp. Physiol., 59:203–214.

Neave, N. M., Dilworth, C. L., Eales, J. G., and Saunders, R. L. 1966. Adjustment of buoyancy in Atlantic salmon parr in relation to changing water velocity. J. Fish. Res. Bd. Can., 23:1617–1620.

Pinder, L. J., and Eales, J. G. 1969. Seasonal buoyancy changes in Atlantic salmon *(Salmo salar)* parr and smolt. J. Fish. Res. Bd. Can., 26:2093–2100.

Qutob, Z. 1962. The swimbladder of fishes as a pressure receptor. Arch. Ned. Zool., 15:1–67.

Saunders, R. L. 1965. Adjustment of buoyancy in young Atlantic salmon and brook trout by changes in swimbladder volume. J. Fish. Res. Bd. Can., 22:335–352.

Scholander, P. F. 1954. Secretion of gases against high pressures in the swimbladder of deep sea fishes. II. The rete mirabile. Biol. Bull., 107(2):260–277.

———. 1957. The wonderful net. Sci. Am., 196:96–107.

———, and van Dam, L. 1954. Secretion of gases against high pressures in the swimbladder of deep sea fishes. I. Oxygen dissociation in blood. Biol. Bull., 107(2):247–259.

Steen, J. B. 1970. The swimbladder as a hydrostatic organ. *In:* Hoar, W. S., and Randall, D. J. (eds.), Fish Physiology, Vol. 4. New York, Academic Press, pp. 414–443.

Tait, J. S. 1956. Nitrogen and argon in salmonoid swimbladders. Can. J. Zool. 34:58–62.

Trewavas, E. 1933. On the structure of two oceanic fishes: *Cyema atrum* Gunther and *Opisthoproctus soleatus* Vaillant. Proc. Zool. Soc. Lond., (3): 601–614.

Wittenberg, J. B. 1958. The secretion of inert gas into the swim-bladder of fish. J. Gen. Physiol., 41:783–804.

13 OSMOREGULATION

Some of the most interesting adjustments that fishes of all kinds must make in their particular environments concern the maintenance of proper water and salt balance in their tissues. Few fishes have internal salt concentrations that closely match the water in which they swim, so they must be able to prevent excessive gain or loss of water physiologically. The body fluids of freshwater fishes have a higher osmotic concentration than their medium, and marine species have more dilute fluids than the sea water surrounding them. Both ecological types, then, are at what appears to be an osmotic disadvantage that had to be overcome in order for fishes to occupy the earth's waters. Regulation of osmotic concentration is accomplished by the kidney, the gills, some special organs and, to some extent, by the integument in its role as a barrier.

A simplified review of the expressions used in discussing salt and water balance might be in order at this point. One gram molecule of a substance in 1 liter of solution is called a molar solution, whereas 1 gram molecule per liter of solvent is referred to as a molal solution. In expressing osmotic activity, which depends on the number of undissociated molecules and ions per unit volume or weight of solvent, the term osmole is used. One gram molecule per liter (kg) of water of a substance that does not dissociate can be said to have an osmolality of 1 osmole/kg and exerts an osmotic pressure of 22.4 atmospheres. Compounds that dissociate have a higher osmolality, corresponding to the degree of dissociation. One mole of sodium chloride/kg has an osmolality close to 2 Osm/kg because the compound dissociates nearly completely in solution In dealing with the body fluids of animals it is convenient to use smaller units than osmoles, so the milliosmole (mOsm) is used.

Osmotic concentration can be expressed also in terms of freezing point depression (Δ_{fp}) of aqueous solutions. A molal solution (1000 mOsm) freezes at −1.86° C, which approximates the freezing point of surface water in temperate seas. Table 13–1 shows comparisons between freezing point depression and salinity, expressed as parts per thousand (ppt).

A solution with a smaller amount of salt per unit of volume than a solution to which it is being compared is said to be hypotonic to the more concentrated solution which, of course, is hypertonic to the less concentrated solution. These terms are often used in treatments of

Table 13–1. FREEZING POINT DEPRESSION OF WATER AT SELECTED SALINITIES

ppt	Δ_{fp} (°C)	m0sm/kg
5	−0.29	155
10	−0.58	312
15	−0.87	444
20	−1.13	608
25	−1.45	780
30	−1.72	925
32 (sea water)	−1.86	1000
35	−2.03	1091
40	−2.35	1263

osmotic regulation, but hyposmotic, hyperosmotic, and isosmotic are preferable.

Although diffusion will be regarded in simple terms for the sake of this treatment, it is a complex process involving such things as the nature of the membranes being penetrated, concentration of solutes, and electrical charges of particles. The discussion here will be mainly concerned with the net results of the process.

Freshwater Fishes

Because the osmotic concentration of typical freshwater fish blood as expressed in mOsm/kg is in the range of about 265 to 325 (Δ_{fp} = −0.5° to −0.61° C), they are hyperosmotic to their medium and tend to gain water by diffusion through any semipermeable surface. If unchecked or uncompensated the inward diffusion would dilute the body fluids to the point that the necessary physiological functions could no longer be carried out, a state referred to by some as internal drowning. Waterproofing the body would appear to be a means of preventing the diffusion, but can be only partially successful. A thick scale covering, or bony armor, or even large amounts of connective tissue in the skin might afford some protection, but any site that maintains circulation of blood near the surface of the skin will be a chink in the armor. The gills, obviously, cannot be waterproofed, and provide a great surface for diffusion of water as well as gases so, overall, there is no way of keeping the water out.

If water in excess of the needs of the organism is driven in by inexorable osmosis, a balance must be maintained by driving the water out through some other means. The task of pumping is accomplished by the kidney. Blood from the dorsal aorta is led to the kidney by the renal artery, where it passes through the capillaries of the glomeruli and then through capillaries surrounding the kidney tubules before leaving via the renal vein. Blood from the renal portal vein joins the network of capillaries around the tubules. The glomerulus is a filter that allows blood plasma containing dissolved materials to pass into the space between the walls of Bowman's capsule and thence into the kidney tubule. Blood cells and large molecules, such as proteins, cannot pass the filter. There would be no osmoregulatory advantage gained by freshwater fishes if the filtrate removed from the blood at the glomerulus were to be excreted with its normal complement of salts. The advantage is gained by excreting urine more dilute than the plasma. (Further physiological advantage is gained in some fishes by the excretion of small amounts of waste nitrogenous compounds in the urine.) As the fluid passes down the tubule, substances are resorbed at specific locations. Glucose is resorbed in the proximal tubule and salts are resorbed in the distal tubule, the walls of which are impermeable to water in many fishes. The urinary bladder appears to function in osmoregulation in teleosts. Its walls are impermeable to water, and it is the site of ion reabsorption. The resultant urine is quite dilute, with osmotic concentrations, in various species and conditions, from about 16 to 55 mOsm/kg ($\Delta_{fp} = -0.03°$ to $-0.09°$ C). The urine contains small amounts of nitrogenous compounds such as uric acid, creatine, creatinine, and ammonia.

Although the urine contains little salt, the copious flow causes a significant amount of salt to be lost. Salts are also lost by diffusion from the body. These losses are balanced by salt intake in food, and by active absorption through the gills.

Lampreys in Freshwater. Although lampreys have not been the subject of as much physiological study as elasmobranchs and bony fishes, their capabilities as osmoregulators have been explored. Osmotic concentration of the body fluids of lampreys in fresh water is about 230 to 280 mOsm/kg. About 70% of the body fluid is intracellular, about 7% is in the plasma, and the remainder is interstitial. The skin and gills of lampreys are permeable to water, but research findings on the extent of permeability have been variable. *Lampetra fluviatilis*, the river lamprey, is reported to absorb water at a rate equivalent to about one third the body weight per day; laboratory tests have shown that the skin allows water to move through at a rate of about 2.5 $\mu l/cm^2/hr$. *L. planeri*, the brook lamprey, apparently absorbs water at about twice the rate of the river lamprey.

Normal urine flow in freshwater lampreys is apparently not well known because of difficulties with laboratory methodology. Many values shown in the literature, ranging up to 360 ml/kg/day for river lamprey, are considered high by some authorities, but all agree that flow of the very dilute urine is copious. The amount of sodium excreted in the urine is about 117 μmole/hr; chloride is excreted at about 5 μmole/hr. Only small amounts of nitrogenous substances are excreted at the kidney. Most nitrogenous excretion is at the gills.

Salts lost by lampreys in the urine and through the body surface are balanced by salts in food and by direct absorption from the water. The gills actively remove chlorides from the water and release them

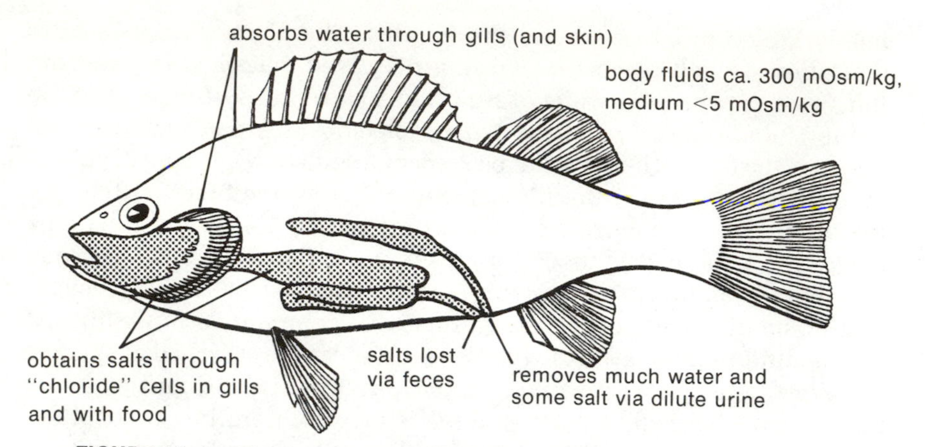

FIGURE 13–1. Diagram summarizing osmoregulation of freshwater teleost.

into the blood. Sodium, potassium, and calcium enter through the gills when made available as chlorides.

Freshwater Elasmobranchs. The matter of invasion of fresh waters by a few sharks and rays, most of which are essentially suited to marine waters, will be presented along with discussion of the marine elasmobranchs.

Freshwater Bony Fishes (Fig. 13–1). Permeability of bony fishes to water is generally less than that of lampreys, although some of the former have been noted as absorbing up to one third of their body weight per day. Scales and other armor in the bony fishes aid in retarding water uptake, as armor must have done in the extinct relatives of the lampreys. The eel, *Anguilla*, is often singled out as having a nearly impervious skin. Eel skin is reported to be so thick that it equals about 10% of the body weight. One investigator estimated that 5 years would be needed to pass 1 ml of water through 1 cm^2 of eel skin at a pressure of 1 atm. For lamprey skin, comparable time would be 91 days. For bony fishes, there is evidence that most of the water absorbed comes through the gills. The body water of freshwater teleosts makes up about 70 to 75% of body weight. Sturgeons, paddlefishes, gars and the bowfin generally have a similar proportion of water. In freshwater teleosts, according to T. B. Thorson, intracellular water accounts for about 60% (55 to 63%) of the total body water; about 12 to 16% is extracellular water, and plasma accounts for about 2% of body water. Figures given by G. Parry are somewhat higher — 74 to 80% as intracellular water, and 2.5 to 3% of the body fluids in plasma.

Osmotic concentration of freshwater fish blood is usually slightly less than 300 mOsm/kg but at least one species, the reedfish, *Erpetoichthys calabaricus*, has a dilute plasma of 199 mOsm/kg. Others, especially anadromous species returned to freshwater, may maintain concentrations around 325 mOsm/kg. Bony fishes move relatively large amounts of water through the kidneys. Urine flow varies with species, temperature, etc., but many determinations have shown between 50 and 150 ml/kg/day. Urine volumes as low as 16 ml/kg/day have been measured for the pike, and as high as 330 ml/kg/day for the goldfish. Osmotic concentration of bony fish urine is usually between 30 and 40 mOsm/kg.

Composition of urine of freshwater fish varies greatly among species and conditions, but commonly the following solutes appear in the amounts given (mmole/liter): sodium, 5 to 17; potassium, 1 to 5; calcium, 0.5 to 1; chloride, 2 to 10. Small amounts of nitrogenous compounds are excreted in the urine of freshwater fishes. In addition to urea and ammonia, which are excreted mainly through the gills, the urine contains creatine, amino acids, uric acid, and creatinine.

Mention should be made of the African lungfishes, *Protopterus* spp., which produce urea during aestivation and store it in the blood, producing no urine for several months.

As in lampreys, loss of salts in the urine and by diffusion is compensated by uptake from ingested food and by absorption through the gills. Uptake of sodium is in part related to the excretion of ammonia at the gills. Ion exchange mechanisms operating in branchial cells facilitate exchange of ammonia for sodium and bicarbonate for chloride. The exchange rate for sodium in freshwater species is very small compared with the exchange that occurs in saltwater species. Only about 1% of the exchangeable sodium in the body of the threespine stickleback (*Gasterosteus aculeatus*) is exchanged per hour in freshwater. The rate is 20 times that for the same fish in sea water. Total exchange of sodium for goldfishes has been reported to be 27 μ equiv/100 g/hr.

Saltwater Fishes

The osmotic concentration due to salts in marine fishes (except hagfishes) is less than that of sea water. Bony fishes adapted to marine life maintain blood at the osmolality of about 380 to 470 mOsm/kg ($\Delta_{fp} = -0.7°$ to $-0.87°$ C). The salt content of elasmobranch blood is slightly higher than that, but forms only a part of the blood's osmolality, the remainder being due to retention of nitrogenous substances. Thus, maintenance of proper water and salt balance in salt water requires different mechanisms from those in fresh water. There are at least three methods evident among marine fish-like vertebrates: hagfishes are essentially osmoconformers; elasmobranchs and *Latimeria* regulate by retaining urea; and the remaining bony fishes and lampreys osmoregulate by special mechanisms.

Lampreys in Salt Water. While in salt water equivalent to one half to full-strength sea water, river lampreys (*L. fluviatilis*) and sea lampreys (*P. marinus*) can maintain body fluids at about 285 to 330 mOsm/kg. As the body fluids are hyposmotic to the medium, diffusion tends to remove water from the body through the gills and skin. To make up the loss, lampreys swallow sea water (50 to 220 ml/kg/day), which is absorbed from the gut.

Also absorbed from the gut are monovalent ions, mainly Na^+ and Cl^-, which are in excess of the lamprey's needs. Divalent ions from the sea water are not absorbed by the gut and are eliminated with feces. Divalent ions are excreted in the meager concentrated urine also. The excess chloride is excreted through the gills, apparently through special excretory cells. These cells will be mentioned later in the treatment of bony fishes.

Hagfishes. These are characterized by a very permeable skin and virtual lack of the ability to regulate sodium chloride, so that exposure to water of higher or lower salinity than sea water results in rather rapid changes in osmotic concentration of body fluids. The body fluids of hagfishes are practically isosmotic to their medium, although some investigators claim a slight hypertonicity. The ionic content of hagfish plasma differs from that of sea water in that Ca^{2+}, Mg^{2+}, and SO_4^{2-} are in lower concentration and Na^+ is in higher concentration. The concentration of Cl^- in hagfish plasma has been reported as lower than that of sea water by some and as higher by others. The concentration of K^+ is reported to be nearly the same as in sea water. Urea appears to be present in very small amounts. Divalent ions and K^+ are excreted in the scant urine.

Elasmobranchs (Fig. 13–2). As mentioned earlier, the osmoregulation of elasmobranchs involves bolstering the osmotic concentration due to salts in the blood by retention of urea and smaller amounts of other nitrogenous compounds. Urea, an end product of nitrogen metabolism, produced in the liver, is excreted only in relatively small amounts via the urine of sharks and rays. As the glomerular filtrate passes along the kidney tubule, special segments resorb much (70 to 90%) of the urea so that the blood contains about 350 mmole/liter urea in a typical marine elasmobranch. The urea content of the blood is usually given as 2 to 2.5%. The urine of marine elasmobranchs usually contains about 100 mmole/liter urea. Trimethylamine oxide (TMAO),

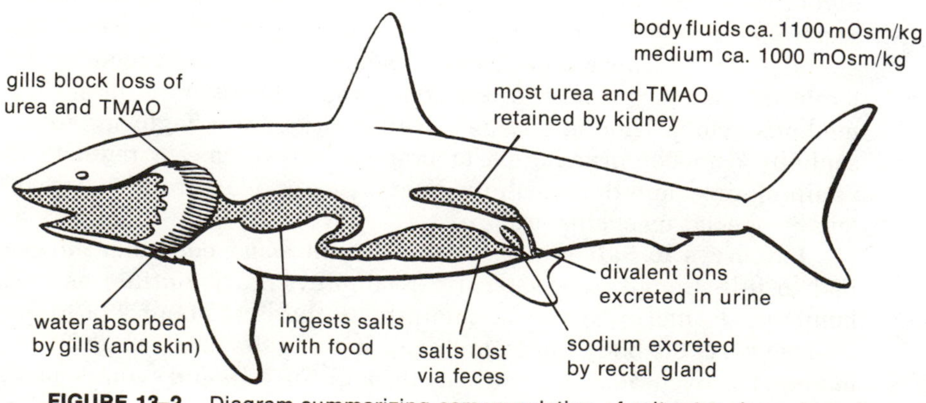

FIGURE 13–2. Diagram summarizing osmoregulation of saltwater elasmobranch.

another nitrogenous waste product, appears in the blood at about 70 mmole/liter and is therefore of secondary importance to the osmolality of the blood. It is reabsorbed in the kidney tubule and its concentration in the urine is about 10 mmole/liter. The gills of elasmobranchs are relatively impermeable to urea and TMAO.

The concentration of the various solutes, mostly sodium, chloride, urea, and TMAO, in the blood of elasmobranchs combine for an osmolality of about 1000 to 1100 mOsm/kg in animals living in sea water of about 930 to 1030 mOsm/kg. Sharks and rays typically have an osmotic concentration in the blood about 50 to 100 mOsm/kg higher than the concentration of the medium. Being hyperosmotic to sea water, elasmobranchs tend to gain water by diffusion. Influx is mainly through the gills, as the skin is nearly impervious. This excess water is excreted as urine, the flow of which is typically from 1 to 1.5 ml/kg/hr. Salts enter the marine elasmobranchs by diffusion as well as via ingested food because their body fluids have a more dilute concentration of salts than the medium. These are excreted via two main pathways.

The urine is the most important medium for excretion of divalent ions, the concentration of magnesium and phosphate in urine being more than 30 times the concentration in plasma, and the concentration of sulfate being nearly 140 times that of plasma. Sodium and chloride account for most of the osmotic concentration of the urine, but appear in nearly the same concentrations found in plasma (Table 13–2).

Sodium is excreted by the rectal gland, an organ that was of unknown function until about 1960. The normal concentration of NaCl in rectal gland fluid is nearly twice that of plasma, although the osmolality of the two fluids is the same. Small amounts of potassium and even smaller amounts of calcium and magnesium are excreted by this gland. The urea concentration of rectal gland fluid is about 4% of the concentration in the plasma. Volume of flow from the rectal gland is variable, usually 0.5 to 0.8 ml/kg/hr. When the rectal gland is ex-

Table 13–2. CONCENTRATIONS OF SELECTED SOLUTES IN PLASMA AND URINE OF THE SPINY DOGFISH (*SQUALUS ACANTHIAS*)*

Solute	Concentration (mmole/kg)		
	Sea water	Plasma	Urine
Sodium	440	250	240
Potassium	9	4	2
Calcium	10	3.5	3
Magnesium	50	1.2	40
Chloride	490	240	240
Sulfate	25	0.5	70
Phosphate	0	0.97	33
Urea	0	350	100
TMAO	0	70	10
Osmolality	930	1000	800

*Adapted from Hickman and Trump, 1969.

perimentally rendered nonfunctional, the kidneys appear to compensate by releasing more copious urine with greater chlorinity.

Appreciable amounts of sodium are excreted from the gills of elasmobranchs, but this elimination is small compared to the amounts released by the kidneys and rectal gland.

A number of elasmobranchs enter water of low salinity or even fresh water, and the freshwater stingrays *(Potamotrygon)* of the Amazon are confined to fresh water. Obviously, a great strain would be placed on the kidneys of freshwater elasmobranchs if they maintained the same osmotic concentration in the body fluids as that maintained by their marine counterparts. The excess water that would enter the body by osmosis would be a tremendous burden. Some marine species can be acclimated to about half-strength sea water. They adjust by reducing the amounts of chloride, urea, and TMAO in the blood until the osmotic concentration is reduced to, say, about 600 mOsm/kg. At this level the urine flow is increased considerably, up to five- to six-fold. The sawfish (*Pristis pectinata*) is commonly found in fresh water in large tropical rivers. This species and the bull shark *(Carcharhinus leucas)*, sometimes called the freshwater or Nicaragua shark, reduce the osmotic concentration of the blood to from 485 to 550 mOsm/liter, retaining only about 100 to 180 mmole/liter urea in the blood. Urine flow in *Pristis* has been measured at an average of 10.4 ml/kg/hr. The rectal gland of *C. leucas* regresses by decreasing the number of glandular tubules while in fresh water. *Potamotrygon* has nearly broken the urea habit after its long history as a freshwater genus; the blood contains slightly more than 1 mmole/liter of urea.

Holocephali. Chimaeras retain a high content of urea in the blood and maintain their internal osmotic pressure at, or slightly above, that of the surrounding sea water. They differ from sharks and rays in the relative proportions of solutes in the blood, as more salt and less urea and TMAO are retained (Table 13–3).

A discrete rectal gland is not recognized in holocephalans, but is considered by various investigators to exist in primitive form. Secretory cells similar to those of rectal glands are present along with ducts entering the rectum.

Latimeria. This ancient marine fish osmoregulates in much the same manner as the two preceding groups. It is tolerant of urea and

Table 13–3. CONCENTRATION OF SELECTED SOLUTES IN SERUM AND URINE OF *HYDROLAGUS**

Constituent (mmoles/liter)	Serum	Urine	Sea Water
Na^+	300	162	400
Cl^-	306	268	476
Urea	245	51.6	–
TMAO	5.5	–	–
Osmolality (mOsm/kg)	897	844	892

*Adapted from Read, 1971.

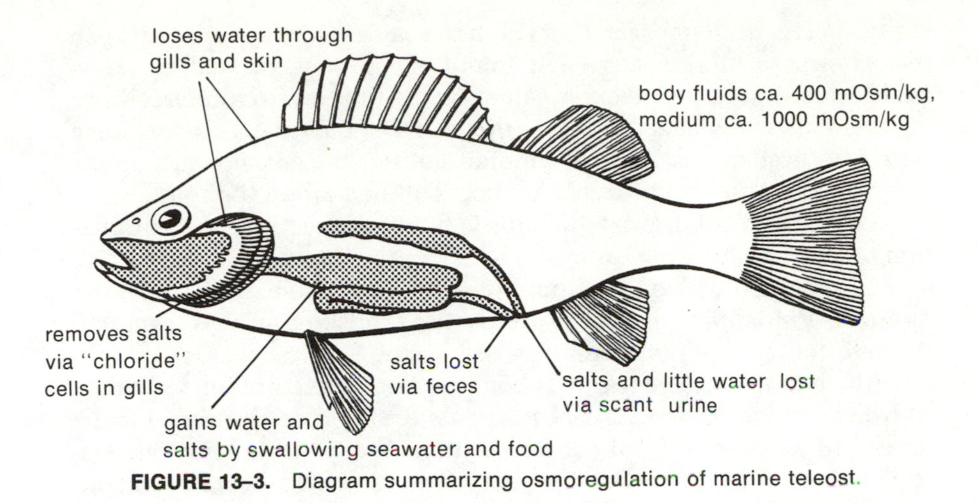

FIGURE 13–3. Diagram summarizing osmoregulation of marine teleost.

TMAO, retaining these in the blood so that the nitrogenous solutes are relatively more important at keeping a high osmotic concentration than they are in sharks, rays, or chimaeras. Sodium chloride concentration in the blood of *Latimeria* is less than that in the cartilaginous fishes. Examples of concentrations in serum of *Latimeria* are Na, 197 mmoles/liter, Cl, 187 mmoles/liter, and urea, 377 mmoles/liter. Osmolality of *Latimeria* serum has been reported in the range of 923 to 1181 mOsm/liter. The lower figure is hyposmotic and the upper figure slightly hyperosmotic to sea water. Apparently the coelacanth is normally slightly hyposmotic to its environment. *Latimeria* has a rectal gland, but its function is not known.

Marine Teleosts (Fig. 13–3). These have slightly less total body water than freshwater teleosts, and have a blood osmolality of 380 to 450 mOsm/kg in an external environment of 800 to 1200 mOsm/kg, so that water constantly diffuses from them to the medium. Some species have been shown to lose from 30 to 60% of their water intake by osmosis. This loss is made up largely by drinking sea water that is in part absorbed through the gut. Rate of drinking varies with species, and within species it varies with salinity. The higher the salinity, the greater the rate of drinking. Marine species commonly swallow sea water amounting to from 7 to over 35% of their body weight per day. From 60 to 80% of the ingested water is absorbed through the gut, and with it the monovalent ions Na^+, K^+, and Cl^-. Divalent ions remain mostly in the gut; usually less than 20% of those swallowed are absorbed.

The excess monovalent ions imbibed with the water are excreted mainly through the gills via cells called "chloride cells" that resemble the salt-secreting cells of other animals. These cells are concentrated in the gills, but appear in the skin of the head and anterior body in many species. These cells are extremely rich in mitochondria and have a remarkable network of tubular structures throughout the cytoplasm. In many species each ion-secreting cell has a cavity at its apex, opening to the exterior. Chloride has been found to be concentrated in these cavities. Apparently molecules can enter the tubules from intercellular spaces. This knowledge has led to theories that the tubules constitute a transport system that carries anionic substances toward the apical cavities, and that the materials are concentrated along the route.

Loss of sodium through the gills is not tied directly to the elimination of chloride. Sodium excretion balances the intake through the gut, and appears to involve a sodium-potassium exchange. Studies on the flounder *Platichthys flesus* showed that no sodium was excreted through the gills in potassium-free sea water.

Much remains to be learned about the mechanisms of ion exchange in fishes. Ranges of amounts of materials excreted or absorbed can be measured and verified, but processes are not yet well understood. The use of the electron microscope has done much to bring about an appreciation of the stucture of the secretory cells, and biochemical studies have led to the knowledge that certain enzyme systems, especially adenosinetriphosphatase, seem to play a strong role in ion exchange.

The kidneys of marine teleosts, as in freshwater fish, are the major site for excretion of divalent ions, but their role as a water pump is diminished because of the extrarenal losses of water. Consequently, the glomeruli are generally smaller and fewer than in freshwater fishes and elasmobranchs (Table 13–4). Several species of marine teleosts have lost the glomeruli, and have minimized renal loss of water. Among

Table 13–4. NUMBER AND SIZE OF GLOMERULI IN SELECTED FISHES*

Species	Weight (g)	Number of Glomeruli in One Kidney	Average Diameter of Glomeruli (μ)	Relative Volume (mm^3) of Glomeruli per m^2 Body Surface
Freshwater Teleosts				
Ictalurus nebulosus	89	18,160	100	126.5
Cyprinus carpio	221	24,310	82	50.8
Perca flavescens	116	4,870	102	30
Marine Teleosts				
Gadus morhua callarias	670	16,250	37	1.49
Lutjanus griseus	544	31,860	55	10.99
Pseudopleuronectes americanus	160	5,300	50	3.14
Elasmobranchs				
Mustelus canis	485	4,400	185	60.93
Raja erinacea	1060	1,200	190	10.38

*Data adapted from Nash, 1931.

these are species of gulpers (Saccopharyngidae), seahorses and pipefishes (Syngnathidae), dragonets (Callionymidae), scorpionfishes (Scorpaenidae), poachers (Agonidae), sculpins (Cottidae), puffers (Tetraodontidae), toadfishes (Batrachoididae), clingfishes (Gobiesocidae), anglerfishes (Lophiidae), frogfishes (Antennariidae), and batfishes (Ogcocephalidae). In addition to diminution, degeneration, or loss of glomeruli, marine fishes lack the distal segment of the kidney tubule.

Urine flow in marine teleosts is scant, usually amounting to 1 to 2% of the body weight per day. In some species most of the water in the urine enters through the kidney tubule. In others water is filtered out by the glomerulus and most is subsequently reabsorbed by the tubule. The osmotic concentration of urine is slightly less than that of the blood.

Principal divalent ions in the urine of marine fishes are, in decreasing order of concentration: magnesium, at 50 to more than 100 times the plasma concentration; sulfate, at concentrations up to more than 300 times that of the plasma; calcium, at 4 to 10 times the plasma concentration. Phosphate occurs in concentrations less than that of calcium. There apparently is active secretion of magnesium, sulfate and, sometimes, phosphate into the tubule. In some marine species calcium and magnesium are precipitated as salts in the tubular lumen and thus no longer influence the osmotic gradient. Chloride may be virtually absent from the urine in some species, but can appear in concentrations up to 85% of the plasma concentration in others.

Diadromous and Other Euryhaline Fishes

Many species are capable of living in both fresh and salt water. Anadromous fish are hatched in fresh water, subsequently move to the sea to feed and grow, and then return to fresh water to spawn. Catadromous species reverse the two media. Many other species have a wide tolerance for salinity and can move freely between fresh and salt water. All of these must be able to adjust osmoregulatory mechanisms more or less rapidly depending upon the speed with which they change habitat. Diadromous species generally undergo progressive changes that may alter appearance as well as physiology. At a given time, depending upon the stage of life history, these fishes are suited mainly to one medium or the other, and do not usually change back rapidly. On the other hand, some euryhaline species can make excursions rapidly through a range of salinity.

The young of several anadromous species of salmonids, upon attaining a characteristic minimum size (often 10 to 15 cm in length), undergo a transformation from the parr stage to the salt-tolerant smolt stage. The critical size range, which in nature might not be reached for one, two, or more years, depending upon the species, can be reached in only a few months of rearing in a hatchery. The fishes change to a silvery color, become slimmer in form, and have a tendency to migrate downstream into the ocean. If they are restrained from migrating the salt-tolerance of some of the species (e.g., *Salmo gairdneri*) regresses after several weeks. Chum salmon (*Oncorhynchus keta*) and pink salmon (*O. gorbuscha*) migrate seaward at a small size, and are salt-tolerant soon after the yolk sac is absorbed. This salt tolerance does not normally regress.

When the diadromous species are adapted to salt water, the osmotic concentration of blood is higher than in the same species adapted to fresh water. Examples of a few species follow. Concentrations are given in mOsm/kg, first that of the animal adapted to fresh water, then a dash and the corresponding figure for salt water adaptation: (1) *Petromyzon marinus,* 280 — 312, (2) *Oncorhynchus tshawytscha,* 304 — 350; (3) *Salmo salar,* 328 — 344; (4) *Anguilla,* 350 — 430.

Adaptation of euryhaline fishes to salt water generally requires drinking of the medium. Fishes such as rainbow trout and eels that swallow little or no water while in fresh water, may drink about 4 to 15% of their body weight per day in salt water. *Tilapia mossambica* may swallow nearly 30% of body weight per day. Many species respond to the saline medium by changes in kidney function. The glomerular filtration rate may diminish dramatically, and the tubular reabsorption of water usually increases. Urine flow decreases to 10% or less of the flow in fresh water. Precipitation of calcium and magnesium salts in the tubule has been noted in coho salmon. Chloride secretory cells in the gills become active. Usually the blood of a fish physiologically ready for hypoosmoregulation stabilizes within 1.5 to about 5 days.

The osmoregulatory capacity of some species is truly remarkable. *Galaxias maculatus* lives in salinities from less than 1 to 49 ppt and can be acclimated to 62 ppt. *Tilapia mossambica* has been acclimated to 69 ppt. Many members of the Poeciliidae and Cyprinodontidae can tolerate high salinities. The sheepshead minnow, *Cyprinodon variegatus,* has been found alive at 142.4 ppt.

In the tropics especially, many members of typically marine families move into and out of fresh water, some at certain life history stages, but some nearly randomly. These must be able to switch abruptly from conserving water to filtering out large volumes through the kidney, and must turn from excretion of excess salts to conservation. Several marine families have given rise to vicarious freshwater forms. A few of these forms seem to be unlikely candidates for a freshwater life inasmuch as they have few or no glomeruli. For instance, there are freshwater species of pipefishes and toadfishes. These species apparently are quite impermeable and can increase urine flow through a mechanism that is not well known. A marine toadfish, *Opsanus tau,* which can adapt to low salinity, has been noted as having plasma osmolarity of 392 mOsm/kg in sea water and 250 mOsm/kg in fresh water.

Eggs and Larvae

Eggs of fishes, at the time of deposition, are essentially isosmotic to the body fluids of the female. In elasmobranchs, urea is present in sufficient quantity so that there is no quick change of medium encountered by the egg upon spawning. Hagfish eggs are nearly isosmotic to the environment. On the other hand, eggs and sperm of teleosts are generally placed in environments with either higher or lower osmotic concentrations. Under conditions normal for the species involved, exposure of the gametes to a medium differing in osmolality exerts no ill effects. Furthermore, results of experimentation with gametes of salmon, herring, and flounders have shown that fertilization in some species can occur over a wide range of salinity. Percentage of successful fertilization decreases at low (<15 ppt) salinity in the marine spawners, and at high (>24 ppt) salinity in the salmon. Most freshwater species have low rates of fertilization and hatching in saline water.

Fertilized eggs, although they imbibe water through the chorion, are somehow able to regulate their salt concentration to a great extent. There appears to be some effect of the salinity at which fertilization takes place on the subsequent ability of the developing egg to develop optimally. This seems to be due to some initial influence on the physical properties of the perivitelline fluid by the salinity of the medium.

There are some examples of incubation and hatching at unusual salinities. The sheepshead minnow *(Cyprinodon variegatus)* has been known to hatch at 110 ppt, the desert pupfish *(C. macularius)* and the fourbeard rockling *(Enchelyopus cimbrius)* at 70 ppt, and the herring *(Clupea harengus)* at 60 ppt. Chum salmon *(Oncorhynchus keta)* eggs, when transferred at the eyed stage to full-strength sea water, showed 50% survival through hatching, although the alevins did not survive. The eggs of a few primary freshwater teleosts are known to incubate normally and hatch well at salinities up to 5 ppt.

Newly-hatched larvae of some marine species (herring, flounder) tolerate wide ranges of salinity, but reduce this tolerance with age. Alevins of chum salmon can survive in quarter-strength sea water one day after hatching, but do not adapt well to full-strength sea water until the yolk sac is absorbed at 60 or more days after hatching. Cells rich in microtubular structures, resembling the "chloride-secreting" cells of mature fishes appear in the epidermis of larval herring. Possibly osmoregulation is begun in such cells prior to the full development of gills, gut, and kidneys.

Endocrine Secretions and Osmoregulation

Although there is a probability that hormones are of great importance to the osmoregulatory powers of lampreys, hagfishes, and elasmobranchs, details and proof are largely lacking. That state of affairs is not due entirely to lack of attention by physiologists. Results of research with those animals have not been clearcut. Partial or complete removal of endocrine tissues in elasmobranchs has not produced rapid, significant changes in osmoregulation. On the other hand, much knowledge is accumulating concerning the relationship of the endocrine secretions to osmoregulation in bony fishes, especially teleosts. As expected, there have been differences found in the responses of various species to such conditions as experimental administration of hormones and the blockage of endocrine secretion, but there appears to be firm evidence that certain hormones affect osmoregulation.

Secretions of the pituitary are probably both directly and indirectly involved in control of salt and water balance. Prolactin has been experimentally implicated in retention of sodium in fresh water. Removal of the pituitary from a fish living in fresh water results in sodium loss that proves fatal unless prolactin is administered. Hypophysectomized marine or euryhaline fishes suffer no upset of sodium metabolism while in salt water. Arginine vasotocin influences kidney function and sodium permeability in both marine and freshwater species. Growth hormone is thought to influence the onset of salt tolerance in the young and anadromous salmonids. It shows effects on salinity preference as well as tolerance, but the exact role is not known.

Adrenocorticotropin and thyrotropin, secreted by the pituitary, may influence osmoregulation by stimulating secretions of the respective target glands. The thyroid is known to be active in migrating diadromous species, but the specific effect of thyroxin on water and ion balance is poorly known. The interrenal tissue produces adrenocortical steroids that act on renal and extrarenal systems to aid in regulating the body fluids in both marine and freshwater species but, again, details of activity are largely lacking. Cortisol may be the most important steroid involved.

Removal of the corpuscles of Stannius leads to defective water balance, but the secretions responsible and the mode of action are still being identified. Knowledge of the influence of the urophysis on osmoregulation is still imperfect, although the secretions of that tissue appear to be involved.

References

Agarwal, S., and John, P. A. 1975. Functional morphology of the urinary bladder in some teleostean fishes. Forma Functio, 8(2):19–26.

Anderson, B. G., and Loewen, R. D. 1975. Renal morphology of freshwater trout. Am. J. Anat., 143(1):93–114.

Bentley, P. J. 1971. Zoophysiology and Ecology, Vol. 1. Endocrines and Osmoregulation: A Comparative Account of the Regulation of Water and Salt in Vertebrates. Berlin, Springer-Verlag.

Black, V. S. 1957. Excretion and osmoregulation. *In:* Brown, M. E. (ed.), The Physiology of Fishes, Vol. 1. New York, Academic Press, pp. 163–205.

Burger, J. W., and Hess, W. N. 1960. Function of the rectal gland in the spiny dogfish. Science, 131:670–671.

———. 1962. Further studies on the function of the rectal gland in the spiny dogfish. Physiol. Zool., 35:205–217.

Burton, R. F. 1968. Cell potassium and the significance of osmolarity in vertebrates. Comp. Biochem. Physiol., 27:763–773.

Butler, D. G. 1966. Effect of hypophysectomy on osmoregulation in the European eel (*Anguilla anguilla* L.). Comp. Biochem. Physiol., 18:773–781.

Chessman, B. C., and Williams, W. D. 1975. Salinity tolerance and osmoregulatory ability of *Galaxias maculatus* (Jenyns) (Pisces, Salmoniformes, Galaxiidae). Freshwater Biol., 5(2):135–140.

Conte, F. P. 1969. Salt secretion. *In:* Hoar, W. S., and Randall, D. J. (eds.), Fish Physiology, Vol. 1. New York, Academic Press, pp. 241–292.

______, and Wagner, H. H. 1965. Development of osmotic and ionic regulation in juvenile steelhead trout, *Salmo gairdneri*. Comp. Biochem. Physiol., 14:603–620.

DiJulio, D. H., and Brown, G. W. 1975. Urea in the coelacanth. 1974. Research in Fisheries; Annual Report of College of Fisheries, Seattle, University of Washington (Cont. No. 415), p. 52.

Goldstein, L., Forster, R. P., and Fanelli, G. M., Jr. 1964. Gill blood flow and ammonia excretion in the marine teleost, *Myoxocephalus scorpius*. Comp. Biochem. Physiol., 12(4):489–499.

Griffith, R. W., Umminger, B. L., Grant, B. F., Pang, P. K. T., and Pickford, G. E. 1974. Serum composition of the coelacanth, *Latimeria chalumnae* Smith. J. Exp. Zool., 187(1):87–102.

Henderson, I. W., and Jones, I. C. 1973. Hormones and osmoregulation in fishes. Ann. Inst. Michel Pacha, 5(2):69–235.

Hickman, C. P., and Trump, B. F. 1969. The kidney. *In:* Hoar, W. S., and Randall, D. J. (eds.), Fish Physiology, Vol. 1. New York, Academic Press, pp. 91–239.

Hoar, W. S. 1976. Smolt transformation: evolution, behavior, and physiology. J. Fish. Res. Bd. Can., 33:1234–1252.

Holliday, F. G. T. 1969. The effects of salinity on the eggs and larvae of teleosts. *In:* Hoar, W. S., and Randall, D. J. (eds.), Fish Physiology, Vol. 1. New York, Academic Press, pp. 293–311.

Holmes, W. N., and Donaldson, E. M. 1969. The body compartments and the distribution of electrolytes. *In:* Hoar, W. S., and Randall, D. J., (eds.), Fish Physiology, Vol. 1. New York, Academic Press, pp. 1–89.

Johnson, D. W. 1973. Endocrine control of hydromineral balance in teleosts. Am. Zool., 13:799–818.

Kashiwagi, M., and Sato, R. 1969. Studies on the osmoregulation of the chum salmon, *Oncorhynchus keta* (Walbaum). I. The tolerance of eyed period eggs, alevins and fry of the chum salmon to seawater. Tohoku J. Agr. Res., 20(1):41–47.

Knutson, S., and Grav, T. 1976. Seawater adaptation in Atlantic salmon (*Salmo salar* L.) at different experimental temperatures and photoperiods. Aquaculture, 8:169–187.

Krogh, A. 1939. Osmotic Regulation in Aquatic Animals. Cambridge, Cambridge University Press.

Lahlou, B., Henderson, I. W., and Sawyer, W. H. 1969. Renal adaptations by *Opsanus tau*, a euryhaline aglomerular teleost, to dilute media. Am. J. Physiol., 216:1266–1272.

Lutz, P. L. 1975*a*. Osmotic and ionic composition of the polypteroid *Erpetoichthys calabaris*. Copeia, 1975(1):119–123.

———. 1975*b*. Adaptive and evolutionary aspects of the ionic content of fishes. Copeia, 1975(2):369–373.

Maetz, J. 1969. Seawater teleosts: evidence for a sodium-potassium exchange in the branchial sodium-excreting pump. Science, 166:613–615.

Miles, H. M. 1971. Renal function in migrating adult coho salmon. Comp. Biochem. Physiol., 38A:787–826.

———, and Smith, L. S. 1968. Ionic regulation in migrating juvenile coho salmon. *Oncorhynchus kisutch*. Comp. Biochem. Physiol., 26:381–398.

Morris, R. 1960. General problems of osmoregulation with special reference to cyclostomes. *In:* Jones, I. C. (ed.), Hormones in Fish. London, Symposium of the Zoological Society of London, (1):1–16.

———. 1972. Osmoregulation. *In:* Hardisty, M. W., and Potter, I. C. (eds.), The Biology of Lampreys, Vol. 2. London, Academic Press, pp. 193–239.

Nagahama, Y., Nishioka, R. S., Bern, H. A., and Gunther, R. L. 1975. Control of prolactin secretion in teleosts, with special reference to *Gillichthys mirabilis* and *Tilapia mossambica*. Gen. Comp. Endocrinol., 25(2):166–188.

Nash, J. 1931. The number and size of glomeruli in the kidneys of fishes, with observations on the morphology of the renal tubules of fishes. Am. J. Anat., 47(2):425–446.

Natochin, Y. V., and Gusev, G. P. 1970. The coupling of magnesium secretion and sodium reabsorption in the kidney of teleosts. Comp. Biochem. Physiol., 37(1):107–111.

Oguri, M. 1964. Rectal glands of marine and freshwater sharks: comparative histology. Science, 144:1151–1152.

Oikari, A. 1975. Seasonal changes in plasma and muscle hydromineral balance in three Baltic teleosts, with special reference to the thermal response. Ann. Zool. Fenn., 12(3):230–236.

Parry, G. 1966. Osmotic adaptation in fishes. Biol. Rev., 41:392–444.

Prosser, C. L. 1973. Water: osmotic balance; hormonal regulation. *In:* Prosser, C. L. (ed.), Comparative Animal Physiology. Philadelphia, W. B. Saunders, pp. 1–78.

———, and Kirschner, L. B. 1973. Inorganic ions. *In:* Prosser, C. L. (ed.), Comparative Animal Physiology. Philadelphia, W. B. Saunders, pp. 79–110.

Read, L. J. 1971. Body fluids and urine of the holocephalan, *Hydrolagus colliei*. Comp. Biochem. Physiol., 39A:185–192.

Robertson, J. D. 1963. Osmoregulation and ionic composition of cells and tissues. *In:* Brodal, A., and Fänge, R., (eds.), The Biology of *Myxine*. Oslo, Universite Tsforlaget, pp. 504–515.

Thorson, T. B. 1961. Partitioning of body water in Osteichthyes: Phylogenetic implications in àquatic vertebrates. Biol. Bull., 120:238–254.

Vickers, T. 1961. A study of the so-called "Chloride-Secretory" cells of the gills of teleosts. Q. J. Micros. Sci., (60):507–518.

Wagner, H. H. 1974. Seawater adaptation independent of photoperiod in steelhead trout. Can. J. Zool., 52:805–812.

———, Conte, F. P., and Fessler, J. L. 1969. Development of osmotic and ionic regulation in two races of chinook salmon, *Oncorhynchus tshawytscha*. Comp. Biochem. Physiol., 19:325–341.

14
FEEDING AND NUTRITION

Feeding

The actions of feeding are carried out daily by most fishes, and may be the most frequent of voluntary activities. Flight from enemies, reproduction, migrations, and many other activities might be occasional or periodic, but feeding is usually part of the daily routine and in some species may require extended periods of time. Most fishes are designed especially well for food gathering. Although many structures serve several functions in the life of the fish, feeding may be an important, if not the principal, use for them. Although some may be of greater importance in certain fishes than in others, all the senses can be instrumental in the detection and selection of food items. The locomotory powers and structures of fishes are used in great measure for finding and gathering food. In earlier chapters there have been many references to specialized structures involved in feeding, mostly concerning the head and mouth, but including lures for attracting prey. A brief review of some of these with regard to their function is appropriate here.

Detection of possible food items at a distance can be through physical or chemical senses so that the eyes, the auditory organs, the lateralis system, or the electrical sense may serve to orient the fish toward the food source. Some specimens are "sight feeders," and approach food mainly on visual cues; examples are large-eyed pelagic predators such as scombroids. In many species of fish the position of the eyes in the head serves the particular feeding habits. Topminnows and other surface feeders have the eyes oriented upward, whereas bottom feeders, including certain catfishes (such as *Pangasius*), have the eyes on the lower part of the head looking downward.

The lateralis sense is of great importance in food detection in most fishes that feed on active prey. Even species that are strongly sight-oriented can continue to locate moving prey when forced to rely on the lateral line in experimental situations. The lateralis sense, as previously described, reaches a remarkable stage of development and great acuity in blind cave fishes and deepsea fishes. Nocturnal fishes and even some daytime predators such as some sharks depend heavily on the lateralis sense. There are a few instances of specialized lateral line structures, as in certain topminnows (*Aplocheilus*) that have modified lateralis organs on the top of the head, which is usually held at the surface film. Prey causing minute ripples at the surface can be detected easily.

Olfaction and taste play important roles in the feeding of fishes of all kinds, with olfaction better suited to detecting food at a distance. Many species that appear to feed mainly by sight are known to have acute olfaction. Bottom fishes such as eels, the spiny eels of the Mastacembelidae, and others have well developed olfactory systems, as do sharks, rays, and many deepsea fishes. No doubt many species get the first signal that food is in the vicinity via the sense of smell and are able to follow a chemical gradient to the source.

Taste, along with the tactile sense, appears to be of great importance in final selection of food and its retention for preparation and swallowing. The distribution of taste buds over the skin of a wide variety of species is well known. Such an arrangement must give a broader area for the selection of food than if the taste receptors were only in the mouth or on the head. The greatest concentration of external taste buds is usually in the region of the mouth, on barbels, the snout, and lips, where final selection is generally made.

Not all materials taken into the mouth while feeding are swallowed. Objects taken by sight only, such as a twig snatched from the surface by a rising trout, must be tested for suitability by sensory receptors inside the mouth. Usually a trout will reject an unsuitable object soon after taking it, but many bottom-feeding species take various extraneous materials with their food and subject them to a selection process inside the mouth and pharynx. Taste buds are well distributed within the mouth, especially on the oral valves, palatal organs, and tongue. Isolated chemoreceptor cells occur in the epithelium of the palate of fishes.

In many fishes food appears to be separated from unwanted detritus at the level of the pharynx. Taste receptors are abundant on the gill arches, gill rakers, epibranchial organ, and in the tissue surrounding the pharyngeal teeth. In many instances material is rejected through the gill openings after having been subjected to a final test by sensory facilities in the pharynx. Large particles are usually ejected through the mouth with a "coughing" action. Gill rakers, pharyngeal teeth and bristles, and epibranchial organs all serve mechanical functions in retaining or rejecting ingested material.

The amounts of food ingested per day and the times of day that feeding is performed depend on many factors. Active predators, with their high metabolic rates, require more food energy than do sluggish fishes. If the predator feeds habitually on small organisms, a relatively great amount of time must be spent gathering prey in order to acquire this energy. However, if the predator can catch and swallow large organisms, it might satisfy itself with one or two captures per day. Because metabolic rate varies with temperature, cold-water predators such as salmon should require less food than warm-water predators such as tunas. Deepsea predators, in constant cold water, probably require infrequent meals. Daily and seasonal temperature fluctuations affect food intake in most species. The predator, feeding on the rich bodies of other animals such as insects, crustaceans, squids, and other fishes, is ingesting high-protein, high-calorie food that provides nutritional requirements without much bulk. On the other hand, the herbivore or detritivore must ingest large quantities of less concentrated foodstuffs, sometimes including a large proportion of undigestible material. Consequently, feeding activity in these fishes must require a greater period of time per day.

Some species feed mainly by sight and are active by day, although peaks of feeding activity occur in morning and evening. Other fishes that depend more on chemical senses can feed effectively in twilight or at night, so they may be most active in early morning and late evening. Other differences in feeding activity are tied to seasons, cycles of migratory or reproductive activity, or age and size.

The results of many studies on feeding of various fishes have shown that small individuals consume more per day in relation to their body weight than large individuals. Small fishes of 2 to 5 g were found to eat the equivalent of 6 to 10% of their body weight per day, while fishes of 30 g or more consume food at about 2 to 3% of body weight. Many fishes show an average daily food intake despite fluctuations when feeding is measured over an extended period. In addition, food intake appears to be adjusted to the nutritive value of available food, so that greater amounts of less concentrated foods are eaten.

Preparation of Food for Digestion

Many predatory fishes, such as the largemouth bass (*Micropterus salmoides*), have cardiform teeth in relatively broad pads both in the mouth and on the pharyngeal bones. These teeth act to catch and hold prey and to aid in swallowing it, but do not prepare it for the digestive process. Other predators, such as salmon or trout, have rows of sharp teeth that commonly tear or break the skin of the prey at the time of capture or during the swallowing process. The squawfishes (*Ptychocheilus* spp.) have no teeth in the mouth but are equipped with strong pharyngeal teeth that tear the prey as it is forced into the esophagus. A few predators, such as sharks and piranhas, bite pieces out of the prey.

Although a number of fishes (lungfishes, chimaeras, wolffishes, croakers, etc.) have tooth plates or molariform teeth in the mouth, and can grind or mash food with them, pharyngeal teeth appear to be the major apparatus used for mastication throughout the teleosts. Some of the greatest development of throat teeth can be seen in species that habitually feed on various shellfishes or corals. Many of these, such as the surfperches (Embiotocidae), have pharyngeal mills that can triturate the food organisms past recognition. Herbivorous fishes must tear or grind plants extensively so that digestive enzymes can act upon the cell contents, and molariform pharyngeal teeth are common among the plant eaters. The grass carp (*Ctenopharyngodon idella*) has relatively long, rough-edged pharyngeal teeth that intermesh while tearing the soft plants upon which the species feeds.

Mucous cells or glands are present in the mouth and pharynx of most fishes, but the greatest mucus production is usually in the esophagus. Mucus facilitates the process of swallowing by lubricating large particles, and may aid in holding finely divided particles together. Butterfishes (Stromateidae) and their close relatives have expanded esophageal sacs that serve to store food, or in those species that have teeth in the lining of the sac, to triturate the food. The sacs are well equipped with mucous glands. The digestive process appears to begin in the esophagus of certain mullets (Mugilidae) and sculpins (Cottidae) that have gastric glands in the posterior section.

Digestion

In most fishes the digestive process begins in the stomach, which differs from the esophagus in the composition of the walls and in the types of glands in the mucosal lining. In addition to secreting a protective mucus and pepsin, a protease, the glands of the stomach secrete hydrochloric acid, which maintains the pH of the stomach contents in a range suitable for the action of pepsin. Pepsin shows a peak of activity at a pH of about 2. pH values of about 1.5 to 4 are common in the stomachs of predatory or insectivorous species. Flow of gastric juices is initiated by the act of feeding and especially by distension of the stomach wall. Secretion is to some extent under the control of the vagus.

The stomach generally acts to store food and initiate digestion by mixing the ingesta with the gastric juices. Depending on the food habits of the species, the organ can be large and distensible or small and capable of passing small food items along during an extended feeding period each day. The stomachs of mullets (Mugilidae), the milkfish *(Chanos)*, some herrings (Clupeidae), and characins (Characinidae), all microphagous fishes, are modified into gizzards. In many of these the gizzard involves only the pyloric part of the stomach, and the secretory function of the stomach is lessened. In the gizzard shad *(Dorosoma cepedianum)* the gizzard is divided into cardiac and pyloric parts. Many microphagous fishes have unspecialized stomachs or none at all. There are great differences in the shape and development of the stomach, even among closely related species or those with similar food habits. Motility of the stomach is in many instances related to the degree of fullness, so that ingesta is removed more rapidly from a full stomach than from one that is partially full.

A variety of fishes lack the stomach. Stomachs are not recognized in chimaeras (Holocephali), lungfishes (Dipnoi), some gobies (Gobiidae), and the minnows and suckers, for example. There are several hypotheses explaining the loss of the stomach in various kinds of fishes. At the low concentrations of chlorine available in fresh water, the ability to digest food in a completely alkaline system would seem to be advantageous as the fishes would be freed of the burden of produc-

Table 14–1. RATIO OF INTESTINE LENGTH (I) TO BODY LENGTH (B) IN SELECTED FISH SPECIES

Species	I/B	Remarks
Atlantic salmon *(Salmo salar)*	0.73–0.80	Carnivorous
Cod *(Gadus morhua)*	1.05–1.50	Carnivorous
Silver carp *(Hypophthalmichthys molotrix)*	4.6–7.1	Herbivorous
Tui chub *(Gila bicolor)*	1.0–1.3	Omnivorous
Northern squawfish *(Ptychocheilus oregonensis)*	0.7–0.9	Carnivorous
Calbasu *(Labeo calbasu)*	3.75–10.0	Herbivorous
—*(Labeo horie)*	15.0–21.0	Detritivorous
Flagfish *(Jordanella floridae)*	2.5–2.7	Herbivorous
Largemouth bass *(Micropterus salmoides)*	0.7–0.9	Carnivorous

ing hydrochloric acid. Stomachless fishes of predatory habits usually have an expanded portion of the intestine in which large morsels can be stored while undergoing digestion. These expansions are often mistaken for stomachs, inasmuch as one of the important functions of the stomach is being served.

Among the teleosts the intestine varies in length and conformation with food habits and, to some extent, with individual diet. Piscivorous species usually have rather straight intestines somewhat shorter than the body, whereas herbivores or detritivores have guts much longer than the body. Fishes that habitually ingest a large proportion of indigestible material with their food appear to have the longest relative gut length, and in these fishes the gut is usually flexed or coiled in an elaborate manner. The added length increases the retention time of the food and allows for more efficient digestion of materials that are hard to digest. Table 14–1 shows a comparison of the relative gut lengths of several species. In many species the length, and thus the volume and absorptive area of the gut, increases by a factor greater than the increase in body length, so that large adult individuals have a markedly greater gut length to body length ratio than do small individuals.

Digestion proceeds in the intestine in a neutral to alkaline medium. Enzymes involved are secreted by the pyloric caeca, pancreas, and intestinal mucosa. Trypsin, secreted by the pancreas, is one of the most important proteases in fish digestion, although other digestive enzymes appear to be present in various species. Carbohydrases, including amylase, saccharase, lactase, and maltase, have been identified in the intestine of fishes. Lipase activity has been demonstrated in many species. The pancreatic tissue is probably the source of most of the enzymes, but the diffuse nature of the pancreas in most fishes presents physiologists with great difficulty in locating the exact source of secretions. The liver secretes emulsifiers, carried to the intestine in bile, that aid in fat digestion. The types and amounts of enzymes present in the digestive system of a given species are related to the general food habits of the species.

Rates of digestion are variable, depending on type of foodstuff, species of fish, temperature, and amount of food ingested. There is

Table 14–2. RATES OF GASTRIC DIGESTION IN SELECTED FISH SPECIES

Species	Type of Food	Temperature (°C)	Range, Mean or Median Time (hr)	Digestion (%)
Florida gar *(Lepisosteus)*	Fish	24	42	100
Rainbow trout *(Salmo)*	Oligochaetes	15	12	70
Rainbow trout *(Salmo)*	Oligochaetes	15	36	99
Pike *(Esox)*	Fish	22.5	50	100
Silver carp *(Hypophthalmichthys)*	Algae	–	10	~100
Goldfish *(Carassius)*	"Pellets"	24.5	7–24	~100
Squawfish *(Ptychocheilus)*	Fish	17–20	1	14
Squawfish *(Ptychocheilus)*	Fish	24	1	40–50
Black bullhead *(Ictalurus)*	Amphipods	24	6	100
Cod *(Gadus)*	Shrimp	5	60	100
Cod *(Gadus)*	Shrimp	15	15	100
Bluegill *(Lepomis)*	Mixed (natural)	22.5	20–26	100
Black crappie *(Pomoxis)*	Mixed (natural)	22.5	19	100

some indication that small fishes of a given species digest food more rapidly than larger individuals. Studies of the rate of gastric plus intestinal digestion have presented some problems, but satisfactory results have been obtained in a number of experiments. These have usually involved timing the passage of a meal from ingestion to defecation of the waste resulting from the meal; this has been accomplished by feeding and subsequent observation following a fasting period, or by feeding materials colored by inert dyes. Gastric digestion has been measured more directly, usually by feeding a measured amount of food and studying the rate of disappearance from the stomach. Fish can be sacrificed and dissected at intervals, the stomachs can be pumped, or X-ray techniques can be used in these studies.

Because various investigators have used different test species, different methods, and different test foods, there is some difficulty in comparing results of digestion rate studies. Temperatures have usually been reported for the experiments, but these have not always been discussed in relation to the ecology of the test species. Temperature influences such phenomena as the rate of secretion and the activity of digestive enzymes, the absorption rate of the digested food, and the muscular activity of the digestive tract. The amount of food fed has an effect on the rate of digestion; usually a large meal is digested at a more rapid rate than a small one. Some results of digestion rate studies are shown in Table 14–2.

Nutrition

Nutritional Requirements

Fishes, like other animals, require the common components of foods — proteins, carbohydrates, fats, minerals, vitamins, and water.

Specific requirements and optimum levels of these in diets have been studied in only a few species, and research on these is not complete. Most information available on fish nutrition pertains to species that are reared in captivity. Included are several members of the Salmonidae, the Japanese eel *(Anguilla japonica)*, some members of Cyprinidae, the red tai *(Pagrus major)*, and the channel catfish *(Ictalurus punctatus)*. Because fish culturists are continually seeking means of producing more and better fish at lower costs, there is a practical value in studying fish nutrition. Even though species may be specific in their requirements, knowledge gained from studying one species can provide a start for the feeding of others.

Fish differ from warm-blooded animals in that their metabolism is directly influenced by temperature, and that some species are adapted to cold waters and others to warm waters, each showing characteristic changes in metabolic rate over their tolerance range. Furthermore, fish of various species, especially carnivores, can utilize proteins and fats for energy sources more readily than mammals. Most species subjected to intensive aquaculture are known to require high dietary protein levels. Protein requirements of salmonid fishes have been studied to some extent, the chinook salmon having been shown to require about 50% protein in artificial diets for maximum growth. This differs with temperature; less protein is required at low temperatures than at higher ones. Natural foods of trout have as low as 12% protein, and appear to be more efficient than hatchery foods. Channel catfish need somewhat less protein in their artificial diets than salmonids, between 25% and 40%. Predatory marine species are reported to require more than 50% protein in the diet. Age has an effect upon the protein requirement of fishes, young fish needing more than older fish. Also, in salmonids, greater amounts of protein are needed in salt water than in fresh water.

Quality of protein is of great importance. Fishes need ten indispensable amino acids for proper growth, as do most animals. These are arginine, histidine, isoleucine, leucine, lysine, methionine, phenylalanine, threonine, tryptophan, and valine. Each is required in a minimum amount, so a proper diet must provide all of them at the proper rate. In certain salmonids some methionine can be spared by additional cystine in the diet.

Inclusion of carbohydrates in the diet can spare some protein for use in growth rather than for energy expended in activity. Carbohydrates are usually much cheaper than proteins, so there is an economic advantage if they can be fed to cultured fishes. Omnivorous species such as carp can digest carbohydrates better than trout or other carnivores, and can have a higher percentage in the diet. Carnivorous species, on the other hand, may show nutritional disorders if fed an excess of carbohydrate. Salmonids usually respond to high levels of dietary carbohydrate by depositing an excess of glycogen in the liver. Accordingly, hatchery diets for trout contain less than 10% digestible carbohydrates. Trout are naturally diabetic and retain a high level of blood sugar long after being fed an excess of carbohydrate. The same condition has been noted in the red tai (*Pagrus major*), which tolerates only low amounts of carbohydrate in formulated diets. Diets containing as much as 20% dextrin were found to retard the growth of the red tai. Channel catfish are not diabetic, and have responded to test diets containing 30% of either starch or dextrin, making good growth.

Fats provide an energy source for fishes but can be used only in limited amounts because, if fed in excess, fat will infiltrate the liver, and possibly cause death. Fats differ greatly in digestibility, those with high melting points being difficult for fishes to digest. When digestible fats are used in balanced diets, some of the dietary protein is spared for growth or other purposes in addition to energy. Rancid fats are harmful in fish diets; fishes fed oxidized fats have been noted to develop fatty degeneration of the liver, particularly if the diet is low in vitamin E. Some naturally occurring oils are known to contain substances toxic to some fishes. For instance, cottonseed oil contains cyclopropene fatty acids that are harmful to trout if fed above a certain level. As in higher animals, there seems to be a requirement for essential fatty acids in fish diets. The requirements differ among fishes, but the linolenic acid series is essential for salmonids. Fats can constitute as much as 20% of properly compounded diets for salmon and trout; this level approximates the highest level of lipids usually found in natural diets of trout. Most trout foods contain 15% or less of fats.

Vitamins are needed in the diets of all the fish species for which this aspect of nutrition has been studied. The most thorough studies have been made of salmonids, especially trouts (*Salmo gairdneri, S. trutta, Salvelinus fontinalis*) and the chinook and coho salmons (*Oncorhynchus tshawytscha, O. kisutch*). Essential vitamins for trout include: ascorbic acid (C), thiamine (B_1), riboflavin (B_2), pyridoxine (B_6), vitamin B_{12}, biotin (H), choline, folic acid, inositol, nicotinic acid (niacin), pantothenic acid, *p*-aminobenzoic acid, tocopherol (E), and vitamins K and A. The requirements of chinook salmon are similar but no requirement for *p*-aminobenzoic acid has been shown for that species. Channel catfish require essentially the same vitamins, with the possible exception of biotin. Generally studies with the above species and with other cultured species such as carp, Japanese eel, red tai, and yellowtail have shown similar qualitative needs for vitamins, but deficiency symptoms differ from species to species, and quantitative requirements may differ.

Vitamin requirements are usually determined by feeding experimental lots of fish diets from which specific vitamins have been withheld, and noting deficiency symptoms that may occur in the experimental fish but not in the control groups. Experiments may involve restoring the missing vitamin to the diet and noting any possible recovery of the deprived fish Deficiency of ascorbic acid in trout can cause abnormal spinal curvature (scoliosis, lordosis) and internal bleeding (Fig. 14–1). Poor or no growth is a consequence of withholding several vitamins, and a high mortality rate accompanies a deficiency of tocopherol, biotin, thiamine, and especially pyridoxine. An ex-

FIGURE 14–1. Rainbow trout (*Salmo gairdneri*), showing severe scoliosis resulting from ascorbic acid-free diet.

cess of vitamin A (hypervitaminosis A) causes pathological changes in trout.

Some examples of daily dietary requirement of vitamins for young trout and salmon, expressed in mg/kg of body weight, are: ascorbic acid, 3–5; thiamine, 0.15–0.20; riboflavin, 0.5–1.0; vitamin B_{12}, 0.0002–0.0003; and choline, 50–60.

Mineral requirements in fish diets are not well known. Presumably fishes need the same elements necessary in the body chemistry of other animals. Iodine is essential in the diet, and its lack can cause goiter and reduced growth. Cobalt supplementation in carp diets has been shown to increase growth and food consumption. Sodium and potassium are important in osmoregulation, iron is necessary for the blood, calcium and phosphorus for bone formation and various metabolic functions, and other elements play their familiar roles. There has been some difficulty in assessing the dietary requirements of minerals and other trace elements in fishes because of their ability to absorb elements directly from the water. Calcium, chloride, cobalt, phosphorus, strontium, and sulfate can all be taken out of the water by trout, and probably by other species as well.

Diets for Fish Culture

An economic advantage can be gained by the fish culturist who produces healthy market-sized fish in a short time, and much of the ability to maximize growth depends on food. The culturist must provide proper space, sufficient water of high quality and correct temperature, and protection from infectious diseases so that the fish can utilize the food to best advantage. The food must supply all the materials that the fish are unable to absorb from the water in sufficient quantity. Vitamins, minerals, and other food elements concerned with metabolism must be present, but the culturist is especially concerned that the diets provide energy for basal metabolism, activity and growth. The energy, expressed in kilocalories (kcal), that can be used for growth depends not only on the amount but also on the kind of food given to the fish. As noted, most commercially reared fish species do not tolerate large amounts of fats and carbohydrates in the diet. The type and balance of protein with other ingredients can make a considerable difference in the amount of food that will produce good growth. For instance, in one study with brook trout (*Salvelinus fontinalis*), natural food that contained about 640 kcal/kg proved to be more than twice as efficient as a compounded dry diet containing 1540 kcal/kg; with the dry diet, 4600 kcal were required to produce 1 kilogram of trout whereas only 2000 kcal were required with the natural food.

Only a portion of the energy ingested is available for growth, for several reasons. First of all, not all the energy may be in materials digestible by the fish, or some materials may otherwise escape assimilation. This portion of the food energy is passed out of the alimentary canal as feces. The materials absorbed through the intestinal wall contain some energy in nitrogenous compounds that cannot be metabolized and are excreted at the gills or kidneys. Estimates of energy in fecal and other wastes are usually in the range of 15 to 20% of energy ingested. The remaining metabolizable materials are available for necessary metabolism and growth. Needs of the organism that must be met before energy in appreciable amounts can be used for growth include standard metabolism of the resting animal, any swimming or other activity over the resting condition, and the energy of what is called specific dynamic activity (SDA). SDA includes energy used in deamination of amino acids that are not used in growth, as well as energy utilized in the digestion and assimilation of food. SDA has been studied in a few fishes. The amount of energy attributed to it in the bioenergetics of the cutthroat trout has been estimated to range from about 14 to about 47% of energy consumed by individual fish. Usually SDA in fishes ranges from 10 to 18% of energy consumed.

An energy budget that takes the fate of all ingested energy into consideration can be expressed as:

$$Q_c - Q_w = Q_g + Q_s + Q_d + Q_a$$

where Q_c = energy in food consumed; Q_w = energy in waste (feces, urine, etc.); Q_g = energy in materials added to body (growth); Q_s = energy of standard metabolism; Q_d = energy of SDA; and Q_a = energy used in activity over that of standard metabolism. If the three latter quantities are combined and termed respiration, Q_r, the equation can be simplified: $Q_c - Q_w = Q_g + Q_r$.

In trying to maximize Q_g while minimizing the other quantities, the fish culturist must plan diets with great care, providing the right balance of all the required vitamins, minerals, fats, carbohydrates, and high-quality protein for growth—all at the least possible cost. At one time, various animal foods from slaughterhouses were available, and many trout hatcheries fed fish on liver, spleen, lungs, and other viscera, as well as day-old calves. These products could be refrigerated or frozen and the daily ration could be ground, mixed with wheat middlings or other grain products, and fed fresh. These diets were cheap and often effective, although nutritional problems were common. As mink ranches and the pet food industry began to compete for these meat products and other, more profitable uses were found for some of them, the supply available to fish hatcheries at low prices dwindled. Hatcherymen, nutritionists, and fishery biologists had long been experimenting with fish diets, trying to eliminate nutritional disorders and formulate inexpensive dry diets that would provide complete nutrition and require a minimum of storage cost. During the 1950's and 1960's dry, pelletized fish foods largely supplanted fresh diets. These foods were formulated from a variety of fish meals and other meals of animal origin such as blood meal, plus vegetable products such as soybean meal, cottonseed meal, and alfalfa meal. Dried milk products, by-products from the brewing and distilling industries, vitamin mixes, and mineral mixes are other ingredients.

Moist pellets, made up of a mixture of typical dry ingredients plus marine fish, pasteurized salmon viscera, tuna viscera, etc., in the ratio of about 60% dry ingredients to 40% wet ingredients, have been very successful in culture of salmon. Moist pellets are quick-frozen immediately after manufacture and can be held frozen for a reasonable length of time without loss of quality.

Diets for trout usually contain from 38 to 50% protein, up to 12% carbohydrate and from 5 to 8% fat. Diets for channel catfish generally contain about 30% protein, up to 25% carbohydrate, and about 5 to 8% fat. Some of the best catfish food contains about 1870 kcal/kg. Trout of about 15 cm held in 14° C water are fed dry diets at about 2% of body weight per day. Smaller fish are fed more and larger fish less; more food is given in warmer water and less in colder water. Channel catfish held in 25° C water are fed at the rate of 2.5 to 3% of body weight per day. Conversion of food into fish is usually from 1.6 to 2.5 kg of dry food to 1 kg of fish produced.

Growth

Individuals of a given fish species have a genetic potential for reaching a characteristic maximum size under the most favorable circumstances. Maximum size is reached in a relatively short time in short-lived species, but may be attained only after decades in long-lived species. Under less favorable environmental conditions, fish reach a size smaller than the maximum physiologically attainable for the species. The sigmoid pattern of increase in size with age generally exhibited by fishes is illustrated in the theoretical curve in Figure 14–2. The curve represents growth from hatching to the maximum possible

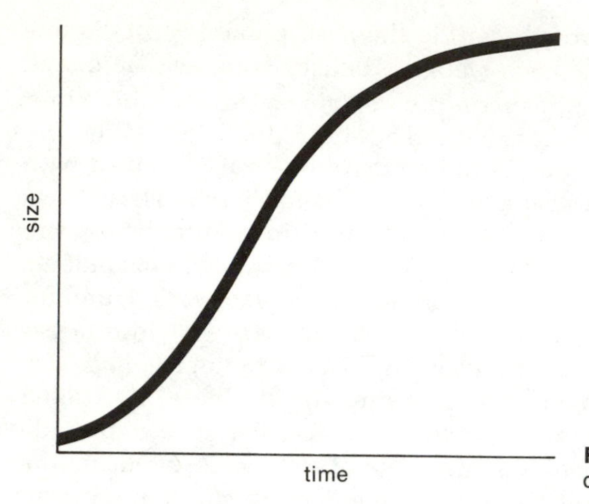

FIGURE 14–2. Idealized sigmoid growth curve.

in a given environment. Actually, fish growth is more irregular than shown in the idealized curve. Growth is usually greater in warm weather than in cold, and may decrease during migrations or spawning, sometimes even being negative when metabolic demands exceed the food energy intake (Fig. 14–3). In addition to annual fluctuations, fish growth may occur in "stanzas" during the normal life history, with each stanza being defined by a sigmoid curve showing a slowing down of growth before resumption of more rapid growth with entry into the next stanza. Growth stanzas generally result from physical, physiological or ecological changes. A migratory fish might end its first growth stanza, begin its second by moving from a stream into a lake or the ocean, and begin a third when it is large enough to feed on other fishes rather than small invertebrates

Factors involved in the irregular growth or in the limitation of

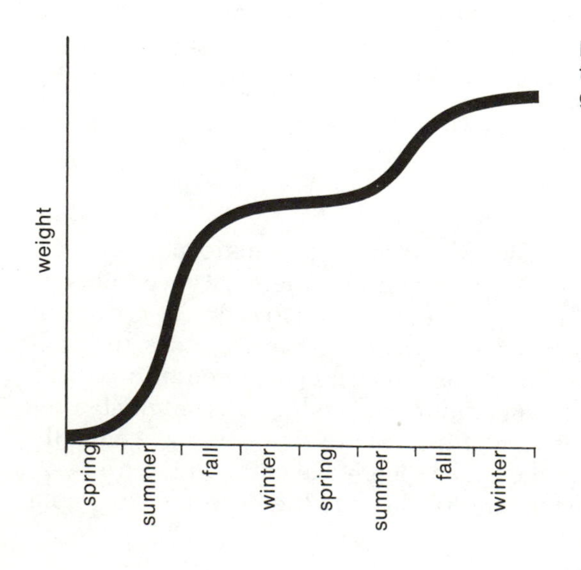

FIGURE 14–3. Growth curve of fish in temperate climate, showing cessation of growth in winter.

growth may be environmental or may be concomitants of the fish's physiology. Environmental factors include temperature and light, both of which change with the seasons. These can affect the fish directly or can influence another all-important factor — the abundance (and availability) of food. Generally, food and temperature favorable for growth occur during spring, summer, and fall in temperate areas, but most species do not grow at a constant rate during this period. Summer-spawning species such as the bluegill may put on most of the year's growth in the spring, and then grow slowly through the spawning period and the hottest part of the summer. Spring spawners grow best following spawning but growth may slow during the hot summer. Some, such as the largemouth bass, feed heavily in late summer and early fall.

Growth is regulated by hormones, with the pituitary growth hormones being of foremost importance. Physiological changes influenced by hormones affect growth rate during migrations, spawning, and wintering, all natural segments of the life cycle. Hormonal influence can be seen in the differential growth of the sexes. In some species males are of much smaller maximum size than females. Among other species, this occurs in many having internal fertilization. The most extreme examples are the ceratioid anglerfishes. In those species in which the males are larger than females, size appears to be related to function, and the males are involved in building or guarding nests, or in other activities that require size and stamina. Hormones are involved in the response of fishes to stressful situations. Most fish grow at a slow rate under stresses such as overcrowding. Even when sufficient food is made available, fish held in overcrowded ponds grow poorly.

The study of growth is not only of great scientific interest, but is also of considerable practical importance in the management of fisheries. Growth may be studied in terms of nutritional bioenergetic considerations, with the expectation that the knowledge gained will aid in growing fish faster on less expensive rations than currently used. Or growth may be considered in relation to age for various species or stocks within species in order, for instance, to gain information on the influence of population size on growth so that catch regulations making the best use of the resource may be set. Comparing size at first maturity among genetically similar stocks can provide insight into possible limitations on growth in various bodies of water. Various relationships of length to weight are used by fishery managers in order to compare the general conditions of fish from separate stocks, of fish of the same stocks at different times of the year, or from different bodies of water.

Growth is studied by several methods. Age of fish can be determined by interpretation of marks left in scales and other hard structures as a result of seasonal growth. Then, measurement of specimens representing several age groups can provide an estimate of the rate of growth from year to year. Frequent sampling of known age groups during the year can aid in estimating growth from season to season.

Individual fish can be captured, measured, tagged for later identification, released, and hopefully recaptured and remeasured, thus giving actual change in size over a period of time.

References

Barrington, E. J. W. 1957. The alimentary canal and digestion. *In*: Brown, M. E. (ed.), The Physiology of Fishes, Vol. 1. New York, Academic Press, pp. 109–161.

Brocksen, R. W., and Bugge, J. P. 1974. Preliminary investigations on the influence of temperature on food assimilation by rainbow trout *Salmo gairdneri* Richardson. J. Fish Biol., 6:93–97.

———, Davis, G. E., and Warren, C. E. 1968. Competition, food consumption and production of sculpins and trout in laboratory steam communities. J. Wildl. Mgt., 32:51–75.

Castell, J. D., Sinnhuber, R. O., Wales, J. H., and Lee, D. J. 1972. Essential fatty acids in the diet of rainbow trout (*Salmo gairdneri*): growth, feed conversion and some gross deficiency symptoms. J. Nutr., 102:77–86.

Chang, W. 1971. Studies on feeding and protein digestibility of silver carp, *Hypophthalmichthys molotrix* (C. & V.). Chinese-American Joint Comm. on Rural Reconstr. Fish., Ser. 11:96–114.

Darnell, R. M., and Meierotto, R. R. 1962. Determination of feeding chronology in fishes. Trans. Am. Fish. Soc., 91:313–320.

Davis, G. E., and Warren, C. E. 1965. Trophic relations of a sculpin in laboratory stream communities. J. Wildl. Mgt., 29:846–871.

———. 1968. Estimation of food consumption rates. *In*: Ricker, W. E. (ed.), Methods for Assessment of Fish Production in Fresh Waters. Oxford, Blackwell Scientific Publications, pp. 204–225.

Dupree, H. K. 1976. Studies on nutrition and feeds of warmwater fishes. Proc. 1st Int. Conf. Aquaculture Nutr. Univ. Del., pp. 65–84.

Furukawa, A. 1976. Diet in yellowtail culture. Proc. 1st Int. Conf. Aquaculture Nutr. Univ. Del., pp. 85–104.

Gaudet, J-L. 1971. Report of the 1970 Workshop on Fish Feed Technology and Nutrition. Washington, D. C., United States Fish and Wildlife Service, Bureau of Sport Fish and Wildlife, Resource Publ. 102.

Groot, S. J. De. 1971. On the relationships between morphology of the alimentary tract, food and feeding behavior in flatfishes (Pisces: Pleuronectiformes). Netherl. J. Sea Res. 5:121–196.

Halver, J. E. (ed.). 1972. Fish Nutrition. New York, Academic Press.

———. 1976. Formulating practical diets for fish. J. Fish. Res. Bd. Can., 33:1032–1039.

Hunt, B. P. 1960. Digestion rate and food consumption of Florida gar, warmouth and largemouth bass. Trans. Am. Fish. Soc., 89:206–211.

Kapoor, B. G., Smit, H,, and Voringhina, I. A. 1976. The alimentary canal and digestion in teleosts. *In*: Russell, F. S., and Yonge, M. (eds.), Advances in Marine Biology, Vol. 13. London, Academic Press, pp. 109–203.

Keast, A., and Webb, D. 1966. Mouth and body form relative to feeding ecology in the fish fauna of a small lake, Lake Opinicon, Ontario. J. Fish. Res. Bd. Can., 23:1845–1874.

Lee, D. J., and Putnam, G. B. 1973. The response of rainbow trout to varying protein/energy ratios in a test diet. J. Nutr., 103:916–922.

Leitritz, E., and Lewis, R. C. 1976. Trout and Salmon Culture (Hatchery Methods). Sacramento, California Fish Bulletin, 164.

McDonald, P., Edwards, R. A., and Greenhalgh, J. F. D. 1966. Animal Nutrition. Edinburgh, Oliver and Boyd.

Molnar, G., Tomassy, E., and Tolg, I. 1967. The gastric digestion of living predatory fish. *In*: Gerking, S. D. (ed.), The Biological Basis of Freshwater Fish Production. Oxford, Blackwell Scientific Publications, pp. 135–149.

National Academy of Sciences—National Research Council. 1973. Nutrient Requirements of Trout, Salmon, and Catfish. Nutrient Requirements of Domestic Animal Series. Washington, D.C., National Academy of Sciences, No. 11.

Pandian, T. J. 1967. Intake, digestion, absorption and conversion of food in the fishes *Megalops cyprinoides* and *Ophiocephalus striatus*. Mar. Biol., 1:16–32.

Phillips, A. M., Jr. 1969. Nutrition, digestion and energy utilization. *In*: Hoar, W. S., and Randall, D. J. (eds.), Fish Physiology, Vol. 1. New York, Academic Press, pp. 391–432.

Randall, J. E. 1967. Food habits of reef fishes of the West Indies. Proc. Int. Conf. Trop. Oceanogr., Univ. Miami, Stud. in Trop. Oceanogr., 5:665–847.

Seaberg, K. G., and Moyle, J. B. 1964. Feeding habits, digestion rates and growth of some Minnesota warmwater fishes. Trans. Am. Fish. Soc., 93:269–285.

Shulman, G. E. 1974. Life Cycles of Fish: Physiology and Biochemistry. New York, Halstead Press.

Steigenberger, L. W., and Larkin, P. A. 1974. Feeding activity and rates of digestion of northern squawfish. J. Fish. Res. Bd. Can., 31(4):411–420.

Suyehiro, Y. 1942. A study of the digestive system and feeding habits of fish. Jap. J. Zool., 10:1–303.

Tyler, A. V. 1970. Rates of gastric emptying in young cod. J. Fish. Res. Bd. Can., 27:1177–1189.

Warren, C. E. 1971. Biology and Water Pollution Control. Philadelphia, W. B. Saunders.

———, and Davis, G. E. 1967. Laboratory studies on the feeding, bioenergetics, and growth of fish. *In*: The Biological Basis of Freshwater Fish Production. Gerking, S. D. (ed.), Oxford, Blackwell Scientific, pp. 175–214.

Weatherly, A. H. 1976. Factors affecting maximization of fish growth. J. Fish Res. Bd. Can., 33:1046–1058.

Winberg, G. G. 1956. Rate of metabolism and food requirements of fishes. Lenina, Minsk, Nauchnye Trudy Belorusskovo Gosudarstrennovo Universiteta im. V. I. Trans. Ser. 194, Fish. Res. Bd. Can.

Windell, J. T. 1967. Rates of digestion in fishes. *In*: Gerking, S. D. (ed.), The Biological Basis of Freshwater Fish Production. Oxford, Blackwell Scientific, pp. 150–173.

———, and Norris, D. O. 1969. Gastric digestion and evacuation in rainbow trout. Progr. Fish-Cult., 31:20–26.

Yone, Y. 1976. Nutritional studies of red sea bream. Proc. 1st Int. Conf. Aquaculture Nutr. Univ. of Del., pp. 39–64.

15
REPRODUCTION

Fishes are well known for their high potential fecundity, with most species releasing thousands to millions of eggs annually. The world would be full of fish if the environment did not take its toll of eggs and hatched young. Thus, the minimum requirement of reproduction, if a species is to maintain itself in stable numbers, is eventual replacement of each spawning pair by an equally successful pair. Stability of population numbers is seldom actually achieved, and numbers wax and wane depending on the pressures of environmental factors. Fluctuations may be erratic or cyclic, depending on these factors, and not all species in an area will be equally affected by the same environmental changes.

Fish species have evolved reproductive methods and an attendant physiology that allows them to be successful under a great variety of conditions. The entire combination of habits, physiology, and behavior, the overall approach to reproduction, will be termed the reproductive strategy. Strategies may require great numbers of eggs, as mentioned, or fewer eggs with greater opportunity for survival. Strategies must ensure survival of a portion of the eggs — through force of numbers, concealment, protection of nests, or retention in the body; strategies must place the earliest feeding stage of the young in the proximity of ample and suitable food, and must ensure that the juvenile fish have eventual access to the living space of the adults. Time and location are generally of great importance.

Placement of eggs or young in the right place at the right time is due to the response of the endocrine system to environmental cues such as temperature and light, so that gametes are matured and spawning migrations are undertaken. In some species reproductive readiness may be achieved by the influence of long photoperiod and warm temperatures, and spawning may be brought on by shortened photoperiod and decreasing temperature. In others the converse is true. Some are mainly responsive to temperature and others mainly responsive to light. Water flow and flooding, availability of food, salinity and, very possibly, other environmental factors may have reproduction-related impact on the endocrine system. The proximal pars distalis of the pituitary secretes gonadotropic hormones that promote the development of eggs and sperm, and stimulate the production of androgenic and estrogenic steroids that control sexual behavior and development of secondary sexual characteristics. The gonadotropic cells of the pituitary are under the control of the hypothalamus and possibly the pineal

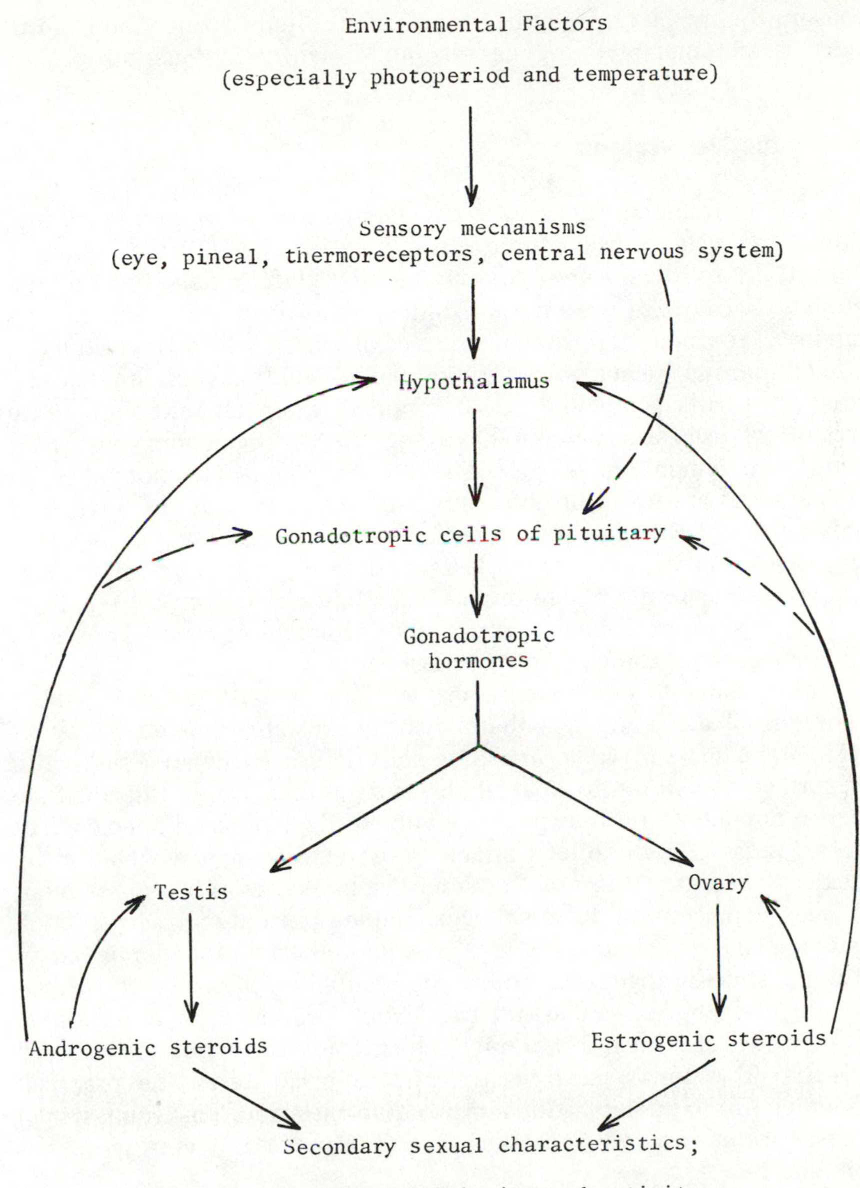

FIGURE 15–1. Possible relationships among environmental factors, receptors, endocrine organs, and reproductive activity.

body. Possible pathways and control mechanisms are shown in Figure 15–1.

Fish culturists have learned to influence or control the maturation of brood stock in hatcheries or fish farms. By controlling photoperiods, temperature, and, for some species, water flow, fishes can be brought into spawning condition earlier or later than the normal reproductive season. Spawning time can be directly controlled by injection of pituitary extract, mammalian gonadotropin, or synthetic gonadotropin.

Reproductive Strategies

There are numerous ways to categorize fishes as to their reproduction — for instance, the simple breakdown into egg-layers (oviparous condition) and live-bearers (ovoviviparous, viviparous)—but this obviously is only the beginning. Egg-layers are involved in numerous modes of reproduction, some examples of which will be mentioned.

Oviparous Fishes. Eggs of oviparous bony fishes are usually small, typically between 1.5 and 3 mm in diameter, although some species of trout and salmon have eggs exceeding 5 mm, and ariid catfishes commonly have eggs of from 15 to 25 mm. The elongate eggs of hagfishes are up to 30 mm long, and oviparous sharks, rays, and chimaeras deposit eggs in cases up to 300 mm long. Some of the larger egg cases contain two eggs. Elasmobranch egg cases are of various shapes, from spindle-like to purse-like, as shown in Figure 15–2. Eggs of bony fishes are typically round, but elongate eggs are known in anchovies, some gobies, and clownfishes.

Most fishes lay eggs that are heavier than water (demersal eggs) but many produce buoyant eggs that may be hydrostatically adjusted by oil inclusions, imbibed water (in a large perivitelline space), or a high ratio of surface to volume to float at the surface or at some intermediate depth, depending on the species involved. Eggs of some species drift freely; those of some others attach to each other or to vegetation by means of tendrils (Fig. 15–3). Tendrils, hooks, or other attachment devices also occur on demersal eggs, and are present on elasmobranch and hagfish eggs. Demersal eggs may be adhesive and deposited in clumps, sticking together through the incubation period, or they may be attached singly to some substrate. Some are temporarily adhesive, like the eggs of trout and salmon. Such eggs adhere to the substrate and to each other for a short period, and then separate. The practical advantage of temporary adhesion for a species that constructs gravel nests in running water and takes several minutes to cover the eggs is obvious.

Some fishes engage in mass spawning, with no pairing. Numerous males and females release gametes together in a suitable environment. Numbers of eggs are high, and they may be left to drift in the open waters, to be carried and buoyed up by the turbulence of a stream, allowed to settle on the substrate, or released so that they may adhere to vegetation. Spawning often ensues after migration to a suitable site, often against a current that will carry eggs and larvae back to a nursery area. Some herrings are exemplary of open sea mass spawners. Others migrate to shore areas and some, like the shad, are anadromous.

Polyandrous spawning is exhibited by many species in both fresh and marine situations. Males position themselves on the spawning

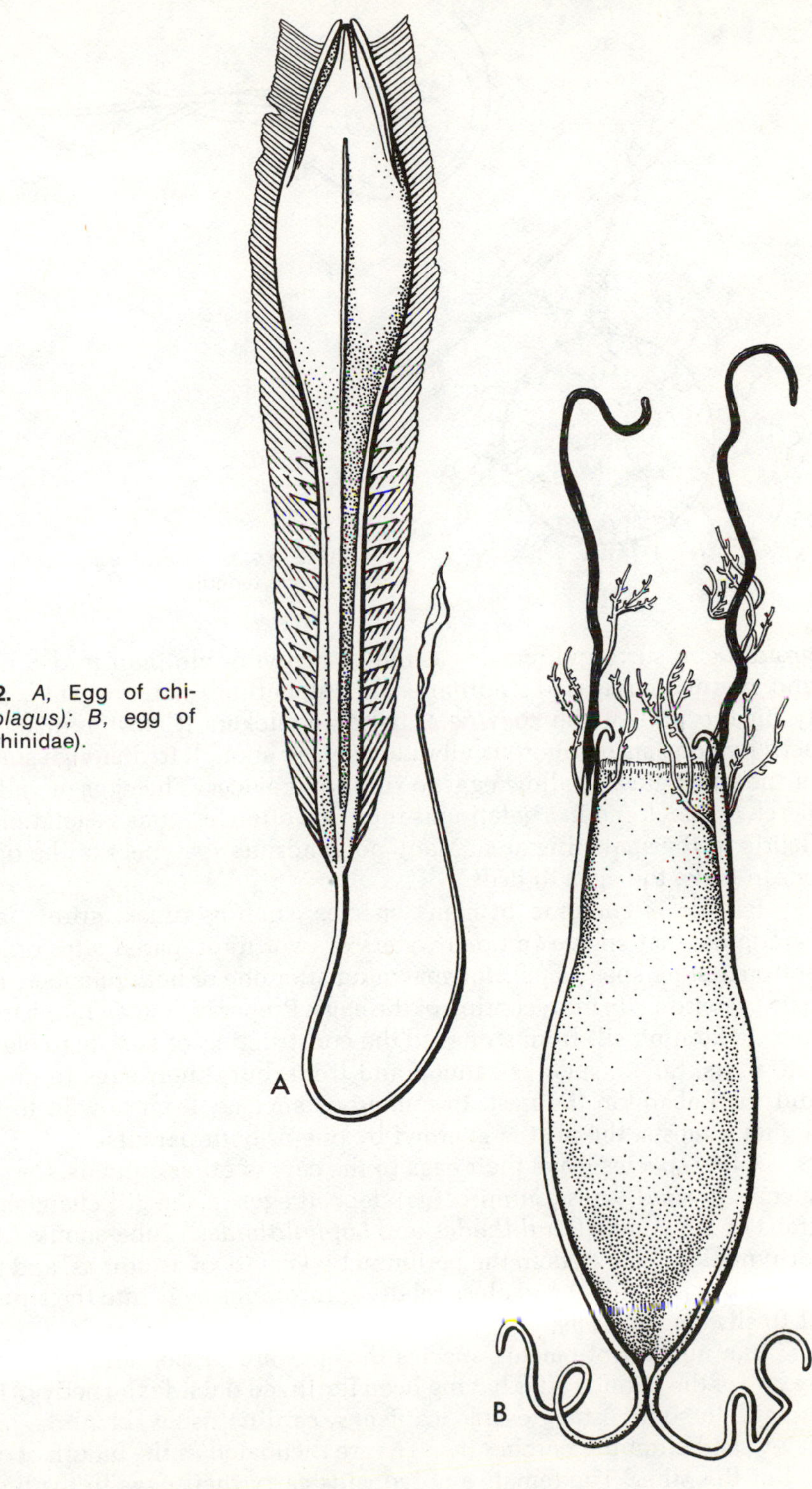

FIGURE 15–2. *A,* Egg of chimaera *(Hydrolagus); B,* egg of shark (Scyliorhinidae).

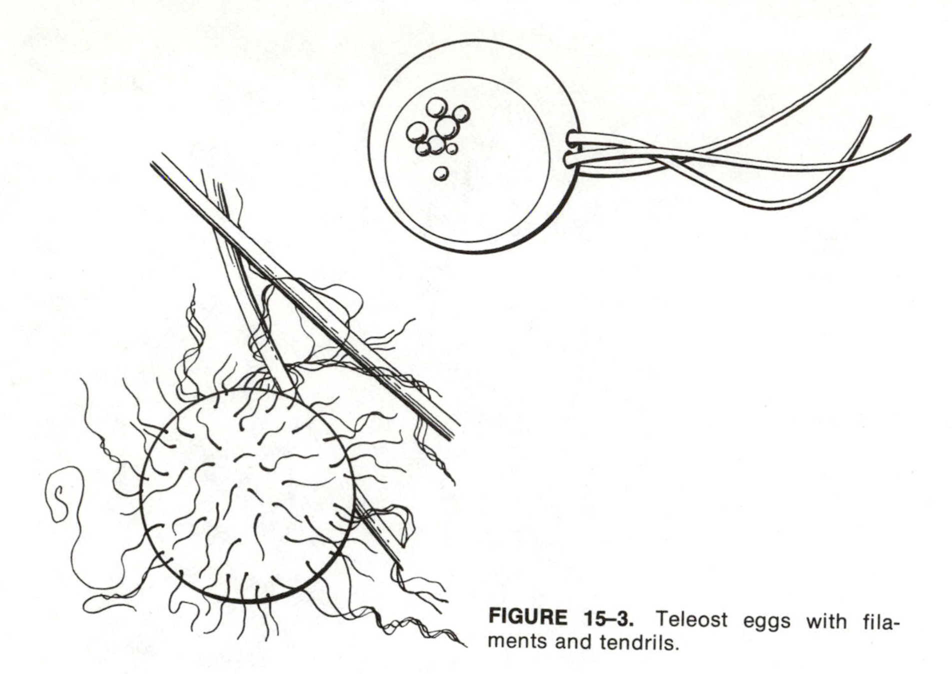

FIGURE 15–3. Teleost eggs with filaments and tendrils.

ground and surround females as the latter swim into their midst. Eggs and sperm are released simultaneously, sometimes with violent activity on the part of the spawners. In some suckers (Catostomidae) the activity consists of vigorous vibration strong enough to dislodge stones of the substrate and allow eggs to sift into crevices. The eggs of yellow perch are enclosed in a gelatinous rope that often festoons vegetation or debris in the spawning area. Many polyandrous spawners of the open ocean leave the eggs to drift.

Pairing is common in many species, such as tunas, grunts, and carangids, that spawn in open waters or over unprepared sites on the bottom. Other species pair for spawning after one or both members of a pair prepare a site for reception of the eggs. Preparation may range from merely fanning silt from stones to the construction of simple to elaborate nests. Some, such as salmon and trout, bury their eggs in gravel and then abandon the nest. In numerous species, from bowfin to the higher teleosts, the nest is guarded by one or both parents.

A few species leave their eggs in the care of other animals. Certain species of snailfishes (*Careproctus*) deposit eggs in the gill chamber of crabs of the genera *Paralithodes* and *Lopholithodes*. Tube-snouts (Aulorhynchidae) lay eggs in the peribranchial cavity of ascidians, and the bitterlings (*Rhodeus*) and close relatives introduce eggs into the siphon of freshwater mussels.

In a number of pairing species the eggs are carried and protected by one of the parents after having been fertilized outside the body of the female. In some osteoglossids, catfishes, cardinalfishes, cichlids, jawfishes, and climbing perches the eggs are incubated in the mouth of one sex or the other. The female amblyopsids carry their eggs in the branchial cavity. Female bunocephalid catfish and some loricariid catfishes carry the eggs embedded in the skin of the lower surfaces of the body and fins. Males of some pipefish and seahorse species bear eggs on the skin, but most species carry eggs in brood pouches. In their close relatives, the Solenostomidae, the female is equipped with a brood pouch formed by the pelvic fins. The males of *Kurtus* have a hook-like structure projecting forward from the forehead, from which the clustered eggs hang during incubation.

Some oviparous fishes fertilize the eggs internally. This is encountered, for example, among skates, chimaeras, many sharks, some characins, catfishes, cyprinodonts, and rockfishes. Egg retention for a short period, after fertilization takes place in the follicle or ovarian lumen, naturally leads into the conditions of ovoviviparity (internal incubation) and viviparity.

Egg Retention, Internal Incubation, and Viviparity. There is a continuum of conditions leading from the deposition of internally fertilized eggs in the cleavage stage to the release of large, well nourished juveniles or young adults. Several possible advantages are conferred by these patterns. The first of these is protection. The eggs and embryos are safe from predators except those large enough to overwhelm the female. They are protected from adverse water conditions; desiccation, anoxia, and injurious temperatures are not dangerous unless the female is unable to escape those conditions. Young are protected against loss by drifting, as eggs or larvae, away from suitable rearing areas. Another possible advantage might be conservation of energy, although careful study would be necessary to confirm energy-saving in specific instances. Certainly nest-building is unnecessary, and the size of the male can be reduced. There is no need for large numbers of eggs, and generally no necessity to make long migrations to a specific breeding site. Usually fertilization is ensured, so that few eggs are wasted. Viviparity (in its broad sense) may be of some advantage in species dispersal, as a single pregnant female might be able to accomplish an extension of range. Similarly, survival of one pregnant female after a catastrophe might allow continued survival of the species.

Because there is some difficulty in fitting uncomplicated definitions to complicated natural processes that grade into each other, there have been several versions of definitions for ovoviviparity and viviparity. Consideration of viviparity in a broad sense to include all conditions in which hatched young are liberated from the female might be most suitable, but the term ovoviviparous is useful for emphasizing those species that incubate eggs and liberate live young without providing any maternal source of nourishment other than that in the egg.

As mentioned earlier, some oviparous species set the stage for ovoviviparity in fertilizing eggs internally and releasing them in an early stage of development. Internal fertilization requires modification of behavior and the development of the means of introduction of

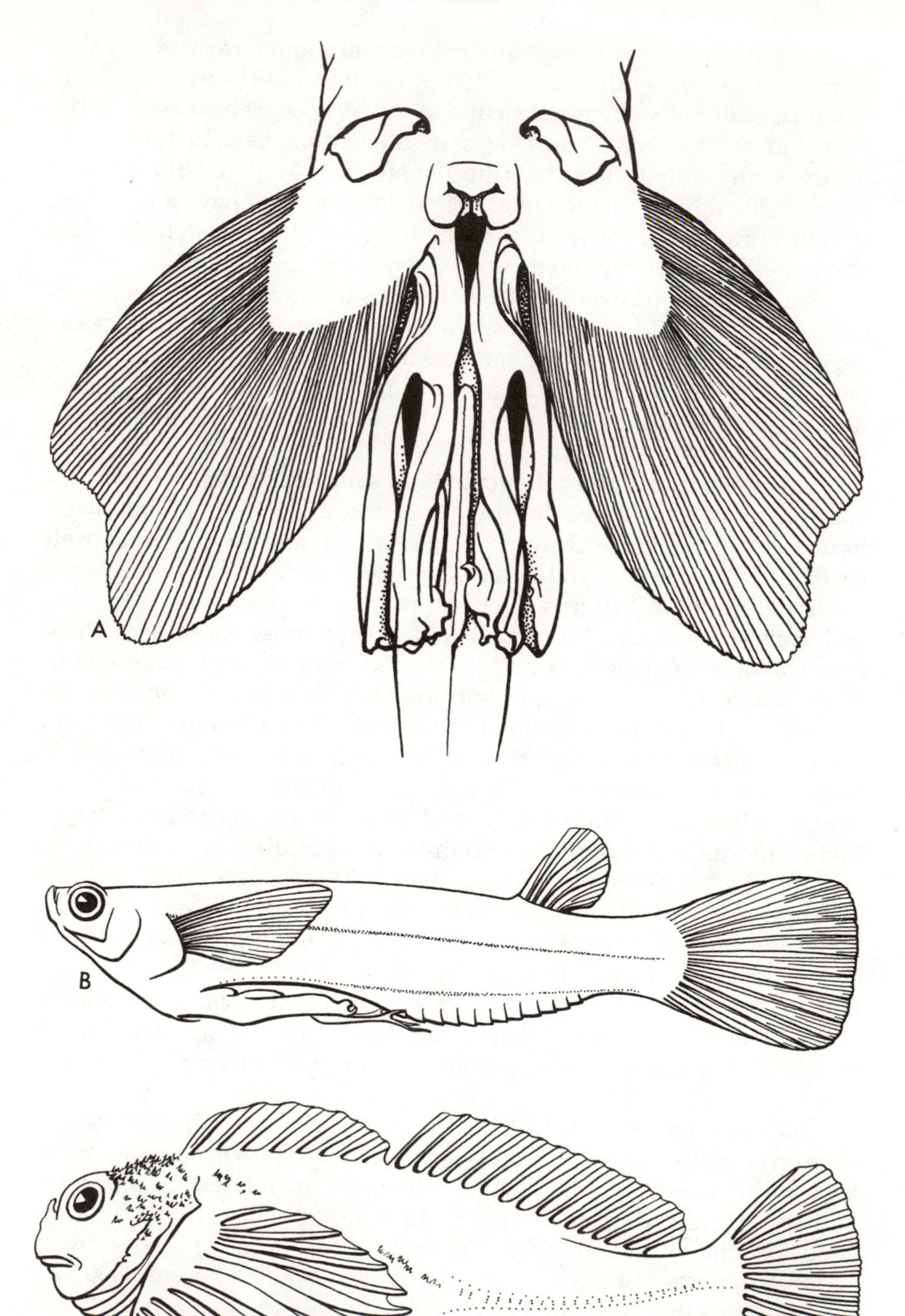

FIGURE 15–4. Examples of male oviparous fishes with large intromittent organs. *A*, Pelvic claspers of a chimaera *(Hydrolagus)*; *B*, *Horaichthys*; *C*, *Clinocottus*.

spermatozoa into the genital orifice of the female. Spermatophores have developed in Poeciliidae and Horaichthyidae; those of the latter are barbed for sure attachment. Some of the most remarkable intromittent organs among fishes are seen in oviparous species (Fig. 15–4). Chimaeras, skates, and oviparous sharks have pelvic "claspers," while the anal fin is modified into a gonopodium in *Tomeurus* (Poeciliidae) and in Horaichthyidae. Phallostethidae and Neostethidae have elaborate structures developed for use in clasping and inseminating the female. Some sculpins have fleshy genital papillae that reach a large size.

From the short-term retention of eggs to incubating and hatching them internally is only a short step. Many species of rockfishes (Scorpaenidae) are ovoviviparous, releasing newly hatched larvae. The rockfishes show little sexual dimorphism and have no particular specialization of the ovary, where the eggs remain during incubation. In a brotulid genus, *Dinematichthys*, eggs hatch from the ovarian follicles and develop into advanced larvae in the lumen of the ovary. The term "larviparous" has sometimes been applied to the condition of giving birth to larvae.

Another type of ovoviviparity is seen in most Poeciliidae, in which the young are retained until the juvenile stage is reached. Young remain in the follicles of the ovary, with sufficient yolk for development. In some species more than one brood can be present at once (superfetation). In certain species, including those of *Gambusia* and *Xiphophorus*, the pericardial sac expands into a hood-like or strap-like structure that folds over the head; this structure is thought to have a respiratory function. Many elasmobranchs retain egg capsules in the oviducts until young are fully developed. In sawfishes, guitarfishes, and some others the capsules may each contain two to four eggs. In some elasmobranchs the egg capsule is not well developed and the young, bearing the yolk sac, are free in the uterine portion of the oviduct. Some of these receive a small amount of nourishment from ovarian or oviducal secretions. Examples are some electric rays, *Acanthias*, and *Squalus*. The coelacanth, *Latimeria chalumnae*, has the largest eggs among the bony fishes — 8.5 to 9.0 cm. in diameter. Because of the egg size, and because of evidence of live-bearing in fossil coelacanths, some scientists believed that *Latimeria* would be shown to be ovoviviparous. This was proven in 1975 by the discovery of five advanced young, up to 33 cm long, in the oviduct of a specimen at the American Museum of Natural History.

Viviparous fishes nourish developing embryos through a variety of adaptations, most involving the secretion of nutritive materials by the female, but including formation of pseudoplacentae. In some of the nonplacental species, the developing young subsist exclusively on the yolk for a time and then the yolk is supplemented or supplanted by secretions ("uterine milk") of the female and, in a few species, material from dead eggs and embryos. In several species of elasmobranchs, the nutritive secretions are taken in through the mouth or the spiracles and swallowed. In some, the villi (trophonemata) that secrete the nutrient fluid are elongate and extend through the spiracles of the embryo to the gut. Among the viviparous teleosts supplemental structures are usually developed to serve as absorptive surfaces. Examples are the expanded pericardial sac of some Poeciliidae, expansions of fin membranes in Embiotocidae, and branching structures called trophotaenia growing from the anal region of Goodeidae. The viviparous eelpout, *Zoarces viviparus*, has no special absorptive structures, and apparently absorbs directly through the skin. Many of these species give birth to young in an advanced state of development. The males of the shiner perch, *Cymatogaster aggregata*, are born sexually mature.

Placentae are formed in a few shark species by the close apposition and eventual interdigitation of thin tissues of the yolk sac and the uterus. Yolk sac placentae are present in *Mustelus canis, Carcharhinus falciformis, Sphyrna* spp., and a few others. Pseudoplacentae of bony fishes are of various types, involving the gut in Anablepidae, the follicle of the ovary in *Heterandria formosa*, and extensions of ovarian tissue in Jenynsiidae.

The varied reproductive methods of fishes have inspired few attempts to classify them from ecological or behavioral standpoints. Dr. E. K. Balon has proposed a classification that includes 32 reproductive

Table 15–1. ECOETHOLOGICAL GUILDS OF FISHES*

Guild	Examples
A. Nonguarders	
A.1. Open substratum spawners (eggs placed in open water or openly on substrate)	
A.1.1 Pelagophils (eggs and larvae drift free or attached to floating material)	Typical of pelagic species, also many marine benthic and some freshwater species
A.1.2 Litho-pelagophils (eggs deposited on rocks, larvae drift)	Some sturgeons, whitefishes, smelts
A.1.3 Lithophils (eggs deposited on rocks, larvae remain on bottom)	Many freshwater species, minnows, suckers, pike-perch, etc.
A.1.4 Phyto-lithophils (eggs deposited on submerged plants, rocks, logs, or other material)	Herring, minnows, perches, etc.
A.1.5 Phytophils (adhesive eggs on aquatic plants, not on bottom)	Carp and other minnows, gar, pike, etc.
A.1.6 Psammophils (eggs on sand)	Few minnows, smelts
A.2 Brood hiders (eggs buried or otherwise hidden)	
A.2.1 Lithophils (eggs in natural or constructed hiding places)	Salmon, trout, minnows
A.2.2 Speleophils (eggs hidden in caves)	Various cavefishes
A.2.3 Ostracophils (eggs hidden in shells of live invertebrates)	Bitterlings, some snailfishes
A.2.4 Aero-psammophils (eggs incubate in sand above waterline)	Grunion, one puffer
A.2.5 Xerophils (eggs endure dry seasons in mud, sod, or sand, etc.)	Certain cyprinodonts (*Aphysemion, Nothobranchius,* etc.)
B. Guarders (one or both parents attend eggs, may both guard and aerate them)	
B.1. Substratum choosers (no nest construction)	
B.1.1 Lithophils (eggs attached to rocks)	Certain gobies, snailfish, puffers, sculpins
B.1.2 Phytophils (eggs attached to, or scattered among, plants)	Bichir, certain catfishes and sunfishes, etc.
B.1.3 Aerophils (eggs attached to surfaces above water)	*Copeina arnoldi*
B.1.4 Pelagophils (buoyant eggs in cluster at surface)	*Channa, Anabas*

guilds. The guilds are arranged in three sections. Section A, *Nonguarders,* has two subsections, open substratum spawners that leave their eggs exposed and brood hiders that conceal the eggs. Section B, *Guarders,* includes substratum choosers and nest spawners. Section C, *Bearers,* includes those that bear eggs externally or internally. Although the full reproductive habits of only a small fraction of the fishes are known, and although Balon's classification may be incomplete or may not provide for some exceptional or versatile species, the guilds form a framework useful in presenting extensive information on breeding habits. Table 15–1 is based on Balon's work.

Sexual Differentiation and Differences

With important exceptions to be covered later, the individuals of most species of fish function as either male or female throughout their adult life. That is, most species are bisexual, or gonochoristic, as opposed to hermaphroditic, a condition in which an individual pro-

Table 15–1. ECOETHOLOGICAL GUILDS OF FISHES* *Continued.*

Guild	Examples
B.2. Nest spawners	
B.2.1 Lithophils (nests in gravel, on rock, etc.)	Certain sculpins, cichlids, minnows, etc.
B.2.2 Phytophils (nest built of, or among, plants)	Bowfin, certain sunfishes and percids, etc.
B.2.3 Psammophils (nests in sand)	*Abbottina rivularis, Cichlasoma nicaraguense*
B.2.4 Aphrophils (nests built of bubbles or froth)	Anabantoids, some armored catfishes
B.2.5 Speleophils (nests in natural or constructed cavities)	Several minnows, catfishes, sculpins, etc.
B.2.6 Polyphils (nests built of various materials)	Some osteoglossoids, sculpins, sunfishes, etc.
B.2.7 Ariadnophils (nest material bound together with kidney secretions)	Sticklebacks
B.2.8 Actinariophils (nests made next to sea anemones)	*Amphiprion* spp.
C. Bearers	
C.1. External bearers	
C.1.1 Transfer brooders (eggs carried by various means after extrusion, then deposited)	Some armored catfishes, some cyprinodonts
C.1.2 Forehead brooders (eggs carried on hook at forehead)	*Kurtus*
C.1.3 Mouth brooders (eggs carried in mouth)	Marine catfishes, cardinalfishes, etc.
C.1.4 Gill-chamber brooders (eggs carried in gill chambers)	Amblyopsidae
C.1.5 Skin brooders (eggs attached to skin)	Certain catfishes and pipefishes
C.1.6 Pouch brooders (eggs carried in cutaneous pouch)	Certain catfishes, pipefishes, and seahorses
C.2. Internal bearers	
C.2.1 Ovi-ovoviviparous (internal fertilization, external incubation)	Skates, chimaeras, some sharks, characins, priapiumfishes, and several others
C.2.2 Ovoviviparous (internal fertilization and incubation)	Many sharks, rays; *Latimeria.* live-bearers, etc.
C.2.3 Viviparous (young retained, nourished in ovary or oviduct)	Certain sharks, live-bearers, Jenynsiidae, Goodeidae, etc.

*As proposed by E. K. Balon (1975). Jour. Fisheries Res. Bd. of Canada (used by permission).

duces both eggs and sperm at some stage of its development. Establishment of sex depends upon the sex chromosomes, designated X and Y for most fishes. Usually XX individuals are females and XY individuals are males, but there are exceptions among some of the Poeciliidae, for instance, in which the homogametic individual is male. Chromosomes designated W and Z are recognized in some of these fishes.

The time at which the sex of an individual becomes established differs among gonochoristic fishes. The genetic material that causes the production of sex-inducing substances may not be activated until some species are in or past the larval stage, so that the very young fish may have undifferentiated gonads. Some geneticists postulate the production of male (M) factors and female (F) factors which, by their balance, point the undifferentiated gonad one way or the other.

Although the sexes have very similar appearances in many species, sexual dimorphism or dichromatism is common in fishes and may be especially well marked in those species with internal fertilization or elaborate reproductive behavior. The differences between the sexes may involve secondary sex characters necessary for the accomplishment of copulation, oviposition, or incubation (requisite characters), or may be so-called accessory characters, which may not be directly involved in the mechanics of reproduction but are important to recognition, courtship, or other reproductive behavior. Requisite secondary sex characters include the claspers of elasmobranchs and the various gonopodia of the males of phallostethiforms and cyprinodontiforms, and the modified anal fins of other fishes with internal fertilization, such as hemirhamphids and embiotocids. Even some oviparous species with external fertilization, such as sculpins, have large genital papillae. Brood pouches and specialized ovipositors are requisite characters.

Accessory secondary sex characters are many and varied, and are usually sexually dimorphic. Many structures and colors change with the reproductive state of the individual, while others are more or less permanent throughout the year. Males of many species are more brightly colored than females, and may have larger fins and bolder markings. Sexual dichromatism is seen in salmon and trout, in which the males are more colorful; in the bowfin, in which the male has a caudal ocellus; and in many minnows, characins, cichlids, and others. Longer fins are characteristic of the males of many fishes. Suckers, gobies, dragonets, and climbing perches are examples. The color and larger fins of males can be significant in courtship or aggressive displays toward rival males, or the fins can, in certain species, aid in holding the spawners together. This is especially true in species having nuptial tubercles or contact organs on fins or body. These structures are prominent on the fins and scales of suckers, and reach large sizes on the heads of certain minnows (Fig. 15–5). They are most common in species that spawn in flowing water. Nuptial tubercles are made up of epidermal cells and are of two types, keratinized and nonkeratinized. The horny caps of the former type are often quite pointed. Whitefish, grayling, smelts, ayu, retropinnids, kneriids, phractolaemids, most cy-

FIGURE 15–5. Nuptial tubercles on *A*, scales, anal fin, and caudal fin of sucker *(Catostomus)*; *B*, scales, head, and pectoral fin of chub *(Mylocheilus)*.

prinoid families, a few characoids, the mochokid catfishes, and percids have nuptial tubercles. Contact organs are small, bony, spine-like structures usually associated with scales or fin rays; they are present in needlefishes, certain cyprinodontoids, characins, and sculpins.

Size differences between the sexes is evident in many species. The reproductive requirements of the species determines which sex is the larger, but commonly the female, carrying the bulky eggs, is larger than the male. The greatest disparity in size occurs in the ceratioid anglerfishes, in which the female can be many times larger than the male. In certain species the male ceratioid grasps the female's skin with his teeth, literally grows to her, and becomes a testis-filled parasite, available for service at spawning time.

Hermaphroditic fishes have attracted attention as a rich resource of physiological and genetic information, as well as for their general interest. Occasional hermaphrodites are found in many gonochoristic species as an abnormality, but there are numerous species that are normally hermaphroditic, some even capable of self-fertilization. Synchronous (or simultaneous) hermaphrodites have ripe ovaries and testes at the same time, but usually spawn with one or more other individuals, alternately taking the role of male and female. *Serranus subligarius*, a marine species of Florida, has fertilized its own eggs in captivity. *Rivulus marmoratus* can fertilize its eggs internally prior to oviposition. Synchronous hermaphrodites are known from the following families: Chlorophthalmidae, Bathypteroidae, Alepisauridae, Paralepididae, Ipnopidae, Evermannellidae, Cyprinodontidae, Serranidae, Maenidae, and Labridae. Individuals of other families may exceptionally be hermaphrodites. A familiar example is the striped bass (Percichthyidae), in which occasional hermaphrodites are seen.

Consecutive hermaphrodites are either first male (protandrous) or first female (protogynous). Many of the species begin life with undifferentiated gonads that contain both male and female elements. The protandrous condition is known in members of the Gonostomatidae, Serranidae, Sparidae, Maenidae, Labridae, and Platycephalidae. Protogynous hermaphrodites have been noted in the Synbranchidae, Serranidae, Maenidae, and Labridae. In the latter family there are species in which some males do not pass through the female stage (primary males), as well as those secondary males that change to male from female. The secondary males are brightly colored, but the primary male may be dull, or at least colored like the females. In one species the two types of males exhibit different spawning behavior, the primary males spawning in groups with a single female and secondary males pair-spawning with females in turn.

Sex reversal and hermaphroditism are controlled by the endocrine system, which is genetically programmed in normally hermaphroditic species to act on the gonads in response to the proper stimuli, whether the stimuli be internal, external, or both. Experimentally, administration of androgens to genetic female fish has changed them into functional males, and administration of estrogens has changed genetic males into functional females. The breeding of sex-reversed, functional males (with no Y chromosome) to normal females can result in all-female progeny. This can be significant to fishery management, especially in situations where no reproduction of early-maturing pond fishes is desired or where a useful but potentially troublesome herbivore, such as the grass carp, *Ctenopharyngodon idella*, is to be stocked.

Embryological and Early Development of Fishes

Development of the fish egg normally begins upon fertilization by a spermatozoon, but has been experimentally induced by various chemical and physical techniques. The artificially activated eggs do not usually develop normally, but there are recorded instances of a fungus infection of ovaries of live-bearers bringing about activation of eggs with subsequent production of all-female broods. In the live-bearer genera *Poecilia* and *Poeciliopsis* there are all-female species that mate with males of other species but do not retain any of the male genetic material (genome) past activation or some very early stage of development.

In normal fertilization only one sperm enters the micropyle, except in elasmobranchs but, of course, only one sperm is involved in fertilization. The great majority of fishes release the sperm into the water in the vicinity of the eggs. Freshwater species usually bathe the eggs in a heavy concentration of seminal fluid as the eggs are being extruded, for the sperm, inactive until they come in contact with the water, have an effective life of from a few seconds to a few minutes. Fish sperm can be held alive for a short time in physiological (0.9%) saline solution. Sperm of marine fishes have a longer active life than those of the freshwater species. Fish culturists are quite interested in methods of long-term storage, as are used in the livestock industry, but definitive techniques have not yet been perfected. Sperm frozen in appropriate supporting media at the temperature of liquid nitrogen has been used with limited success in fertilization of fish eggs.

The development of the fertilized eggs follows much the same pattern as in other vertebrates. Fish eggs are telolecithal, with a relatively good supply of yolk, but there are wide differences among the fishes regarding the actual amount of yolk. A few species have only a small egg and a small enough supply of yolk so that the entire egg divides during cleavage (holoblastic cleavage). This is the case in lampreys, some sturgeons, and lungfishes. The cleavage is unequal, with larger cells at the vegetal pole, where the yolk is concentrated. Some fishes have eggs with the cytoplasm thinly distributed around the relatively large yolk, while others have cytoplasm concentrated at the animal pole. The cytoplasm forms a polar cap at the site of the nucleus following fertilization, and this begins to divide, forming the embryo on the surface of the yolk (meroblastic cleavage). Some stages in the development of a bony fish are shown in Figure 15–6.

Teleosts and lampreys share a peculiarity of embryology in that the central nervous system forms by the hollowing of a medullary keel instead of by the formation of medullary folds. Another interesting peculiarity in the development of bony fishes is the diapause undergone by eggs of certain annual cyprinodontids that deposit fertilized eggs in the bottoms of drying ponds. These eggs do not complete development until the ponds hold water again. *Nothobranchius* and

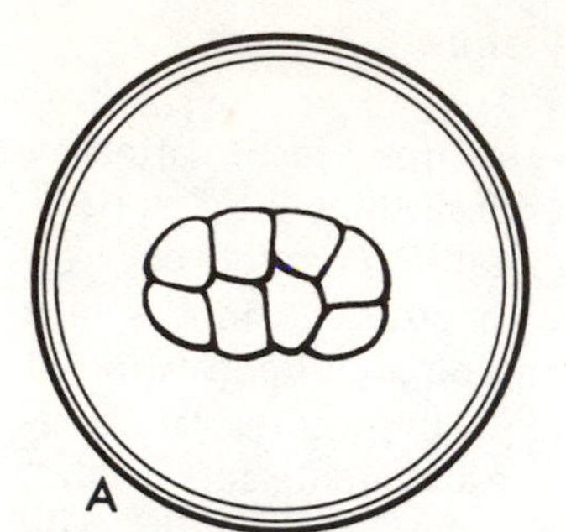

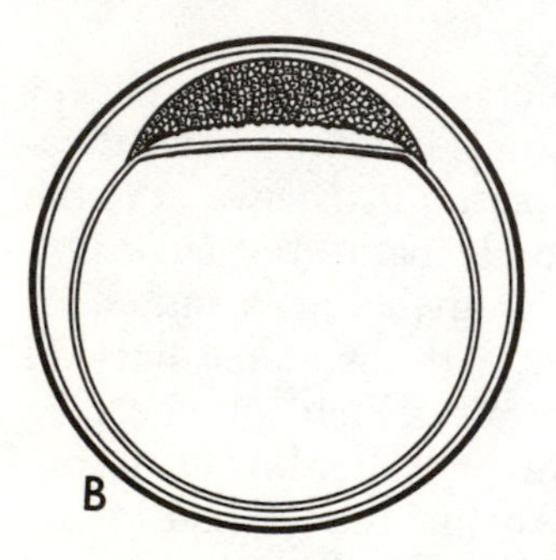

FIGURE 15–6. Examples of embryonic stages in teleosts. *A*, Cleavage; *B*, blastula (sectioned); *C*, embryonic shield; *D*, organogeny.

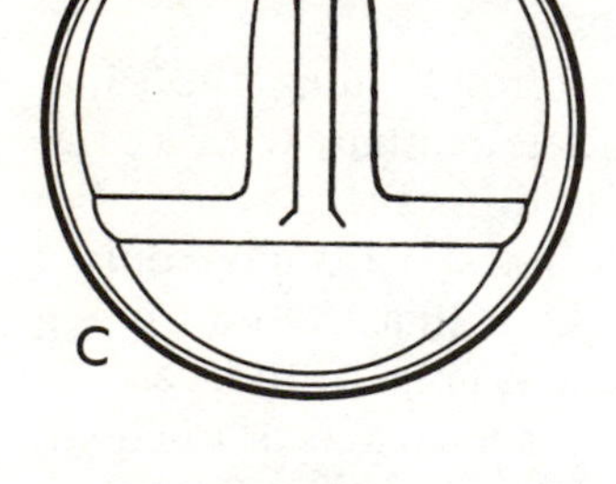

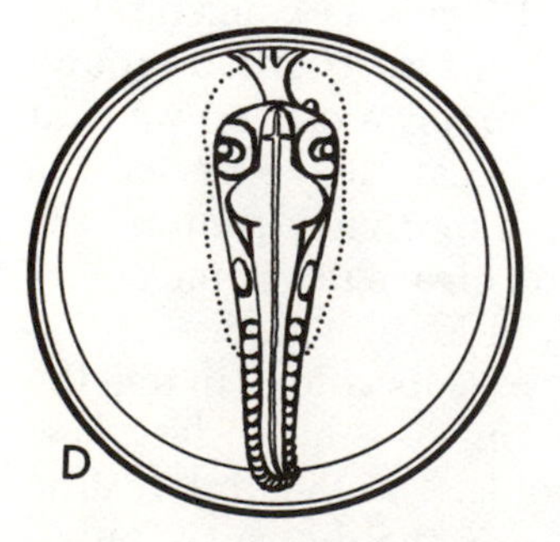

Aphyosemion are genera with annual species in which eggs might not hatch for several months.

Usually the length of the incubation period is governed by temperature. Within the optimum range for normal development, the period shortens as the temperature increases. For instance, trout and salmon eggs will hatch in about 50 days at 10° C but, at 2° C, incubation requires about 6 months. The eggs of the common carp incubate normally at temperatures of from 15° to 30° C, with hatching occurring in about a week at the lower temperature and 2 days or a few hours less at the higher temperature. Within the genetic capability of the species incubation temperature influences the meristic features of the individual. These are the features primitively tied to segmentation of the body, especially vertebrae, scales, and fin rays. As a general rule, those individuals of a selected batch of eggs incubated at low temperatures will have more vertebrae, scale rows, and fin rays than their siblings incubated at higher temperatures. There are a few instances in which the opposite has been shown to be true.

Hatching in many bony fishes is aided by secretions of special glands on the head or inside the mouth. The secretions are generally enzymatic in nature and weaken or even liquefy the chorion.

Newly hatched fishes of oviparous species may be tiny, unformed creatures destined to undergo considerably more development, as in the lampreys, or may be essentially small replicas of the adults, as in sharks, rays, hagfishes, and some bony fishes. Discussions of early life history stages of fishes have suffered somewhat because of lack of standard terminology, and the diverse nature of very young fishes seems to prevent the ready acceptance of any set of terms. Generally, early development is divided among embryonic, larval, and juvenile periods, but some workers consider the embryonic period to last until hatching or parturition, and others believe it to last until the young begin to feed for themselves. In one of the least complicated sets of terms, proposed by Dr. C. L. Hubbs in 1944, larval stages are those that are beyond the embryo but well differentiated from the juvenile, which is essentially like the adult. Two larval stages are recognized: the prolarva, which retains a yolk sac, and the postlarva, which has absorbed the yolk sac but is still unlike the juvenile stage. The upright-swimming larvae of flatfishes or the leptocephali of eels are examples of postlarva. If yolk-bearing larvae transform directly into a juvenile, as is the case in many salmonids and certain sculpins, these larvae are called alevins. They are the "sac-fry" of salmonid culturists.

A terminology suggested by Dr. E. K. Balon in 1975 divides the embryonic period into three phases — cleavage egg, embryo, and eleutheroembryo (free embryo), which is free from the egg and is equivalent to prolarva or alevin. The early larval phase, with undifferentiated fin folds, is called protopterygiolarva. The later larval phase, with fin rays forming, is called pterygiolarva.

A marvelous array of adaptations to the environment is seen among fish larvae. Some of these involve structures and shapes entirely unlike those in the juvenile stage and require an extensive metamorphosis. Lampreys, for example, undergo great internal changes as well as some obvious external modifications. They lose the functional gall bladder and bile ducts, grow a new esophagus as the respiratory tube disconnects from the alimentary canal, gain functional eyes, lose the oral hood and filtering sieve, replacing them with an oral disc set with horny teeth, and acquire larger fins. Eels change from the toothy, leaf-like, transparent leptocephalus to transparent "glass eels" to the juvenile or elver. Both lampreys and eels shrink considerably from larva to juvenile.

Examples are endless, including: the flatfishes, which change the entire architecture of the head to get both eyes on the same side; the headfishes, which lose a fine set of spines and most of the caudal region of the body during metamorphosis; and the swordfishes, which as larvae have prolonged, toothed jaws and extremely long opercular spines.

Many larvae show special adaptations for respiration, ranging from highly vascular fin folds, pectoral fins, or yolk sacs, to the feather-like external gills of *Polypterus, Protopterus,* and *Lepidosiren.* Gill filaments project from the gill openings of larval loaches. Attachment or adhesive organs are present on larvae of *Protopterus, Lepidosiren, Amia,* and *Lepisosteus.*

Many pelagic larvae are specially modified for flotation so that they maintain a specific depth or range of depth. Oil globules are effective in conferring hydrostatic balance and are common in drifting larvae. Inclusion of a large proportion of water in the flesh is another common flotation device, found among leptocephali. In the larvae of many fishes sinking is retarded by a high ratio of surface to volume. In these, the fins may be of exceptionally large size, the fin rays may extend into long trailing filaments, or the body may be covered with spines.

In exceptional instances, fishes with larvoid characteristics mature sexually and reproduce. This phenomenon, known as neoteny or paedogenesis, is common in the icefishes (Salangidae) of the coasts of China and Korea. Certain sauries and needlefishes are neotenic, and a tendency toward this condition is seen in some brook lampreys in which the ammocoetes go a considerable way toward developing gonads prior to transformation. Perhaps the best example of a neotenic fish is *Schindleria praematurus* of the Hawaiian Islands. These tiny transparent fishes reach about 20 mm in length and retain several larvoid characters, including a functional pronephros, opercular gills, and a nonfolded heart. The atrium is behind the ventricle instead of being folded over it.

References

Ahlstrom, E. H. 1968. Review of: development of fishes of the Chesapeake Bay region, an atlas of egg, larval, and juvenile stages. Part 1. Copeia, 1968 (3):648–651.

Amoroso, E. C. 1960. Viviparity in fishes. Symp. Zool. Soc. Lond., 1:153–181.

Atz, J. W. 1964. Intersexuality in fishes. *In*: Armstrong, C. N., and Marshall, A. J. (eds.), Intersexuality in Vertebrates Including Man. New York, Academic Press, pp. 145–232.

Balon, E. K. 1971. The intervals of early fish development and their terminology. Lab. Fish. Res. Slovak Agr. Acad., 35(1):1–8.

———. 1975. Reproductive guilds of fishes: A proposal and definition. J. Fish. Res. Bd. Can., 32:821–864.

Blaxter, J. H. S. 1969. Development: eggs and larvae. *In*: Hoar, W. S., and Randall, D. J. (eds.), Fish Physiology, Vol. 3. New York, Academic Press, pp. 177–252.

Breder, C. M., and Rosen, D. E. 1966. Modes of Reproduction in Fishes. Garden City, N.Y., Natural History Press.

Burns, J. R. 1976. The reproductive cycle and its environmental control in the pumpkinseed, *Lepomis gibbosus* (Pisces: Centrarchidae). Copeia, 1976(3):449–455.

Cherfas, B. I. (ed.) 1969. Genetics, Selection, and Hybridization of Fish. Academy of Sciences of the U.S.S.R., Ministry of Fisheries of the U.S.S.R., Ichthyological Commission. Jerusalem, Israel Program for Scientific Translations.

Collette, B. B. 1966. *Belonion,* a New Genus of Freshwater Needlefishes from South America. New York, American Museum Novitates No. 2274.

Donaldson, E. M. 1973. Reproductive endocrinology of fishes. Am. Zool., 13:909–927.

Gandolfi, G. 1969. A chemical sex attractant in the guppy *Poecilia reticulata* Peters (Pisces, Poeciliidae). Monitore Zool. Ital., 3:89–98.

Gosline, W. A. 1959. Four new species, a new genus, and a new suborder of Hawaiian fishes. Pacific Sci., 13:67–77.

Gupta, S. 1974. Observations on the reproductive biology of *Mastacembelus armatus* (Lacèpede). J. Fish Biol., 6:13–21.

Harrington, R. W. Jr. 1971. How ecological and genetic factors interact to determine when self-fertilizing hermaphrodites of *Rivulus marmoratus* change into functional secondary males with a reappraisal of the modes of intersexuality among fishes. Copeia, 1971(3):389–432.

Hoar, W. S. 1969. Reproduction. *In*: Hoar, W. S., and Randall, D. J. (eds.), Fish Physiology, Vol. 3. New York, Academic Press, pp. 1–72.

Hubbs, C. L. 1943. Terminology of early stages of fishes. Copeia, 1943(4):260.

Kille, A. 1960. Fertilization of the lamprey egg. Exp. Cell. Res., 20:12–27.

Koenig, C. C., and Livingston, R. J. 1976. The embryological development of the diamond killifish. Copeia, 1976(3):435–449.

Lauman, J., Pern, U., and Blum, V. 1974. Investigations on the function and hormonal regulation in the anal appendages in *Blennius pavo* (Risso). J. Exp. Zool., 190(1):47–56.

Liley, N. R. 1969. Hormones and reproductive behavior in fishes. *In*: Hoar, W. S., and Randall, D. J. (eds.), Fish Physiology, Vol. 3. New York, Academic Press, pp. 73–116.

Long, W. L., and Ballard, W. W. 1976. Normal embryonic stages of the white sucker, *Catostomus commersoni.* Copeia, 1976(2):342–351.

Macey, M. J., Pickford, G. E., and Peter, R. E. 1974. Forebrain localization of the spawning reflex response to exogenous neurohypophyseal hormones in the killifish, *Fundulus heteroclitus.* J. Exp. Zool., 190(3):269–280.

Manner, H. W. 1975. Vertebrate Development. Dubuque, Kendall Hunt.

Matthews, L. H. 1955. The evolution of viviparity in vertebrates. Mem. Soc. Endocrinol., 4:129–148.

Mead, G. W., Bertelsen, E., and Cohen, D. H. 1964. Reproduction among deep sea fishes. Deep-Sea Res., 11:569–596.

Moser, H. G. 1967. Reproduction and development of *Sebastodes paucispinis* and comparison with other rockfishes off Southern California. Copeia, 1967 (4):773–797.

Myers, G. S. 1931. The primary groups of oviparous cyprinodont fishes. Stanford Univ. Publ. Biol. Ser., 6(3):243–254.

Neilsen, J. G., Jeppersen, A., and Munk, O. 1968. Spermatophores in Ophidioidea (Pisces, Percomorphi). Galathea Rep., 9:239–254.

Ott, A. G., and Horton, H. F. 1971. Fertilization of chinook and coho salmon eggs with cryo-preserved sperm. J. Fish. Res. Bd. Can., 28:745–748.

Richards, W. J. 1976. Some comments on Balon's terminology of fish development intervals. J. Fish. Res. Bd. Can., 33:1253–1254.

Rosen, D. E. 1962. Egg retention: pattern in evolution. Nat. Hist., 71(10):46–53.

———, and Bailey, R. M. 1963. The poeciliid fishes (Cyprinodontiformes), their structure, zoogeography, and systematics. Bull. Am. Mus. Nat. Hist. 126:1–176.

Schreck, C. B. 1974. Control of Sex in Fishes. Blacksburg, Virginia, Virginia Polytechnic Institute, No. SG–74–01.

———, and Scanlon, P. F. 1977. Endocrinology in fisheries and wildlife. Fisheries, 2:20–27.

Soin, S. G. 1968. Adaptational Features in Fish Ontogeny. Jerusalem, Israel program for Scientific Translations, translated from Russian 1971, pp. 1–72.

Sundararaj, B. I., and Vasali, S. 1976. Photoperiod and temperature control in the regulation of reproduction in the female catfish. J. Fish. Res. Bd. Can., 33:959–971.

Svardson, G. 1949. Natural selection and egg number in fish. Rep. Inst. Freshwater Res. (Drottningholm), 29:115–122.

Turner, C. L. 1936. The absorptive processes in the embryos of *Parabrotula dentiens*, a viviparous, deep-sea brotulid fish. J. Morphol., 59:313–325.

———. 1946. Male secondary sexual characters of *Dinematichthys ilucoeteoides*. Copeia, 1946 (1):92–96.

———. 1947. Viviparity in teleost fishes. Sci. Monthly, 65:508–518.

Waal, B. C. W. van der. 1974. Observations on the breeding habits of *Clarias gariepinus* (Burchell). J. Fish Biol., 6:23–27.

Weibe, J. P. 1968. The reproductive cycle of the viviparous seaperch, *Cymatogaster aggregata* Gibbons. Can. J. Zool., 46:1221–1234.

Wiley, M. L., and Collette, B. B. 1970. Breeding tubercles and contact organs in fishes: their occurrence, structure, and significance. Bull. Am. Mus. Nat. Hist., 143:143–216.

Yamamoto, T. 1969. Sex differentiation. *In*: Hoar, W. S., and Randall, D. J. (eds.), Fish Physiology, Vol. 3. New York, Academic Press, pp. 117–175.

Yamazaki, F. 1976. Applications of hormones in fish culture. J. Fish. Res. Bd. Can., 33:948–958.

16 BEHAVIOR

No doubt man studied animal behavior in an informal and practical way long before he defined it, for a knowledge of the habits of the species hunted by early man and of those that hunted him was of daily importance. Mass migrations of birds, mammals, insects, and fishes are conspicuous examples of behavior that must have held the attention of the ancients in the same way that they command attention today—perhaps more so, for early man must have been more dependent on the mass appearances for food and other commodities that could be taken at those times.

Attaching a simplistic definition to a complex subject such as behavior may seem easy, but not when one begins to examine the ramifications of the definition and to recognize the tremendous body of knowledge necessary to understand or explain even small problems in the field. Fish behavior can be defined as all that a fish *does*, but obviously the actions of the fish are in consequence of everything the fish *is*. This, of course, depends on all the attributes of the particular species—attributes that have been formed in the crucible of evolution—plus any individual traits carried, and selection of traits on the level of the individual through time governs the evolutionary direction in which the species will be taken.

Perhaps a good definition of behavior is that it is the sum of all the motor responses of the organism to all the external and internal stimuli acting upon it. At once behavior appears as a dynamic phenomenon, but it includes the quiescence of the lurking stonefish and the aestivation of the lungfish, as well as other responses that require no conspicuous movements. Because of the immensity of the field of ethology, as the study of behavior is called, and the endeavors of those engaged in studying behavior of fishes, the coverage here will be limited to a general review of a few important aspects. Fishes are excellent subjects for ethological study, and articles on their actions and patterns of behavior can be found in a great variety of books and journals.

Paramount aspects of the behavior of fishes are feeding and reproductive behavior, one immediately involved with the existence of the individual, and the other involved with the existence of the species. The remainder of the behavior is mostly subservient to feeding and breeding, maintaining the individual in a suitable state between the times given over to these essential activities. There may be additional activity that does not fit into these three kinds of behavior, for such "play" activity as needlefishes performing backflips over flotsam would seem, without proper research, to have little to do with the serious business of perpetuating a species.

Because the ability of a fish to move is of primary importance to most aspects of life, locomotion has been studied from many standpoints, including behavioral responses to stimuli of many kinds. Locomotory behavior will be thus covered first, because of its general relationship to other behavior. Social aspects will then be covered in general, followed by sections on reproductive behavior, feeding behavior and, finally, a brief mention of behavior modification. Throughout this discussion, bear in mind the variety of fishes, and remember that only relatively few have been subjected to ethological study. Furthermore, behavior of fishes is species specific, even though closely related forms may have similar responses to a given stimulus. Generalizations are therefore tenuous if not hazardous.

Locomotory Responses to Stimuli

Fishes, as well as many other animals, have long been known to orient themselves, either in position or direction of movement, in response to a variety of stimuli. In most instances the responses are more easily studied in the young, but the adults of many fishes can be seen to react predictably when presented with a stimulus that requires orientation. The simplest type of orientation and movement in response is called a kinesis; this involves essentially random activity when stimulated. An example is the activity of lamprey ammocoetes under a bright light. Although blind, ammocoetes have light-sensitive cells in the skin of the tail. When these are exposed to light, the larvae respond with burrowing or swimming movements that will eventually remove them from the illumination. The larvae of some pelagic bony fishes exhibit photokineses but, overall, kineses are more typical of invertebrates than of vertebrates.

The type of directed movement seen in vertebrates is called *taxis* (plural *taxes*), which is a forced response to, or from, the source of a stimulus. Taxes were formerly known as tropisms, but this term is now used to describe the reactions of plants (and sometimes sessile animals). There are elaborate ways to classify taxes, but they will be considered here in regard to the specific stimulus. For instance, a movement in response to light may be called a phototaxis, whether it is toward (positive) or away (negative) from the light source. Orientation in response to current is called rheotaxis, that involving a chemical, chemotaxis, and a movement toward contact with some object is positive thygmotaxis. The examples given are all of importance in fishes.

Geotaxis. The posture of most fishes is governed by reaction to gravity (geotaxis) mediated through the membranous labyrinth, and reaction to light. Most swim with the dorsal surface up, but will alter the exact angle at which they hold the body if the light source is moved because of a response known as the dorsal light reaction (Fig. 16–1). As

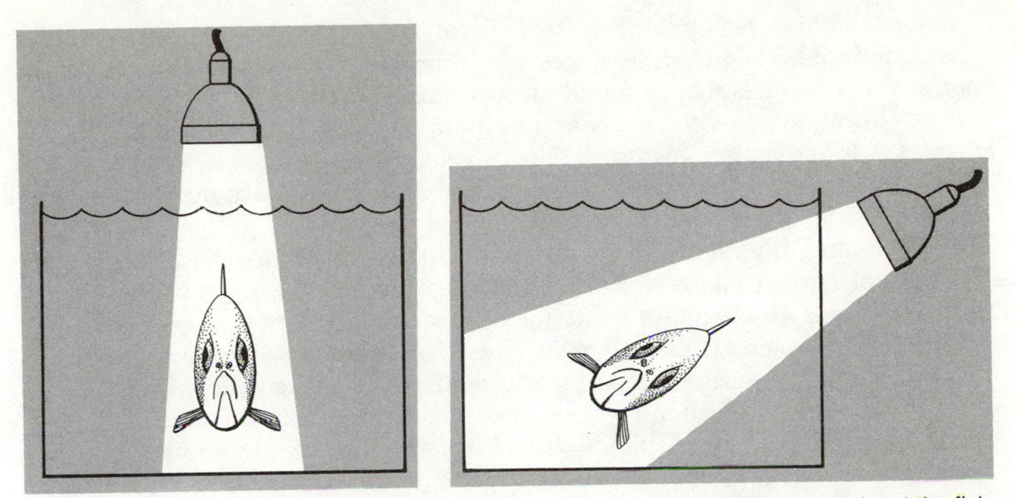

FIGURE 16–1. Representation of dorsal light reaction. When light is directly overhead the fish assumes a normal upright position, but as the light is moved to the side the fish maintains the same relative position in regard to the light by assuming an oblique posture.

long as the inner ear is intact there is a limit beyond which the fish will not incline but, if the labyrinth is removed, some species will present the dorsal surface toward a light at the bottom of the tank.

Phototaxis. Positive phototaxis is evident among many diurnal pelagic fishes in that they will swim toward lights at night. Some species will approach lights displayed under the water or above it but some, such as the Pacific saury, are more attracted to lights above the surface. Fishermen of many nationalities have taken advantage of the phototactic response to devise means of harvesting pelagic species. Methods have ranged from erecting a vertical net and displaying a torch to catch flyingfishes to elaborate banks of lights and special nets and pumps for the mass capture of sauries, mackerels, herring-like species, and others. Some species are more attracted to blue lights than to white, and experience has taught fishermen which light intensity to employ. A typical method is to attract schools of fish to one side of the vessel while a lift net is moved into place on the dark side, then the attractant lights are extinguished while lights over the net are lit. When the fish move into position over the net, it is lifted.

There are negatively phototactic species as well as those that are attracted to light. Sandrollers *(Percopsis)*, sculpins *(Cottus)*, and warmouth *(Lepomis gulosus)* are examples of nocturnal fishes that retreat from light and spend their days under cutbanks, rocks, vegetation, or other suitable cover. There have been a few attempts to guide fish away from hazards by the use of bright lights.

Electrotaxis. Although the study of this subject is in an early state of development, attention is being given to electrotaxis (galvanotaxis). The knowledge that most fishes will swim toward the positive electrode of a direct current field has been used in a variety of ways.

Some of the high seas fishing systems involving lights also employ a direct current system to draw fishes to the mouth of a pump or into a net. Fishery biologists have found electrofishing by means of portable direct current systems to be easier and more effective than netting in many kinds of population studies. The current is usually more effective if it is pulsed, and there are obvious differences in the responses of species to the frequencies of the pulses.

In natural situations, there are many instances of responses of fishes to electric or magnetic fields. Additional research will probably reveal reliance on magnetic fields for long-range navigation in many wide-ranging species, and will refine the knowledge of the electrical capabilities of species that have no recognized electroreceptors, but nonetheless seem to respond well to electrical fields. As in the case of light, reaction to electrical fields differs from one species to another, and within a species various strengths of stimuli will elicit varying responses. Fishes that can be made to swim toward an electrode of a pulsed direct current field can be repelled by alternating current or by different applications of direct current. There has been some success in guiding desirable fishes past traps set to capture migrating sea lampreys, which are not repelled by the electrical field used in the guidance system. The ability of sharks to seek out dipole fields, whether from animate or inanimate sources, has been mentioned earlier (Chap. 11). Avoidance of natural electrical fields has been demonstrated in some species; certain salmonids, for instance, have shown avoidance of the field set up around the head of the sea lamprey.

Thygmotaxis. Most fishes probably avoid contact with other individuals or with objects in their environments, except perhaps during the spawning season, but cryptic species may require such contact, being positively thygmotaxic. These species live in vegetation, in bottom debris, among stones, or in many other materials that can give them cover. Pipefishes and seahorses, morays, worm eels, flatfishes, and loaches are among many examples. The need for contact is highly developed in the freshwater sculpins (*Cottus*). When placed into a rectangular glass aquarium with no hiding place, they will seek out the corners where contact can be made with three planes at once. If the tank has a frame that darkens the corners, individuals will orient vertically in the corners, making contact mainly with the two sides, but taking advantage of the darkest areas as well. In a round glass container, contact can be made with the bottom and the side only and, without corners to congregate in, sculpins will usually pile up, making contact with each other. These aggregations can be composed of more than one species and can include those that are commonly preyed upon by other species represented in the pile. If a hiding place formed from a bent piece of sheet lead is provided in an otherwise bare tank, it becomes the site of a clump of individuals that may lift it off the bottom or perhaps even turn it over.

Rheotaxis. Reaction to a current of water is commonly observed among stream fishes and is also recognized in some still-water forms. Some species that will cruise as a school around a tank without a current, proceeding at seemingly random directions, will break up the school and orient into the current as individuals when water is made to flow through the tank. This positive rheotaxis is common in salmonids, and in anadromous species serves to keep the young in streams until they transform to smolts and begin a downstream movement. Even on the downstream migration some species may orient with the head upstream and drift backwards. This type of drift has been reported for species that move mainly at night and retain a position by a visual fix during the day. In the catadromous *Anguilla*, a positive rheotaxis is evident in the elver stage when the newly transformed individuals are entering fresh water and making their way to suitable living space. After the freshwater period of the life history is nearly over, and the eel transforms to the silver stage, a negative rheotaxis causes a downstream movement to the sea. In the ocean, there are many species of fish that appear to move along currents at certain stages of life history. Various herrings are examples of fishes that undertake contranatant migrations to reach spawning grounds from which the eggs and larvae can drift with the current back into the rearing areas.

Practical use of rheotaxis in carp is seen in some European fish farms where, by lowering the water level in a rearing pond and letting water flow in from another compartment, the carp can be induced to swim against the current to the new compartment. At some installations in Poland carp are harvested by inducing them to swim through a long canal into a trap. The length of the canal, through which river water is directed, ensures that the carp are "freshened" by the time they reach the trap. Many salmonid hatcheries make use of the rheotactic response to guide and sort fish. An example is the collection of breeders, in which the positive response is heightened. These will swim through grates or over barriers that fish in a less advanced stage of gonad development will avoid because of the swiftness of the current.

Reaction to water current is employed in guiding fish at dams and other installations where water is diverted from streams. Reluctance of young salmonids to make a sharp turn and enter accelerating downstream current at the same time has led to the construction of giant louver-like screens that are placed diagonally in front of water intakes and deflect downstream migrants past the danger.

Optomotor Response. Related to rheotaxis is the phenomenon called "optomotor response." This involves the visual fix mentioned previously. Fish of many species appear to maintain position in a current by lining up with visible objects on the stream bottom or side. In some fishes, contact is an additional factor, and position is sometimes held by keeping fins or a part of the body lightly against the bottom. Optomotor response is best seen in experimental situations, where it is often used to induce fishes to swim at a given rate of speed. A typical test device consists of a circular tank with a transparent outside wall and a circular partition inside to form a narrow swimway in which the fishes can move adjacent to the side. If a circular curtain marked with vertical light and dark stripes is rotated around the tank, test fish of many species will predictably move with a given stripe or will move at a slightly different speed from the speed of the stripes, but will move in the direction in which the stripes move. There are many variations in apparatus to fit particular research requirements; some have stripes moving on the bottom and next to the inside wall as well as on the outside, and some consist of elongate rectangular tanks with stripes moving on an endless screen along one wall.

The optomotor response is especially well developed in stream fishes, but is evident also in pike, perch, and others that are quite at home in still water. Marine fishes, including certain herrings, cods, and smelts, show the response also. Species differ in the speeds at which they will follow the moving background; pike, for instance, will move at slow speeds only, whereas herring tend to gain on the stripes. Bottom-living fishes do not usually respond by swimming with the stripes in experimental situations. Young fish show a stronger response than adults.

Researchers have made use of the optomotor response in several ways. One of the most common is the exercise of experimental fish in studies of active metabolism. Use of the optomotor response to guide fishes into traps or entangling nets has been suggested. Sequential switching of underwater lights on and off causes the response in some species and has been employed in attempts to guide fishes into bypasses at dams. Many commercial species are known to show the response or a similar following response. These will move along or above the sea bottom as a trawl is drawn along, adjusting their speed to the speed of the trawl and avoiding it. Countermeasures suggested by fishing gear specialists include using trawls that do not contrast with the background, and using devices ahead of the trawl to muddy the water and reduce visibility.

Reactions to Chemicals. Responses to chemical substances in the water have great importance in the lives of fishes. They are important in the location of food, in breeding and parental care, in avoidance of danger, and in orientation to home ranges and migration routes. Various species of minnows and poeciliids have been shown to react with characteristic fright activity, including violent swimming, darting to the surface, "freezing," etc., when presented with the odor of pike. The reaction appears to be innate, and can be elicited in specimens with no previous experience with predatory fishes. At least in certain parts of their ranges, Pacific salmons show a strong escape response in the presence of the odor of bears, seals, and man. When extracts of mammalian skin or solutions of an amine that appears to be the active component are placed in fishways or natural streams containing migrating adult salmon, many of the fish will abandon their positions and retreat downstream, returning later when the odor has been diluted.

Perhaps the most remarkable fright reaction is that shown by certain cypriniform species at the release of a "fright substance," or *Schreckstoff*, by injury to their own or closely related species. This material is produced in special cells in the skin and is released when the skin is torn. It can be sensed in very small quantities and usually causes immediate flight from the area by the individuals attuned to the odor. The phenomenon is known in one form or another in minnows, loaches, catfishes, and characins. It is not universal in its development or effect. The material produced by some is species-specific, but others produce a substance that is widely recognized among cypriniforms.

Fish are known to avoid some chemicals that appear in water as pollutants. In some instances this may be due to the irritating effect of the foreign material, but in others may be due to a fortuitous built-in reaction. Of course, there are harmful materials that are not avoided.

Homing Response. Many studies have shown that fishes of widely different habitats, systematic placement, and modes of life have home ranges in which they spend most of their existence. In some the home range may be small, consisting of a single pool in a creek, or it may encompass a given section of coastline or lake shore. That fish of certain species return to the home range after being displaced has been demonstrated by experiments in which individuals have been carried by boat, by towing in a cage, or otherwise transported to other places in the same body of water. After the marked fish are liberated they are able to orient to the direction of their home range properly and many complete the journey home. The cues by which they orient and navigate must include a combination of physical and chemical entities, and probably include some random search. Searching along the shore of a small lake would eventually take a fish home, but this does not account for movement toward the home range across a large lake or from one rock reef across deeper ocean water to the home reef. Homing behavior has been shown to involve sun-compass orientation, a response to the learned position of the sun in relation to the familiar area. There are hints that homing fishes can orient to the earth's magnetic field.

The homing response has been considered to be the return to a place that an individual formerly occupied instead of to other similar places. It is demonstrated by species that normally show a great restriction of movement, as well as by salmon that swim thousands of miles away from a locality they occupied for only a short time as young. Displacement experiments with marine shore fishes have usually involved short distances, but the cunner (*Tautogolabrus adspersus*) will return to a home site from as far as 4 km, and the yellowtail rockfish (*Sebastes flavidus*) will find the previously occupied site from a distance of 22.5 km, some successfully crossing deep water. Perhaps a more remarkable aspect than homing from a distance is the ability of some species to remember the home site and return to it after being held in artificial surroundings for a matter of months. This has been demonstrated for the cunner and the yellowtail rockfish, and has been suggested for the tidepool sculpin (*Oligocottus maculosus*).

Aggregations, Schools, and Loners

The social aspects of fish behavior have attracted a considerable amount of attention from both scientists and laymen. Interactions of fish, easily observed in the natural habitat or in the aquarium, hold our interest for a variety of reasons. Knowledge of the schooling habits can aid in increasing the catch; fishery biologists base management and conservation plans on the social behavior of some species; many aquarists select species for display because of social interactions, and have learned to breed some species only after gaining an understanding of certain kinds of intraspecific social behavior. Ichthyologists have many direct and peripheral interests in social activities of fishes, but among the most important is the relationship of social behavior to speciation.

Gatherings of fishes are usually referred to as schools or shoals, but behaviorists recognize a certain kind of association as a true school, and provide different terms for other groupings. Acceptance of a limited definition of schooling aids in its study by focusing attention upon essential aspects so, although there may be no harm in calling any group of fish a school for general purposes, we will consider that a true school meets the criterion that the fish in a school are together because of a social attraction for each other. In a polarized school fish share the same general orientation and maintain, within limits, a uniform spacing and speed. Fish in a school can be nonpolarized.

The use of the term aggregation has been variable. It has been used in an inclusive sense that encompasses schools as well as other groupings, and on the other hand for groups excluding schools. The term will be used here in the latter sense for the sake of convenience. Aggregations can form because of an attraction of a particular environment for members of a species.

Some fishes school throughout their lives and are not normally found in a solitary condition. Many of the herrings and herring-like fishes (Clupeidae, Engraulidae) are such obligate schoolers, as are certain mackerels (Scombridae), silversides (Atherinidae), mullets (Mugilidae), and others. Other fishes gather in schools only at given times in their lives, as in migrations, during spawning season, or as part of their early life, when they move in schools as juveniles. Actually, most fish species school as juveniles but, as adults, only about 20% of the species school. Schooling species include fresh- and saltwater fishes both large and small, from tiny minnows and characins to giant tunas. Numbers range from a very few individuals to the millions observed in herring or menhaden schools. Even in schools that can stretch for miles the members of a given group will be of the same approximate size. The sorting by size is apparently related to the swimming speed of the different-sized fish; large fish are able to sustain a faster cruising speed than small individuals of the same species. In general, schools are made up of members of a single species, but there are special instances that bring more than one species into a school.

The tendency to approach and orient to other members of the species appears early in many fishes. This behavior (sometimes referred to as biotaxis) is dependent upon visual stimuli, according to the findings of many investigators. Size, shape, color, and patterning of conspecifics attract the very young of schooling species to each other. Although fishes are well equipped with physical senses other than vision, and have keen chemical senses, vision appears to be vital to the formation and integrity of schools. Cues to speed and directional changes come especially from the lateral visual fields. Several species have been shown to cease schooling at certain light intensities. Blinded individuals of most species tested school poorly if at all, but blinded saithe (*Pollachius virens*) have schooled in laboratory experiments. Physical senses other than sight can aid in maintaining distance between individuals, and may be active in keeping dispersed members of a school within the same vicinity in the dark. Some species are known to produce sounds while dispersed but not while schooling, and electric fishes send and receive information in dark or turbid waters.

Schooling behavior would surely seem to bring some kind of advantage to the fishes that display it, but what this advantage is and how it is conferred has been debated. Probably adaptive advantage accrued to the individuals seeking company of others in a school, and advantage to the species has evolved concomitantly. For the most part, observation, field and laboratory experiments, and mathematical modeling have shown that prey species gain advantage from schooling, and that other aspects of life might also benefit. Some of the ideas of how schooling helps protect potential prey have been tested to some extent and others have not, and counterarguments to the following will be easy to devise. Obviously, a closely packed school could give the impression of a large organism and this could intimidate or discourage casual predators. Some predatory fishes have been noted to consume more prey fish when these are presented singly or in small numbers than if made available as a group. Even when a few prey are introduced at one time predators delay in selecting and striking, whereas a single fish will be victimized immediately. Dense packs of prey species not only might confuse predators but might also make predation physically difficult, as could be deduced from observation of the herring "balls" that occasionally form when herring are under attack by predators such as Pacific salmon. The herring crowd into a writhing mass so dense that a portion of it can be above the surface, while the predators appear to strike only at individuals that detach from the mass.

Not only the prey species school. There are many examples of schooling species that feed on other fishes, so protection must not be the only benefit obtained from schooling. Location of food sources can be facilitated by fishes moving as a group, for when food is discovered on one side of a school, individuals on the other side are soon aware of it. In some species, color pattern changes give a signal that food is in view. In laboratory experiments, young fish allowed to feed in a social group tended to grow faster than those eating by themselves. In learning experiments, fish are more rapidly conditioned in groups than as individuals; they not only learn faster but remember longer when taught in groups.

Reproduction can be facilitated by schooling. Many fishes are communal spawners, and both sexes are represented in a single school, so at the proper season and place they have only to release gametes en masse in order to achieve a high rate of fertilization. If a species schools by sex between spawning seasons, the situation is complicated by the necessity of intermingling on the spawning area, but seeking an individual mate and pairing off is not necessary.

Schooling appears to confer a hydrodynamic advantage. Oriented schools move through the water more efficiently than individuals, in that vortices set up by lead fish allow for easier progress by those behind. Because the schools have no permanent or habitual leaders, and the lead fish are constantly changing with every change of direction as well as by continual shifting of position within the school, the effort of trailbreaking is shared.

Change of direction by a school seems to be a concerted, instantaneous action on the part of all members, and there was early speculation on how thousands of individuals could be of the same mind at one time. By means of cinematography and other techniques, researchers have found that direction change can be initiated by individuals well within the mass of fish. The entire school surrounding the initiator responds within a fraction of a second. Experimentation can provide stimuli that bring about changes in direction of schools but, in nature, the cues that cause schools to change shape and direction can be so subtle that some of them cannot yet be recognized.

Mention was made earlier of schools containing more than one species. An example is in rivers of the Columbia drainage, where schools of small fish about 2 cm long, seen streaming through gaps between stones in very shallow water, feeding along the bottom, prove to be made up of speckled dace *(Rhinichthys osculus)*, redside shiners *(Richardsonius balteatus)*, and suckers *(Catostomus macrocheilus)*. Northern squawfish *(Ptychocheilus oregonensis)* and chiselmouths *(Acrocheilus alutaceus)* can be represented also. These species do not associate as adults. Certain reef fishes, including snappers (Lutjanidae), goatfishes (Mullidae), and grunts (Pomadasyidae), have been reported to form multispecies schools when they are not actively feeding. The species observed in the schools are nocturnal feeders, thought to school together in appropriate places for mutual protection during the day.

The two examples given may not truly be schools in the restricted sense but spacing, orientation, and similarity of size of individuals make the groups seem more like schools than aggregations that form because of the attraction of the immediate environment. Nonschooling fishes often gather to take advantage of resources that might be available in some part of the habitat. Food, spawning sites, and cover for resting or for wintering may all bring typically solitary fish together for periods of time but, although they may be in close proximity, each fish (or spawning pair) is making its own use of the resource without being socially attracted to the other individuals.

Nonschooling fishes by no means ignore each other. In fact, some of the most interesting social behavior known in fishes results from interplay between or among individuals in modes of contact other than schools. The following will introduce some of these relationships.

Social hierarchies have been observed and studied in a large number of fishes. In these arrangements an order is established whereby certain individuals, usually males, become dominant over the remainder of that species in a given section of habitat. The order of dominance is often related to size, so that the largest fish can chase the rest, and the next largest can chase all but the dominant (or alpha) fish and so on down the scale of size. In large groups the direct size-related dominance must break down somewhere along the line, so that there may be a number of individuals of equal social rank and reduced aggressiveness. Even in small groups there may be codominant individuals. Social orders in which each fish knows its place have some adaptive importance in reducing excessive fighting and in stopping other unproductive encounters short of actual combat.

Territoriality

Hierarchies and the matter of dominance and submission among fishes are usually related to such important aspects of life history as feeding and breeding, and most can be related to the phenomenon of territoriality. Home range was mentioned earlier as the area through which an individual habitually moved in its search for living requirements. A special part of the home range is the territory, a space usually concerned with nesting, feeding, or resting, which is defended against members of the same species. Some species will defend territories against other species, especially those with similar requirements. Adaptive advantages of territoriality are probably numerous, considering the many species of diverse habits in which the behavior is developed. Regardless of the function that defense of a space serves in a given species, overall advantages concerned with population control, selection of the most vigorous as breeders, reduction of fighting, and other benefits, can accrue to the species.

A species that exemplifies the relation between population regulation and territoriality is the coho salmon (*Oncorhynchus kisutch*). Although the young of this salmon will form schools in hatchery ponds and in other still water, the usual habitat is flowing streams, where drifting insects and other small organisms are available as a food supply. In a small spawning stream many thousands of alevins may survive incubation, hatching, yolk-sac absorption, and emergence from the gravel. For a short time the young parr may cruise in the shallow margins of the stream, and most may find adequate food. In some streams there may be a downstream relocation of numbers of the fish not long after emergence. These might find living space in larger stream areas or, as in the case of some short streams in Alaska, drift to the sea and disappear. For those that stay in the upstream areas the survival of the most aggressive members of the population is favored.

As the parr resort to deeper water and larger food items, certain spots in the stream become favored as feeding stations because of such factors as abundance of food, cover for protection from predators, and relative ease of maintaining position in the current. A good position, for instance, might be a declivity in a large rock not far from the water's surface. From the vantage point of the eddy in the hollow, a small fish can command a view of a relatively broad stretch of the swifter water flowing above, as well as the surface of the rock itself. If two small coho attempt to occupy the station at the same time, one will displace the other through aggressive displays or actual combat. The winner is often larger, perhaps a few days or weeks older, or more brightly colored, but sometimes the contest is won by a small fish with a superior show of aggressive action, especially if the smaller fish is the resident despot in a given area.

As in many other species, the territorial and hierarchical behavior of the juvenile coho includes a repertoire of standard acts and signals of pigmentary and morphological nature. The parr marks are of a distinctive size, shape, and spacing, and the dorsal and anal fins have elongate rays at the leading edges. These anterior edges are white, bordered posteriorly by a black stripe, and the ground color of the remainder of the fin is a shade of orange (Fig. 16–2). The length of the dorsal and anal "spurs" and the brightness of the fin color varies with locality, the fins being shorter and of a duller hue in parts of Alaska than on the coast of Oregon. When erected to the full height, the fins constitute a showy signal that accompanies threatening postures, either frontal or lateral, both of which may be offensive or defensive in nature, depending on circumstances and the vigor with which the display is executed. Submission is shown by lowering the fins and assuming a head-down posture. When territories and hierarchies are being established, threatening encounters may be followed by nipping of the loser by the dominant individual, or fighting may be resorted to in order to establish dominance. Vigorous chasing of a defeated or submissive fish sometimes occurs.

The overall effect is that a social order is established with strong and aggressive individuals commanding territories in favored parts of the stream, and less dominant fish holding such stations as they can defend. The territories of the latter may be small, and are likely to be in water shallower or swifter than optimum for normal feeding and growth. As the parr grow, and territories shift and enlarge, individuals at the lower end of the hierarchy are displaced, perhaps by inability to make a living in their station, or perhaps by more doiminant fish enlarging territories. If the displaced fish cannot locate and hold a territory downstream, they cannot feed and develop normally. Coho

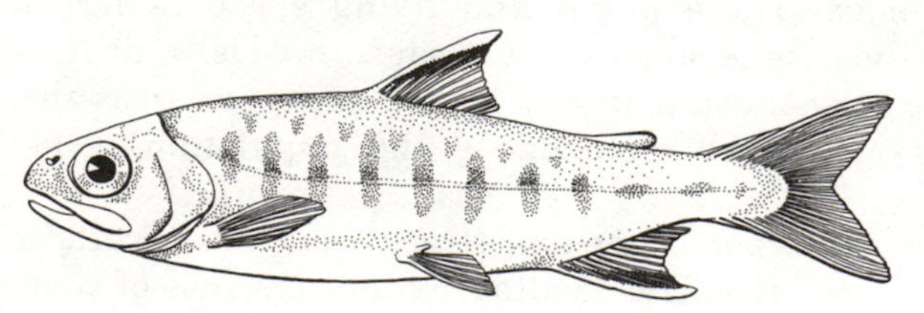

FIGURE 16–2. Juvenile coho salmon *(Oncorhynchus kisutch)*, showing specially shaped and colored dorsal and anal fins.

remain in fresh water for about a year after hatching in the southern part of their range, and for as much as two to three years in Alaska. The effective downstream migrants are those that have grown to the proper size for transformation to smolts prepared for entry into salt water. Those that leave early, prior to reaching the right physiological state, have little chance of survival. Counts made of smolts migrating from small streams over a number of years have shown that even though egg deposition and fry emergence may fluctuate widely, numbers of smolts are remarkably constant from year to year. Much of this constancy in smolt populations is due to limited resources of the stream partitioned through the medium of the behavior of the fish.

The young of other species of salmon and trout show similar agonistic behavior, territoriality, and hierarchical social arrangements, but activities, functions, and adaptive advantages do not correspond exactly with those of the coho. Some of the other salmonid species coexist in streams with coho and interact with that species. Usually, differences in habitat selection cause a separation of sympatric species so that most of the agonistic behavior is intraspecific, but sometimes the coho select territories at the expense of other species. Coho and steelhead (*Salmo gairdneri*) that occupy slightly different parts of the stream during the summer may live in the same pools in the winter, but individual steelhead usually seek hiding places under stones and other cover, while the coho swim in groups above the bottom. Young coho are known to direct interspecific aggression at juvenile chinook (*Oncorhynchus tshawytscha*) and set up territories at the heads of feeding areas, so that coho get first chance at the food and grow faster.

Species of several other families display territoriality with or without the formation of hierarchies. Studies have been made of these aspects of behavior in swordtails (Poeciliidae), sunfishes (Centrarchidae), wrasses (Labridae), parrotfishes (Scaridae), and many others. In some species aggression is displayed in competition for food, not for space, as in the medaka (*Oryzias latipes*). This species shows the greatest degree of aggressive behavior at intermediate population densities and develops hierarchies if food is limited. Localized limited food is defended as a territory. In experiments, socially dominant medaka grew faster than subordinates. Food-related territoriality has been observed in the field in the threespot damselfish (*Pomacentrus planifrons*), which defends territory against many of the other reef fishes, but protects larger areas if the intruder is a potential competitor for food. The largest areas are defended against members of the same species.

Signals and Social Behavior

Most of those fishes mentioned, and undoubtedly most fishes, depend mainly upon the sense of vision for the cues, signs, and signals that trigger and maintain social behavior, but other senses can be significant or even more important than sight. Optical signals in fishes can range from those as passive as the usual, constant outline or ground color of a given species to those of a very active nature, such as the flashing of a luminescent structure. Signals involving color can be accompanied by meaningful movements — dances, approaches, flights, or postures. Some of these movements show off specially colored parts, such as brightly colored or significantly marked fins, bellies, opercular and branchiostegal membranes, and mouth linings. In some species these colors or patterns are present only at breeding season or at other significant times of the year. In others the color may be present but may not be displayed until required by breeding or other behavior.

Although significant colors and patterns are a seasonal phenomenon in some species, other species change signals to suit the activity or "mood" of the moment, paling in fright, darkening in anger or sexual motivation, or changing from stripes to spots according to the stage of feeding activity. Some cichlids can display a fright pattern within a second of receiving the stimulus. Certain species may have several patterns involving stripes, ocelli, bars, and such. The oriental fish *Badis badis* has about a dozen patterns that are used to signal behavior, but sometimes activity changes faster than the patterns, so there might be a lag in the signals.

Various reef fishes, wrasses (Labridae), and damselfishes (Pomacentridae), for example, show strikingly different colors and patterns in given stages of their life histories.

Mention has been made (Chap. 11) of the significance of sounds in the life of fishes. Threats, warnings, and signals to prospective mates can be produced by vibrations of the gas bladder, movement of one hard part against another, or by other mechanisms. Electrical signals during certain behavior are known in fishes with active electrical systems; gymnotoids include bursts of electricity in their repertoires of aggressive activity, along with head butts, serpentining, bites, etc.

Chemical signals may be part of most behavioral systems, but have not been studied to the extent that has been accorded other means of communication. The work of Dr. John H. Todd and his associates has revealed the great importance of chemical communication in the bullheads *(Ictalurus)* and a few others, and has paved the way for further advances in this line of research. Significance of olfaction in the social behavior of bullheads is truly amazing, although visual signals such as the open mouth threat are part of their complex behavioral repertoire. Social status of individuals can be recognized by olfaction alone, but not just by individual odor. Pheromones or other substances advertising the status of the individual appear to be released and recognized by others in the population. In laboratory experiments, Dr. Todd found that subordinate individuals would rush to the territory of the dominant fish and be tolerated there when a strange fish or its odor entered a tank in which a stable order had been established. After the stranger had been driven off or the odor had dissipated, the dominant animal would force the subordinates out of his living area. If a dominant suffered a defeat, even outside the presence of the others in his tank, he would no longer be recognized as fit to hold the top position in the order. Appraisal of his status was by smell, and could be sensed immediately upon his return after being removed and trounced by a star gladiator. Incidentally, a defeated dominant usually failed to grow normally.

In natural situations bullheads live together in large numbers quite peacefully. Chemical signals advertising amicable relationships apparently are released by such groups, for Todd was able to use water from a peaceful tank to break down agonistic behavior in tanks to which the odor-carrying water was introduced.

Olfaction is important in recognition of young in cichlids. The importance of chemical cues in reproductive behavior, such as initiating nest building in several species, will be discussed later in this chapter.

Interspecies Relationships

There are numerous examples of social relationships involving two or more species of fish. Pilotfish *(Naucrates ductor)* are often seen in the company of sharks, swimming ahead of, or along with, the larger animals. Pilotfish will accompany boats and turtles as well, but in these instances might not receive the benefits of food and protection that have been suggested by some authors as deriving from the association with the sharks. Remoras and suckerfishes (Echeneidae), with a powerful adhesive device on the top of the head, are usually found attached to large fishes, including sharks, or large turtles. The remoras get a free ride from the larger fishes, and no doubt have an opportunity to eat food particles that escape the jaws of the predators that they often accompany.

Some fishes have established associations with other species of different feeding habits and take advantage of food that becomes easily available because of the activity of the other species. Certain sea basses (Serranidae) follow herbivorous fishes as they seek algae among the rocks, and feed on suitable small fish and crabs that the herbivores dislodge or frighten. Other sea basses follow octopi or morays, and catch small fish and other animals that avoid those fearsome predators.

Some of the best known and most interesting relationships among fishes involve cleaning symbiosis, an association bringing mutual benefit to the individuals involved. Cleaner-fishes are allowed by larger fishes to explore the surface of the body and even the mouth and gill cavity in search of parasites and loose or injured skin, which the cleaner devours. Although cleaning has been studied mainly in tropical reef areas, where it apparently reaches its best development, there are cleaning relationships in fresh water, some involving species that are usually predator and prey. The bluegill *(Lepomis macrochirus)* acts as a cleaner; the juveniles clean larger bluegills and the adults clean the largemouth bass. This is truly the lion lying down with the lamb, for the success of typical farm fish pond management depends upon the use of the bluegill as food for bass. In the cleaning relationship the bluegills establish cleaning stations where the largemouth bass can present themselves for service, taking a position with the head down. The bluegill respond to the invitation by approaching and cleaning the larger fish, and by displaying a dark coloration, which is thought by K. J. Sulak, who reported the behavior, to be a signal of submissiveness.

In marine waters, cleaner-fishes are known from a number of families, mostly tropical, although there are a few cleaners in temperate waters. Of course cleaning may be more prevalent in temperate and cold waters than is realized, but the relationships may not be as obvious as in the clear waters of the reefs with their multiplicity of species. A member of the temperate family Embiotocidae, the kelp perch *(Brachyistius frenatus)*, is a cleaner, but does not seem to enjoy the relative immunity from predation seen in most tropical cleaners. Most cleaners in warm seas are in the families of wrasses (Labridae), gobies (Gobiidae), and butterflyfishes (Chaetodontidae).

In a typical cleaning relationship the cleaner, which is usually distinctively colored, displays itself habitually at a rock, sponge, or other place where most cleaning activity is carried out. This is the cleaning station, although the individual may have a larger range over which it will serve its customers. One species has been described as having a "focal point" of cleaning activity within the cleaning station. Some cleaners have a territory within the home range; this is defended against all intruders. In some species the size of the cleaning ranges served is dependent upon the size of the individual and, if the ranges are contiguous, intraspecific aggression is prominent.

Usually, fish present themselves for cleaning singly, signaling the cleaner with a particular posture, and often erecting the fins and opening the mouth and opercula. In some instances entire schools will surround a cleaner. Some cleaners operate singly only but other species may operate in pairs or in small groups. There are recorded instances of two species cleaning the same host simultaneously. The quiescent attitude of the host is often an invitation not only to the cleaner, but to other small fishes that feed upon the skin or fin membranes of the larger species. Some remarkable instances of mimicry are known among the wrasses, in which skin biters mimic the cleaners of similar genera rather well. As the cleaners are not often preyed upon, some measure of protection can be gained by other small species that mimic their pattern, configuration, and activity. Usually cleaners are quite aggressive toward mimics and other species that enter the bounds of the cleaning station to take advantage of the hosts.

Not only fishes, but shrimp as well, are involved in cleaning, and fish approach the stations of the tiny crustaceans with a characteristic behavior. They move close to the station and allow the shrimp to climb aboard and feed, showing no inclination to feed upon them. As mentioned earlier (Chap. 8), there are many instances of fishes associating commensally with members of other phyla, having adapted their behavior to obtain shelter, protection, or some other benefit from the host.

Reproductive and Parental Behavior

The subject of reproduction is one of great complexity and embodies many aspects closely related to behavior. A general review of some of the requisites can help introduce a brief treatment of the relationships of reproduction and behavior. First of all, a given fish must come into breeding condition simultaneously with fish of the opposite sex, and at a time propitious to the survival of the eggs and young. These potential breeders must be in, or move to, a suitable site for mating, spawning, and rearing the young. The site can be used as it is found, or can be prepared by one or more of the breeders. If the preparation is elaborate, and a nest is constructed, guarding the nest will reserve it for the builder. When the fish are in the right place at the right time there must be recognition of the opposite sex of the same species, so that one (usually the male) can induce the other to participate in the proper acts of courtship, mating, and spawning. From spawning on, the behavior may consist simply of leaving the scene, expiring, or beginning a complex set of acts involved in ensuring the survival of the eggs and offspring.

Ethologists, ecologists, fishery biologists, and aquarists have learned much about reproduction in fishes, and much of the effort has been directed toward behavior. Numerous accounts are available in many books and journals, outlining examples of sexual and parental behavior among fishes. Referral to selected examples will be made in order to illustrate points in the discussion below.

Some treatment of spawning migrations has been given in Chapter 9. The gathering and movement of numbers of fish at a given time of the year are among the most conspicuous behavioral manifestations among those involved with the reproductive cycle, and are noted by the general public as well as by fishermen. In general, the reproductive migration is synchronous with maturation of the gonads, so that fish arriving at the spawning area are "ripe" and ready to spawn. The environmental stimuli that influence sexual readiness can be suspected to influence migratory behavior, either directly or indirectly through a series of endocrine pathways. Typically, fishes will migrate and gather on the spawning grounds and complete the reproductive sequence under a given set of stimuli. For instance, rainbow trout, over much of the range, will migrate and spawn in the spring, under influence of increasing photoperiod and rising water temperature. On the other hand, the related brown trout migrates and spawns in the fall under the influence of decreasing light and temperature. There are numerous examples available among the Salmonidae to show temporal separation between migration and spawning — situations that might call for a separation of the stimuli that trigger migration from those that bring on final maturation of the gonads and spawning. The example of the spring chinook salmon might be sufficient. In this race, migration from the ocean to the streams occurs in the spring months, at a time of increasing photoperiod and warming temperatures. Once in the streams the fish are attracted to deep pools where they spend a few to several months in a state of relative inactivity, although they often jump as if in play, and make circular, follow-the-leader sashays around the pools. Development of the gonads proceeds during these months, and spawning occurs in late August and September during a time of decreasing day length and cooling temperatures.

Whether reproductive migrations are short or long, they require marked changes in behavior. Fishes must turn from their usual pursuits, whether feeding or wintering, and take heed of cues that orient and move them in the direction of spawning grounds. The world of the fish is full of potential cues that can be sensed, and any or all may be of significance in orientation and migration. Water movement, especially current and the rise and fall of the tides, solutes in the water, temperature, light from the sun or other celestial source, the earth's magnetic field, sounds, and configuration of shore and bottom all may be involved in stimulating and guiding migrating fish. Behavioral reactions to these stimuli might, in some instances, seem rather simple, as in some species that move short distances against or with a current to arrive at their destination. In actuality most migrating species are involved in a complex series of reactions to environmental features and gradients. In homing species the bases for these reactions were laid at an earlier time with the imprinting of the young, sometimes only a few weeks old, with the characteristics of its habitat and migration route. Information on sights, smells, tastes, degrees of azimuth, relationships of the magnetic field, and probably other modalities are locked in the nervous system of the fish until recalled by signals from the endocrine system as the fish is guided back along its early migration path.

At the end of the migration or other movement to a reproductive area, fish select the actual site of the reproductive activity. According to the habits and life histories involved, the site may be simply a stretch of open water from which the eggs and larvae can drift in a general direction that will allow reasonable survival, or the site may be a complex situation encompassing current, cover, substrate, depth, intragravel flow of water, and so on. Availability of nest-building materials may be the most important requisite for some species. In the selection process, fishes test the environment in various ways, mouthing potential nest material, exploring crevices and burrows, settling on the bottom to feel it, or making preliminary digging or cleaning efforts. Sometimes a partially constructed or excavated nest will be abandoned in favor of a new one nearby for reasons that may have something to do with suitability.

Mate selection and some courtship activity may precede or accompany the preparation of the spawning site, or may take place after the site is prepared. In many species breeding is communal or polygamous, with no distinct pairing and nest preparation. In others males may wait upon the spawning site and attend en masse any female of the species approaching them. In this instance recognition of the species and sex is exercised with no particular process, depending mainly upon visual signals. Colors, dark and light patterns, configuration of body and fins, or placement of photophores might be involved as signals in various species. In those fishes in which pairing takes place, or in which the male entices and spawns or copulates with a few to several females in succession, stimuli that release the behavior of courtship and mating may be those visual ones mentioned, or those coupled with ritualized actions or postures, or may be due to auditory, olfactory, or possibly other modalities.

Chemical cues are important in triggering nest-building in certain members of the Belontiidae, a family containing many bubble nest-builders. Olfaction plays a part in the ritual of the courting stickleback, and apparently is involved in the reproduction of other fishes. Cods, croakers, toadfishes, and others employ sounds in their reproductive activity, and the midshipman adds the light of photophores to his nuptial noises. Posturing and dancing serve as sign stimuli in many pairing fishes, and have become so ritualized that they are recognized as important mechanisms that ensure reproductive isolation of closely related species.

Courting behavior usually begins with some kind of stylized approach, the male sidling toward the female or otherwise displaying his form, color, and posture. If the female is "impressed" and is ready to spawn or copulate she responds by approaching or allowing the male to approach. Nuzzling, butting, lateral contact, "dancing," or other requisite acts may occur before the mating is consummated. If a nest or other prepared site is involved, the male must cause the female to follow him to it. The famous zigzag dance of the male stickleback is one of the best examples. Many fishes pair off before the nesting site is prepared, and both sexes may work at building or cleaning a spawning site, as in the Cichlidae. In the Salmonidae, the females usually dig the redd while the males discourage intruders.

Territoriality is generally strongly manifested during the breeding season, especially in nest-building species. Territorial males are usually brightly colored and are quick to threaten conspecifics that come into the space containing the nest or females. Protection of the reproductive site may be as ritualized as the mating behavior, usually beginning with some kind of direct and threatening approach, and followed by threat displays of mouth lining, opercle-flaring, fin-spreading, sidling, and so on, depending upon the species involved (Fig. 16–3). Actual contact is usually avoided but, if an intruder persists, nips, butts, or biting may be employed. Locking of jaws with attendant pushing, tugging, or twisting is a common test of strength and purpose seen in male-to-male agonistic encounters over a great range of fish species. (Mouth-to-mouth contact is often employed in mate selection, and even jaw-locking is part of that process in cichlids.) Destructive fighting is usually avoided, but does occur in encounters between well matched males. Males of the genus *Betta*, or Siamese fighting fish, commonly fight to death. In some species of fighting fish, the male will attack and even kill the female after all the eggs have been placed in the bubble nest if she does not leave the scene. Pacific salmon engage in fights on the spawning grounds in defense of territories. A favorite ploy is to grasp the opponent's caudal peduncle

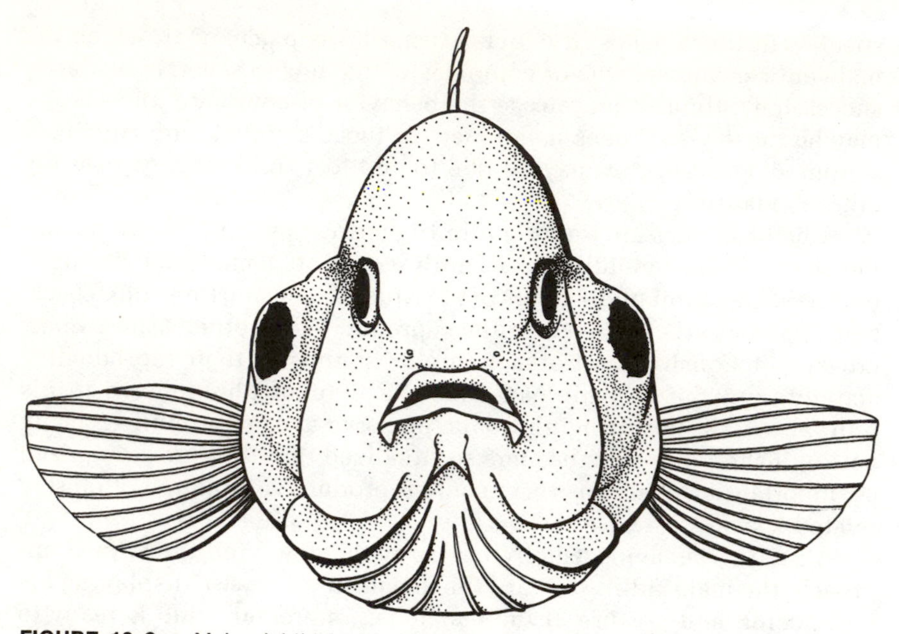

FIGURE 16–3. Male cichlid in frontal display — opercula flared and pectoral fins spread.

and twist violently. The kype of salmon with its arching curve and large recurved teeth can inflict much damage (Fig. 16–4). A part of agonistic behavior that should be mentioned is the so-called "displacement activity," in which a beaten or frustrated animal will carry out, at least partially, a behavioral act such as nest-building, unrelated to the fight, but which allows some kind of relief.

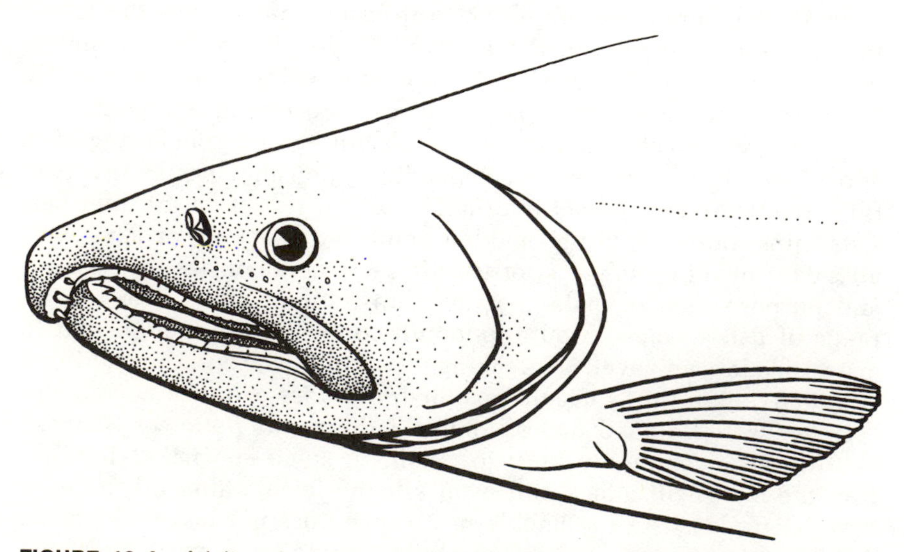

FIGURE 16–4. Adult male coho salmon (*Oncorhynchus kisutch*), showing curved jaws (kype).

The act of spawning can be triggered by a variety of releasing mechanisms acting alone or in concert. Tactile stimulation of the female by the attentions of the male is possibly the most common mechanism. Head butts, nips, lateral contact with quivering action, and passes underneath with the male's dorsal surface in contact with the vent region of the female have all been noted in one or more species. Visual stimuli are important to some fishes; the sight of the male, or of the particular place where the eggs are to be deposited, might be enough to induce the spawning sequence. Ethologists have implicated olfaction as a cue in the spawning of various fishes; secretions from either sex can have a stimulatory effect on others of the same or opposite sex.

Parental care of eggs and young is common among fishes and may be simple, as in the covering of a nest with gravel or other materials, or complex, as in the carrying or guarding of the eggs or even the schooling young. Numerous substrate spawners remain with the eggs, guard them, supply them with oxygenated water by fanning with the fins, and remove dead eggs or those infected with fungus. Nesting fish perform the same functions, but the bubble nesters, by the nature and location of the nest, have automatically provided a supply of oxygen. Pipefishes and seahorses carry eggs in brood pouches; cardinalfishes, some cichlids, sea catfishes, and others incubate the eggs in the buccal cavity; still others carry eggs on the body (Chap. 15). Care of the young, as well as care of the eggs, is known over much of the phylogenetic range of fishes. *Protopterus* guards the larvae as they are held in the nest by their adhesive organs; *Polypterus* swims with the young for a time; *Amia* males fiercely guard the young. Assiduous attention to the welfare of the young is seen in catfishes, sunfishes, cichlids, and many others. In some instances cichlids appear to recognize their own young by odor. Some species provide food in the form of mucal secretions.

Species derive obvious benefits from the specific behavior that accompanies their reproductive processes. Although much of the behavior is tied to launching a great number of young of the right size at the right time, other behavior helps to prevent the flooding of a given area with young beyond the capacity of the resources of the habitat. Some types of behavior insure that the most suitable males, presumably with genetic material most advantageous to the species, have the best chance of spawning. Other behavior protects the integrity of the species, preventing hybridization, or preventing the waste of gametes in infertile alliances. Some postspawning behavior that takes the adults away from the vicinity of the hatching young prevents intraspecific predation and competition. Intraspecific competition may be avoided by distinctly different feeding behavior in adults and young.

Feeding Behavior

The finding and consumption of food is governed by physical constraints as well as by behavior. The two are closely related, and some fishes are so extremely specialized physically that food selection is limited and feeding behavior is usually specific. On the other hand, species with generalized structure can be rather flexible in feeding behavior, and can exploit a wide range of resources. Locomotory capabilities are of great importance in feeding, obviously. The ability to swim faster than anything else serves some predators well, but others make a living by slow stalking of prey. The serpentine mode of movement seen in many elongate fishes can allow search for food in the water, on the bottom, in the bottom materials, or even out of the water in swamps and on reefs. Some fishes profit by sustained swimming over long distances, but others require short, swift bursts with tight turns and sudden stops. Even a little observation of fishes in an aquarium will serve to establish an idea of the important relationships between feeding and locomotion.

No less important than movement, and of course more important in some instances, is the feeding mechanism. Many fishes without protrusible jaws, and with unspecialized opercular and branchiostegal pumps, simply put the mouth in contact with or around the food item and either bite a piece off, as do many sharks and some characins, or capture the item in the jaws, as is done by a wide variety of species, from trout to tunas. Combining a sucking action with grasping increases the range somewhat; even species with a fixed premaxilla can bring food to the jaws from a short distance away by suddenly opening the mouth and enlarging the buccopharyngeal cavity. Development of protrusible jaws allows extension of the mouth beyond the usual profile of the head, so that the sucking in and grasping of prey becomes more efficient. Fishes with greatly extensible jaws or with long tubular mouths depend primarily on the sucking action to bring food past the jaws well into the mouth cavity. Referral has been made earlier to the specializations in dentition, gill rakers, and other adaptations for food gathering in fishes.

Although some fishes are primarily vision oriented in their feeding and others depend mainly on olfaction, the marvelous sensory arsenal of fish generally allows them a large degree of flexibility in their feeding behavior. Certain environments and habitats favor the use of some senses over others, but usually most senses are functional to some degree and available for use. A fish might be alerted to the presence of potential food through the acoustico-lateralis system, move toward it by following a scent, and finally capture it by seeing, tasting, feeling, or sensing it by electrolocation.

Quests for food may involve active searching over a large area, as in pelagic predators, or may be confined to a restricted home range. Some species habitually take stations and wait for food to come by; anglerfishes and a few others increase the chances of the approach of prey by displaying a lure. For the most part, fish do not seek food continuously every hour of the day. Usually there is a definite period (or periods) during the day when feeding activity reaches a peak. Some species are nocturnal, feeding only or mainly at night; others are diurnal, feeding in the bright of day only. Twilight or crepuscular feeding is a common periodicity; fishes on this rhythm feed once or twice a day in dim light. The daily pattern of activity is fixed in some species, so that individuals show a circadian rhythm in activity and physiology even when kept in a constant environment. There are cycles of feeding activity attuned to the seasons and life histories of fishes. Fishes of temperate waters usually feed most during seasons of optimum water temperatures, reducing food intake during the winter and the hottest part of the summer. The seasonal cycle is further complicated in mature fish by the reproductive season, when feeding diminishes or ceases for a time.

Search for food is affected by social relationships among individuals. Feeding is a solitary experience in some species, and a good feeding station or area might be guarded as a territory. Other species feed in aggregations or schools, in which signals of changing color patterns or certain actions can advertise that food has been located. Feeding actions of conspecifics serve to attract members of schools and aggregations to a food source. The same can be said for solitary fishes also, but habitual loners generally regard other fishes with the same or similar food habits as competitors.

Conditioning or habituation can have a marked effect on food selection in fishes. Long association with food of a certain size or color, for instance, can cause individuals to seek and select food items of the same nature out of an abundant supply of varied food items, at least until the learning process changes the habit. An interesting large-scale field study illustrating conditioning to a certain food involved young steelhead trout reared in earth-bottomed ponds where they had access not only to prepared hatchery food but to midge larvae that grew in abundance in the pond bottoms. After these fish were released to a river they continued to seek out midge larvae, although wild fish of the same size fed on mayfly and stonefly nymphs, which were quite abundant and constituted larger and more easily obtained food items.

Obviously search for food can be modified by the abundance of a habitual or preferred food. Distances traveled and the particular part of the water volume searched are related to the presence or absence of the preferred items. If the food item is not present, the fish might leave the vicinity or shift to an alternate food source. Regardless of preference or habit, some species are known to respond to bonanzas in which great amounts of a seasonal or unusual food become suddenly available. For instance, tui chub *(Gila bicolor)* that feed mainly on snails, algae, midge larvae, and other benthic organisms will quickly turn to feeding on surface foods, as during the swarming of flying ants.

Recognition and selection of food depends on the sensory assessment of eligible items. Size of particles may be of importance to some species, as shown in one study in which yellow perch and rainbow trout would eat only *Daphnia* that were 1.3 mm and larger. In others, size makes little difference, as in cod that would take morsels from 2 mm up to a size too large to swallow. In experimental situations various species have responded to real or dummy food of size ranges characteristic for them. Shape is of importance, also; fish of given species snap more often at dummy food of certain shape than at others. Star-shaped objects were seldom chosen in one extensive study — in fact, the star seemed to cause avoidance in some species. Fish apparently learn to associate certain colors or sheen with acceptability of food, as experiments with fish brought in from the wild showed them to prefer colors closely related to natural foods. Motion may be of importance of itself, or may combine with color and sheen to make an object attractive to certain fishes. Otherwise peaceful fishes will often attack others of the same species that twist and turn following an injury, ignoring others of the same size and normal activity. Manufacturers of fishing tackle take heed not only of the foibles of the fisherman but of the idiosyncrasies of the fish in preparing lures for the market. Size, color, sheen, shape, and motion can be combined to make effective lures. Because some fishes seem to be attracted to sounds and other vibration, some lures are made to rotate, wiggle, rattle, or otherwise set up vibrations. Electric fields, odor, taste, and touch are all involved in food selection.

The final capture of an item of food after it has been recognized may involve a simple approach culminating in the grasping of the item, or may involve pursuit, dislodgement, digging, or even herding. Schools of predators may round up schooling prey, preventing escape and making predation easier. Cooperation in capture of food has been reported for species that dig food out of the bottom, or dislodge it from under stones. Several individuals rooting in the same spot can turn over rocks that one fish would have no chance at moving. Once food is obtained, it might require some positioning or preparation prior to swallowing. Predators may swim about with prey in the mouth, shifting it until it is positioned with the head toward the throat so it can be easily swallowed. Molluscs might be shaken about until dislodged from the shell, or shell and all might be broken up and the shell rejected. Some species chew or crush prey before swallowing. Plankton feeders generally consolidate masses of organisms before swallowing. At times food too large to swallow can be set upon by two or more fish, each struggling to get its share or more with the result that, through a kind of greedy cooperation, the food is reduced to pieces that can be ingested.

Modification of Behavior

Modification of behavior through learning has been mentioned several times in the previous section. Obviously, learning is involved in the return of migrating or displaced fish to spawning areas or home ranges. Recognition of mates, young, cleaners, predators, and other biologically significant individuals or kinds of fishes requires some behavior modification. Learning and memory are necessary in the acceptance of new food items and feeding sites over the annual cycle. Even general observation of the lives of fishes can lead to the conclusion that fish learn quickly and have reasonably good memories. The appreciation of that aspect of fish behavior has led to the use of learning in both practical and scientific endeavors with fish.

The knowledge that anadromous salmonids return to the stream from which they migrated to the ocean, and not to the stream to which their parents migrated, is of great importance to fishery management. Knowing that the young migrants imprint on the migration route by storing away olfactory (and other?) cues as they move downstream, and then react to these cues on their subsequent upstream travels instead of following some genetically fixed urge to go to the same place their parents went, has made possible the use of a few strategically located hatcheries to supply downstream migrants for many streams, instead of having one hatchery for each stream's stock. The quick and positive response of many kinds of fishes to new foods has been of practical importance to fish culture in that the fishes can be taught to eat artificially formulated diets instead of more expensive natural diets. Not only do the fishes learn to accept unfamiliar foods, but learn rapidly where and at what time the food will be given. The ayu (*Plecoglossus altivelis*) normally feeds by scraping diatoms off submerged rocks, but readily learns to take dry food from the surface. Channel catfish (*Ictalurus punctatus*) are normally subsurface feeders, but are fed floating pelletized food in fish farms. This is advantageous to the fish farmer in that little food sinks to the bottom where it might possibly be wasted, and overfeeding and underfeeding can be avoided by balancing the food given against the feeding activity of the fish. There have been a few experiments on the rearing of salmonids in open waters after training them to concentrate for feeding at the stimulus of some appropriate underwater sound.

The study of learning in fishes has proven to be very rewarding from the scientific standpoint, and studies of increasingly sophisticated designs are being conducted by ethologists and experimental psychologists. These studies not only provide knowledge that can aid us in understanding more about fishes, but elucidate some of the learning processes of vertebrates in general. The field is becoming specialized, and in many laboratories the fishes under study are being researched not because of a specific interest in fish but because they are convenient subjects for the study of certain brain functions. The conditioned response is well known in fishes, and is employed widely in a variety of studies. By presenting the experimental animal with a conditioned stimulus that can be readily appreciated through one of the senses and, at the same time, subjecting it to punishment or reward, the subject can be made to show such obvious responses as flight or feeding reactions, or more subtle responses such as changes in heart or respiratory rate.

Conditioning paves the way for study of the ability of the fish to discriminate among colors, small differences in sounds, odors, temperature, visual patterns, and the like. Fishes can be trained to carry out tasks that require a series of actions. Once a subject is conditioned, memory can be studied, and once good information on memory is obtained, the investigator can study the effects of chemicals, elapsed time, or other factors, such as removal of parts of the brain, on retention. Removal of the forebrain does not make experimental fish incapable of learning or remembering, although some functions are slowed or altered. Certain chemicals appear to prolong memory in goldfish, but others are known to interfere with retention — for instance, DDT seems to reduce retention time in salmonids.

One of the most obvious ways to modify behavior in a species of fish is through selective breeding, and the selection can be deliberate or incidental to the maintenance of fish in an artificial hatchery or aquarium environment. A simplified theoretical example concerning trout can illustrate this point. In the wild, trout that are unwary of objects moving on the bank or overhead have a greater chance of falling victim to predators, whereas the wary ones that retreat from such disturbances are more likely to grow to reproductive age. In the hatchery, however, there is little premium placed on wariness. In fact, the wariest trout might grow poorly and be subject to greater stress than less wary ones, because they hide when the hatchery attendant appears with food. The least wary fish are generally better fed and less subject to panic and injury brought on by reaction to disturbances. Hatchery strains of trout that have been under domestication for 40 or 50 years have behavior patterns quite different from those of wild trout of the same species. Fish only one generation removed from the wild have been seen to exhibit changes in behavior.

References

Agranoff, B. W., and Davis, R. E. 1967. Further studies on memory formation in the goldfish (abstr.). Science, 158:523.

Alekseev, A. P. (ed.) 1971. Fish Behavior and Fishing Techniques. Jerusalem, Israel Program for Scientific Translations.

Anderson, J. M., and Peterson, M. R. 1969. DDT: Sublethal effects on brook trout nervous system. Science, 164:440–441.

Aronson, L. R. 1971. Further studies on orientation and jumping behavior in the gobiid fish, *Bathygobius soporator*. Ann. N.Y. Acad. Sci., 188:378–392.

Assem, J. 1967. Territory in the three-spined stickleback, *Gasterosteus aculeatus* L., an experimental study in intraspecific competition. Behav., suppl. 16:1–164.

Atema, J., Todd, J. H., and Bardach, J. E. 1969. Olfaction and behavioral sophistication in fish. *In:* Pfaffmann, C. (ed.), Olfaction and Taste, III. Proc. 3rd Int. Symp. N.Y., New York, Rockefeller University Press, pp. 241–251.

Baerends, G. P. 1971. The ethological analysis of fish behavior. *In:* Hoar, W. S., and Randall, D. J. (eds.), Fish Physiology, Vol. 6. New York, Academic Press, pp. 279–370.

———, and Baerends-van Roon, J. M. 1950. An introduction to the study of the ethology of cichlid fishes. Behav., suppl. 1:1–243.

Barlow, G. W. 1963. Ethology of the Asian teleost *Badis badis*. II. Motivation and signal value of the colour patterns. Anim. Behav., 11:97–105.

Beukema, J. J. 1968. Predation by the three-spined stickleback (*Gasterosteus aculeatus* L.): The influence of hunger and experience. Behaviour, 31:1–126.

Breder, C. M., Jr. 1959. Studies on social groupings in fishes. Bull. Am. Mus. Nat. Hist., 117:393–482.

Brock, V. E., and Riffenburgh, R. H. 1960. Fish schooling: a possible factor in reducing predation. J. du Conseil Cons. Perm. Int. pour l'Expl. de la Mer, 25:307–317.

Brown, F. A., Jr. 1972. The "clocks" timing biological rhythms. Am. Sci., 60(6):756–766.

Cane, V. R. 1963. Some ways of describing behavior. *In*: Thorpe, W. H., and Zangwill, O. L. (eds.), Current Problems in Animal Behaviour. Cambridge University Press, pp. 361–388.

Carlson, H. R., and Haight, R. E. 1972. Evidence for a home site and homing of adult yellowtail rockfish, *Sebastes flavidus*. J. Fish. Res. Bd. Can., 29:1011–1014.

Chapman, D. W. 1962. Aggressive behavior in juvenile coho salmon as a cause of emigration. J. Fish. Res. Bd. Can., 19:1047–1080.

Clark, E., Aronson, L. R., and Gordon, H. 1954. Mating behavior patterns in two sympatric species of xiphophorin fishes: their inheritance and significance in sexual isolation. Bull. Am. Mus. Nat. Hist., 103:141–225.

Crone, R. A., and Bond, C. E. 1976. Life history of coho salmon, *Oncorhynchus kisutch*, in Sashin Creek, southeastern Alaska. Fish. Bull. U.S., 74:897–923.

Cushing, D. H., and Harden Jones, F. R. 1968. Why do fish school? Nature, 218:918–920.

Ehrlich, P. R. 1975. The population biology of coral reef fishes. Ann. Rev. Ecol. Syst., 6:211–247.

Frankel, G. S., and Gunn, D. L. 1961. The Orientation of Animals. New York, Dover.

Gilbert, P. W. 1962. The behavior of sharks. Sci. Am., 207(1):60–68.

Gleitman, H., and Rozin, P. 1971. Learning and memory. *In*: Hoar, W. S., and Randall, D. J. (eds.), Fish Physiology, Vol. 6. New York, Academic Press, pp. 191–278.

Goodyear, C. P., and Ferguson, D. E. 1969. Sun-compass orientation in the mosquito-fish, *Gambusia affinis*. Anim. Behav., 17:636–640.

Green, J. M. 1971. High tide movements and homing behaviour of the tidepool sculpin, *Oligocottus maculosus*. J. Fish. Res. Bd. Can., 28:383–389.

———. 1975. Restricted movements and homing of the cunner, *Tautogolabrus adspersus* (Walbaum) (Pisces: Labridae). Can. J. Zool., 53:1427–1431.

Hamilton, W. D. 1971. Geometry for the selfish herd. J. Theor. Biol., 31:295–311.

Hara. T. J. 1975. Olfaction in fish. Progr. Neurobiol., 5 (Part 4):271–335.

Harden Jones, F. R. 1963. The reaction of fish to moving backgrounds. J. Exp. Biol., 40(3):437–446.

Hartman, G. F. 1965. The role of behavior in the ecology and interaction of underyearling coho salmon (*Oncorhynchus kisutch*) and steelhead trout (*Salmo gairdneri*). J. Fish. Res. Bd. Can., 22(4):1035–1081.

Hasler, A. D. 1966. Underwater Guideposts. Madison, University of Wisconsin Press.

———. 1971. Orientation and fish migration. *In*: Hoar, W. S., and Randall, D. J. (eds.), Fish Physiology, Vol. 6. New York, Academic Press, pp. 429–510.

———, and Larsen, J. A. 1955. The homing salmon. Sci. Am., 193(2):72–76.

———, and Wisby, W. J. 1958. The return of displaced largemouth bass and green sunfish to a "home" area. Ecology, 39:289–293.

Hemmings, C. C. 1966. Olfaction and vision in fish schooling. J. Exp. Biol., 45:449–464.

Hunter, J. R. 1969. Communication of velocity changes in jack mackerel (*Trachurus symmetricus*) schools. Anim. Behav., 17:507–514.

Limbaugh, C. 1961. Cleaning symbiosis. Sci. Am., 205(2):42–49.

Lyon, E. P. 1904. On rheotropism. I. Rheotropism in fishes. Am. J. Physiol., 12:149–161.

Magnuson, J. J. 1962. An analysis of aggressive behavior, growth, and competition for food and space in medaka *(Oryzias latipes* [Pisces, Cyprinodontidae]). Can. J. Zool., 40:313–363.

Marler, P., and Hamilton, W. J., III. 1966. Mechanisms of Animal Behavior. New York, John Wiley and Sons.

Mayr, E. 1963. Animal Species and Evolution. Cambridge, Mass., Belknap Press.

McFarland, W. N., and Moss, S. A. 1967. Internal behavior in fish schools. Science, 156:260–262.

Myrberg, A. A., Jr. 1972. Ethology of the bicolor damselfish, *Eupomacentrus partitus* (Pisces: Pomacentridae): a comparative analysis of laboratory and field behavior. Anim. Behav. Monogr., 5:197–283.

Pfeiffer, W. 1962. The fright reaction of fish. Biol. Rev. Cambridge Phil. Soc., 37:475–511.

Pitcher, T. J. 1973. The three-dimensional structure of schools in the minnow, *Phoxinus phoxinus* (L.). Anim. Behav., 21:673–686.

———, Partridge, B. L., and Wardle, C. S. 1976. A blind fish can school. Science, 194:963–965.

Radakov, D. V. 1973. Schooling in the Ecology of Fish. New York, Halstead Press.

Robertson, D. R., and Choat, J. H. 1974. Protogynous hermaphroditism and social systems in labrid fish. Proc. 2nd Int. Coral Reef Symp, 1:217–225.

Roe, A., and Simpson, G. G. (eds.). 1958. Behavior and Evolution. New Haven, Yale University Press.

Schutz, D. C. 1969. An Experimental Study of Feeding Behavior and Interaction of Coastal Cutthroat Trout *(Salmo clarkii clarkii)* and Dolly Varden *(Salvelinus malma)*. M.S. Thesis, University of British Columbia.

Schwassmann, H. O. 1971. Biological rhythms. *In*: Hoar, W. S., and Randall, D. J. (eds.), Fish Physiology, Vol. 6. New York, Academic Press, pp. 371–428.

Shaw, E. 1970. Schooling in fishes: critique and review. *In*: Aronson, L. R., Tobach, E., Lehrman, D. S., and Rosenblatt, J. S. (eds.), Development and Evolution of Behavior (Essays in Memory of T. C. Schneirla). San Francisco, W. H. Freeman, pp. 452–480.

———, and Tucker, A. 1965. The optomotor response of schooling carangid fishes. Anim. Behav., 13:330–336.

Stasko, A. B. 1971. Review of field studies on fish orientation. Ann. N.Y. Acad. Sci., 188:12–29.

Storm, R. M. (ed.). 1967. Animal Orientation and Navigation. Corvallis, Oregon State Univ. Press.

Sulak, K. J. 1975. Cleaning behaviour in the centrarchid fishes, *Lepomis macrochirus* and *Micropterus salmoides*. Anim. Behav., 23:331–334.

Tavolga, W. N. 1969. Principles of Animal Behavior. New York, Harper & Row.

Tinbergen, N. 1952. The curious behavior of the stickleback. Sci. Am., 187(6):22–26.

Todd, J. H. 1971. The chemical languages of fishes. Sci. Am., 224(5):98–108.

———, Atema, J., and Bardach, J. E. 1968. Chemical communication in social behavior of a fish, the yellow bullhead *(Ictalurus natalis)*. Science, 158:672–673.

Verheijen, F. J., and Reuter, J. H. 1969. The effect of alarm substance on predation among minnows. Anim. Behav., 17:551–554.

Wallace, R. A. 1973. The Ecology and Evolution of Animal Behavior. Pacific Palisades, N.J., Goodyear.

Westby, G. W. M. 1975. Further analysis of the individual discharge characteristics predicting social dominance in the electric fish, *Gymnotus carapo*. Anim. Behav., 23:249–260.

Whitney, R. R. 1969. Schooling of fishes relative to available light. Trans. Am. Fish. Soc., 98:497–504.

Wickler, W. 1967. Specialization of organs having a signal function in some marine fish. Stud. Trop. Oceanogr., Miami, 5:539–548.

Williams, G. C. 1957. Homing behavior of California rocky shore fishes. Univ. Cal. Publ. Zool., 59:249–284.

———. 1964. Measurement of consociation among fishes and comments on the evolution of schooling. Publ. Mus. Mich. State Univ. Biol. Ser., 2:349–384.

Winn, H. E. 1958. Comparative reproductive behavior and ecology of fourteen species of darters (Pisces-Percidae). Ecol. Monogr., 28:155–191.

17 FISHES AS A RESOURCE: THE RELATIONSHIPS BETWEEN FISHES AND MAN

In considering the relationships between fishes and man, one automatically thinks of the use of fishes as food, for this is the greatest and most obvious relationship. This section will devote attention to the widespread fisheries and products made from fishes, but other relationships will also be examined. Most of these will emphasize positive aspects of interaction between fishes and man, such as recreational fisheries and the use of fishes for aquarium display, but attention will first be given to some negative or more unpleasant relationships. Fortunately, the incidence of these negative experiences is minor in comparison to the many important positive associations man has with fishes. However, some may receive undue attention; one nonfatal shark attack is sure to get more newspaper coverage than the capture of enough fish to feed a large city.

Negative Aspects

Poisonous and Venomous Fishes

Throughout the many orders of fishes there are species known to be poisonous to man. Some can cause illness or death when eaten; others have stinging spines that introduce venoms. Together they constitute genuine hazards to divers, fishermen, and bathers in the natural habitat of the fishes, and to aquarists and diners who might come into contact with fishes or flesh in areas far removed from their origin.

A considerable vocabulary of terms designates the toxins of the fishes and the conditions they cause in human beings. A few of the important terms are:

Ichthyotoxin — generally, any poison originating from fishes.

Ichthyosarcotoxin — poison found in flesh of fishes, excluding poisons due to bacterial action.

Ichthyohemotoxin — poison found in blood of fishes.

Ichthyootoxin — poison found only in roe of fishes.

Ichthyoacanthotoxin — poison secreted at the site of a venom apparatus such as spines, stings, or teeth of fishes.

Ciguatera — a particular ichthyosarcotoxism caused by eating various marine fishes of tropical and subtropical areas.

Scombroid poisoning — an ichthyosarcotoxism caused by eating improperly preserved scombroid fishes.

Tetrodotoxin — the poison in viscera of puffer fishes.

In this treatment the adjective "poisonous" will be used for fishes containing poisons (ichthyosarcotoxins, etc.) that affect man after ingestion of the fish. "Venomous" will be used for those fishes that introduce toxins by means of stings, spines or teeth. This usage roughly follows that of Bruce W. Halstead.

Poisonous Fishes. The most widely known type of poisoning caused by fishes is ciguatera, a type of intoxication that causes nausea, vomiting, abdominal pain, reversal of hot and cold sensation, and numbness of the mouth. Various other symptoms may include headache, muscular aches, dizziness, and, occasionally, blistering and loss of skin on hands and feet. A great variety of tropical marine fishes cause this type of poisoning, although some species may be toxic in some geographical areas and not in others. Usually large individuals of a species are poisonous whereas small ones are not. The toxin appears to be obtained through the food chain, originating in a marine alga of the genus *Diplopsalis*, which grows on detritus and plants around coral reefs. The plant is eaten by herbivorous fishes that then become the food of carnivores. Large carnivores tend to concentrate the poison to the point that they become dangerous to eat. Fishes most often implicated in ciguatera are morays (Muraenidae), barracuda (Sphyraenidae), snappers (Lutjanidae), groupers (Serranidae), and jacks and their close relatives (Carangidae). All these families contain excellent food fishes, which are habitually eaten in tropical areas.

There have been reports of poisonings from ciguatera dating back to 1600 A.D. These usually tell of incidents involving a ship's crew or a family of islanders, but some list 50 or 60 persons taken ill and at least one report tells of 1500 persons affected. Relatively few deaths (less than 10%) are reported in most instances, but there is a report of more than 400 deaths among Marshall and Caroline islanders during 1940 and 1941. There is no effective remedy at this time. Cooking does not destroy the toxin.

Other than the families mentioned earlier as being among the most common ciguatoxic fish, there are many families of fishes that have been implicated in one or more poisonings. Some of these are herbivores or plankton feeders. Families include bonefishes (Albulidae), milkfish (Chanidae), tarpons (Elopidae), herrings (Clupeidae), anchovies (Engraulidae), lizardfishes (Synodontidae), conger eels (Congridae), flyingfishes (Exocoetidae), squirrelfishes (Holocentridae), surgeonfishes (Acanthuridae), butterflyfishes (Chaetodontidae), mackerels and tunas (Scombridae), plus a number of other perciform fishes and some of the tetraodontiforms (filefishes and relatives).

Other types of fish poisoning are known, but do not appear to be as common as ciguatera. Lampreys and hagfishes have been implicated in what is called cyclostome poisoning, and several families of sharks have toxic livers or flesh. Elasmobranch poisoning might be due to three different kinds of toxins (including ciguatera). The flesh of the Greenland shark *Somniosus macrocephalus* often contains a toxin that can be removed by drying or thoroughly washing strips of the meat. Many tropical sharks have toxic livers, a condition not related to hypervitaminosis A.

Oilfishes (Gempylidae) have been called "purgativefishes" by seafarers because of the diarrhea caused by eating their oily flesh. Flesh of tunas and their relatives, when improperly stored, can become toxic owing to the presence of histamines and histamine salts. The name "saurine" has been applied to one of these toxic agents. Histamines have appeared in toxic levels after storage of tuna flesh at 20° C for four days. Tunas and swordfishes are known to concentrate mercury through the food chain so that unacceptable levels have been noted. Landed fish are routinely tested, and those with excessive mercury content are not used for human consumption.

Much has been written about puffer poisoning (tetrodotoxism) caused by eating the viscera of various tetraodontiform fishes, which are common table fare in Japan. The toxin, tetrodotoxin, is particularly strong, and the fatality rate is usually over 50% of those intoxicated. Nonetheless, the flesh is of such high quality that the Japanese maintain a fishery for the species and license chefs to prepare them so that only the nontoxic portions reach the table. The ovary and liver are the most toxic parts of the fish, with the stomach and intestines being nearly as virulent. The eyes and kidneys are toxic as well. The skin, subcutaneous tissue, and testes are poisonous in some species.

Puffers (Tetraodontidae), porcupinefishes (Diodontidae), and molas (Molidae; Fig. 7–23C) have been implicated in tetrodotoxism. These are widely distributed in warm and warm temperate waters and have been the cause of illness and death throughout their ranges, especially in the Orient. The poison found in puffers and their relatives has been isolated and its chemical structure determined. It is identical to tarichatoxin, a poison found in newts of the genus *Taricha*.

A few fishes are known to produce toxins from skin glands not associated with any kind of a stinging structure. These poisons, called ichthyocrinotoxins, are harmful if ingested by man. Fishes known to have poisonous skin or slime or to secrete toxins upon being disturbed include hagfishes, lampreys, morays, soapfishes (Grammistidae), puffers, and their near relatives.

Venomous Fishes. A number of fish families contain species capable of inflicting painful stings that combine mechanical injury with release of venom. The stingrays of Dasyatidae, Potamotrygonidae, Urolophidae, and Gymnuridae are among the best known of stinging fishes. Their generally larger relatives — eagle rays (Myliobatidae), cownose rays (Rhinopteridae), and mantas (Mobulidae) — have stings and venom-producing tissue, but are not often implicated in injury to humans. The stings of the rays are stiff, dagger-like spines with recurved teeth on the edges. In the integumentary sheath, especially along the ventrolateral grooves of the spine, there is a layer of venom-producing glands. When the spine is thrust into the flesh of a victim, the skin sheath is broken and the venom released into the wound.

Spines of stingrays are usually situated close to the thick, muscular base of the tail (Fig. 6–8C). When the ray is stimulated to defend itself—for example, by a person stepping on the disc—the tail is curled quickly over the back and the spines thrust at the offender. The wounds resulting from these stings are dangerous and painful. The victim is usually in danger from secondary infections, tetanus, and gangrene, as well as the effects of the venom.

Other cartilaginous fishes with venomous spines are the dogfish shark (*Squalus*) (Fig. 6–7C) and the chimaeras (Fig. 6–4). In these the venom glands are along the dorsal spines, and the danger of being stung is not as great as with the stingrays. Most punctures come from careless handling. The venom of sharks and chimaeras is not as dangerous as that of stingrays.

Among the bony fishes certain catfishes, weevers, surgeonfishes, scorpionfishes, stargazers, rabbitfishes, toadfishes, and the sabertooth blennies are known to be venomous, and several other groups including Carangidae and Scatophagidae are reported to have venom associated with spines.

Catfishes have venom glands in the skin sheathing the dorsal and pectoral spines, and some groups have axillary venom glands that supply their secretions to the exterior of the pectoral spine. Most catfish stings are painful but not dangerous, but some species cause edematous swelling and gangrene at the wound. The family Plotosidae, an Indo-Pacific group containing both fresh- and saltwater members, contains the most dangerous venomous catfishes. Some of the catfishes commonly kept in home aquaria can inflict severe stings that occasionally result in numbness and shock. Included are the shovelhead catfishes of Pimelodidae, a South American family, the air-breathing Clariidae and Heteropneustidae, and the armored Doradidae. Other venomous catfishes are the Bagridae, Siluridae, the marine Ariidae, and the familiar North American Ictaluridae.

Catfishes react to being grabbed or restrained by lashing violently from side to side. Usually the pectoral and dorsal spines are locked in an erect position during this activity so that they can pierce the attacker. Care should be exercised when handling live catfish of any kind. Usually they can be grasped directly behind the pectoral fins with reduced danger of being stung.

Fishes of the family Scorpaenidae are widespread along tropical and temperate shores. Members of the family generally have venom glands in grooves along the dorsal, anal, and pelvic spines. In some genera, such as *Sebastes*, the venom is not virulent and the spines mainly inflict a painful wound. In other genera, such as *Scorpaena* and the tropical *Pterois*, the venom is more powerful, that of *Pterois* sometimes killing human beings. Species of *Pterois*, called turkeyfish, lionfish, etc., are extremely colorful and are sometimes kept in home aquaria.

The family Synanceidae (stonefishes) have very large venom glands associated with the dorsal, anal, and pelvic spines. The ducts of the spines run in a groove to a point near the tip of the spine. Stonefish venom is extremely dangerous and, as the fishes are so well camouflaged, there are many recorded instances of people stepping on them and being severely affected by the venom. There are records of several deaths caused by stonefishes. The family lives in the warm Indo-Pacific region.

Weevers (Trachinidae; Fig. 7–24C) are among the fishes that have venomous opercular spines as well as venomous fin spines. Weevers are found from the North Sea south into the Mediterranean, and have considerable contact with fishermen and divers throughout their range. They are said to attack when disturbed, striking with the blade-like opercular spine. Both fin and opercular spines are deeply grooved, as

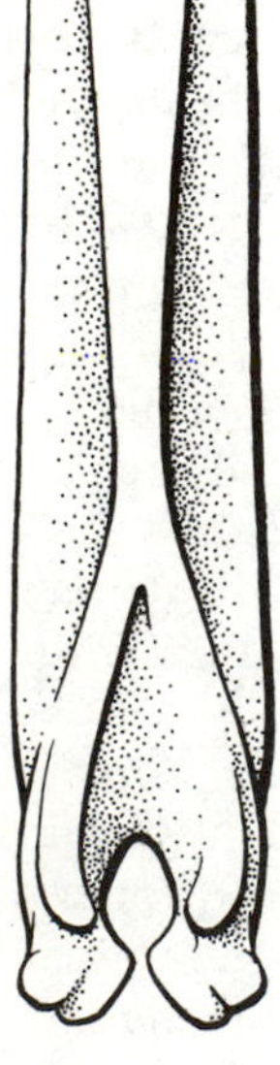

FIGURE 17–1. Grooved dorsal spine of a venomous toadfish.

in most venomous fishes (Fig. 17–1). The venom-producing tissue is in the grooves. Toadfishes (Batrachoididae; Fig. 7–29*A*) are found in warm coastal waters of most seas. Like the weevers, they are small bottom fishes equipped with venomous opercular and fin spines. They differ in having hollow spines with large venom glands surrounding them, so they tend to inject venom into wounds caused by the spines. Though their apparatus might seem more efficient than some of the other venomous fishes, their venom is not as dangerous as that of the weevers.

Very few fishes appear to have a venomous bite. Morays have been reported to have poison fangs, but that has been disproved. There may be some venom-producing tissue in the skin of the palate, according to some authorities. The sabertooth blennies, *Meiacanthus*, have large canine teeth in the lower jaw. The teeth are grooved and are associated with venom gland, so that the bite is very painful.

Other Dangerous Fishes

Among the fishes that are injurious to man there are electrogenic species such as *Torpedo, Electrophorus,* and *Malapterurus* (electric ray, eel, and catfish; Fig. 10–10). These can all cause pain and temporary numbness, and the electric eel is large and powerful enough to be fatal. Other nonbiting dangerous fishes are the billfishes and sawfishes. Both types are large and armed with a long extension of the rostrum. Occasionally billfishes ram boats, doing considerable damage to the small craft; swordfishes have been recorded as splintering several small boats. Sawfishes inhabit shallow water and habitually enter tropical or subtropical rivers, so they often come in contact with bathers or fishermen. Serious injuries and some deaths have been caused by sawfishes in various areas, especially in India where the Ganges can be crowded with pilgrims. A sawfish 4 or 5 meters long blundering into a crowd of bathers and turning swiftly is certain to do damage.

Needlefishes (Belonidae; Fig. 7–20*A*) can leap out of the water at great speed, their sharp-pointed beaks and arrow-shaped bodies combining to make them dangerous projectiles. They occasionally strike boaters, swimmers, or surfboard riders and cause serious effects. In Hawaii during the summer of 1977 at least three persons were struck. Two were hit in the leg, the beak passing through the calf of one. The third, a 10-year-old boy, was fatally injured when a meter-long needlefish struck him in the eye.

The candiru *(Vandellia)* is a tiny South American catfish that parasitizes larger fishes, living in their gill cavities. It is apparently attracted to urine and is known to enter the urethra of humans. Surgery is required to remove it because of its recurved spines.

Fishes that bite are probably feared more than all others. Most predatory fishes and some others, such as the coral-eating parrotfishes, can deliver painful or injurious bites when mishandled after capture. There are stories of the wolf-herring (*Chirocentrus dorab*; Fig. 7–3C) attempting to reach and bite its captors when boated, but most bites from captured fish are probably accidental. The pike eels (*Muraenesox*) are also dangerous to capture. There are some fishes, however, that have made direct biting attacks on bathers and shipwrecked individuals. The deep-set, knife-like teeth of the barracuda are capable of tearing out great chunks of flesh. Morays occasionally make unprovoked attacks on swimmers. One of the most feared groups of fish are the piranhas (*Serrasalmo, Rooseveltiella*). These South American characins have strong jaws and sharp triangular teeth, and commonly feed on animals larger than themselves. They can attack and severely wound or kill wading or swimming mammals, including man. The wound from a single piranha bite is said to be a hemispherical hollow.

Sharks, of course, are the fishes that draw most attention as maneaters. Some are of large size and have voracious appetites, seeking large prey. These apparently attack man as a source of food as they would attack a seal or a large fish. Smaller sharks, too, have been known to feed on man, but some authorities believe that one half or more of known shark attacks have not involved feeding sharks.

The Office of Naval Research has sponsored the accumulation of data on more than 1600 shark attacks in a file at the Smithsonian Institution. Data from over 1100 attacks were subjected to computer analysis at the Mote Marine Laboratory, Sarasota, Florida, in an attempt to discover important factors related to shark attacks; some of that laboratory's findings are summarized here. In general, shark attacks occur where large numbers of swimmers and divers frequent waters containing large numbers of sharks. Usually swimmers are found in water over 20° C and sharks are seldom found in water over 30° C, so most attacks occur between those temperatures, with the peak range between 21° and 24° C. The great white shark has attacked men at temperatures as low as 10.5° C.

Male victims of shark attack outnumber females by about 9 to 1, probably a reflection of the relative swimming and diving activities of the sexes rather than the taste preference of the sharks. Nearly a fifth of the more than 1000 victims for which such information was available were spearfishermen, with many known to have been carrying captured fish. About two thirds of the attacks recorded took place within 200 feet of shore, which, of course, is the area where the greatest numbers of people are found. Actual possibility of attack is probably greater farther from shore. The fatality rate among shark victims is about 35%.

In about 270 attack cases sharks of eight families were identified. The worst offender appeared to be the great white shark (*Carcharodon carcharias*), a giant that wanders well into temperate waters. The tiger shark (*Galeocerdo cuvieri*) was a close second. Sharks of the genus *Carcharhinus*, some of which enter tropical rivers, often attack man. The mako shark and its relatives (*Isurus*) have been implicated in many attacks, as have the hammerheads (Sphyrnidae).

Fishes as Carriers of Parasites and Diseases

One negative aspect of man's relationships with fish that is important in some geographical areas is the ability of various species to harbor parasites that can affect man directly. The parasites are usually of concern only in areas where freshwater fishes are eaten raw or without sufficient processing, although there are some parasites that can be transmitted to man by marine fishes.

Most of the parasites involved are worms — nematodes (roundworms), cestodes (tapeworms), and trematodes (flukes). One potentially dangerous nematode is the kidney worm, *Dioctophyma renale*, which is known mainly from the Orient. Another type occasionally found in man, especially in regions where raw herring is consumed, is represented by *Anisakis* and its relatives. These worms can cause illness if a person becomes infested with enough of them.

Probably the best known fish-borne cestode affecting man is the broad tapeworm, *Diphyllobothrium latum*, common in some northern European countries and in parts of Asia. In North America it is known primarily from the Great Lakes area. Those infested with this worm generally suffer from anemia. The worm is transmitted to man through uncooked freshwater fishes.

Among the trematodes are a few that are known to be transmitted from fish to man. Some are intestinal parasites but one family, Opis thorchidae, contains liver parasites that can cause serious effects. Infestations are known mainly from Asia. An interesting relationship involving a snail, a fluke, and one of our most highly prized domestic animals, the dog, is found in coastal portions of the Pacific Northwest. A snail, *Oxytrema silicula*, harbors the early stages of the fluke, *Nanophyetus salmonis*, the cercariae of which are carried mainly by salmon and trout. The adult fluke is found in various carnivores including skunks, raccoons, and others. Dogs and other canids are susceptible to the fluke, which of itself is not a dangerous parasite, but carries a rickettsial disease that is extremely dangerous to members of the dog family. The fluke's cercariae are so common in salmonids of the region that before the true nature of the disease was discovered it was generally believed that salmon were poisonous to dogs. Methods of prevention and treatment are now known, so the mortality rate of dogs that eat raw salmon has been considerably reduced.

Fishes Out of Place

One definition of a weed is that it is a plant out of place, and the same might be said of some fishes. Regardless of their adaptation to the environment in which they evolved, fishes, when introduced to other areas, can have adverse effects on the new environment or upon other species there. Adverse effects may materialize through competition, predation, or hybridization. In some instances where fish managers wish to enhance populations of game fishes, the presence of even the native nongame fishes is undesirable. A good example of an introduced fish considered a "weed" or "trash fish" in many areas is the carp (*Cyprinus carpio*). In Europe and in Asia, where this species is native, the carp is esteemed as food, is subject to fisheries, and apparently has ecological checks and balances, so that it does not dominate. In North America there are only minor fisheries on the carp, and populations in many places run unchecked. The carp is such an efficient forager that few native fishes can compete with it, especially as the carp can uproot and destroy aquatic vegetation that serves as cover for fish and fish food organisms. Some of the greatest difficulty with carp is on certain migratory waterfowl refuges, where the fish destroys the aquatic vegetation upon which the waterbirds feed.

Some native North American minnows have bad reputations among trout managers, especially when they are introduced to lakes in which only trout are desired. Tui chub (*Gila bicolor*) and redside shiner (*Richardsonius balteatus*) are two minnows that seem to compete with trout, in some situations to the point of near exclusion. Often management agencies resort to poisoning the entire lake with rotenone or other ichthyocides and begin the management over again. In some instances such control operations are not detrimental, but in others there may be some danger to rare fishes or to susceptible invertebrates.

Perhaps the greatest fish control campaign of all was the effort directed at the sea lamprey (*Petromyzon marinus*) after it invaded the upper Great Lakes, contributing to the decline of the commercial and sport fisheries. While a specific toxin for the lamprey was being sought, electrical barriers were devised that decreased the numbers of adult lampreys reaching the spawning grounds. After a specific poison was discovered and developed for use, a systematic effort was made to eradicate the larval lampreys in all the tributary streams. The effort was sufficiently successful so that salmonid fishes, especially the introduced coho salmon (*Oncorhynchus kisutch*) are now the objects of fisheries.

Wherever fishes or other aquatic organisms are grown in monoculture, species other than the chosen object of culture will be removed, even though they might be prized for culture in another application.

Positive Aspects

Fishes as Biological Control Agents

The varied feeding habits of fishes sometimes lead to the use of certain kinds of fishes to control organisms considered undesirable. Probably the best known of the insect-controlling species are the various members of the family Poeciliidae, especially the genus *Gambusia*. One species, *G. affinis*, aptly called the mosquitofish, has been introduced into many areas outside its native range in the Mississippi and adjacent drainages. Its reputation as a mosquito eater is so great that it even has been taken to other continents where there probably are native top-feeding fishes of equal efficiency.

There are several fishes that feed upon freshwater snails, thereby destroying intermediate hosts of parasites of man. Of special note are certain African cichlids and the black or mud carp of Asia, *Mylopharyngodon piceus*.

Another Asiatic carp, the grass carp (*Ctenopharyngodon idella*), feeds almost exclusively on aquatic plants, ingesting prodigious amounts daily. Although only a portion is digested, the vegetation is shredded by the carp's pharyngeal teeth and passed through the alimentary canal, becoming, in effect, a fertilizer for other plant growth. Usually phytoplankton growth is promoted by the activity of the carp, so that rooted plants are destroyed not only by being eaten but by the shading due to phytoplankton blooms. The grass carp is an excellent table fish, and has been introduced to many countries, where it serves double duty as a food fish and an aquatic weed control agent. Various members of Cichlidae, especially species of *Tilapia* (including *Sarotherodon*), feed upon filamentous algae and soft vascular plants. Some are efficient at removing vegetation from irrigation systems, and are used for weed control in ditches in the southwestern United States. In some areas they cannot survive the cool winters, but can be planted annually in the spring. Reproduction during the summer increases populations sufficiently to control the problem plants.

In some instances predatory fishes are used to control populations of other fishes. In *Tilapia* culture in rice fields, for example, a few snakeheads (*Channa*) are often added to crop some of the excess juveniles. The walleye (*Stizostedion vitreum vitreum*) is sometimes used in managed lakes as a control for the yellow perch (*Perca flavescens*), but success is not always achieved. The northern pike (*Esox lucius*) have been successfully employed as a control for sunfishes.

Scientific Uses of Fishes

In general, fishes are excellent subjects for the study and demonstration of anatomy, physiology, ecology, evolution, and other aspects of science. Because of their ready availability and their representation of typical structure of lower vertebrates, the dogfish (*Squalus acanthias*) and yellow perch (*Perca flavescens*) have become standard dissection laboratory animals. These two species are the basis of an important trade and are the subjects of numerous laboratory manuals. Lampreys and their larvae are dissected routinely in college laboratories as representatives of Agnatha, and there is a good trade in gars and bowfins as primitive bony fishes.

The hagfishes, with their extra hearts, have aided physiologists in studies of cardiac pacemaker cells. Rainbow trout have been used extensively in the study of relationships of various food components to hepatoma and other cancers. Many species, including the rainbow trout, are used in studies of water pollution and the effects of various pesticides and waste products on aquatic life. The fathead minnow (*Pimephales promelas*), the bluegill (*Lepomis macrochirus*), and the goldfish (*Carassius auratus*) are favorite bioassay animals and, along with the mummichog *(Fundulus heteroclitus)*, are also used in physiological experiments. Many species have been used in research in ethology and experimental psychology, including the goldfish, various sticklebacks, cichlids, and others. The medaka *(Oryzias)* and the zebra danio *(Danio rerio)*, both of which reproduce well in captivity, are among fishes that are useful in the study of embryology.

Recreational Fisheries

Imagining how fishing for fun originated from hook-and-line fishing for food fishes is easy. Many fishes have qualities of speed, stamina, and leaping abilities that allow them to put up noble struggles to prevent capture, and the lightening of tackle to give the fishes a fighting chance can be seen as a natural development transforming a serious matter of food gathering into a form of play. Usually, anglers utilize their catches as food so, for many, sport fishing serves a dual purpose. There are, however, some sport fisheries, based on the use of artificial flies as lures, in which no fish are killed, but are returned to the water to be caught and released another day. Types of fishes sought by anglers range from the sharks through the lower bony fishes to the higher bony fishes. The size range runs from trouts and others of a few centimeters in length to the giant tunas, swordfishes, and sharks. Sport fishing is geographically widespread, but it reaches its greatest importance in affluent societies where the demand for outdoor recreation transcends the need for commercial fisheries.

Freshwater angling in the Northern Hemisphere is based on a variety of fish families, with the Salmonidae ranking high in popularity and in money spent in pursuit of sport. Salmon and trout are found in abundance in the northern part of the hemisphere and are represented at high altitudes south to Turkey, Morocco, and Pakistan. Many species support recreational fisheries. The arctic char and its relatives of the genus *Salvelinus* are limited to the coldest waters, so are best distributed in the far north and high altitudes. The genus *Salmo* contains two of the best known sport fishes, *S. gairdneri*, the rainbow trout, and *S. trutta*, the brown trout. These highly prized fish have been introduced to many areas outside their original distributions. Other famous salmonids are the Atlantic salmon *(S. salar)* and two of the Pacific salmons, the chinook *(Oncorhynchus tshawytscha)* and the coho *(O. kisutch)*. Various whitefishes (Coregoninae), the grayling *(Thymallus)*, the huchen *(Hucho)*, and others are locally important as game fishes.

Pikes (Esocidae) and perches (Percidae) are other popular northern game fishes. Members of these families range from cold to temperate regions and include both large and small species. The pike *(Esox lucius)* is holarctic in distribution and provides sport throughout its wide range. Other large pike are the muskellunge *(E. masquinongy)* and the Amur pike *(E. reicherti)*. The pikeperches *(Stizostedion)*, the largest members of Percidae, are found in both North America and Eurasia. The walleye *(S. vitreum)* is a noted game fish in the U.S. and Canada. The perch of Europe *(Perca fluviatilis)* and the yellow perch

(P. flavescens) of North America are locally numerous in rivers, lakes, and reservoirs, and have good reputations as panfishes.

Minnows (Cyprinidae) are distributed in both cool and warm waters of the Northern Hemisphere. In Europe several species are used for hook-and-line fishing, including many smaller species as well as large ones such as the carp *(Cyprinus carpio)* and the tench *(Tinca tinca)*. In India the very large minnow, the mahseer *(Barbus tor)*, is a favorite of sport fishermen. The mahseer and related species of the Tigris *(B. schejki)* may reach nearly 100 kg. A few North American anglers fish for the introduced carp, but other minnows are not generally sought. Warm-water sport fishes of North America include many of the catfish family, Ictaluridae. Three of these are large fishes — the flathead catfish *(Pylodictus olivaris)*, and blue catfish *(Ictalurus furcatus)*, and the channel catfish *(I. punctatus)*. Other members of *Ictalurus* are smaller but are popular panfishes. In eastern Europe the huge catfish *Silurus glanis* is sought by anglers. Other favorite warm-water game fishes are included in the North American sunfish family, Centrarchidae. Chief among these are the black basses of the genus *Micropterus*, but the crappies, *Pomoxis*, and the sunfishes, *Lepomis*, have many devotees.

Many of the gamefishes and panfishes of the Northern Hemisphere have been introduced to areas where they were not natively found. For instance, centrarchids, ictalurids, and the rainbow trout have been established in Europe, and the carp and brown trout have been brought to North America. There has been a strong tendency for anglers to take their favorite sport fishes with them wherever they go, and many northern fishes are now established in the Southern Hemisphere. Rainbow and brown trout are now present in suitably cold streams and lakes in Africa, South America, Australia, and New Zealand. The carp is now present in many southern areas, and a few centrarchids, notably the largemouth bass, have been introduced to Africa.

Native Southern Hemisphere sport fishes include some large and interesting species. The arapaima *(Arapaima gigas)* of the Amazon is sought with hook and line. This species is reported to reach a weight of over 200 kg, and specimens of about 100 kg have been taken by anglers. The dorados of South America (*Salminus* spp.) are large, colorful fishes reputed to be among the world's best game species. *S. maxillosus* reaches a weight of about 23 kg. Some of the cichlids of South America are colorful and sporting. One, the oscar or peacock bass *(Astronotus ocellatus)*, has been tried as a game fish in Florida.

African game fishes include the huge Nile perch *(Lates niloticus)* and the voracious tigerfish *(Hydrocynus goliath*; Fig. 7–14B). The former reaches a length of more than 2 meters, the latter slightly less. In Australia a large member of the Percichthyidae, the Murray cod *(Maccullochella macquarienensis)*, is a favored game fish. It reaches about 1.8 meters in length.

Just as there are hundreds of species of freshwater fishes sought by anglers, there is also a tremendous variety of saltwater species that provide sport fishing. The so-called big game fishing is pursued mainly in tropical and subtropical oceans, with large scombroid fishes as the targets. The marlins (*Makaira, Tetrapturus*), swordfish (*Xiphias*), and bluefish tuna (*Thunnus thynnus*) are some of the most highly prized big game fishes, but others, such as the white shark (*Carcharodon*) and the mako shark (*Isurus*) are also popular. Other large marine game fishes include the tarpon (*Megalops atlantica*), the sailfishes (*Istiophorus*), jewfish (*Epinephelus itajara*), giant sea bass (*Stereolepis gigas*), barracuda (*Sphyraena*), and cobia (*Rachycentron*).

Most saltwater anglers seek smaller quarry, including members of several common families such as temperate basses (Percichthyidae), cods (Gadidae), sea basses (Serranidae), jacks (Carangidae), snappers (Lutjanidae), dolphins (Coryphaenidae), porgies (Sparidae), drums (Sciaenidae), surfperches (Embiotocidae), rockfishes (Scorpaenidae), greenlings (Hexagrammidae), and flounders (Pleuronectidae).

The importance of sport fishing is growing in many areas. In North America the sport catch of some species exceeds the commercial take, and some species and fishing areas are reserved exclusively for sport angling. Considerable economic value is placed on recreational fisheries in developed countries. Besides being an excellent use of leisure time, angling promotes a great circulation of money. Anglers purchase such items as fishing tackle, boats, outboard motors, and camping gear. They expend money for travel, food, and lodging to the extent that some small coastal communities depend upon good fishing for continued prosperity. In addition, there is great nutritional value in the fish caught, even if the angler's cost per pound in catching them greatly exceeds the price per pound in the fish market.

The Aquarium Trade

Color, form, motion, and habits of many fish species are enjoyed by aquarists in many parts of the world. For instance, home aquaria and ornamental pond enthusiasts are estimated to number more than 20 million in the U S A and about 2 million each in Canada and Japan. There are records of imports of ornamental fishes by most of the nations of Europe as well as some Asian nations. Freshwater species are used much more than marine species because of the relative ease of culture, care, and shipping, but marine species are becoming more popular as the technology of maintaining small saltwater systems advances.

There are probably over 1000 species of freshwater fishes available to aquarists in the U.S.A. Most belong to the following few families: the characins, tetras, etc. (Characidae); the minnows, carps, and relatives (Cyprinidae); the loaches (Cobitidae); the topminnows (Cyprinodontidae); the livebearers (Poeciliidae); various armored catfishes (Callichthyidae, Loricariidae, and Doradidae); the cichlids (Cichlidae); and the gouramis (Belontiidae). Members of the foregoing and many other tropical freshwater families are imported to temperate climates in what has been until recently a relatively unrestricted trade.

Saltwater species constitute only a small percentage of the commerce in aquarium fishes, but represent a growing segment of the trade, generally commanding high prices. Usually reef fishes such as butterflyfishes (Chaetodontidae), damselfishes (Pomacentridae), surgeonfishes (Acanthuridae), triggerfishes (Balistidae), squirrelfishes (Holocentridae), and cardinalfishes (Apogonidae) are the most popular. Over 20 families are commonly stocked by wholesalers.

Some of the major exporters of aquarium fishes are Brazil, Colombia, Guyana, Singapore, Hong Kong, and Thailand. Usually fishes are captured by means of traps, seines, or dip nets by individuals who then sell them at a collecting point from which they are transported to exporters. For instance, in Thailand they are trapped from creeks flowing into the Chao Phya River, then taken by small boat to a buyer who sorts the species into liveboxes attached to a barge. When enough are accumulated the barge is taken downriver to Bangkok where the fishes are sold to an exporter who ships to Singapore or Hong Kong for transshipment to other countries.

The export trade in ornamental fishes from Hong Kong is worth about 4 million dollars per year, that from Singapore about 4.5 million dollars per year. Aquarium-related trade in the U.S.A., including fishes and accessories, is estimated to have an annual value of about 700 million dollars. This includes sales of fishes and other aquarium animals, fish food, fish health aids, aquaria, pumps, filters and other accessories, books, and magazines.

Restrictions are being placed on the transport of aquarium fishes by various countries and states for a variety of reasons. Some countries are realizing that unrestricted export of wild native fishes will eventually deplete the supplies of certain species. Other countries prohibit export or import of species declared endangered or especially rare. In addition, there are species considered potentially harmful to humans or to the environment of a country where they might be introduced, and are therefore restricted from free trade. Certain stingrays, stonefishes, lionfishes, and weevers are listed as undesirable in the U.S.A. Various states prohibit piranhas, walking catfishes, snakeheads, parasitic catfishes, the electric eel, and others.

In addition to the enjoyment afforded by private aquaria there is the pleasure of public aquaria where both native and exotic fishes of all sizes can be displayed. Many large aquaria have facilities for keeping both fresh and saltwater species. Some feature giant circular tanks that accommodate large sharks and rays and active fishes such as tunas. Such aquaria are often set up in conjunction with museums or public parks, or may be attached to commercial ventures where marine shows are staged.

Fishes as Items of Commerce

Although fishes and fisheries may be of considerable importance and value from the standpoints of hobbies, recreation, biological control, and other interests, the greatest value from a worldwide standpoint arises from commercial use of fishes as industrial products and food. The industrial uses to which fishes can be put are numerous and varied, but the most important appears to be the production of fish meal, mainly for agriculture. Protein-rich fish meal is a basic ingredient in the food of poultry, trout, catfish, pigs, and other domestic animals. Some grades of meal have been used in the fertilizer industry. Fish oil is another important product of reduction plants, and is used in many manufacturing processes and in various foodstuffs.

Other than meal and oil, fishes are used in manufacture of glue, as a source of leather, as a source of silver pigment (guanine) for certain paints, and as a source of many other minor items or products. Fishes have been especially important in some primitive societies, furnishing, for instance, spines for needles and awls, skin for leather, and in some areas large stinging spines or large teeth for spear or arrow points or for making club edges more formidable.

The greatest use of fishes is as food. Species of all sizes and habitats contribute to the table fare of people in most parts of the world. Small species may be of great local importance. For example, larval fishes are harvested in New Zealand and Asia as they make their way upstream. In the Philippines and Southeast Asia tiny gobies and larvae of other fishes are used in the manufacture of bagoong or fish sauces of various types. Sardines, sprats, and other small marine fishes are harvested en masse for use as food for man. Although the relatively small herrings (Clupeidae) and anchovies (Engraulidae) account for a great share of the world's fish catch — historically 20 to 30% (22% in 1975) — most of the tonnage landed of those families is used for industrial purposes. Most food fishes are medium-sized, weighing a kilogram or more, but a few species are giants, such as the great tunas and the swordfishes. Table 17–1 lists the major groups of food fishes, with an indication of their relative importance. The impact of the fisheries on the world food supply can be appreciated from the total landings. The estimated world catch of fishes (not including shellfish) in 1976 was about 65 million metric tons, of which 20 million metric tons were used for industrial purposes (and in part cycled through farm animals to produce higher-priced protein). Fish provides about 50% of all animal protein consumed in southeast Asia, and is known to be extremely important in the protein supply in many other areas.

Table 17–1. MAJOR GROUPS OF COMMERCIAL FISHES, WITH AVERAGE CATCH, 1971 –1975 *

Species Group	Average Catch (Metric Tons)
Herrings, sardines, anchovies	14,536,488
Cods, hakes, haddocks	11,716,161
Tunas, billfishes, mackerel-like fishes, cutlassfishes	5,322,566
Salmon, trout, smelt	2,509,829
Flatfishes (flounders, halibuts, etc.)	1,258,967
All freshwater fishes	9,277,836

*From FAO Yearbook of Fishery Statistics, Vol. 40. Published by Food and Agr. Org. of the United Nations, Rome.

Table 17–2. LEADING FISHING NATIONS SHOWING ESTIMATED CATCHES FOR 1975*

Country	Catch (Metric Tons)
Japan	10,508,000
U.S.S.R.	9,876,000
China	6,880,000
Peru	3,447,000
U.S.A.	2,799,000
Norway	2,550,000
India	2,328,000
Republic of Korea	2,133,000
Denmark	1,767,000
Spain	1,533,000

*Figures represent the total landings, including fishes, molluscs, and crustaceans. From FAO Yearbook of Fishery Statistics, Vol. 40. Published by Food and Agr. Org. of the United Nations, Rome.

The World Fisheries

The fisheries are pursued in both fresh and salt water, with the great bulk of landings near shore on the continental shelf and slope. Freshwater fisheries account for about 8 to 9 million metric tons annually, according to best estimates. The remainder of the catch, about 53 million metric tons, is taken from marine waters. The marine catch comes mainly from areas where upwelling or other nutrient-rich currents promote growth of plankton, which in turn promotes the growth of fish communities. A notable example is the area off Peru, where the impingement of the cold Humboldt current on warmer water is the basis of a tremendous fishery on the anchovetta, *Engraulis ringens*. Because of this species Peru has become one of the leading fishing nations, with average landings of about 5 million metric tons from 1971 through 1975. Table 17–2 shows some of the prominent fishing nations.

Other great fishing areas are the grounds off New England and the maritime provinces of Canada, the North Sea, the Bering Sea, and the continental shelf and slope along Alaska, British Columbia, and south to Mexico, and the waters from the East China Sea north past the Japanese islands to Kamchatka and the Kuriles. In all these areas cold currents and warm currents meet.

Most of the world's fisheries involve capture of wild stocks of fishes wherever they can be found. This, of course, is a form of hunting, for finding the fishes in large lakes, rivers, and the seas may account for much of the effort expended by fishermen. Methods of hunting range from the subsistence fishermen groping along the bottom of a swamp to the employment of aircraft or the use of sophisticated sonic detectors. Few modern fishing vessels of any size are without some means of detecting fish.

Methods of fishing employed in fisheries include the entire arsenal from the spears, traps, hand lines, and small nets of the traditional subsistence fisherman to the tremendous trawls and purse seines of the modern, high seas fishing fleets. Trawls, seines, and related gear are responsible for most of the tonnage of fish caught, as they can effect the mass capture of various schooling fishes such as cods, herrings, tunas, and salmons. Fishing gear is obviously devised and employed to take advantage of the habits, behavior, and movements of the target species. Ground fishes of the families Gadidae (cods), Merlucciidae (hakes), Pleuronectidae (flounders), Scorpaenidae (rockfishes), and others are taken by means of various trawls which are dragged along the bottom at shallow and moderate depths. Some slow-moving schooling species that live well off the bottom can be taken with midwater trawls. Pelagic schooling species, such as the families Engraulidae (anchovies), Clupeidae (herrings), Salmonidae (salmons), and Scombridae (tunas), are vulnerable to capture by purse seines or other encircling nets.

Migratory species or others having a definite pattern of movements can be captured by entangling gear such as trammel nets or gill nets, or can be led by long weirs or fences into pound nets or traps. Beach seines of various types can be utilized to catch species that migrate close to shore.

Hooks and lines have their places in modern commercial fisheries. Trollers rigged to tow several lures at various depths seek salmon and tuna. Halibut are caught by long lines stretched along the bottom; tuna are taken by long lines at the surface or by lures on short poles and lines. In this type of fishing live bait is released into a school of tuna, and the lures are cast among the frenzied feeding fish. The hooked tunas are heaved onto the deck of the boat where they are easily freed from the barbless hook. If the tuna are large, two or more poles are used to each lure, the fishermen working together throw the fish onto the deck.

Modern fishing vessels are equipped with cold storage or freezing facilities so that catches from trips of many weeks or months can be kept until delivered ashore. Fleets of some countries feature factory ships or "mother ships" that receive and process fish captured by trawlers. Other ships bring supplies, transport products, and even shuttle crewmen on rotation, so that the fleet can stay on the fishing grounds for long periods of time and high quality fish products can reach the home country.

Fish Culture

For thousands of years man has kept fishes in enclosures or impoundments, either to hold them alive until needed for food or to allow them to grow, thus producing more food. The practice has been traditional in China and other parts of Asia and was used by the ancient Romans. Fish culture of several kinds probably predates history, and many traditional methods practiced today have no doubt persisted for a thousand years or more.

Fish culture is taking on a new importance in modern times,

supplying about 3 to 5 million metric tons annually. As the capture fisheries have reached the limit of their production in some areas, and since the expansion of fisheries to other areas farther from population centers or to areas with lower fish production tends to raise prices, more attention has been given to culture of desirable fishes at competing prices. There are, however, many other reasons for the growing popularity of fish culture in addition to the obvious advantages of producing protein for subsistence or high quality fish for luxury prices. For instance, fish culture can make better use of some lands than the farming of terrestrial animals, or can be a means of recycling farm wastes to produce a useful product. Sometimes pig, chicken, or duck farms are built over fish ponds in Asia so that fertilization is continuous and direct. Fish culture can be used to replace ruined or inaccessible spawning grounds or to supplement natural populations. Obviously, depending on the worth of the species cultured, some types of culture will warrant greater effort and financial expenditure than others, so that various levels of intensity can be recognized. A low level of intensity might consist of transferring young fish from the wild into ponds where they could be left without care until harvested. In high intensity culture the entire life cycle of the stock is controlled. Brood fish are kept in special ponds, induced to spawn or spawned artificially, the eggs cared for, and the young provided with prepared food until they reach harvestable size.

Fishes of many families are cultured — from sturgeons to puffers — because of the combination of availability, demand, and adaptability to artificial surroundings. A high demand can justify, economically, at least, the culture of high-cost carnivorous fishes such as trout and salmon (Salmonidae), porgies (Sparidae), the yellowtail (*Seriola quinquiradiata*), eels (Anguillidae), and others, even though they must be fed high-protein food made in part from other fishes. The channel catfish (*Ictalurus punctatus*) is another fairly high-priced fish fed high-protein food.

Greater bioenergetic efficiency, but not necessarily greater profits, can be realized by rearing omnivores, planktivores, or herbivores. These are closer to the base of the food web than the carnivores and can be cultured with less outlay for food. Fishes of the family Cyprinidae are extremely popular for fish culture in Asia, Europe, and parts of Africa. Polyculture systems can be set up with a number of Chinese carps so that fresh vegetation can be fed the grass carp (*Ctenopharyngodon idella*), which passes a large proportion of the vegetation through its digestive tract shredded but undigested, thus fertilizing the water. Plankton resulting from this fertilization is eaten by such fishes as the silver carp (*Hypophythalmichthys molotrix*) and bighead carp (*Aristichthys nobilis*). Bottom organisms can be eaten by the common carp (*Cyprinus carpio*) and others. Several ecological niches are filled in such a pond and greater production is realized than if only one species were used. Polyculture systems are used in India, utilizing fish of the genera *Catla*, *Cirrhina*, and *Labeo*.

The common carp is one of the most important cultured fishes in the world. It is an omnivorous species that has been bred for rapid growth, great efficiency in utilization of food, good body conformation, and other desirable attributes. At present it does not enjoy great popularity in North America, but the time may come when this efficient animal will be instrumental in providing protein to an ever-growing population of Americans.

Another important family containing some herbivorous fishes used in fish culture is Cichlidae. African species of the genus *Tilapia* (including *Sarotherodon)* are used in subsistence fish culture both in Africa and Asia, and are subject to culture in other areas, including North America. *Tilapia (Sarotherodon) mossambica* has been widely introduced and used with varying success in many warm countries. One of the greatest disadvantages of that species is its fast reproductive rate, which often leads to overpopulation and stunting, with consequent production of fish unattractively small for table use. Other species, such as *T. nilotica, T. melanopleura,* and *T. zillii,* are also cultured. The native cichlids of India *(Etroplus* spp.) are of minor use in fish culture. In South and Central America native fishes of the genera *Cichlasoma, Astronotus,* and *Cichla* are subject to culture.

Important marine fishes from the standpoint of fish culture are the herbivorous *Chanos chanos* or milkfish, and mullets of the genus *Mugil,* which are generally detritivores. The milkfish is cultured in brackish water ponds in the Indo-Pacific region, especially the Philippines, Indonesia, and southern Taiwan. Mullets are cultured in brackish or marine waters in many regions having tropical to warm temperate climates.

Although many other species and families are used in fish culture, those mentioned are among the most prominent. Research efforts are constantly being directed at discovering additional species to culture, developing new and more efficient methods, better food, and so on. A recent development in the North Pacific is the concept of "ocean ranching" in which Pacific salmon are reared to migratory age and size in a hatchery, then released to the ocean after being held for a time at a place that serves both as the release and capture site. While being held prior to release, they are behaviorally imprinted with the odor of the release site, so that after they have grown to maturity in the ocean they return to the same site and are harvested.

Management of the Fisheries

When only a few fishermen were fishing, with primitive and inefficient gear, the waters must have seemed inexhaustible. Subsistence fisheries in fresh waters and near the shores of the ocean traditionally have been pursued with the idea of harvesting as much usable protein as possible with the least amount of effort. Generally, the history of such a fishery is that, as more fishermen enter the activity, and as better gear is devised, the capacity of the fish populations to produce large numbers of large fish is diminished. This becomes evident in the smaller average size of the fish landed, in the increased amount of effort needed to harvest the catch, and possibly in an absolute decrease in the weight of the annual harvest.

These are symptoms of overfishing, and they can be recognized even in the absence of environmental degradation — not only in traditional fisheries in small or confined fishing grounds, but also in modern, large-scale commercial and recreational fisheries. An unexploited or lightly used fish stock often has a significant proportion of its biomass in the slow-growing older brood stock. These large individuals may represent, because of the peculiarities of fish population dynamics, a deterrent to the recruitment of great numbers of smaller fish into the usable size classes. Also, these mature fish constitute a bonanza for the fishermen who first exploit the stock, and may attract the fishing pressure that can eventually result in reducing the number of spawners to the point where optimum numbers of recruits cannot be produced. Fishery scientists have determined a hypothetical relationship of numbers of spawners to numbers of young fish entering the fishery (Fig. 17–2).

Fishermen and scientists generally recognize that the aquatic ecosystems are not inexhaustible, because they now recognize factors that limit biological productivity. They also recognize that man's activities can alter natural processes either through impact upon the populations themselves or upon the environment upon which the fish populations depend. Most alterations in the environment that attend advancing civilization can have deleterious effects upon fisheries. Such activities as land clearing, logging, grazing, land drainage, diversion of water for irrigation, obstruction of streams by dams, and industrial and domestic pollution can all diminish the productivity of water. Because productivity of future fisheries may be contingent upon current treatment of fish populations and the ecosystems that produce them, management of the fisheries to protect their capacity for producing a sustained yield is highly desirable. What constitutes the "yield" will differ among

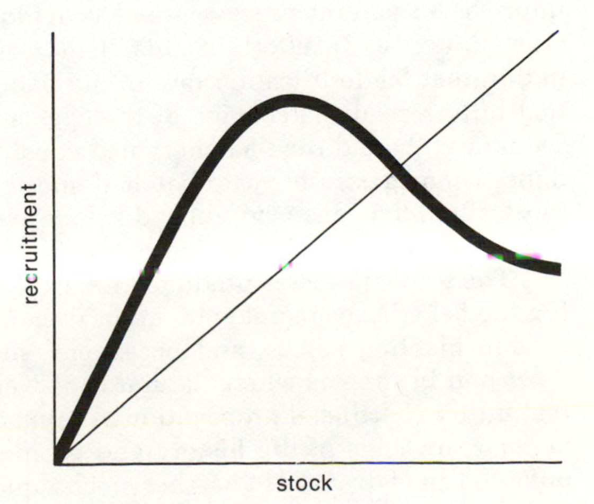

FIGURE 17–2. Hypothetical relationship between a fish stock and the young fish recruited to the stock.

fisheries, depending upon what a community or nation expects from a particular fishery. There can be considerable differences in what fishery scientists believe should be taken from a given fishery. "Maximum sustainable yield," "optimum sustainable yield," "maximum economic efficiency," and other concepts are argued in relation to commercial fisheries. In recreational fisheries the yield sought may be different. Numbers of fish caught may be important in some sports fisheries but, in others, merely the opportunity to try for trophy-sized fish may be the return that devotees wish to perpetuate.

Regardless of the beneficial product expected, fishery managers are devoted to using all available knowledge and technology to perpetuate, distribute, and enhance the fisheries for man's greatest advantage. The managers are backed by scientists in many disciplines — mathematicians involved in studies of population dynamics, physiologists, pathologists, ecologists, statisticians, and many more — as well as that combination of many of these disciplines known as the fishery biologist.

Historically, fishery management seems to have begun with laws and regulations. Artificial propagation may have been the next method, followed by control and improvement of environmental factors. Laws and regulations usually were instituted to protect breeding populations or undersized fish in order to ensure recruitment of adequate numbers of usable-sized fish. These regulations restricted length of season or size of fish taken, or limited the type of fishing gear used. In recent times the amount of gear entering a fishery may be limited, or overall catch quotas for an entire fleet may be set. In recreational fisheries there may be special applications of regulations. There may be both upper and lower size limits to protect both undersized fish and the large brood stock, or some sport fisheries may operate on a "catch and release" basis so that a single fish might produce sport for a number of different fishermen during its life.

Artificial propagation, the incubation and hatching of eggs stripped from female fish from wild or domesticated stocks with subsequent release of young fish, was instituted with the idea that it would improve on natural processes and provide more recruits to the fisheries. Much early effort in this field was predicated on the false notion that the fertilization rate in the natural situation was poor and that only a small percentage of the eggs survived and hatched. Consequently, the old-time hatcheryman considered that his job was well done when eggs in his care hatched and the unfed fry were released. Most often, the eggs were robbed from a wild run of fish trapped at a weir and not allowed to spawn naturally.

The great successes predicted for artificial propagation in enhancing the fisheries were not evident for decades. The practice was useful in transplanting stocks, and occasional success was experienced in lakes and in streams where natural runs were cut off by dams, but the real utility of artificial propagation of salmonids was not realized until better knowledge of life history and ecology of cultured species was obtained and advances in hatchery techniques were made. Better nutrition, disease control, and selective breeding of hatchery-generated brood stock has allowed fishes to be reared economically for several months to a year before release. Fishes may be released for several purposes. Trout may be released in streams or lakes for immediate capture, with those placed in lakes usually giving a much better return to the angler. Others may be released in autumn into rich lakes where good winter survival is expected. Anadromous species may be reared to the smolt phase and released with the expectation that they will immediately descend to feeding grounds in the ocean or in large lakes. There has been notable success with the artificial propagation of the coho salmon, and research in progress on steelhead trout and certain stocks of chinook salmon promises better culture of these species in the future.

In freshwater fisheries, protection and improvement of the environment has become important. Many streams in their natural state have falls, log jams, or other barriers that prevent migration of fishes, or these streams may not have a distribution of riffles and pools that provide for a viable combination of optimum food production and nesting and hiding places. Such situations can often be improved by construction of fish ladders around falls and barriers, destruction of log jams, and placement of check dams and other devices to alter the character of the stream. Often the capacity of streams to support normal populations of fishes is degraded by activities such as logging, farming, construction of large dams, and release of pollutants into the waters. Some productivity can be restored through proper watershed management, pollution control and abatement, and the provision of passage facilities at the dams.

References

Baldwin, N. L., Millemann, R. E., and Knapp, S. E. 1967. "Salmon poisoning disease." III. Effect of experimental *Nanophyetus salmincola* infection on the fish host. J. Parasitol. 53:556–564.

Bardach, J. E., Ryther, J. H., and McLarney, W. O. 1972. Aquaculture. New York, John Wiley and Son.

Borgstrom, G. (ed.). 1961. Fish as Food, Vol. 1. New York, Academic Press.

Childers, W. F., and Bennett, G. W. 1967. Experimental vegetation control by largemouth bass-tilapia combinations. J. Wildl. Mgt., 31:401–407.

Conroy, D. A. 1975. An evaluation of the present state of world trade in ornamental fish. FAO Tech. Pap., (146):1–128.

Costlow, J. D., Jr. (ed.). 1969. Marine Biology, Vol. 5. New York, Gordon and Breach.

Cross, D. G. 1969. Aquatic weed control using grass carp. J. Fish Biol., 1(1):27–30.

Cushing, D. H. 1968. Fisheries Biology. Madison, University of Wisconsin Press.

———. 1971. Upwelling and the production of fish. Adv. Mar. Biol., 9:255–324.

Everhart, W. H., Eipper, A. W., and Youngs, W. D. 1975. Principles of Fishery Science. Ithaca, N.Y., Cornell University Press.

Fishelson, L. 1974. Histology and ultrastructure of the recently found toxic gland in the fish *Meiacanthus nigrolineatus* (Blenniidae). Copeia, 1974 (2):386–392.

Foo, L. Y. 1976. Scombroid poisoning. Isolation and identification of 'saurine.' J. Sci. Food Agric., 27(9):807–810.

Gerberich, J. B., and Laird, M. 1968. Bibliography of papers relating to the control of mosquitoes by the use of fish. FAO Fish. Tech. Pap., (75):1–70.

Gulland, J. A. 1971. The Fish Resource of the Ocean. Surrey, Fishing News (Books) Ltd.

Halstead, Bruce W. 1967. Poisonous and Venomous Marine Animals of the World, Vols. 2 and 3. Vertebrates. Washington, D.C., U.S. Government Printing Office.

Hanson, Joe A. (ed.). 1974. Open Sea Mariculture: Perspectives, Problems, and Prospects. Strandsberg, Pa., Dowden, Hutchinson and Ross.

Hardy, R., and Smith, J. G. M. 1976. The storage of mackerel (*Scomber scombrus*). Development of histamine and rancidity. J. Sci. Food Agric., 27(7):595–599.

Helm, T. 1976. Dangerous Sea Creatures. New York, Funk and Wagnalls.

Hoffmann, G. L. 1967. Parasites of North American Freshwater Fishes. Berkeley and Los Angeles, University of California Press.

Holt, S. J. 1969. The food resources of the ocean. Sci. Am., 221(3):178–182, 187–194.

Laevastu, T., and Hela, I. 1970. Fisheries Oceanography. London, Fishing News (Books) Ltd.

Lagler, K. F. 1956. Freshwater Fishery Biology. Dubuque, W. C. Brown.

McCarraher, D. B. 1959. The northern pike-bluegill combination in north-central Nebraska farm ponds. Progr. Fish. Cult., 21(3):188–189.

Milne, P. H. Fish and Shellfish Farming in Coastal Waters. London, Fishing News (Books) Ltd.

Ospuszynski, K. 1972. Use of phytophagous fish to control aquatic plants. Aquaculture, 1(1):61–74.

Reichenbach-Klinke, H., and Elkan, E. 1965. The principal diseases of the lower vertebrates. New York, Academic Press.

Rounsefell, G. A. 1975. Ecology, Utilization and Management of Marine Fisheries. St. Louis, C. V. Mosby.

Royce, W. F. 1972. Introduction to the Fishery Sciences. New York, Academic Press.

Russell, F. E. 1965. Marine toxins and venomous and poisonous marine animals. Adv. Mar. Biol., 3:255–384.

——— 1969. Poisons and venoms. *In* Hoar, W. S., and Randall, D. J. (eds.), Fish Physiology, Vol. 3. New York, Academic Press, pp. 401–449.

Ryther, J. H. 1969. Photosynthesis and fish production in the sea. Science, 166:72–76.

———, *et al.* 1975. Physical models of integrated waste recycling marine polyculture. Aquaculture, 5:163.

Schuytema, G. S. 1977. Biological Control of Aquatic Nuisances — A Review. Research Report, Ecological Research Series, U.S. Environmental Protection Agency, No. EPA–60013–77–084. Technical Information Staff, Cincinnati, Ohio.

Shapiro, S. (ed.). 1971. Our Changing Fisheries. Washington, D.C., U.S. Department of Commerce.

Tressler, D. K., and Lemon, J. Mc. 1951. Marine Products of Commerce. New York, Reinhold.

Vesey-Fitzgerald, B., and LaMonte, F. 1949. Game Fish of the World. New York, Harper.

Woolner, F. 1972. Modern Saltwater Sport Fishing. New York, Crown.

World Health Organization. 1974. Fish and Shellfish Hygiene. Geneva, W.H.O. Technical Report, Series 550.

GLOSSARY

The following is a list of defined words, the meanings of which may not be explicit in the text. The definitions are restricted to the usage in this book.

Acanthopterygian – Referring to spiny-rayed teleosts.
ACTH – Adrenocorticotropic hormone.
Aestivation – The dormant state of certain animals during periods of drought or high temperatures.
Autonomic nervous system – Those efferent motor fibers and their ganglia that regulate bodily functions not under voluntary (conscious) control. The system is composed of two antagonistic parts: parasympathetic and sympathetic.
Biomass – The total weight of all members of a species (taxon) or group of species (taxa) in a given area at an instant in time. ("Total weight of all organisms in a particular habitat or area; the term is also used to designate the total weight of a particular species or group of species.")
Bowman's capsule – The expanded proximal end of the kidney tubule surrounding the glomerulus.
Circadian – Pertaining to 24 hour biological cycles.
Commensalism – A form of symbiosis in which one species benefits and the other neither suffers nor benefits.
Commissure – A linking or connecting of parts of the nervous or lateral line systems, usually from one side to the other.
Convergence – The attainment of functionally similar structures by distantly related taxa.
Cristae – Patches of sensory cells (neuromasts) at the juncture of the semicircular canals and the utriculus.
Cryophylic – Referring to organisms thriving at relatively low temperatures.
Demersal – Referring to aquatic organisms living on or in close association with the substrate (bottom).
Derived – Modified relative to the primitive condition.
Diapause – The state of suspended development.
Diastole – The dilation phase of the heart action.
Diverticulum – Any blind sac or pouch connected to a larger cavity.
Emmetropic – Referring to normal ocular vision, i.e., not near- or far-sighted.
Endemic – Native or confined (restricted) to a particular geographical region.
Endogenous – Originating within. Produced from within the body, an organ, or a geographical area.
Endolymph – Fluid contained within the inner ear (membranous labyrinth).
Epithelium – The thin layer of tissue covering internal and external body surfaces.
Fecundity – An organism's capacity to produce offspring. (In fish, often expressed as the number of eggs produced per female.)
Fenestra – An aperture in a bone or a transparent portion of a membrane.
Fimbria – A bordering fringe.
Follicle – Any small sac or pit.
Fontanelle – A membrane-covered opening in a bone.
Foramen – A small opening or perforation in any body structure.
Gametes – Mature, haploid, male or female reproductive cells.
Ganglion – A concentration of nerve cell bodies located outside the central nervous system.
Glomerulus – A knot of small blood vessels contained within Bowman's capsule, at the proximal end of a kidney tubule.
Glycoproteins – Organic molecules composed of carbohydrates and proteins.
Gnathostomes – Jawed vertebrates.
Gonadotropic hormones – Hormones, secreted by the anterior lobe of the pituitary, that induce development of gametes.
Gonopodium – A modified anal fin that functions as a copulatory organ.
Hemopoietic – Referring to production of blood cells.
Homologous – Referring to characters, in different taxa, that are structurally similar owing to common evolutionary origin.

Hormones — Chemical substances that are released from endocrine glands, are transported via the circulatory system, and regulate a wide range of physiological functions.

Illicium – Modified first dorsal fin ray of angler fishes, used to attract prey.

Intergrades – As used here, individuals that are, for whatever reason, intermediate between two species (or other taxa).

Intromittent organ – Male copulatory organ.

Maculae – Patches of sensory cells (neuromasts) in the utriculus, sacculus, and lagena.

Meninges – Protective membrane enclosing (surrounding) the brain and spinal cord.

Metamorphosis – The stage in development during which an animal undergoes a radical change in form and function.

Microphagous – Referring to organisms that feed on relatively small food items.

Micropyle – An aperture in the vitelline membrane of an egg, through which the sperm enters.

Mimicry – Imitation of another organism or object in the environment (in form, color, and/or behavior).

MSH – Melanophore-stimulating hormone; can cause either dispersion or aggregation of pigment.

Myelinated – Referring to nerves ensheathed by a fatty membrane.

Myoid – Contractile segment of visual cell.

Neuromasts – Mechanoreceptors of the acoustico-lateralis system.

Nictitating membrane — The third "eyelid" present in many vertebrates (aids in cleaning and protecting the eye).

Nuptial – Referring to breeding.

Ocellus – An eyelike marking.

Oligotrophic – Referring to bodies of water, especially lakes, with low biological productivity.

Otoliths – Calcareous nodules in the utriculus, sacculus, and lagena of the membranous labyrinth.

Oviparous – Referring to egg-laying animals.

Ovoviviparity – Condition of retention and incubation of eggs within ovary or oviduct. Young receive no (or very little) nourishment from the female.

Parabranchial cavity – Cavity bounded by the operculum and branchiostegal membrane; receives water that has passed through gills.

Parr – That stage in anadromous salmonids between yolk sac absorption and transformation to the smolt stage prior to seaward migration.

Pelagic – Referring to organisms that inhabit open waters of the oceans (or large lakes).

Peritoneum – Membrane lining the coelom.

Phagocytosis – The process by which certain cells engulf other cells or foreign particles.

Photophores – Light-producing organs.

Phylogeny – Evolutionary development of a species (or other taxon).

Placenta – An intimate association of embryonic and maternal membranes through which gases, nutrients, and wastes are exchanged.

Polyandrous – Referring to species in which one female mates with more than one male.

Polyculture – The use of two or more species to utilize two or more trophic levels within a culture system. (The same effect can be obtained in some instances by the use of distinct life stages of one species.)

Polyphyletic – Referring to a taxon in which the members are not of the same immediate line of descent.

Primary production – The rate at which plants and other autotrophs accrue biomass, or energy and nutrients. Expressed as weight/area/time or Calories/area/time.

Sexual dimorphism – Morphological variation within a species correlated with the sex of the individual.

Smolt – The seaward migrating stage of anadromous salmonids.

Steroids – A class of organic compounds composed of four interlocking carbon rings. Included are cholesterol and male and female sex hormones.

Symbiosis — An intimate living arrangement (relationship) between two species. (One or both species may benefit from the association.)

Sympatry – Temporal and spatial overlap of the ranges of two or more species.

Systole – The contraction phase of the heart.

Telolecithal – Referring to eggs in which the yolk is concentrated at the vegetal pole.

Trophic – Referring to nutrition or to ecological levels or mode of feeding.

Velum — In hagfishes, a scroll-like pharyngeal membrane that acts as a respiratory pump. In adult lampreys, the fleshy, tentacled flap guarding the opening to the respiratory tube.

Viviparity – The condition in which fertilized ova (eggs) are retained within the female and derive nourishment from the female via placentae or from secretions.

Zymogen – The inactive precursor of an enzyme.

INDEX

Page numbers in *italics* indicate illustrations; those followed by (t) indicate tables.